Lecture Notes in Morphogenesis

Series editor

Alessandro Sarti, CAMS Center for Mathematics, CNRS-EHESS, Paris, France
e-mail: alessandro.sarti@ehess.fr

For further volumes:
http://www.springer.com/series/11247

Giovanna Citti · Alessandro Sarti
Editors

Neuromathematics of Vision

Springer

Editors
Giovanna Citti
Dipartimento di Matematica
Università di Bologna
Bologna
Italy

Alessandro Sarti
CAMS Center for Mathematics
CNRS-EHESS
Paris
France

ISSN 2195-1934 ISSN 2195-1942 (electronic)
ISBN 978-3-662-51005-6 ISBN 978-3-642-34444-2 (eBook)
DOI 10.1007/978-3-642-34444-2
Springer Heidelberg New York Dordrecht London

Printed on acid-free paper

Springer is part of Springer Science+Business Media (www.springer.com)

To my mother Maria and my father Roberto

To my mother Gabriella and my father Vittorio

Preface

This book tells the story of an adventure, a scientific adventure moved by the desire to understand the geometrical structures of the visual brain. For geometry we attend here not the anatomical geometry of the brain shape, but the differential geometry of the connectivity between neural cells. This connectivity builds and shapes the hidden brain structures underlying visual perception. The story of the problem runs over the last 30 years, since the discovery of Hubel and Wiesel of the modular structure of the primary visual cortex, and slowly cams towards a theoretical understanding of the experimental data on what we now know as functional architecture of the primary visual cortex.

Experimental data comes from several domains. The two classical sources of data, neurophysiology and phenomenology of perception, are nowadays more and more sustained by neurocognitive imaging. Imaging techniques like functional MRI and diffusion tensor MRI allow us to deepen the study of cortical structures at the mesoscale, completing the scale range already well covered by neurophysiology at the microscale and by psychophysics at the global scale. Due to the variety of sources of experimental data, neuromathematematics deals not just with modelling of cortical structures but also with modelling of perceptual spaces, in the Spinozist tradition where "res cogitans" and "res extensa" are just the two sides of the same sheet of paper.

From the mathematical point of view, neuromathematical structures are forged by classical differential geometry and Lie groups, but are more and more demanding for new instruments to pure mathematicians: research in sub-Riemannian geometry is important to model horizontal connectivity, harmonic analysis in non commutative groups is fundamental to understanding the pinwheels structure, as well as non-linear dimensionality reduction is at the base of many neural morphologies and possibly of the emergence of perceptual units. But what is at the center of the neurogeometrical problem is overall the art to harmonize contemporary mathematical instruments with neurophysiological findings and phenomenological experiments towards the construction of a theoretical model of vision. The constributions to this book are devoted to this task and come from the very founders of the discipline.

Jan Koenderink in the '80s started using differential geometry to study perceptual spaces, and William Hoffman proposed a model of the visual cortex as a fiber bundle equipped with a contact structure. Almost at the same time Steven Zucker was observing the relation between the measurement of Euclidean curvature and the role of end-stopping cells. A new idea of the brain as "geometric machine" was being raised, and it started to complement the theories on neural dynamics, which were already challenging the "cybernetic brain" of traditional artificial intelligence.

In the '90s a number of phenomenological models of vision appeared in the mathematical framework of calculus of variations and parabolic (possibly degenerate) partial differential equations. The number of models is overwhelming and we cite here just some pioneeristic papers, the segmentation model of Mumford - Shah, the multiscale analysis of Alvarez, Lions, Morel and the first developements of applied wavelet analysis. Even if these models are mathematically elegant and their impact in image processing has been very important, they remained at that time purely phenomenological, and they lacked of the double neuro-psycho nature of neuromathematical models.

An important contribution in this sense has been carried out by David Mumford with its elastica curves modelling perceptual illusory contours and paying attention to both their psychophysical and neural basis. These curves are still at the center of contemporary research, but rethought in new kinds of spaces and in a renewed mathematical setting. A first stochastic model of illusory contours in his space of position and orientation was proposed by Williams and Jacobs in '95. At the end of the '90s a fundamental contribution of Jean Petitot and Yannick Tondut reprehends the model of the cortex as a contact bundle of William Hoffman with which authors compute the geodesic curves of the non-integrable structure, observing that they are able to model perceptual association fields, measured by Fields, Heyes and Hess. In this way, they were able to predict the shape of an illusory contour, given its inducturs. Petitot explicitly introduced the word Neurogeometry to denote the inner geometry of cortical connectivity. Also Steven Zucker reprehends the model of the cortex as a fiber bundle to describe the pinwheel organization and proposed a model of horizontal connectivity between pinwheels in terms of Frenet frames. In 2003 Govanna Citti and Alessandro Sarti observed that the cortical structure is not defined in the Heisenberg group but in the $SE(2)$ Euclidean group of rotation and translation equipped with a sub-Riemannian metric. In this fiber bundle structure, they proposed a model of image completion, which is currently used, expressing cortico-cortical propagation in terms of sub-Riemannian diffusion and $SE(2)$ invariant advection-diffusion PDE.

Bressloff and Cowan are interested in the dynamics of neural population in a model of the cortex defined in the group $SE(2)$. An impressive result is obtained when the activation map is computed in absence of an external input and the effect of chemical drugs is simulated. In this case the resulting activation distribution corresponds impressively to visual hallucinations, as reported in classical literature. A similar model is proposed by Olivier Faugeras and Pascal Chossat, but in the hyperbolic geometry, in relation to texture perception.

Remco Duits works in the $SE(2)$ group from an image processing point of view. He considers lifting and propagation of the visual signal in the $R^2 \times S^1$ space, but, by using invertible kernels (invertible scores), is able to reproject results in the 2D image plane without loss of information. When these kernels are fundamental solution of subelliptic PDE, their expression is provided a formal series with Fourier transform. Results are impressive when applied to medical image processing.

Close to the neuromathematical paradigm is the entire activity of Jean-Michel Morel and his group on computational gestalt, looking at the problem of geometry of vision from a phenomenological point of view. His proposition to individuate in the Helmoltz principle the basis of the classical theory of Gestalt is theoretically deep and computationally fruitful. Daniel Bennequin is interested in a variety of neuromathematical problems, all expressed in terms of invariance, symmetry and ambiguity: from orientation maps distribution to neural structures in Lie groups. He formalized these problems with principles of "information topology" theory, a new co-homology theory of information based on probability and entropy theory, partially inspired by quantum field theory.

Jim Bednar is oriented in reproducing the geometric morphologies of the visual cortex on the base of brain plasticity principles. He recovers the main characteristics of the functional architecture of the cortex simply by a process of learning of a suitable set of stimuli. His work allows a possible reconciliation between an entire ensemble of studies based on a differential geometry description of structures and new trends towards organization of forms best adapted to their environment.

We feel great honor and privilege to have been part of this human and scientific adventure. The present book contains the contributions of many protagonists of this story, and on the other hand it is far from exhaustive, since many others are missing. The book is conceived to testify the past and to look towards contemporary challenges in the understanding of the architecture of vision.

Finally we owe a huge debt to the various people who have supported us at many levels. We would like to thanks particularly Davide Barbieri, Giacomo Cocci and Gonzalo Sanguinetti.

Bologna, Paris
August 2013

Giovanna Citti
Alessandro Sarti

Contents

List of Contributors

Jean Petitot
École des Hautes Études en Sciences Sociales, Center of Mathematics (CAMS), 190-198 avenue de France, 75244 Paris, France
e-mail: petitot@ehess.fr

Jan Koenderink
Laboratory of Experimental Psychology, Katholieke Universiteit Leuven, Tiensestraat 102, 3000 Leuven, Belgium
e-mail: jan.koenderink@ppw.kuleuven.be
and
Universiteit Utrecht, Faculteit Sociale Wetenschappen, Afdeling Psychologische Functieleer, Postbus 80140, 3508 TC Utrecht, The Netherlands
e-mail: J.J.Koenderink@uu.nl

Andrea van Doorn
Universiteit Utrecht, Faculteit Sociale Wetenschappen, Afdeling Psychologische Functieleer, Postbus 80140, 3508 TC Utrecht, The Netherlands
e-mail: A.J.vanDoorn@uu.nl
and
Department of Industrial Design, Delft University of Technology, Landbergstraat 15, 2628 CE Delft, The Netherlands
e-mail: A.J.vanDoorn@tudelft.nl

Steve Zucker
Department of Computer Science, Yale University
e-mail: steven.zucker@yale.edu

Giovanna Citti
Università di Bologna, Dipartimento di Matematica, Piazza di Porta San Donato 5, 40126 Bologna, Italia
e-mail: giovanna.citti@unibo.it

Alessandro Sarti
École des Hautes Études en Sciences Sociales, Center of Mathematics (CAMS), 190-198 avenue de France, 75244 Paris, France
e-mail: alessandro.sarti@ehess.fr

Remco Duits
IST/e, Eindhoven University of Technology, IST/e, Den Dolech 2, NL-5600 MB, The Netherlands
e-mail: R.Duits@tue.nl

Arpan Ghosh
IST/e, Eindhoven University of Technology, IST/e, Den Dolech 2, NL-5600 MB, The Netherlands
e-mail: A.Ghosh@tue.nl

Tom Dela Haije
IST/e, Eindhoven University of Technology, IST/e, Den Dolech 2, NL-5600 MB, The Netherlands
e-mail: T.C.J.Dela.Haije@tue.nl

Yuri Sachkov
Russian Academy of Sciences, Program Systems Institute, Pereslavl-Zalessky 152020, Russia
e-mail: sachkov@sys.botik.ru

José Lezama
École Normale Supérieure de Cachan, Cachan, France
e-mail: lezama@cmla.ens-cachan.fr

Samy Blusseau
Centre de Mathématiques et de Leurs Applications - ENS de Cachan Paris, France
e-mail: sblusseau@gmail.com

Jean-Michel Morel
École Normale Supérieure de Cachan, Cachan, France
e-mail: morel@cmla.ens-cachan.fr

Gregory Randall
Instituto de IngenierÃa Eléctrica, Universidad de la República, Montevideo, Uruguay
e-mail: randall@fing.edu.uy

Rafael Grompone von Gioi
Instituto de Ingegneria Eléctrica, Universidad de la República, Montevideo, Uruguay
e-mail: grompone@cmla.ens-cachan.fr

Daniel Bennequin
Institut de Mathématiques, Université Denis Diderot - Paris VII, Paris
e-mail: danielbennequin@gmail.com

James A. Bednar
Institute for Adaptive and Neural Computation, The University of Edinburgh, 10 Crichton St, EH8 9AB, Edinburgh, UK
e-mail: jbednar@inf.ed.ac.uk

Chapter 1
Landmarks for Neurogeometry

Jean Petitot

Abstract. We present a historical survey of the way Neurogeometry links the modelling of the functional architectures of primary visual areas with sub-Riemannian geometry.

1.1 Introduction

In the 1990s, we coined the expression "neurogeometry of vision" to refer to geometrical models of the functional architecture of primary visual areas. The very particular connectivity of the functional neuroanatomy of these areas explains the geometry of percepts and must therefore be implemented in the synaptic weights of the neural nets used for modelling.

The term "neurogeometry" (of vision) presents two complementary aspects.

1. The geometry of (visual) perception described, but not modelled, since the times of Goethe, Helmholtz, Hering, Brentano, Poincaré, Husserl and Gestalttheorie (von Ehrenfels, Wertheimer, Stumpf, Koffka, Köhler, Klüver, etc.), from Kanizsa to Marr in psychology, from Evans to Peacocke or McDowell in philosophy of mind. We will use the expression "perceptual geometry" for its description. It must be emphasised that, until the 1960s, a relevant *mathematical* perceptual geometry was essentially lacking.
2. Strictly speaking the term "neurogeometry" concerns mathematical models for the neural algorithms processing perceptual geometry.

Now, we will see that these mathematical models of neural implementation are also geometrical but in a sense deeply different from that of perceptual geometry. The difference is quite similar to that found in computer sciences between high

Jean Petitot
École des Hautes Études en Sciences Sociales, Center of Mathematics (CAMS),
190-198 avenue de France, 75244 Paris, France
e-mail: petitot@ehess.fr

G. Citti and A. Sarti (eds.), *Neuromathematics of Vision*,
Lecture Notes in Morphogenesis,
DOI: 10.1007/978-3-642-34444-2_1,

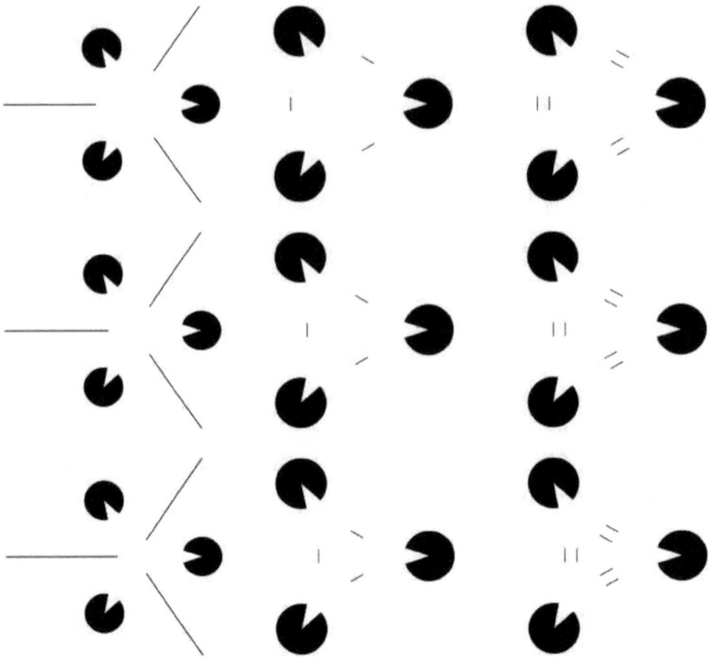

Fig. 1.1 Examples of curved Kanizsa triangles with three methods for determining the exact position of the illusory contour. One has to put a mark (the end of an orthogonal line, a small segment, the axis of a small stripe) as exactly as possible on the extremum of the contour. The good positions are shown on the middle line.

level logic and low level λ-calculus. Neurogeometry is, so to speak, "internal", neural, and "immanent", while perceptual geometry is "external", ideal, and "transcendent".

To give an example (but, as we say in French, "l'exemple est la chose même"), let us consider Kanizsa illusory contours (see the celebrated Kanizsa's *Grammatica del Vedere* [73]). On the one hand, at the level of perceptual geometry, they have been analyzed by a lot of psychophysicists, and we know quite well their properties as gestalts. They are particularly interesting when they are curved. Figure 1.1 shows different curved Kanizsa triangles with three methods for determining the exact position of the illusory contour.

On the other hand, at the level of neurogeometry, we will see that illusory contours result from the *functional architecture* of the primary cortical areas. Now, to take only the case of $V1$, the first of these areas, we will see that its functional architecture implements the contact structure of the fibre bundle of 1-jets of planar curves. There exists a natural metric, called sub-Riemannian, on this contact structure and, mathematically speaking, Kanizsa illusory contours are "geodesics of a sub-Riemannian geometry defined on the contact structure of the fibre bundle of 1-jets of planar curves". The fact that such a formulation appears highly esoteric and even non-sensical reveals the gap between perceptual geometry and the neurogeometry in which it is implemented. The gap will be filled in section 1.4.10. In fact, we will see that this seemingly complicated neurogeometrical model deepens in terms of the functional architecture of $V1$ a model proposed in 1992 by David Mumford for computer vision [90].

We will evoke briefly in this survey the theory of caustics in optics. It is interesting to parallel them with our theory of illusory contours. Caustics are well known since antiquity as envelopes of optical rays. As we will see, in their current definition, they are "singularities of the projections on $\mathbb{R}^3$ of the Lagrangian solutions of the Hamiltonian system defined on the cotangent fibre bundle $T^*\mathbb{R}^3$ by the Hamiltonian H associated to the wave equation". The formulation is as esoteric and non-sensical as that of illusory contours and reveals the long way between the phenomenological description of caustics and their physical explanations. The main difference is that, in physics, such "long ways" are completely usual while in cognitive sciences they are completely lacking. The ambition of neurogeometry is to introduce a "mathematical turn" analogous to that found in theoretical physics.

Concepts such as jets, contact structures, sub-Riemannian metrics or geodesics belong to differential geometry in the sense of Elie Cartan, Hermann Weyl, René Thom or Misha Gromov, and it is therefore this type of geometry which is "internal", neural and "immanent" in neurogeometry and explains the "external", ideal and "transcendent" perceptual geometry. It is crucial to understand this *twofold* aspect of geometry in neurogeometry. The situation is quite similar to what one finds in particle fundamental physics. One observes complex trajectories of particles in space-time M. And to explain them in the framework of quantum field theory, one has to consider fibre bundles over M, Cartan connections, curvatures of connections, etc. All these deep geometrical structures model a physical immanence, which accounts for the observed empirical trajectories.

Since the editors of this volume thought that it could be of interest to present some historical and biographical backgrounds for neurogeometry, we will give in this survey a few landmarks, first for perceptual geometry, then for neurogeometry.[1]

1.2 Perceptual Geometry Since the 1970s

1.2.1 Thom's Models

The first elements of mathematical perceptual geometry were worked out by René Thom at the end of the 1960s as an aspect of his mathematical models for morphogenesis (see [130] and [131]). He used the fundamental tools of singularity theory (of which he was one of the main inventors after Marston Morse and Hassler Whitney) to explain how morphologies and patterns can appear and develop in material substrates M. The key idea was that, at every point a of M, the physical,

[1] Many thanks to Francesco Galofaro and Heather Arielle Griffin for the English translation of the text.

chemical or metabolic properties of the substrate are described by an attractor A_a of an "internal" dynamics X_a and that the dynamics of neighbouring points are coupled. Then for some critical values a_c of a bifurcations can happen, the attractor A being substituted by another attractor B. The subset K of the a_c can be very complex (fractal, Cantor, etc.), but in simple cases it *stratifies* M and breaks its homogeneity. This *symmetry breaking* generates a morphology. In that sense, any morphology is a segmentation of the qualities of a substrate by a set of qualitative discontinuities.

Thom's models constitute a wide expansion of the pioneering *reaction-diffusion* models introduced in 1952 by Alan Turing [134]. The challenge is the same: explaining how "the chemical information contained in the genes" can be "converted into a geometrical form". In Turing, the internal dynamics are systems of (non linear) differential equations modelling the chemical reactions between "morphogens" inside the substrate. The spatial coupling is afforded by diffusion, and the cause of the "patternized" morphologies is the breaking of homogeneity induced by "diffusion-driven instabilities" (see our survey [107]). The great biologists inspiring Turing's and Thom's projects are also the same: Sir D'Arcy Wentworth Thompson and Conrad Hal Waddington.

Thom's models were called "catastrophic" by Christopher Zeeman. We preferred the term "morphodynamical". Morphodynamical models belong to a mathematical universe which experienced an extraordinary development in the 1960s-1970s with masters such as René Thom, Bernard Malgrange, John Mather, Christopher Zeeman, Vladimir Arnold, Stephen Smale, David Ruelle, David Mumford, John Milnor, Martin Golubitsky, Robert MacPherson and many others. Their main tools were the theory of dynamical systems (for the study of internal dynamics), their attractors, their properties of structural stability, and their bifurcations; the theory of critical points of differentiable mappings (for when the internal dynamics are gradient like); the geometrical theory of jet spaces and their stratifications; and universal unfoldings of finite codimension singularities (see Thom [130], [131] and our compiling [98]). They aimed at a mathematical comprehension of morphogenesis and, beyond biological morphogenesis, of morphological *structures* whatever their substrate may be.

Thom's models were inspired by a deep and rather universal mathematical "philosophy" elaborated in the 1950s and the 1960s (see *e.g.* his two classical papers "Les Singularités des applications différentiables" in 1956 [128] and "Ensembles et morphismes stratifiés" in 1968 [129]):

1. The singularities of a space or of a map between spaces concentrate information on its global structure into local morphologies.
2. To analyse locally a differentiable map $f : M \to N$ it is efficient to look at its successive jets modulo a change of coordinates (a diffeomorphism) in M and in N. In general, jets of sufficiently high order can be eliminated, and f can be reduced locally to an algebraic form (codimension of singularities and normal forms).
3. Jet spaces $J^k(M,N)$ of successive order k are manifolds *stratified* by a stratification Σ_k whose strata of increasing codimension (of decreasing dimension) correspond to more and more singular singularities.

4. One of the most fundamental theorem is the *transversality theorem* in jet spaces. The k-jet $j^k(f)$ of f is a map $j^k(f) : M \to J^k(M,N)$ and the theorem says that, *generically*, $j^k(f)$ is transverse on Σ_k. A consequence is that f cannot present generically singularities of codimension $> \dim(M)$.
5. *Morse theory* is a privileged tool for analysing manifolds. If M is a manifold of dimension n and if $f : M \to \mathbb{R}$ is a Morse function – that is a smooth function whose all critical points (points where the gradient $\nabla f = 0$) are non degenerate (the Hessian is of maximal rank n) and whose all critical values ($f(a)$ for a critical) are distinct – then M can be qualitatively reconstructed from f via a "handle representation". In particular a fundamental formula allows computation of the Euler-Poincaré characteristic $\chi(M)$ of M from any of its Morse function: $\chi(M) = \Sigma_{i=0}^{i=n}(-1)^i m_i$ where m_i is the number of critical points of Morse index i (a is of Morse index i if in a suitable system of local coordinates f is of the form $-x_1^2 - \ldots - x_i^2 + x_{i+1}^2 + \ldots + x_n^2$).
6. Morse theory can be generalized to stratified manifolds (MacPherson).

1.2.2 Some Autobiographic Remarks

Perhaps the reader will authorise some more personal remarks aiming at precising briefly the initial interdisciplinary context of the following sections. I became interested in Thom's new ideas for two reasons. First, I began my career at the Centre of Mathematics directed by Laurent Schwartz at the Ecole Polytechnique and worked on singularity theory in differential geometry with René Thom and in algebraic geometry with Heisuke Hironaka and Jean Giraud (a disciple of Alexander Grothendieck working also on Hironaka's ideas). Second, I was also very interested in structural theories in social sciences from Saussure and Jakobson to Lévi-Strauss. Thom's ideas unified these two different interests. So, after having joined the EHESS (Ecole des Hautes Etudes en Sciences Sociales) Mathematical Centre (CAMS) in 1971, I focused on the applications of morphodynamics to cognitive sciences and semiolinguistics and also on their far-reaching epistemological consequences.

In order to elaborate this kind of models, it was essential to maintain close relationships with pure mathematical theories. I could succeed thanks to the colleagues with whom I worked at Schwartz laboratory: Bernard Teissier, Alain Chenciner, Jean-Pierre Bourguignon (who would become director of the IHES, Institut des Hautes Etudes Scientifiques), Jean-Marc Deshouillers (an arithmetician interested in cognitive sciences), and later on some younger colleagues such as Daniel Bennequin and Marc Chaperon. Moreover, the weekly seminar by Thom at the IHES in Bures-sur-Yvette on the theory of singularities allowed me to follow the avant-garde of the research in this field and to learn many things attending the talks of a number of important geometers and physicists (*e.g.*, Stephen Smale on complex dynamical systems or David Ruelle on strange attractors, chaos and turbulence).

During the 70s an extraordinary interdisciplinary interest arose on the theory of singularities. It started in mathematics (including mathematical physics: caustics in optics, defects in ordered media, critical phenomena and phase transitions) with important symposia such as the *Liverpool Singularities Symposium* organized and edited by C.T.C Wall in 1971 or the Summer Schools of the Institute of Scientific Studies in Cargèse organized in 1973 and 1975 by Frédéric Pham. The theory has then been developed in an extremely rich scientific context in which different traditions converged. Not without controversies, the morphodynamical models met with the dissipative structures which came from thermodynamics and chemistry at the Bruxelles school of Ilya Prigogine, Grégoire Nicolis, and Isabelle Stengers, with the works on self-organization by Henri Atlan, Jean-Pierre Dupuy and Francisco Varela, with synergetics by Hermann Haken and Scott Kelso, and so on. A real new scientific paradigm had been deployed to understand the emergence of morphologies in the fields of physics, chemistry, and biology.

The philosophical echoes were sensational. The problem of dynamics of forms had been the great loser of the Galileo-Newtonian revolution which allowed the development of mechanics of forces during the XVIIth century: such dynamics implied to conserve certain Aristotelian teleological concepts such as "entelechy". From a philosophical point of view, many great authors were aware of this point. First of all, Leibniz, who investigated the problem during his whole life, Diderot (cf. his debate with d'Alembert), Kant (see the *Critique of Judgment*), Geoffroy Saint Hilaire, Goethe (the inventor of modern structural morphology), Brentano as psychology is concerned, Husserl and phenomenology, Gestalt theory, D'Arcy Thompson, Waddington which I already quoted. By showing how a *mathematical* dynamics of forms compatible with physics was possible and how it could be extended to psychological and social sciences thanks to its transphenomenal nature, René Thom ignited a philosophical breakthrough blowing up the traditional frontiers between natural and human sciences.

At those times I decided to focus my researches on this *unification* of the problem of form and morphogenesis in different empirical disciplines. In order to detail its dimension and implications, in 1982 I organised in honor of René Thom the Cerisy Symposium *Logos and Catastrophe Theory* in which Christopher Zeeman and many mathematicians participated as well as physicists, for example David Ruelle and Michael Berry; biologists as Yves Bouligand or Brian Goodwin, and experts in morphogenesis; philosophers of science, semiolinguists and experts in Aristotle.

1.2.3 Perceptual Geometry and Phenomenology

Within perceptual geometry, five problems seemed particularly relevant in this context. First of all, the link between morphological models and phenomenology. The similarities are strong between the definition of forms as systems of qualitative discontinuities on spatial substrates and Husserl's eidetic descriptions of perception, in particular in *Ding und Raum*. During the 70s, it seemed strange how morphological models allow to relaunch phenomenology by "naturalizing" it. Today things have changed, and neurocognitive sciences converge with phenomenology on different points, as it is shown, *e.g.*, by the works of Alain Berthoz and Jean-Luc Petit

Physiologie de l'Action et Phénoménologie [18]. The interested reader can consult the volume *Naturalizing Phenomenology* [93].

Another problem in perceptual geometry is to construct objects in a tridimensional space, starting from two-dimensional retinal sensations. From a geometrical point of view, it is the problem of the *apparent contours* (AC) of objects which appear as surfaces S in $\mathbb{R}^3$. The deformations and the transformations of these AC in the perceptual temporal flux allow to reconstruct the $3D$ objects. An AC is a singular locus. It is the singular locus $S_{\pi,\delta}$ of the projection of S on the π plane in the direction parallel to δ (transverse to π). If S is smooth, its projection is a map between manifolds of dimension 2, and, thanks to a theorem by Whitney, we know the singularities it can present generically: fold lines, transversal intersections of fold lines, and cusp points. The set of the (π,δ) is called a Grassmannian, so a $3D$ object S is equivalent to an infinity of AC $S_{\pi,\delta}$ parametrised by a Grassmannian. The direct problem, namely how to construct the AC $S_{\pi,\delta}$ knowing S, is rather simple. On the other hand, the *inverse* problem, namely how to reconstruct S and its metric properties (its hyperbolic domains of negative curvature, its elliptic ones of positive curvature, its parabolic lines of curvature $= 0$), starting from a certain number of AC $S_{\pi,\delta}$ is extremely difficult. However, it is what perception solves in every instant. Also in this case, the similarity with the eidetic descriptions of the phenomenology of perceptual geometry, namely Husserl's theory of *adumbrations* (Abschattungen) is remarkable.

1.2.4 Caustics in Optics

A third fundamental link between the theory of singularities and perceptual geometry is given by the already mentioned theory of caustics in optics. René Thom considered it of a central importance because caustics can realize all the "catastrophes" of codimension ≤ 3 (folds, cusps, swallowtails, hyperbolic umbilics, elliptic umbilics). In geometrical optics, caustics are envelopes of light rays and are manifested as forms on a screen because the intensity of the light diverges on it. In wave optics, what Michael Berry called "diffraction catastrophes" are superimposed to these geometrical skeletons and their modelling as solutions of the wave equation is founded on the theory of oscillatory integrals (see our compilation [98]).

Wave optics looks for solutions $v(q,t)$ of the wave equation $Dv = 0$, where q is a point of $\mathbb{R}^3$ with coordinates (x,y,z), t represents time and D is the differential linear operator of second order $D = \frac{\partial^2}{\partial t^2} - \Delta$ (where Δ is the spatial Laplacian). The separation of spatial and temporal variables leads to the search for stationary solutions of frequency τ, $v(q,t) = e^{i\tau t}u(q)$, where $u(q)$ is an amplitude which satisfies $D_\tau u = 0$, with $D_\tau = \tau^2 + \Delta$ and initial condition a given function $u_0(q)$ on the source surface S_0. This means that one looks for a spatial propagation $u(q)$ of $u_0(q)$ on which a wave pattern $e^{i\tau t}$ is engrafted.

The approximation of geometrical optics corresponds to an infinite frequency τ, *i.e.* to a wavelength $= 0$. However, when $\tau = \infty$, the operator D_τ is not defined any longer. This is the source of the idea of searching for *asymptotic solutions* u_τ, parametrised by τ, of the perturbed equation family $D_\tau u_\tau = \varepsilon_\tau$, where ε_τ is a rapidly decreasing function in τ (*i.e.* a function decreasing more rapidly than every negative power of τ). "At the limit" the function u_∞ will then be a the solution of the equation $D_\infty u = 0$. Then, one looks for spatial solutions of the form $u_\tau(q) = a_\tau(q)e^{i\theta}$, where $\theta = \tau\varphi(q)$ with $\varphi(q)$ a *spatial* phase, and where the amplitude $a_\tau(q)$ admits an asymptotic expansion in τ of the form:

$$a_\tau(q) \sim a_0(q) + \frac{1}{\tau}a_1(q) + \cdots + \frac{1}{\tau^k}a_k(q) + \ldots \text{ with } a_0 \neq 0.$$

When $\tau \to \infty$, every $\frac{1}{\tau^k} \to 0$ for $k \geq 1$. So it is simple to compute $D_\tau u_\tau$ as an expansion in decreasing powers of τ, and since u_τ has to be a solution of $D_\tau u_\tau = \varepsilon_\tau$ where ε_τ is rapidly decreasing, the coefficients of this expansion have to vanish. This yields an infinite number of equations, first of all the *eikonal* equation $1 - |\nabla\varphi|^2 \equiv 0$ (where $\nabla\varphi$ is the gradient of the spatial phase) which says that the level surfaces of S_t of φ, the wave-fronts of the geometrical approximation, are parallel surfaces, and that the gradient lines of φ, the light rays (so-called *characteristics* of the wave equation), are orthogonal to the wave-fronts. Further, the equations (known as "transport" equations) for the a_k coefficients, the first of which is $a_0\Delta\varphi - 2\nabla\varphi \cdot \nabla a_0 = 0$. On a light ray, given that $q(t)$ satisfies $\frac{dq}{dt} = 2\nabla\varphi$, we have $2\nabla\varphi \cdot \nabla a_0 = \nabla a_0 \cdot \frac{dq}{dt} = \frac{da_0}{dt}$, and this equation is reduced to an ordinary differential linear equation $\frac{da_0}{dt} + a_0\Delta\varphi = 0$. On the caustic C the amplitude a_0 diverges.

In order to understand the *geometrical* status of caustics, it is necessary to introduce the conjugated momenta p of positions $q \in \mathbb{R}^3$, to work on the phase space which is the cotangent bundle $T^*\mathbb{R}^3$ and to shift toward an Hamiltonian formalism considering the Hamiltonian $H(q,p) = 1 - |p|^2$ associated to D_τ. Consequently, the eikonal equation will be written $1 - |\nabla\varphi|^2 = 1 - |d\varphi|^2 = H(q,d\varphi) = 0$, since the Euclidean metric of $\mathbb{R}^3$ establishes an isomorphism between $T^*\mathbb{R}^3$ and $T\mathbb{R}^3$ which allows identification of the differential 1-form $d\varphi$ with the gradient field $\nabla\varphi$. Let then Λ_φ be the graph of $d\varphi$ on the open subset U of $\mathbb{R}^3$ where φ is defined. It is simple to verify how, because of the canonic symplectic structure of $T^*\mathbb{R}^3$ defined by the fundamental 2-form $\omega = dq \wedge dp$[2], (i) Λ_φ is a *Lagrangian* sub-manifold of T^*U, *i.e.* a sub-manifold Λ of dimension 3 on which the 2-form ω vanishes, (ii) the Hamiltonian H vanishes on Λ_φ and (iii) Λ_φ is *transverse* at each one of its points to the fibres of the canonical projection $\pi : T^*U \to U$.

The obstruction to the construction of a *global* functional solution φ comes from some *deficits of transversality*. Then one calls a *Lagrangian solution* of the problem

[2] On the cotangent bundle T^*M of a manifold (of local coordinates (q,p)), there exists a *canonical* 1-form $\theta = pdq$. $\theta(q,p) \in T^*_{(q,p)}(T^*M)$ associates to every tangent vector $(\xi,\eta) \in T_{(q,p)}(T^*M)$ the real $p(\xi)$ (well defined, because $p \in T^*_qM$ and $\xi \in T_qM$). The 2-form $\omega = d\theta$ is fundamental in classical mechanics because it is at the base of the Hamiltonian formalism.

a Lagrangian manifold Λ on which the conditions (i) and (ii) are verified but not necessarily the condition of transversality (iii). Λ is a reunion of solutions of the Hamiltonian system defined by H, solutions called *bicharacteristic* curves, and the caustic C is the *apparent contour* on $\mathbb{R}^3$ of the Lagrangian solution Λ, *i.e.* the projection of the critical locus Σ where the condition of transversality (iii) is no longer satisfied.

To transform such a geometrical solution into a functional one, sophisticated tools such as *oscillatory integrals* are necessary. The key idea is due to Maslov (see Maslov [84], Duistermatt [49] and our compilation [98]) and consists in looking for asymptotic solutions $u_\tau(q)$ no longer of the previous type $a_\tau(q)e^{i\tau\varphi(q)}$ but for sums locally finite of oscillatory integrals:

$$I(q,\tau) = \left(\frac{\tau}{2\pi}\right)^{p/2} \int e^{i\tau\varphi(q,\alpha)} a_\tau(q,\alpha)d\alpha$$

where $\alpha \in \mathbb{R}^p$, where $(\tau/2\pi)^{p/2}$ is a renormalisation factor and where $a_\tau(q,\alpha)$ (with compact support in α) is given, as $a_\tau(q)$, by an asymptotic expansion. The relation between such functional representations and Lagrangian solutions is given by the *stationary phase principle*, which says that, due to destructive interferences, the oscillatory integral concentrates when $\tau \to \infty$ on the critical locus V_φ where the phase $\varphi_q(\alpha)$ is stationary, *i.e.* where $\dfrac{\partial \varphi_q(\alpha)}{\partial \alpha} = 0$. A theorem due to Lars Hörmander, says that, at least locally, every Lagrangian solution can be represented that way.

1.2.5 Structuralism and Categorical Perception in Phonetics

The fourth link between morphodynamical models and perceptual geometry, which concerned us for a long time, does not regard vision but what is called *categorical perception* in phonetics: the categorisation of the phonemes is a product of the perception of sounds, when these are recognised as speech sounds. Inside a category, the discrimination ability degenerates (two different allophones of the same phoneme are perceived as identical even if they are acoustically different), while at the boundaries of the categories the discrimination is strong. As phoneticians say, discrimination is subordinate to identification: we can't discriminate correctly some occurrences unless they are identified as occurrences of different phonemes. It is possible to explain this remarkable phenomenon (which is really different, *e.g.*, from the perception of colours) considering that the phonetic percepts are attractors of neuronal internal dynamics X_a parametrised by acoustic cues a. In each language these cues as the *voice onset time* of stop consonants (VOT, the interval between the release of the consonant and the vibration of vocal folds) have central values. Using methods of vocal synthesis, it is possible to let them vary in a continuous way, and then to observe how bifurcations of the phonetic percepts are induced when crossing critical values. For example, a voiced labial \p\ becomes an unvoiced labial \b\.

At the end of the 70s, we developed morphodynamical models of the phonetic categorisations (see [99]) and we emphasised their double interest. First of all, they show how to generalise the models to the cases in which the external space M parametrising the internal dynamics X_a is not the standard space any longer, but a space of control-parameters. Second, as far as phonetics has been a model for modern structuralism, with Prince Troubetzkoy and Roman Jakobson, they show how morphodynamical models are natural for structuralism. On this base we build the "morphodynamical structuralism" program, which goes from phonetics to semiolinguistics (in Greimas' and Eco's sense), and anthropology (in Lévi-Strauss' sense), which in 1982 led to the four volumes of *Pour un schématisme de la structure. De quelques implications sémiotiques de la théorie des catastrophes* and to the three books *Les Catastrophes de la parole. De Roman Jakobson à René Thom* (Maloine, Paris, 1985), *Morphogenèse du Sens* [99] and *Physique du Sens* (Presses du CNRS, 1992).

1.2.6 Mental Dynamics

A fifth aspect of the morphodynamical models of the 70s is already linked with neurosciences, even if in a theoretical and qualitative fashion and not in an experimental and quantitative one. In his 1965 article *Topology of the Brain* (see [144]), Christopher Zeeman introduced the key idea that brain activity must be modelled by dynamical systems X_a on internal configuration spaces $P = I^N$, where $I = [0,1]$ is the range of activity of a neuron, N is the number of neurons of the system under consideration, and the X_a depend on control parameters a, micro-parameters such as synaptic weights and macro-parameters such as behavioural or psychological values. The main hypothesis was to identify mental contents with the topological structure of attractors of the X_a, and the flow of consciousness with a "slow" temporal evolution of the X_a. The strategy for explaining mental phenomena was then to use theorems concerning the general structure of attractors and their bifurcations for drawing empirical conclusions from this dynamical scheme *without knowing explicitly the* X_a. Indeed, if mental contents are modelled by attractors, then their significant changes during mental processes are modelled by bifurcations K observable in the control space M of *relevant* control parameters (the relevance depends of course on the nature of the mental phenomena under consideration). The dynamics X_a are defined on the *very high dimensional* manifold P and are "implicit", while the bifurcation sets K are "explicit". But suppose that, due to theorems of classification as those of Whitney-Thom, we know models of the K that are generated by dynamics Y_a defined on *low dimensional* internal spaces Q. Such a drastic *reduction of dimension* of the internal space (which is very similar to what one meets in statistical physics when one reduces an enormous number of degrees of freedom using what is called an *order parameter*) can then be identified to the passage from the neurodynamical micro-scale to the "psychological" macro-scale.

Thom's and Zeeman's neurodynamical hypothesis could not be confirmed at those times because of the lack of experimental tools. Nevertheless it raised an exciting problem, namely the *mereological* problem of *constituency*. Indeed, a key feature of mental contents is indeed how they are structured into components, what is called their "compositionality" or their "constituent structure". One need only think to the syntactic structure of a sentence to be convinced. If it is possible to model a mental content through a neurally implemented attractor, then this attractor must have a constituent structure on its turn. What could this possibly mean? This is what I proposed to call the problem of an *attractor syntax*. It is particularly difficult. If structures can be modelled in terms of bifurcations of internal dynamics X_a defined on internal spaces P, bifurcations which are unfolded in external spaces M, then one must "internalise" these external unfoldings into higher level internal dynamics defined on the product spaces $P \times M$.

The interested reader could consult our work (written with the help of René Doursat) *Cognitive Morphodynamics* [106] dedicated to this theme in relation to semiolinguistics (Per Aage Brandt, Wolgang Wildgen, Jean-Pierre Desclés, Pierre Ouellet, Jean-Guy Meunier, etc.) and cognitive linguistics (Len Talmy, Ron Langacker, George Lakoff, Paul Smolensky, Peter Gärdenfors, Terry Regier, Rick Grush, etc.).

1.3 The Connections between Perceptual Geometry, Image Processing, and Computational Vision from the 80s

During the 80s, the research inspired by Thom in perceptual geometry has been connected with parallel works by specialists in natural and computational vision and image processing. We will cite some of the most remarkable ones.

1.3.1 Koenderink, Hoffman and the "Singularities of the Visual Mapping"

In the mid-80s, the pioneering works by William Hoffman on the application of differential geometry and Lie group theory to vision, and by Jan Koenderink (and, later, by the Utrecht University group, in particular Luc Florack) on visual perception geometry, interested me, because, for the first time, I met specialists in perceptual psychology using resources coming from differential geometry, and, in particular, from singularity theory. Among the fundamental contributions of these scholars, I would mention four issues:

(i) The thesis according to which the visual cortex is a "*geometric engine*" which implements mathematical structures as fibrations, jet spaces of order 1 and 2, and contact structures (see [77], [64]).

(ii) The structuring role of singularities in perceptual geometry (see [76]).

(iii) The necessity to *integrate* (in the mathematical sense) the neural detections of local structures into global geometrical structures. If the brain can be a "geometric engine", this is because groups of receptive fields of visual neurons detect local features as edge orientations, crossings, inflection points, etc., which can be integrated later on through the functional architectures connecting these feature detectors in a specific way.

(iv) The essential role of *scale*. Perceptual geometry results from the integration of local measures by receptive fields which have a certain width, and it is consequently processed at a certain scale, *i.e.* at a certain resolution. For this very reason, perceptive differential geometry must be *multiscale*, because classical differential geometry corresponds to the idealisation of an infinite resolution (scale $= 0$). Koenderink and Witkin introduced the idea of *scale-space analysis*. It consists in a uniform parametrisation of all the relevant geometrical structures by a scale parameter σ. A constraint called "causality constraint" implies that when σ increases the complexity of the geometrical structures *simplifies*. Generally speaking, this constraint is expressed through a diffusion operator D. This means that if one considers a structure $S(x)$ defined on a background space E, one considers the multiscale background space $E \times \mathbb{R}^+$ endowed with the operator $D(x,\sigma)$ and imposes that $S(x)$ will be the initial value $S(x,0)$ of a solution $S(x,\sigma)$ of the diffusion equation $D(S(x,\sigma)) = 0$. The simplest way to obtain this for $E = \mathbb{R}^n$ is to use the heat equation $\frac{\partial S}{\partial \sigma} = \Delta S$ (where Δ is the spatial Laplacian). This strategy, called *Gaussian blurring*, raises interesting mathematical problems. For example, James Damon rewrote in this new framework Morse theory and the Whitney-Thom-Mather theory of universal unfoldings (see [42] and [104]).

1.3.2 Scale Space Analysis and Anisotropic Diffusion

The main drawback of Gaussian blurring in image processing and computational vision is that it does not respect the morphologies of the image. In fact, as we saw, these are dominated by the perceptual saliency of qualitative discontinuities and, by definition, the isotropic diffusion induced by the heat equation smooths discontinuities. That's why a certain number of specialists founded scale-space analysis on highly *anisotropic* non linear parabolic equations of diffusion, in which the intensity of the gradient of the image *inhibits* diffusion. Along discontinuities the gradient is very strong and even diverges, and therefore there is no diffusion any longer: diffusion only operates transversally to discontinuities and therefore preserves morphologies. The most known of these anisotropic equations is perhaps the one introduced by Jitendra Malik and Pietro Perona in 1990 for images defined by their intensity function $I(x,y;\sigma)$ (if we allow a small initial blurring, we can suppose that I is smooth): $\frac{\partial I}{\partial \sigma} = \operatorname{div}(g(\nabla I).\nabla I)$ where ∇I is the spatial gradient of I, and g is a decreasing positive function so that $g(x) \underset{x\to\infty}{\longrightarrow} 0$. As $\operatorname{div}(\nabla I) = \Delta I$, we retrieve the standard heat equation for $g \equiv 1$. Such PDEs are difficult to integrate numerically, because they imply phenomena of inverse diffusion and deconvolution which make them unstable. Let us also mention the well-known equation (1992) by Pierre-Louis Lions, Jean-Michel Morel, and Luis Alvarez [6] $\frac{\partial I}{\partial \sigma} = \partial^2_{\xi^2} I$ where ξ is a normal coordinate to the gradient, that is to say the tangent to the level line at the considered point. This equation is written :

$$\frac{\partial I}{\partial \sigma} = |\nabla I| \operatorname{div}\left(\frac{\nabla I}{|\nabla I|}\right) = \Delta S - \frac{H(\nabla I, \nabla I)}{|\nabla I|^2}$$

where H is the Hessian of S. It is uniformly parabolic along level curves, but it is completely degenerated in the direction of the gradient. It lets the level curves evolve as fronts with a normal speed equal to their curvature.

In a more general way, considering the level lines of the intensity functions $I(x,y;\sigma)$, it is possible to generate evolutions of the plane curves C_σ, which propagate as fronts in accordance to a law of the kind $\frac{\partial a}{\partial \sigma} = F(\kappa)\nu$ where a is a point of C_σ, ν the (external) normal at a to C_σ and κ the curvature of C_σ at a. The more studied cases have been the propagation at a constant speed, $F(\kappa) = v$, for example $v = 1$ (wave equation): $\frac{\partial a}{\partial \sigma} = \nu$, $\frac{\partial \kappa}{\partial \sigma} = -\kappa^2$. This way we obtain "grassfire" models where the propagation speed is constant as in optics.

The above model corresponds to $F(\kappa) = -\kappa$. Under the title of "curve shortening", "flow by curvature" or "heat flow on isometric immersions", this equation has been studied by geometers as Michael Gage, Richard Hamilton, Matthew Grayson, Lawrence Evans or Joel Spruck.[3] According to Grayson's theorem [60], "curve shortening" convexifies the curves C_σ and makes them metrically converge towards a circle. If $j_\sigma : \mathbb{S}^1 \to \mathbb{R}^2$ is the isometric immersion defining C_σ, we have $\Delta j_\sigma = -\kappa n$ and the diffusion equation $\partial_\sigma j_\sigma = -\kappa n$ is the heat equation $\partial_\sigma j_\sigma = \Delta j_\sigma$. In the functional space $\mathscr{J}$ of immersions $j : \mathbb{S}^1 \to \mathbb{R}^2$, this equation defines the gradient field of the function giving the length of the image curve $C = j(\mathbb{S}^1)$ and the theorem says that if C_0 is an embedded curve (as winding as it can be), then the heat equation contracts it into a circular point. And C_σ becomes convex *without* reaching any singularity.

This result concerning 1-dimensional curves cannot be generalized to *surfaces*. For instance a "dumbbell" like surface can be pinched and disconnected into two spheres. The interested reader could consult *e.g.* Brakke [26], Huisken [69], [70], Ecker, Chen, Giga, Goto [33].

To return to curves, Stanley Osher and James Sethian [96] studied the intermediate cases where $F(\kappa)$ is neither constant nor equal to $-\kappa$ but of a mixed form $F(\kappa) = 1 - \varepsilon\kappa$. If τ is the curvilinear abscissa of the curve C_σ, the curvature κ satisfies an equation of the type $\frac{\partial \kappa}{\partial \sigma} = \varepsilon \frac{\partial^2 \kappa}{\partial \tau^2} + \varepsilon\kappa^3 - \kappa^2$ which is a reaction-diffusion equation where the reaction term $\varepsilon\kappa^3 - \kappa^2$ (which pushes C_σ towards singularities) is counterbalanced by the smoothing effect of the diffusion term $\varepsilon \frac{\partial^2 \kappa}{\partial \tau^2}$.

[3] As a matter of fact, the theory comes from Richard Hamilton which tried to solve much more difficult problems in General Relativity. By using the heat equation, he showed how, if X is a Riemannian compact 3-manifold with a positive Ricci curvature R_{ij}, then X admits a Riemannian metrics with *constant* positive Ricci curvature. Now, the latter have been classified. Hamilton tried also to generate *closed geodesics* starting from any closed curve. Among others, these techniques allowed Grigori Perelman to prove Poincaré's conjecture in 2003.

1.3.3 Segmentation and Illusory Contours in David Mumford

Morphological geometry of images, clearly very close to perceptual geometry, became this way a top-ranking scientific object as much as it was necessary to find specific performing *algorithms* to compute it. This computational approach goes beyond morphological modelling, and it intermediates in some sense between perceptual geometry and neurogeometry insofar as, on the one hand, it is based on explicit algorithms while, on the other hand, these algorithms don't have to be neurally implemented. One could think that this is not a real problem since neural nets are computationally equivalent to Turing machines. But this is not the case, because in neurogeometry the neural hardware is *dedicated* and it is the *specificity* of its functional architecture that causes its activations to be equivalent to a calculus.

Another great mathematician in image processing, who has been an essential inspirer for neurogeometry is David Mumford, 1974 Fields medal for his works in algebraic geometry. Two contributions are of a peculiar importance.

(i) First, the Mumford-Shah segmentation model (see [91]), a variational model which allows to optimize the approximation of an image $I(x,y)$, defined on a window W and potentially very noisy, by a "good" morphological structure (u,K) where K is a segmentation of W partitioning W in open domains W_i (connected components of $W-K$) and u an approximation of I which is regular on the W_i while presenting discontinuities along K. A considerable number of algorithms of this kind have been proposed. In a way or in another, they all consist in a merging of local domains into regular, homogeneous regions limited by crisp edges, that is in "*splitting* and *merging* different parts of the domain W" as Mumford said. The principal theoretical problem which they encounter is that the $2D$ regions and the $1D$ edges are geometrical entities of a different dimension, which compete and interact in a very subtle way (see the synthesis Morel-Solimini [89]). In order to compare the possible segmentations between them and to measure their approximation degree of the image I, it is relevant to introduce an energy. The energy proposed by David Mumford and Jayant Shah is

$$E(u,K) = \int_{W-K} |\nabla u|^2 \, dx + \lambda \int_W (u-I)^2 + \mu \int_K d\sigma \, .$$

It includes that three terms: the first term measures the variation and controls the regularity of u on the open subsets W_i; the second term controls the quality of the approximation of I by u; the third term controls the length, the regularity, the parsimony, and the localisation of the edges K and inhibits the oversegmentation phenomena (by segmenting in sufficiently small regions it is obviously possible to approximate I as much as we want, but such oversegmentations are irrelevant).

This variational model is a particular case of what is called in physics a "free boundary problem". It is extremely difficult to solve and a lot of works have been dedicated to it until now in particular by the Ennio De Giorgi centre at the Scuola Normale Superiore in Pisa (among the others Luigi Ambrosio and Gianni Dal Maso), Jean-Michel Morel, Alexis Bonnet and Guy David (for an introduction see [103] and [105]). If the set of edges K is *fixed*, then it is simply a classic

Neumann problem: $\Delta u = \mu(u - I)$ inside the connected components W_i of $W - K$ and $\frac{\partial u}{\partial \nu} = 0$ along the edges $\partial W \cup K$ (ν is the normal at the edge). If K is *free*, the problem is really different. In the simplest case, where the approximants u are *locally constant*, we have $\nabla u = 0$ and $E(u,K)$ is reduced to a

$$E(u,K) = \lambda \int_W (u-I)^2 + \mu \int_K d\sigma \,.$$

Then, the approximation u is entirely determined by K because, on the connected components W_i of $W - K$, it is simply equal to the average of I. Thus, we have only to find K. Mumford has shown that, in this case, the minimum of $E(u,K)$ exists and is realised through a K which is piecewise C^1, whose curvature is bounded by $8\,\mathrm{osc}(I)^2$ (where the oscillation of I is $\mathrm{osc}(I) = \max I - \min I$) and which has triple points at 120° and edge points at 90° on ∂W as only singular points.

In the general case where the approximations u are not necessarily locally constant, Mumford's conjecture says that it is still the same, except that a third class of singularities of "end point" type could occur (they are called "cracktips" or "fault" ends in reference to physical models). They are end points of the discontinuity lines. The conjecture is still unsolved. First it has been shown that K is a closed and "regular" set of topological dimension 1 (it can't be fractal). Later on, David and Semmes [43] showed a property of uniform rectifiability. Then Alexis Bonnet proved the conjecture for the isolated connected components of K. Then Bonnet and David [21] showed how the "cracktips" are minimisers.

In 2003, Giovanna Citti and Alessandro Sarti proved that the Mumford-Shah model is a limit (in the variational sense of the word "limit") of a model of synchronisation of oscillators, a result which would justify a neural implementation (we will return on this point in section 1.5.1.1).

(ii) David Mumford's second major contribution concerns a variational model proposed in 1992 [90] regarding the illusory contours. This is the model which will be reformulated by neurogeometry. It uses *elastica* curves, introduced by Euler in elasticity theory, which are curves minimising at the same time the length and the square of the integral of the curvature κ of curves, *i.e.* an energy $E = \int_\gamma (\alpha\kappa^2 + \beta)ds$ where γ is a smooth curve in $\mathbb{R}^2$ with element of arclength ds. It is possible to justify this model in this way: the virtual contour will correspond to a chain of triplets (a_i, p_i) (a_i = positions and p_i = orientations at the positions a_i) along which the loss of activity is the weakest possible. But leaks can have a double origin: (a) leaks proportional to the number N of elements of the chain, with a constant factor β; (b) leaks due to curvature and equal to the sum of the deflections of orientation between consecutive elements, with a constant factor α. If θ_i is the angle of the slope p_i, we can take for example $\sum_{i=1}^{i=N-1} (\theta_{i+1} - \theta_i)^2$. At the limit, the first term tends to the length $\int_\gamma ds$ and the sum of the deviations $\Delta\theta$ in the second term tends to $\int_\gamma \kappa^2 ds$ since, by definition, $\kappa = \frac{d\theta}{ds}$. The minimisation of leaks "online" consequently leads to Mumford's variational problem. Its solutions are transcendental curves (*i.e.* non algebraic ones) called *elastica* which could be explicitly represented by elliptic functions.

David Mumford developed a *stochastic* explanation of his model. Let us suppose that the curvature $\kappa(s)$ of the curve γ (parametrised by its arclength s) is a *white noise*. As $\kappa(s) = \dot{\theta}(s)$, this means that $\theta(s)$ is a *Brownian motion*. In other words, at every time, motion is a random Gaussian variable of average 0 and of variance σ. If we further suppose that the length l of γ is a random variable which follows an exponential law $\lambda e^{-\lambda l} dl$ (thus l is constant for $\lambda = 0$), then the probability of a particular curve γ is given by $\Pr(\gamma) = e^{-\int(\alpha\kappa^2+\beta)ds}$ with $\alpha = \frac{1}{2\sigma^2}$ et $\beta = \lambda$. Consequently, elastica are the most probable curves.

1.3.4 Receptive Profiles and Wavelets

Another fundamental progress in image processing which can be used to mediate between perceptual geometry and neurogeometry is the remarkable development of *wavelets* algorithms promoted since the late 80s by analysts such as Yves Meyer, Stéphane Mallat or Ingrid Daubechies. The wavelet transform of a signal is a localised and multi-scale harmonic analysis that allows to easily detect (unlike the Fourier transform) the qualitative discontinuities encoded in the signal. For an introduction, see Stéphane Mallat's *A Wavelet Tour of Signal Processing* [80]. The connection with David Marr's pioneering way to treat the retinal signal by the retinal ganglion cells was quickly established, and it became clear that the retina, the lateral geniculate nucleus, and the primary cortical visual areas perform a series of consecutive wavelet analyses of the optical signal and that this is where the primary geometric formatting of the visual input comes from. In particular, the receptive profiles of the "simple" neurons of $V1$, which detect edge orientations, are assimilable to oriented wavelets modelled on second derivatives (*i.e.* anisotropic Laplacians) of Gaussians (the width of the Gaussian defines the scale of the processing). The zones of the receptive field where the receptive profile is > 0 (resp. < 0) are called ON (resp. OFF) by neurophysiologists.

A lot of fine grained analyses have been conducted on the receptive profiles of the visual neurons, for example by Gregory DeAngelis' team at Duke. In fact, the receptive profiles implement derivatives of Gaussians up to order at least 3. For example, figure 1.2 from [45] shows a receptive profile where the spatial and temporal profiles are non separable: the ON/OFF regions are shifted during time, and for this reason the periphery response of the receptive field is delayed with respect to the central one. This allows the neuron to measure the speed of the edges it detects. Figure 1.3 shows a model using the third derivative of a Gaussian $\frac{\partial^3 G}{\partial u^2 \partial v}$, where the (x,t) plane is obtained through a $\frac{\pi}{10}$ rotation of the (u,v) plane (it is the origin of the non-separability). The agreement with the experimental data is quite remarkable.

This analogy between a wavelet analysis of an image and the processing of the optical signal by fields of receptive profiles which operate in parallel and detect local geometrical features plays a crucial role in neuroscientific theorisation on vision. However, it failed to consider the functional architectures, which allow integration of these local data into global structures.

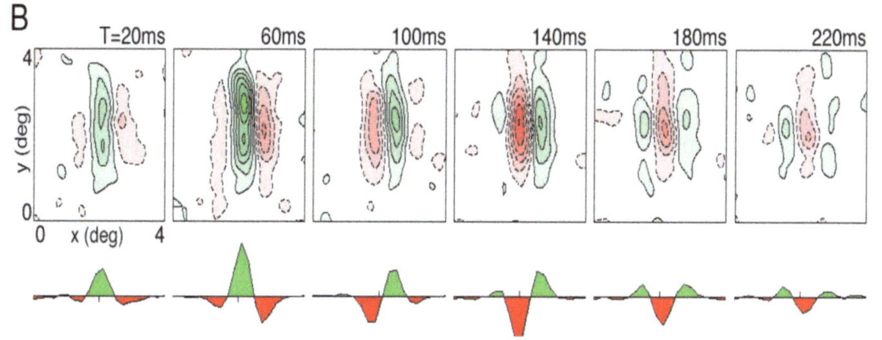

Fig. 1.2 Temporal evolution of a non-separable receptive profile $\varphi(x,y,t)$ from $t = 20$ms to $t = 220$ms. The spatial organisation of the receptive profile changes; this is not the case for the separable receptive profiles. Above, the spatial profile $\varphi_t(x,y) = \varphi(x,y,t)$. Below, its section for $y = 0$. (From DeAngelis *et al.* [45]).

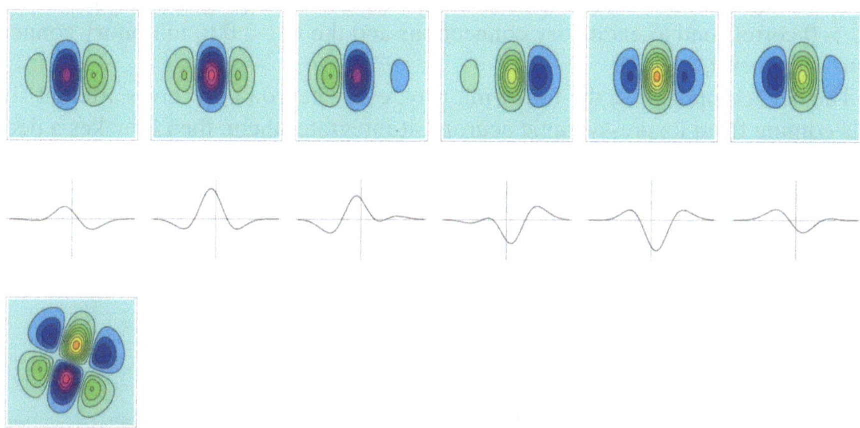

Fig. 1.3 Model of the temporal evolution of the receptive profile $\varphi(x,y,t)$ of figure 1.2. We took a third derivative of a Gaussian $\varphi(x,y,t) = \frac{\partial^3 G}{\partial u^2 \partial v}(u,y,v)$ with the rotation $(u,v) = r_\theta(x,t)$, $\theta = \frac{\pi}{10}$. Above, the evolution of the spatial profile $\varphi_t(x,y) = \varphi(x,y,t)$. At the middle, the sections for $y = 0$. Below, the receptive profile $\varphi(x,0,t)$ on the (x,t) plane for $y = 0$.

1.3.5 Neural Net Dynamics and Attractors

As we saw in section 1.2.6, in the 70s Thom and Zeeman introduced the main idea, according to which a mental content is an attractor of a neural dynamic. However, these dynamics remained *implicit* in their models and if the hypothesis was yet operative, this was because deep theorems on classification of singularities and bifurcations show their *universality*. So, from a methodological point of view, the models were quite special: rather than starting from *explicit* equations and deducing from them the bifurcations, they started from the observed bifurcations and they ascended by *abduction* to some generating dynamics of minimal complexity, a dynamics of which the real implicit dynamics should only be an extension through irrelevant parameters.

The situation changed with the introduction, at the beginning of the 80s, of the neural net models inspired by statistical physics and later developed, after Jack Cowan et Hugh Wilson (1972), by specialists as John Hopfield, Daniel Amit, Haim Sompolinski [123], Misha Tsodyks [132] and, in France at those times, by physicists as Gérard Toulouse, Marc Mézard, Jean-Pierre Nadal, Gérard Weisbuch at the Lab of Statistical Physics of the ENS (Ecole Normale Supérieure), or Manuel Samuelides at Toulouse [48], or Daniel Hansel and Claude Meunier at the Ecole Polytechnique, or also Gérard Dreyfus at the ESPCI. Under its simplest form, a neural net consists of N units u_i whose activation state y_i changes in a certain state space S. The most useful cases are $S = \{0,1\}, \{-1,1\}, [0,1]$. An instantaneous net state is therefore described by the vector $y = (y_i)_{i=1,\ldots,N}$ in the configuration space $P = S^N$ which is a space of great dimension if N is large. The units u_i are little threshold automata connected to each other through connections whose strength is measured by synaptic weights w_{ij}. The w_{ij} determine the program of the net. The $w_{ij} > 0$ correspond to excitatory connections and the $w_{ij} < 0$ to inhibitory connections. In general, $w_{ii} = 0$.

The net "computes" in the following way: every neuron u_i receives afferent signals coming from its pre-synaptic neurons, it integrates them, then it makes a decision and sends an efferent signal to its post-synaptic neurons. One defines in general the input of the unit u_i as the weighted summation of the afferent signals

$$h_i = \sum_{j=1}^{j=N} w_{ij} y_j, \text{ i.e. } h = wy$$

and the local transition laws are of the kind:

$$y_i(t+1) = g\left(h_i(t) - T_i\right), \text{ i.e. } y(t+1) = g\left(h(t) - T\right)$$

where T_i is a threshold and g a highly non linear gain function. Typically we have $g =$ Heaviside function if $S = \{0,1\}$, $g =$ sign function if $S = \{-1,1\}$, $g =$ sigmoid function, for example $1/(1+e^x)$, if $S = [0,1]$. The synaptic weights w_{ij} and the thresholds T_i constitute a control space W. The global dynamics of the net Y_w on P is obtained by aggregating the local transition laws and by iterating them. In the limit of a continuous time, we get a large system of differential equations of the kind:

$$\dot{y} = -y + g(wy - T) .$$

In the limit of a spatial continuum of neurons, we get PDE (on densities) of the kind:

$$\frac{\partial y(x,t)}{\partial t} = -y(x,t) + g\left(\int (w(x,z)y(z,t) - T(x))\, dz\right) .$$

As Daniel Amit emphasised, under the hypothesis of a complete feedback (loop between the inputs and the outputs) the asymptotic states of the system – in particular its *attractors* – are meaningful (see [7]) and define the internal states of the net. The basic dynamic phenomenon is therefore the asymptotic capture of an instantaneous global state y_0 of the net by an attractor A. That's why Daniel Amit proposed the expression "attractor neural networks". ANNs are dynamical computers that bifurcate from attractors to attractors. They make explicit the general neurodynamical models worked out by Thom-Zeeman.

The dynamics that can be obtained this way are in general of a great complexity. For example, in the case (neurobiologically completely unrealistic) of *symmetrical* connections, John Hopfield remarked at the beginning of the 80s that for $S = \{-1, +1\}$ and $g =$ sign function, the equations of the net are analogous to the ones for systems of interacting spins which we meet in statistical physics. The energy minimised by the dynamics is given by:

$$E = -\frac{1}{2}\sum_{i \neq j} w_{ij} y_i y_j + \sum_i T_i y_i \ .$$

In the measure in which the synaptic weights w_{ij} are analogous to the coupling constants and are, in an intricate way, at the same time > 0 et < 0, these systems – which exemplify the *simplest* case of a formal neural net – correspond to the *most complex* case of spin systems, the so called "spin glasses". Their energy presents a considerable number of local relative minima and, in order to find the global absolute minimum, the classical methods of the "gradient descent" kind do not work. It is necessary to use sophisticated algorithms coming from statistical physics such as the so-called *simulated annealing*.

When the synaptic weights become *asymmetrical*, the energy function does not exist any longer, and the dynamics can become of a greater complexity. One finds different classical routes toward chaos and in particular the period doubling route known as Coullet-Feigenbaum-Tresser subharmonic cascade.

Many results of this kind show how it is possible to give a rigorous status to Thom's thesis that mental contents are attractors of dynamical systems implemented in neural nets and therefore that cognitive functions can be naturally conceived in terms of statistical physics and neural dynamics. The example is particularly evident for the fundamental cognitive processes of *categorisation* and *learning*. A sensory input is modelled as an input of the net and the evoked percept is identified with the attractor which captures the input. The categorisation of the inputs processed by the percepts is interpreted as the partition of the configuration space of P in attraction basins $B(A)$ of attractors A – the categories – which work as *prototypes*. The scales which psychologists call "gradients of typicality" are then interpreted as Liapounov functions on the $B(A) - A$ (*i.e.* functions of the " energy" type that strictly decrease on the trajectories when those asymptotically tend towards their limit attractors and vanish on these attractors). We retrieve the morphodymamical models of the 70s, *e.g.* for categorical perception in phonology, evoked in section 1.2.5.

The aspect of learning corresponds to an *inverse* problem. The associated direct problem consists, given a matrix w of synaptic weights, of finding the attractors of the dynamic Y_w. On the contrary, the inverse problem is, some attractors A_i being given, to find a matrix w that generates them. Some algorithms have been developed for this purpose, in particular the one called *backpropagation* which consists of computing, from an initial matrix w_0, the gap between the attractors of Y_{w_0} and the desired attractors and to back-propagate the error by adjusting w_0. Similar algorithms define *slow* dynamics in the control spaces W of the synaptic weights.

In this way, neural nets become privileged models for cognitive processes under the name of *connectionist* models. The fundamental problem of *constituency* in an "attractor syntax" (see section 1.2.6) was raised again but in a more technical way because now the internal dynamics of the models were explicit. It was also raised in a more polemical way, because in 1988 Jerry Fodor and Zenon Pylyshyn [51] attacked Paul Smolenky's article [122] on the "Proper Treatment of Connectionism" (PTC), explaining how connectionist models can't in principle generate cognitive models because their attractors lack an internal structure. The problem was particularly sharp as much as it concerns grammatical relations and semantic roles, constituency and compositionality, but was partially solved by using the rich resources of the theory of singularities and bifurcations, which were unused by standard connectionist models of the PDP type ("parallel distributed processing" in the sense of David Rumelhart and James McClelland) or PTC (Paul Smolensky). We refer to our works [100] and *Cognitive Morphodynamics* [106] for details.

In 1991 and 1992 two Conferences on *Compositionality in Cognition and Neural Networks*, organized by our colleagues Daniel Andler, Elie Bienenstock and Bernard Laks [39], were held in France at the Royaumont Abbey. Then, Bernard Victorri and Catherine Fuchs organized in 1992 a Conference on *The Continuum in Linguistics*. We organized in 1995 with Umberto Eco and Patrizia Violi in San Marino an international Conference *Topology and Dynamics in Cognition and Perception* focusing on Cognitive Grammars (in particular Len Talmy who was there). We edited also a special issue of *Sémiotiques* "Dynamical Models and Cognitive Semiotics". Inspired by the Royaumont workshops, another Conference on *Dynamic Representations in Cognition* was also organized by Robert Port and Tim van Gelder at Indiana University in 1992. They edited then at the MIT Press in 1995 a reference book *Mind as Motion* on dynamical models in linguistics.

1.4 The Neurogeometrical Turn Since the 1990s

1.4.1 Neuromathematical Interdisciplinarity

At the end of the 80s, many instruments were available:

1. A perceptual geometry in a Thom-Koenderink fashion, which mobilised the tools of Morse's theory, jets, singularities, bifurcations, stratifications, and universal unfoldings.
2. Deep connections with some algorithms of multi-scale image processing: local algorithms as wavelet analysis (Mallat-Marr) and anisotropic diffusion, or global algorithms as Mumford-Shah variational segmentation model.
3. Equally important deep connections with neural nets and statistical physics.

In other terms, one could count on suitable elements for morphological geometry and suitable implementation principles for image processing. Many works have been developed in these directions, in particular at the CREA (Ecole Polytechnique), a research centre created in 1982 by Jean-Pierre Dupuy, which I joined in 1986. From 1991, Bernard Teissier invited me for some lectures on these topics at the department of mathematics of the ENS (Ecole Normale Supérieure), and I also dedicated to them some of my master classes. However, what lacked at that time was a technical link with neurosciences founded on precise experimental data. To be sure, since the 70s Zeeman had introduced, in a qualitative way, some neural dynamics, (see section 1.2.6), to be sure Koenderink and Hoffman (see section 1.3.1) or wavelet analysis (see section 1.3.4) constituted a link with data on receptive fields and the organisation of visual areas in hypercolumns, but it lacked a real confluence with neurosciences, a confluence which would allow to connect rich experimental data to non trivial mathematical models.

In my case, this confluence occurred at the beginning of the 90s at two levels. First through the great richness of a new interdisciplinary context, and then through the first neuroimaging data on the functional architectures of the primary visual areas.

The interdisciplinary context was the one of the Master created in 1990 by the prominent specialist in vision Michel Imbert. This original Master which institutionally grouped the EHESS, the ENS and the Pierre et Marie Curie University (Paris VI) was a great success and allowed collaborations with specialists of neurosciences, physiology and psychophysics as, apart from Michel Imbert himself, Alain Berthoz, director of the LPPA (Laboratoire de Physiologie de la Perception et de l'Action) at the Collège de France, Yves Frégnac director of the UNIC (Unité de Neurosciences Intégratives et Computationnelles) at the CNRS (Centre National de la Recherche Scientifique), Jean Lorenceau (UNIC and LPPA), the colleagues of the LENA (Laboratoire de Neurosciences et d'Imagerie Cérébrale) of the CNRS at the hospital "La Pitié-Salpétrière". In this particularly favourable context, many fruitful interdisciplinary exchanges quickly started. A working group organised many conferences at the "Fondation des Treilles" (founded by Anne Gruner-Schlumberger). The 1993-1994 ones on *Geometry and Vision* grouped among others Bernard Teissier, Jean-Michel Morel, David Mumford, Gérard Toulouse, Stéphane Mallat, Yves Frégnac, Jean Lorenceau, Olivier Faugeras, Elie Bienenstock, and the 1998 one was dedicated to *Methodology in Cognitive Science*. There have been many other meetings, in particular in 1995 a symposium on *Mathematics and Brain* at the Institut Henri Poincaré in Paris. Jean-Michel Morel, David Mumford and Bernard Teissier also organised in 1998 a special quarter on *Mathematical Questions on Signal and Image processing* at the Centre Emile Borel of the Institut Henri Poincaré. Then an important symposium, *The Mathematical, Computational and Biological Study of Vision*, was held in 2001 at the Oberwolfach Mathematisches Forschungsinstitut by David Mumford, Christoph von der Malsburg and Jean-Michel Morel.

During the 1990s, another privileged collaboration with neurosciences was established with Francisco Varela, member of the CREA, where he developed his ideas on self-organised and autopoietic systems, and also member of the LENA, where he worked on global neural dynamics which are highly complex non linear dynamics defined on very high dimensional spaces, presenting many chaotic properties but also simpler observable ones. To reconstruct such dynamics from the time series yielded by the measure of suitable neural signals, one can use tools elaborated on the basis of deep mathematical results by Whitney, Thom, Zeeman, Takens, Crutchfield and others. Often, the dynamics result from the interaction of a large number of oscillators of the FitzHugh-Nagumo type[4]. As was emphasised by Wolf Singer, Heinz Schuster and Varela, it is the *synchronisation* of oscillators which is cognitively functional. It is of course complementary of desynchronisation processes, and it is that way the difficult mathematical theory of coupled oscillators (Yoshiki Kuramoto, Hiroaki Daido, Bard Ermentrout, Nancy Kopell, etc.) becomes involved in neurodynamics. What is particularly interesting in global neural dynamics is that they couple many subdynamics related by control, feedback and hierarchical relations. These dynamical hierarchies are structured in synchronisation / desynchronisation temporal flows, which explains the temporal segmentation of consciousness along "syntactically" organized cognitive acts.

On both experimental and theoretical sides, one of the contributions of the new imaging techniques has been to allow the identification of two fundamental mathematical concepts corresponding respectively to the two systems of connections which constitute the functional architecture of $V1$: the concept of *fibre bundle* or *fibration* for the "vertical" retino-geniculo-cortical connections, and the concept of *contact structure* for the "horizontal" cortico-cortical connections (also called "lateral" sconnections).

1.4.2 Efficient Coding, Receptive Profiles and Natural Images

Let us briefly return on the peculiar form of the receptive profiles of visual neurons (from the ganglion cells of the retina to the cortical cells) modelled as derivatives of Gaussians or Gabor patches. Many specialists, for example Joseph Atick and Jean-Pierre Nadal, have shown how these profiles can be retrieved from general hypotheses of information theory. They correspond to an "informational strategy" and to "design principles" which optimise the efficiency of the representation of information. As Atick said ([8], p. 213), "efficiency of information representation in the nervous system potentially has evolutionary advantages". The efficient neural codes of visual information allow to describe the environment in a compact way using an effective "visual vocabulary" of geometrical features. Now, *natural* image statistics is particular because strong correlations exist between the pixels and because of the existence of edges. The different possible statistics of the inputs influence the distribution of the spikes. Yves Frégnac's team, for example, studied four statistics: moving gratings, dense noise, natural images with eye motion, and gratings with eye motion too. They found that the variability of spikes decreases with the

[4] FitzHugh-Nagumo equations are a simplified version of the Hodgkin-Huxley equations for spikes.

complexity of the classes of stimuli and that the temporal precision of their emission time increases. It is a remarkable result.

The "pixelised" representation at the level of photo-receptors does not reflect these statistics and is therefore fundamentally inefficient. Consequently, it had to be improved. From the years 1950-1960 Fred Attneave and Horace Barlow proposed that, in order to be efficient, the neural coding must eliminate as much as it is possible the enormous redundancy of the inputs, in other terms it has to *compress* them. For this reason, the statistical laws of natural images play a fundamental role in the evolutionist explanation of the design of neuronal hardware. It imposes to the neural information processing "ecological" constraints (in James Gibson's sense of the term "ecological") and some "priors" (in the Bayesian sense). As far as the relations between neurons are concerned, the maximisation of the code-efficiency is based upon the possibility of rendering the responses of the different neurons statistically *independent* as much as possible. The problem is difficult.

The leading idea for solving it is to optimise the information and to decorrelate the signal by eliminating increasingly higher correlations. The simplest way is to suppose that it exists a linear filtering (a convolution by a receptive profile) which performs a compression of the signal $I(a)$ by decorrelating first of all its spatial self-correlation $C(a_1,a_2) = \langle I(a_1)I(a_2)\rangle$ (where $\langle\ \rangle$ represents the average). Due to homogeneity and isotropy, the self-correlation depends only on the distance $r = |a_1 - a_2|$, and $C(a_1,a_2)$ has the form $C(r)$. The Fourier transform $\widehat{C}(\omega)$ of $C(r)$ is called the *power spectrum* of the signal: $\widehat{C}(\omega) = \int C(r)e^{-i\omega r}dr$. Now, the natural image statistics shows that they have a spectrum in $\widehat{C}(\omega) = \frac{1}{|\omega|^2}$ (Field's law) and this corresponds to the fact that the spatial self-correlation is scale-invariant (*i.e.* $C(\alpha r) = \alpha C(r)$). Indeed, the inverse Fourier transform is $C(r) = \int \frac{e^{i\omega r}}{|\omega|^2}d\omega$ and therefore, using the change of variable $\omega = \frac{\lambda}{\alpha}$ we get

$$C(\alpha r) = \int \frac{e^{i\omega\alpha r}}{|\omega|^2}d\omega = \int \frac{\alpha^2 e^{i\lambda r}}{|\lambda|^2}\frac{d\lambda}{\alpha} = \alpha C(r)\ .$$

Let $\varphi(a)$ be the receptive profile of the considered linear filters, and let $(I*\varphi)(a)$ be the result of the filtering of the signal I. The decorrelation means that the average of the product $\langle (I*\varphi)(a).(I*\varphi)(a')\rangle$ is $\delta(a-a')$ (where $\delta(x)$ is the Dirac distribution). If we consider only Field's law, then the power spectrum of $I*\varphi$ is flat and the decorrelation is expressed by the "whitening" of the signal. But even if this method is suitable to the spectra in $\frac{1}{|\omega|^2}$, it is indeed very bad for the noise because it amplifies it in the range of high frequencies where it already dominates. It is necessary therefore to associate the decorrelation to a smoothing of the signal suppressing the noise. That's the way in which Atick shows that if we want to correctly decorrelate the signal also when noise is present, we must use receptive profiles in Laplacian of Gaussian ΔG of the type of ganglion cells.

Jean-Pierre Nadal and his colleagues [133] deeply analysed the problem by considering not only Field's law, but also, in an explicit way, the importance of multi-scale *edges* in natural image statistics. They showed how the minimisation of neural code redundancy under the constraint of edge detection leads to oriented wavelets.

In a more general way, we can consider large data bases of natural images and try to analyse them in independent components (ICA) which are neurally implemented. We look for components (the receptive profiles of the neurons) which are at the same time statistically independent and "sparse", where "sparsity" means that, for a given image, the greatest part of the components present a very weak response, only a few of them presenting a meaningful response. Sparse representations have many advantages: they offer a compromise between, on the one hand, representations in which almost all the neurons of the net are implied in the elaboration of every stimulus, and, on the other hand, representations in which, for every stimulus, there would be just a neuron which would selectively respond (the so-called "grandmother" cells); they are composed of basic patterns (the receptive profiles considered as "atoms") which are adapted to the considered set of stimuli (in our case, natural images); they magnify the power of neural nets as associative memories by eliminating the interferences between the activity patterns elicited by the different inputs; they make explicit the structure of the stimuli because their components are relevant features; they allow the net to economise its energy; they can easily be learned through Hebb's learning rules.

The problem to find the optimal sparse representations for a class of stimuli is difficult. It can be formulated in the following way. Let $\Sigma \subset \mathbb{R}^P$ be a class of stimuli I^k, $k = 1,\ldots,S = \#(\Sigma)$, which are images I with $P = p^2$ pixels described as vectors $I = (I_r)_{r=1,\ldots,P}$ in the canonic basis of $\mathbb{R}^P$. We want to find a "good dictionary" Φ of atoms $\varphi_i \in \mathbb{R}^P$, $i = 1,\ldots,N$, which allows us to "well" decompose the $I \in \Sigma$ under the form

$$I = \sum_{i=1}^{i=N} s_i \varphi_i \ .$$

The representation of $I \in \Sigma \subset \mathbb{R}^P$ by $s = (s_i)_{i=1,\ldots,N}$ is the code of I with respect to the "dictionary" Φ which is a matrix $P \times N$ the columns of which are the $\varphi_i \in \mathbb{R}^P$. In the present case, the one of the retina and of $V1$, we have $N \gg P$, *i.e.* what it is called an "overcompleteness". This means that the φ_i generate $\mathbb{R}^P$ without being linearly independent and are on the contrary highly redundant. That's for the same reason that the codes s can be sparse.

The space Σ is not a vector sub-space of $\mathbb{R}^P$ because a linear superposition of natural images it is not a natural image. It has a complex form and the atoms φ_i are a way to locally analyse it at numerous points by means of a kind of tangent structure.

In order to find optimal sparse codes, the basic idea is to minimise an energy of the form

$$E(I,s) = \frac{1}{2}\|I - \Phi s\|^2 + \lambda \sum_{i=1}^{i=N} |s_i|$$

where the first term is the square of the euclidean distance between the stimulus I and its code Φs and the second term is the L^1 norm of the code. The minimisation of the first term guarantees that the representation Φs is a good approximation of I and the minimisation of the second one guarantees that the code is sparse. To precise this guiding idea leads to a lot of difficult problems. We can refer, for example, to the articles of Bruno Olshausen, David Field, Karol Gregor, Yann Le Cun, Eeno Simoncelli [95], [61], [121]. They show how the base functions which one obtains by similar methods are very close to the receptive profiles of the simple cells of $V1$ with their selectivity to orientation, their odd symmetry ($\varphi(-a) = \varphi(a)$) or even symmetry ($\varphi(-a) = -\varphi(a)$), their feature ON or OFF, and their selectivity to spatial frequency (the scale). One observes also detectors of end points.

1.4.3 Orientation Hypercolumns, "Internal" Variables, and Fibre Bundles

Let us return to the functional architecture of the striate area $V1$. After pioneering experiments carried out in the late 50s by Vernon Mountcastle on the somatosensory cortex of the cat, the structure of $V1$ (cortical area 17 of the cat) in orientation hypercolumns has been a major discovery of David Hubel and Torsten Wiesel in the early 60s. It won them the Nobel prize in 1981. In the $V1$ area the "simple" neurons (as opposed to "complex" and "hypercomplex" cells) are sensitive to orientation (and also to phase, resolution, spatial frequency, ocular dominance and colour). If we simplify the situation, we can say that these neurons detect pairs (a,p) of a retinal position $a = (x,y) \in M$ (M is the retinal plane or the visual field, $M \simeq \mathbb{R}^2$) and an orientation p at a. Along a penetration *orthogonal* to the cortical surface, the retinal position a and the prefered orientation p detected by the neurons remain approximately constant. This "vertical" redundancy – which implements a population coding of the position – defines *orientation columns* of about 20μm. As Gregory DeAngelis demonstrated, the *phase* variation dominates inside the columns. Moreover, population coding allows to the system a better resolution than the one of individual neurons. But, along a penetration *parallel* to the cortical surface, the prefered orientation p varies from $0°$ to $180°$ in steps of about $10°$. This "horizontal" grouping of columns defines an *orientation hypercolumn* which is a neuronal micro-module from 200μm to 1mm wide.

This neuroanatomic structure means that, for every position $a \in M$, there exists a functional micro-module P_a which implements the space P of every possible orientation. Moreover, what neurophysiologists call the *retinotopy* of the retino-geniculo-cortical pathway means that the P_a vary in a regular way with a. We recognise in this structure the *fibre bundle* $\pi : \mathbb{V} = M \times P \to M$ which projects the Cartesian product $M \times P$ onto its first factor M, through the identification of the hypercolumn P_a with the fibre $\{a\} \times P$. P can be, according to the case, modelled by the projective line $\mathbb{P}^1$ (the circle of angles θ modulo π), the circle $\mathbb{S}^1$ (the one of θ modulo 2π) or the real line $\mathbb{R}$ (if we take the slope $\tan(\theta)$).

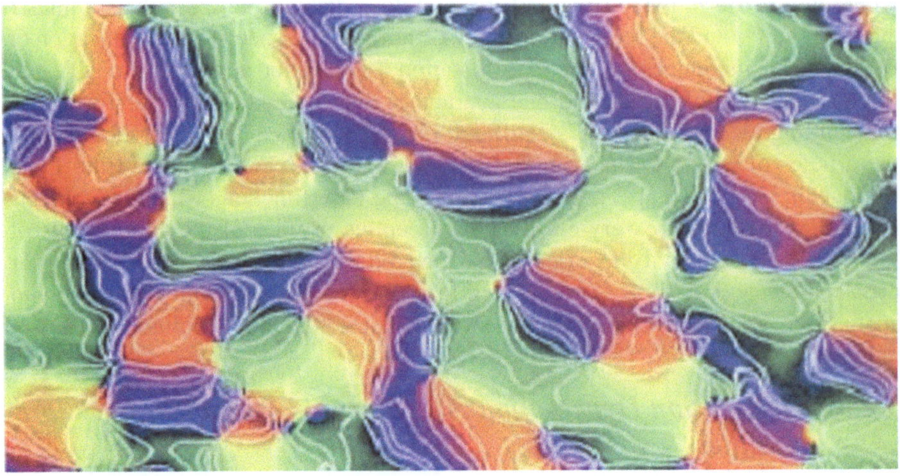

Fig. 1.4 The pinwheels of $V1$. The isochromatic lines are the iso-orientation lines of $V1$. (From [138]).

By the way, this natural geometrical interpretation of the hypercolumns in terms of fibre bundles corresponds to a description given by Hubel himself in 1988 [67] with the concept of "engrafted variables": the base-variables are the variables (x, y) of retinal position and the cortex "engrafts" above them other secondary variables such as the orientation. The "engrafted" variables, *i.e.* the ones which correspond to the fibres of the fibre bundles, are similar to what physicists call *"internal" variables* in field theory.

1.4.4 Pinwheels and in Vivo Optical Imaging

The structuration of $V1$ in orientation hypercolumns has been well understood only after the introduction of revolutionary brain imaging techniques by Amiram Grinvald and Tobias Bonhöffer in the early 90s. This "in vivo optical imaging based on activity-dependent intrinsic signals" used the fact that the metabolic activity of cortical layers change their optical properties (differential absorption of oxyhemoglobin or deoxyhemoglobine whose fluorescence is an index of the local depolarisation of neurons). They enable to acquire images of the activity of the superficial cortical layers. Key experiments show how hypercolumns are geometrically organised in "orientation wheels" called *pinwheels*: the observed cortical layer is reticulated by a lattice of singular points (distant about 1200μm in cats and about 600μm in primates) which are the centres of local pinwheels glued together in a global structure. This pinwheel organisation can be found in numerous species, cat, primate, tupaya (tree shrew, see [25]), prosimian Bush Baby (see [139]), etc. Figure 1.4 from [138] represents $V1$. Orientations p at different positions a are coded by colours and the isochromatic lines are therefore iso-orientation lines.

We notice immediately that there exists a sort of characteristic length Λ of the lattice. It is possible to measure this approximate periodicity by taking the orientation map, by translating it and calculating the correlation between the two maps. The first peak of self-correlation gives Λ. We notice then how the orientations (the colours) are globally distributed in a homogeneous way. We notice also three classes of points. (i) Regular points where the orientation field is locally trivial (the iso-orientation lines are approximatively parallel). (ii) Singular points at the centre of the pinwheels where all orientations converge: they have a positive or negative "chirality" depending on whether, when turning around them, orientations turn in the same direction or in the reverse one; they present opposed chiralities when they are adjacent. (iii) Saddle points at the centre of the domains defined by the lattice, that is points where the iso-orientation lines bifurcate: two iso-orientation lines can start from the same singular point and end to two opposed singular points.

We shall not confuse the iso-chromatic lines which are iso-orientation lines with the integral curves of the orientation field itself. We owe figure 1.5 to Shmuel and Grinvald. White segments represent the preferred orientations coded by colours. We added the field lines of the orientation field near two singularities whose chiralities are opposed. We can see how pinwheels respectively clockwise and counterclockwise are associated to the two kind of generic singularities of direction fields in the plane. This is due to the fact that when a ray of a pinwheel rotates of an angle θ, the associated orientation rotates of $\theta/2$. So, two diametrally opposed rays correspond to orthogonal orientations. If the orientation ψ_θ associated to the ray of angle θ is $\psi_\theta = \alpha + \theta/2$, the two directions will be the same for $\psi_\theta = \alpha + \theta/2 = \theta$, *i.e.* for $\theta = 2\alpha$. Since α is defined modulo π, there's only one solution, and we get an end point. On the contrary, if the orientation is $\psi_\theta = \alpha - \theta/2$, the two directions will be the same for $\psi_\theta = \alpha - \theta/2 = \theta$, *i.e.* for $\theta = 2\alpha/3$. There are then three solutions, and we get a triple point.

It is really important to notice how the pinwheel structure is a *material* structure of dimension 2 which implements an *abstract* structure of dimension 3 (3 degrees of freedom x, y, p). If we compare it to the fibration model $\pi : \mathbb{V} = M \times P \to M$, we see that the *dimensional collapse* consists in selecting a finite number of positions $c_1, \ldots, c_N$ as singular points, taking the fibres P_{c_i} and projecting them onto M as small circles surrounding the c_i. The fibration π will then appear as a limit of such a structure for an infinite number of pinwheels. We will return later (section 1.4.6) to this point.

1.4.5 The Pinwheels as Phase Singularities or Dislocations

We see how two a priori different ways of considering pinwheels appear: either as singularities of an orientation field or as a discretisation of a fibration π. The relation between the two is rather subtle.

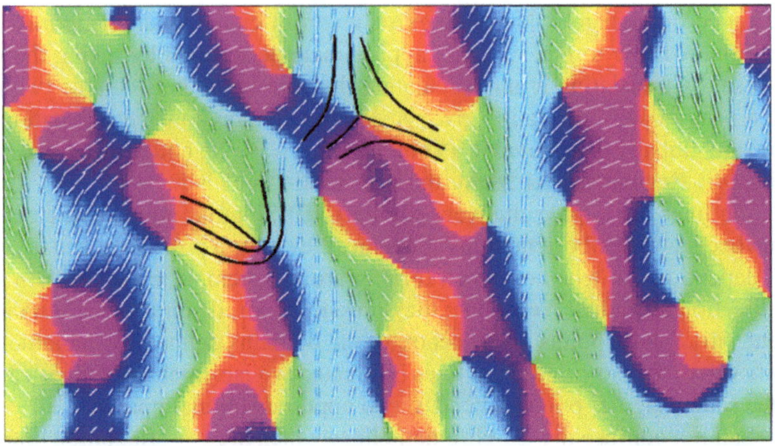

Fig. 1.5 The orientation map and the pinwheels of $V1$ in a tupaya (tree shrew). We observe the relation between the pinwheels (colours) and the preferred orientations (white segments). We represented the field lines (the integral curves) of the orientation field near two singularities whose chiralities are opposed. (Partially from Shmuel [119]).

1.4.5.1 From the Wolf-Geisel Model to Berry's Dislocations

Being a *phase* field, an orientation field can be described by a section $e^{i\varphi(a)}$ (φ modulo π) of the fibre bundle $M \times \mathbb{P}^1 \to M$. To simplify let's take $M = \mathbb{R}^2$. Fred Wolf and Theo Geisel proposed to interpret the *selectivity* to the preferred orientation (in other terms the width of the tuning curve) as a field *amplitude* (see [141], [142] and the synthesis [74]). Under this hypothesis, the orientation field is modelled by a *complex scalar field* $Z : \mathbb{R}^2 \to \mathbb{C}$, $a = \rho e^{i\theta} \mapsto r(a)\, e^{i\varphi(a)}$. The singularities of the pinwheels, which are points where the phase $\varphi(a)$ is not defined, become then the *zeroes* of Z and, as such, they are similar to *dislocations* of phase fields commonly encountered in optical structures (and also in liquid crystals). They are in some sense the dual singularities of caustics because the intensity vanishes instead of diverging. Of course, the function $\varphi(a)$ can also present (if it is differentiable), some singularities in the classical sense, *i.e.* some critical points where the gradient $\nabla\varphi = 0$. These are generically of three kinds: extrema (maxima or minima) and saddles. From an empirical point of view, extrema (where, locally, the isochromatic curves will be concentric circles) seem to be absent while saddles are numerous and occupy the "centres" of the domains defined by the pinwheels. But, dislocations are singularities of a different nature.

According to the general "philosophy" enlightened by René Thom (cf. section 1.2.1), in every situation in which we have to analyse fields, singularities play a structural role which determines and concentrates what is essential in morphological information. As it is remarked by Michael Berry in [16] (p.724) for 3 dimensional waves, this philosophy can be applied to phase singularities: "Wave vortex lines can be regarded as a skeleton, characterizing and supporting the full structure of the wave ψ." A series of concepts from singularity theory (genericity, codimension, bifurcation, unfolding, normal form, etc.) seem therefore relevant for studying pinwheels from a geometrical point of view.

In Cartesian coordinates, the $Z(a)$ field is written $Z(a) = X(a) + iY(a)$, where X and Y are two real functions of the variables (x,y). Since $\mathbb{R}^2 \simeq \mathbb{C}$, it is sometimes convenient to consider the conjugated variables $z = x+iy$ and $\bar{z} = x-iy$ and to express Z as a map $Z : \mathbb{C} \to \mathbb{C}$ which depends on z and $\bar{z}$. We can analyse the geometry of Z by using classical tools as gradient ∇, divergence, vorticity, Laplacian Δ, etc. Near a point $a_0 \in \mathbb{R}^2$ taken as origin $a_0 = (0,0)$, we have up to the first order

$$Z(x,y) \simeq X(0) + x\frac{\partial X}{\partial x}(0) + y\frac{\partial X}{\partial y}(0) + i\left(Y(0) + x\frac{\partial Y}{\partial x}(0) + y\frac{\partial Y}{\partial y}(0)\right)$$

i.e. $Z(a) \simeq Z(0) + a.\nabla_0 X + ia.\nabla_0 Y$ where $\nabla_0 X$ is the value at 0 of the gradient of X, $\nabla X = \left(\frac{\partial X}{\partial x}, \frac{\partial X}{\partial y}\right)$ (idem for Y), and $u.v$ is the scalar product. Thus,

$$|Z(a) - Z(0)|^2 = R^2 = (a.\nabla_0 X)^2 + (a.\nabla_0 Y)^2$$

and the level curves $R = \text{cst}$ are ellipses. They are circles only if $Z(a)$ can be written as a function $Z(z)$ only of z, in other terms if on the one hand $x\frac{\partial X}{\partial x} + iy\frac{\partial Y}{\partial y}$ is proportional to z, which implies $\frac{\partial X}{\partial x} = \frac{\partial Y}{\partial y}$, and on the other hand if $y\frac{\partial X}{\partial y} + ix\frac{\partial Y}{\partial x}$ is proportional to iz, which implies $\frac{\partial X}{\partial y} = -\frac{\partial Y}{\partial x}$. These fundamental conditions, called *Cauchy-Riemann* conditions, express that the gradients ∇X and ∇Y are *orthogonal* and characterise the property of *holomorphy* of Z.

Since $Z = X + iY = re^{i\varphi}$, the dislocations $Z = 0$ are the intersections of the curves of equation $X = 0$ and $Y = 0$. The condition $X = 0$ corresponds to $r\cos(\varphi) = 0$, *i.e.* $\varphi = \frac{\pi}{2} \bmod \pi$ if $r \neq 0$ and $Y = 0$ corresponds to $r\sin(\varphi) = 0$, *i.e.* $\varphi = 0 \bmod \pi$ if $r \neq 0$. If $X = Y = 0$, we necessarily have $r = 0$ because the two conditions on φ are incompatible. Generically, the curves $X = 0$ and $Y = 0$ intersect transversely at isolated points. This means that the points which satisfy the two conditions are of codimension 2 and, as the ambient space $\mathbb{R}^2$ is of dimension 2, they are isolated points (in an ambient space of dimension 3, they would be lines).

As we saw in the previous section 1.4.4, it is necessary to distinguish two fields. The field Z is the field of orientations. As such, it defines a foliation of the plane $\mathbb{R}^2$ through its integral curves. In 1.4.4 we saw, with the models of end points and triple points, the local geometry of these foliations at the singular points. Besides, we have the field of *iso-orientation* or isophase lines, called "wavefronts" by analogy with optics, represented by the isochromatic lines in the pinwheel maps. Let's write it $W(a) = s(a)e^{i\chi(a)}$ and suppose that we could assign a meaning to the amplitude $s(a)$. Unlike the phase φ which is only defined modulo π, the angle χ is defined modulo 2π (if the pinwheels are without distortion, χ can be identified with θ). Now, we saw how in these cases we locally have the relation $\varphi = \alpha \pm \frac{\theta}{2}$, *i.e.* up to a factor, $e^{i\chi} = e^{\pm 2i\varphi} = \left(e^{\pm i\varphi}\right)^2$. This lets us think that there is a relation of proportionality of the type $W \propto Z^2$ between the two fields. We will return to this point later.

Let a_0 be a dislocation of Z taken as the origin $a_0 = (0,0)$. The *topological charge* of this singularity is defined by $q = \frac{1}{2\pi}\oint_\gamma d\varphi$ where $\oint_\gamma d\varphi$ symbolises the integral of the differential 1-form $d\varphi$, *i.e.* the variation of φ, along a little loop γ rotating around a_0 in the forward direction. Since a_0 is an isolated singularity, there exists a γ which surrounds only this singularity, and we can show how the integral is independent from the choice of such a γ. Through the Euclidean structure of $\mathbb{R}^2$, the differential 1-form $d\varphi$ corresponds to the gradient vector $\nabla\varphi$ and, if we interpret differentials as infinitesimal variations, we have $d\varphi = \nabla\varphi.da$ (scalar product) and $\oint_\gamma d\varphi = \oint_\gamma \nabla\varphi.da$ turns to be what it is classically called the "circulation" of the gradient field $\nabla\varphi$ along the loop γ. Then, the topological charge $q = \frac{1}{2\pi}\oint_\gamma \nabla\varphi.da$ can be interpreted as the topological *index* of the field $\nabla\varphi$. For the field W of isophase lines Z, the angle χ varies as $\pm\theta$ and the index is ± 1. For the Z field itself, φ varies as $\pm\frac{\theta}{2}$ and the index is $\pm\frac{1}{2}$.

Near a pinwheel, the isophase lines $\varphi = \text{cst}$ (the wave-fronts) are the rays of the pinwheel. Along these fronts $d\varphi = 0$, and therefore $\nabla\varphi.da = 0$. This means that the field $\nabla\varphi$ is orthogonal to the rays and that its trajectories are therefore locally concentric circles centred on the singular point. In general, the trajectories of $\nabla\varphi$ are orthogonal to the fronts. At a dislocation point the gradient $\nabla\varphi$ is not defined and diverges. In order to regularise this kind of situation, Berry and Dennis [13], [47] consider, as physicists usually do, what is called the *current* $\mathcal{J}$ of the field, *i.e.* the vector (which points in the direction of the gradient $\nabla\varphi$ when it is $\neq 0$) defined by $\mathcal{J} = r^2\nabla\varphi$. We notice how $\mathcal{J} = X\nabla Y - Y\nabla X$ and therefore how $\mathcal{J}$ is well defined also at the singular points of the phase φ. In terms of complex conjugated values Z and $\overline{Z}$ the current $\mathcal{J}$ is written $\mathcal{J} = \text{Im}\left(\overline{Z}\nabla Z\right)$ because

$$\overline{Z}\nabla Z = (X - iY)(\nabla X + i\nabla Y) = X\nabla X + Y\nabla Y + i(X\nabla Y - Y\nabla X) \ .$$

Another vector (in fact a pseudovector) considered by physicists is the *vorticity* Ω of the current $\mathcal{J}$. By definition, $\Omega = \frac{1}{2}\nabla \times \mathcal{J} = \nabla X \times \nabla Y$, where $\times$ represents the exterior product of two vectors of $\mathbb{R}^2$ and consists in taking on an axis orthogonal to $\mathbb{R}^2$ a unitary vector e_3 so that the frame $\{e_x, e_y, e_3\}$ is direct and to put on it the value $\omega = \det\begin{pmatrix} \frac{\partial X}{\partial x} & \frac{\partial X}{\partial y} \\ \frac{\partial Y}{\partial x} & \frac{\partial Y}{\partial y} \end{pmatrix}$ of the determinant of the Jacobian of Z considered as a map from $\mathbb{R}^2$ into $\mathbb{R}^2$. The good interpretation of $\Omega = \omega e_3$ is to make of it a differential 2-form. We will remark that when $\Omega = 0$, either $\nabla X = 0$ or $\nabla Y = 0$, or the real gradients ∇X and ∇Y share the same orientation. This condition, which tells that ∇X and ∇Y are parallel, is opposed to the Cauchy-Riemann conditions, which tell that ∇X and ∇Y are orthogonal. We will notice also how, in terms of complex conjugate values Z and $\overline{Z}$, the vorticity Ω can be written $\Omega = \frac{1}{2}\text{Im}\left(\nabla\overline{Z} \times \nabla Z\right)$.

As we have seen before, to first order in the neighbourhood of a point a_0 taken as origin 0, the module $|Z|$ of Z is given by $|Z(a) - Z(0)|^2 = R^2 = (a.\nabla_0 X)^2 + (a.\nabla_0 Y)^2$. Near a dislocation taken as the origin, the current $\mathcal{J}$ is given up to the first order by $\mathscr{J}(x,y) \simeq \Omega_0 \times a = \omega_0(-y,x)$. This enables to evaluate $|\mathscr{J}| = r^2|\nabla\varphi| \simeq |\omega|\rho$ near the dislocations where $\omega \neq 0$. But locally φ is constant on the rays of such a singular point, so $\nabla\varphi$ is orthogonal to the rays and, since in polar coordinates $\nabla\varphi = \frac{\partial\varphi}{\partial\rho}e_\rho + \frac{1}{\rho}\frac{\partial\varphi}{\partial\theta}e_\theta$ (where e_ρ is the unitary vector of the ray at a and e_θ is the unitary vector orthogonal to e_ρ), we have $\nabla\varphi \simeq \frac{1}{\rho}\frac{\partial\varphi}{\partial\theta}e_\theta$. Whence the formula $r^2\left|\frac{\partial\varphi}{\partial\theta}\right| = \rho^2|\omega|$ which tells that, while r is (locally) constant on the ellipses $r^2 = (a.\nabla X)^2 + (a.\nabla Y)^2$, $r^2\left|\frac{\partial\varphi}{\partial\theta}\right|$ is on its turn constant on the circles $\rho = \text{cst}$. As it has been noticed by Mark Richard Dennis in his thesis supervised by Michael Berry [47] (p.41), it is a sort of "Kepler law" for $r^2\left|\frac{\partial\varphi}{\partial\theta}\right|$ which works as an "angular momentum": "equal area vectors of the core anisotropy ellipse [$r^2 = \text{cst}$] are swept out in equal intervals of phase". We notice how the eccentricity of the ellipses measures the anisotropy of the vorticity. As we saw, isotropy is absent (the ellipses are not circles) unless the Cauchy-Riemann conditions are satisfied.

1.4.5.2 The Helmholtz Equation

When presenting some experimental data on the pinwheels, we saw how they are not intrinsic singularities and appear only when we superimpose the different cortical maps of responses to different orientations. We saw also that there exists a characteristic mesh of the pinwheel lattice. These two empirical facts suggest to consider the Z field as a *superposition* of simpler fields. Now, every field can be considered through its Fourier transform as a superposition of plane waves $Ae^{i\kappa.a}$, where A is a complex amplitude $Ee^{i\phi}$ and $\kappa = (\kappa_x, \kappa_y)$ is a wave vector of wave number $k = |\kappa|$. k is analogous to an impulsion and is associated to the wavelength $\Lambda = \frac{2\pi}{k}$. When they evolve during time, plane waves are of the form $Ae^{i(\kappa.a-\omega t)}$, where ω is an angular frequency (or pulsation) associated to the frequency $\nu = \frac{\omega}{2\pi}$ and the period $T = \frac{1}{\nu} = \frac{2\pi}{\omega}$.

It is easy to verify how the plane waves $U = Ae^{i\kappa.a}$ satisfy *Helmholtz's equation* $\Delta U + k^2U = 0$. And since this PDE is linear, every linear superposition of solutions with different κ but same wave number k is a solution too. That's why it is natural to suppose that the field Z satisfies Helmholtz's equation for a certain characteristic wavenumber k: $\Delta Z + k^2 Z = 0$. Figure 1.6, from a work of Michael Berry [13] on optical currents, shows a superposition of 10 plane waves sharing the same k. We see at what point this phase field is similar to our orientation fields with pinwheels, iso-orientation lines, orthogonal gradient lines and saddle points.

Let's consider now under this hypothesis the divergence of the current $\mathscr{J}$ which is given by

$$\text{div}(\mathscr{J}) = r^2\Delta\varphi + 2r\nabla r.\nabla\varphi$$

where $\nabla r.\nabla\varphi$ is the scalar product and where the Laplacian operator $\Delta\varphi$ is by definition the divergence of the gradient. The Laplacian ΔZ of $Z = re^{i\varphi}$ is given by the formula

$$\Delta Z = \frac{\partial^2 Z}{\partial \rho^2} + \frac{1}{\rho}\frac{\partial Z}{\partial \rho} + \frac{1}{\rho^2}\frac{\partial^2 Z}{\partial \theta^2}$$
$$= e^{i\varphi}\left(\Delta r - r\left|\nabla\varphi\right|^2 + i\left(r\Delta\varphi + 2\nabla r.\nabla\varphi\right)\right)$$

and therefore if $\Delta Z + k^2 Z = 0$ we necessarily have the two equations

$$\begin{cases} \Delta r + r\left(k^2 - \left|\nabla\varphi\right|^2\right) = 0 \\ r\Delta\varphi + 2\nabla r.\nabla\varphi = 0 \end{cases}$$

The second equation expresses that the current divergence vanishes: $\operatorname{div}(\mathscr{J}) = 0$. It is a *law of conservation*. It implies that $\mathscr{J}$ can be written as $e_3 \times \nabla S = \left(-\frac{\partial S}{\partial y}, \frac{\partial S}{\partial x}\right)$ with

$$\omega = \frac{1}{2}\left(\frac{\partial \mathscr{J}_y}{\partial x} - \frac{\partial \mathscr{J}_x}{\partial y}\right) = \frac{1}{2}\left(\frac{\partial}{\partial x}\left(\frac{\partial S}{\partial x}\right) + \frac{\partial}{\partial y}\left(\frac{\partial S}{\partial y}\right)\right) = \frac{1}{2}\Delta S\,.$$

As much as the first equation is concerned, we met it in optics for caustics in section 1.2.4 under a variant where k is integrated to the phase, *i.e.* where $Z = re^{ik\varphi}$. In this case, it is written $\Delta r = k^2 r\left(\left|\nabla\varphi\right|^2 - 1\right)$. When $k \to \infty$, which corresponds to the approximation of geometrical optics, it becomes the eikonal equation $\left|\nabla\varphi\right|^2 = 1$

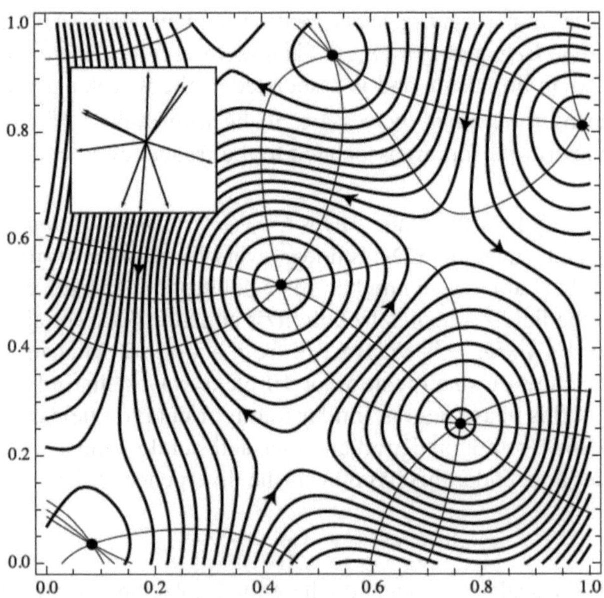

Fig. 1.6 Superposition of 10 plane waves with the same wave number k. The wave vectors κ are given in the upper-left square. (From Berry [13]).

of section 1.2.4, which expresses that the "light rays" (*i.e.* the trajectories of the gradient field $\nabla\varphi$) turn at a constant speed around the singularity, while the wavefronts $\varphi = \text{cst}$ are the rays coming from the singularity (we must be careful here with the lexical confusion between the two meanings of the term "ray": the rays of the dislocation singularity are orthogonal to the trajectories analogous to "light rays").

1.4.5.3 Mesogeometry and Microphysics

The optical analogy suggested by the processing of orientation maps as phase fields and of pinwheels as dislocations of those fields is useful also to better understanding the relations between levels. In optics, there are three levels: a geometric level, a wave level, and a quantum level. In the analogy, the geometric level corresponds to the mesogeometric level of contact, symplectic, and sub-Riemannian structures which we will develop later in this survey. The wave level corresponds to the analysis of pinwheels as singularities of phase fields that we just performed. However, as Michael Berry noticed [13], wave optics is an average of microphysical interactions at the quantum level. In particular, the optical current is an "energy flow" whose trajectories are level lines of S, in a way a "momentum density" which yields the classic force on a little particle placed at a. As much as the phase gradient $\nabla\varphi$, it gives the momentum induced on the particle by the impacts of individual photons. And, as the probability of these impacts is r^2, the average momentum is $\mathscr{J} = r^2\nabla\varphi$.

In this context, it could be relevant to make the hypothesis that there exists a micro-physics of elementary events, in relation to which the geometry of orientation maps is a sort of morphological skeleton. *Spikes* could then play the role of "little particles".

1.4.5.4 Statistics of Pinwheels and Gaussian Fields

Pinwheel maps as phase fields can present great diversity. It is therefore interesting to study them from a statistical point of view starting from certain simplifying hypotheses. It is the purpose of converging studies by Wolf and Geisel (cf. [141] and [142]), Berry and Dennis [14], [15], [47], Daniel Bennequin's workgroup, and also of recent works by Citti, Sarti and their PhD student Davide Barbieri [10].

In his thesis [47], Mark Dennis gives precise results for the superposition of plane waves $Z = \sum_\kappa A_\kappa e^{i\kappa.a}$ with complex amplitude $A_\kappa = E_\kappa e^{i\phi_\kappa}$, in particular in the isotropic case where the E_κ have a distribution which depends only on the wave number $k = |\kappa|$ of the wave vectors and where the spatial phases ϕ_κ are random angular variables uniformly distributed on $[0,\pi)$. If the sampling of the κ in Z is sufficiently fine grained, it is possible to consider that the statistics of the components X and Y of Z and of their partial derivatives are Gaussian distributions, which makes the computations quite accessible. In particular, one defines the energy spectrum $\mathscr{E}(\kappa)$ by $\frac{1}{2}\sum_\kappa E_\kappa^2 = \int \mathscr{E}(\kappa)^2 d\kappa$ and the radial spatial spectrum $\mathscr{R}(k)$ by $\frac{1}{2}\sum_\kappa E_\kappa^2 = \int \frac{\mathscr{R}(k)^2}{2\pi k} d\kappa$. An ulterior simplification consists in considering the monochromatic waves sharing the same k, κ varying then on the circle of radius k. In this case, $\mathscr{R}(u)$ becomes the Dirac distribution $\delta(u-k)$. This hypothesis corresponds to the fact that Z is a solution to the Helmholtz equation.

It is then possible to compute the average density d of phase dislocations. As these are defined by the conditions $X=0, Y=0$, it will be given by the average of $\delta(X)\delta(Y)$ with respect to the measure $dXdY$. With respect to the measure $dxdy$ we must involve the Jacobian of $Z(x,y) = X(x,y) + iY(x,y)$, *i.e.*

$$|\omega| = |\nabla X \wedge \nabla X| = \left| \frac{\partial X}{\partial x}\frac{\partial Y}{\partial y} - \frac{\partial X}{\partial y}\frac{\partial Y}{\partial x} \right| .$$

So we have to calculate the average $\left\langle \delta(X)\delta(Y) \left| \frac{\partial X}{\partial x}\frac{\partial Y}{\partial y} - \frac{\partial X}{\partial y}\frac{\partial Y}{\partial x} \right| \right\rangle$. As X, Y and their partial derivatives are independent random Gaussian variables, we know how to compute them and we find:

$$d = \frac{K}{4\pi} \text{ with } K = \int_0^\infty k^2 \mathscr{R}(k)\,dk = \left\langle k^2 \right\rangle_{\mathscr{R}} \text{ for the measure } \mathscr{R}(k)\,dk .$$

But the wave number k is proportional to the inverse of a wavelength $\Lambda = \frac{2\pi}{k}$ so $\Lambda^2 = \frac{4\pi^2}{k^2}$ and $\frac{k^2}{4\pi} = \frac{\pi}{\Lambda^2}$. Therefore, the density d of singularities is the average $\left\langle \frac{\pi}{\Lambda^2} \right\rangle_{\mathscr{R}}$. This term in π/Λ^2 has been found by Fred Wolf and Theo Geisel too.

Generally speaking, we can consider that the orientation maps are *random* sections of the fibre bundle $\mathbb{R}^2 \times \mathbb{P}^1 \to \mathbb{R}^2$ which satisfy a set of constraints that explain their pinwheel geometrical structure. At every point a in $\mathbb{R}^2$, we therefore consider a random variable $\mathscr{Z}_a$ and this defines a *random field* $\mathscr{Z}$ whose orientation maps $Z(a)$ are samples.

In order to simplify, one supposes in general that the field $\mathscr{Z}$ is *Gaussian, i.e.* that the $\mathscr{Z}_a$ are Gaussians of average $m_a = \mathbb{E}\{\mathscr{Z}_a\}$ ($\mathbb{E}$ = expectation) and variance $\sigma_a^2 = \mathbb{E}\left\{(\mathscr{Z}_a - m_a)^2\right\}$ and that all the joint distributions $\sum_i \alpha_i \mathscr{Z}_{a_i}$ on a finite number of points a_i are Gaussian too. Moreover, it is natural to suppose that the law of the $\mathscr{Z}_a$ is $SE(2)$-invariant. Translational invariance is called *stationarity* and rotational invariance is called *isotropy*. For an introduction to Gaussian random fields, see for example [1].

Random variables $\mathscr{Z}_a$ for different a can't be decorrelated (*i.e.* independent since for Gaussian random variables independence and decorrelation are equivalent) because otherwise there would only be Gaussian noise and no geometrical structures. By the way, the definition of a *continuum* of independent Gaussian random variables raises sensible problems. The field $\mathscr{Z}$ is characterized by the *correlation function* $C(a,b) = \mathbb{E}\{(\mathscr{Z}_a - m_a)(\mathscr{Z}_b - m_b)\}$. If we divide by variances, we obtain the normalized correlation function $\Gamma(a,b) = \frac{C(a,b)}{\sigma_a \sigma_b}$. Stationarity implies that $C(a,b) = C(a-b)$ is a function of $a-b$, and isotropy implies that $C(a,b) = C(a-b) = C(\|a-b\|) = C(r)$ is a function of $r = \|a-b\|$. Averages are all equal, $m_a = m$, variances too, $\sigma_a^2 = \sigma^2 = C(0)$, and $\Gamma(r) = \frac{C(r)}{C(0)}$.

Correlation functions are very particular because they are symmetrical and positive definite.

(i) In the stationarity case, a theorem by Bochner says that they have a spectral representation which is a generalised Fourier transform. This means that we have $\Gamma(a) = \int_{\mathbb{R}^2} e^{i\langle a,\kappa\rangle} dF(\kappa)$ where F is a bounded positive measure on the space of wave vectors κ dual of positions a. We have $F(\mathbb{R}^2) = \Gamma(0) = 1$.

(ii) If the measure F is regular with respect to the Lebesgue measure $d\kappa$, then it has a spectral density $f(\kappa)$ and Bochner's theorem is reduced to the Fourier transform $\Gamma(a) = \int_{\mathbb{R}^2} e^{i\langle a,\kappa\rangle} f(\kappa)\, d\kappa$ with the inverse transform $f(\kappa) = \frac{1}{(2\pi)^2} \int_{\mathbb{R}^2} e^{-i\langle a,\kappa\rangle} \Gamma(a)\, da$.

(iii) If isotropy is present too, then $\Gamma(r) = \int_0^\infty J_0(kr)\, kf(k)\, dk$, where J_0 is Bessel's function.

(iv) If we consider also some solutions of Helmholtz's equations of wave number k_0, then $f(k)$ is proportional to $\delta(k - k_0)$ and $\Gamma(r)$ is proportional to $J_0(kr)\, k_0$.

In this context, the formula which gives us the statistics of the dislocations of the fields $Z(a)$ which realize the Gaussian random field $\mathscr{Z}$ is a particular case of a fundamental formula called *Rice-Kac* formula (see [9]). We want to calculate the average number $d = \mathbb{E}\{\#\{a \in T : Z(a) = 0\}\}$ (# means the cardinal of a set) of zeroes of Z in a unit square T. Let N_T be this number. Then we have $d = \mathbb{E}\{N_T\}$. Rice's formula says that

$$d = \int_T \mathbb{E}\{|\det(\mathrm{Jac}(\mathscr{Z}_a))| : \mathscr{Z}_a = 0\}\, p_{\mathscr{Z}_a}(0)\, da$$

where Jac is the Jacobian and $p_{\mathscr{Z}_a}$ the density of $\mathscr{Z}_a$. The computations effectuated by Mark Dennis [47] rest on a calculus of Gaussian integrals. Let $\mathscr{X}_a$ and $\mathscr{Y}_a$ be the components of $\mathscr{Z}_a$ and $\mathscr{J}_a = |\det(\mathrm{Jac}(\mathscr{Z}_a))|$. One can show how, for a given a, the six random variables $\mathscr{X}_a$, $\mathscr{Y}_a$, $\left(\frac{\partial \mathscr{X}}{\partial x}\right)_a$, $\left(\frac{\partial \mathscr{X}}{\partial y}\right)_a$, $\left(\frac{\partial \mathscr{Y}}{\partial x}\right)_a$, $\left(\frac{\partial \mathscr{Y}}{\partial y}\right)_a$ are independent Gaussian variables, each of them with a law of the form $\frac{1}{\sqrt{2\pi}\sigma} e^{-\frac{\xi^2}{2\sigma^2}}$. As we saw, we must compute an integral of the form

$$\int \delta(X)\, \delta(Y)\, Jpd(X)\, d(Y)\, d\left(\frac{\partial X}{\partial x}\right) d\left(\frac{\partial X}{\partial y}\right) d\left(\frac{\partial Y}{\partial x}\right) \left(\frac{\partial Y}{\partial y}\right)$$

where p is the product of the laws. The two first variables $\mathscr{X}_a$ and $\mathscr{Y}_a$ are of variance 1, and this introduces a factor $\left(\frac{1}{\sqrt{2\pi}}\right)^2 = \frac{1}{2\pi}$ in the integral and the four other variables are of variance $\sigma^2 = \frac{K}{2}$, and this introduces a factor $\left(\frac{1}{\sqrt{2\pi}\sigma}\right)^4 = \left(\frac{1}{2\pi\sigma^2}\right)^2 = \frac{1}{(\pi K)^2}$. The condition $\mathscr{Z}_a = 0$ is translated by the $\delta(X)\, \delta(Y)$ in the integral, but $\int \delta(X)\, e^{-\frac{X^2}{2}} = 1$ and the same about Y. It remains therefore to calculate

$$\frac{1}{2\pi} \frac{1}{(\pi K)^2} \int J e^{-\frac{\|\nabla X\|^2 + \|\nabla X\|^2}{K}} d\left(\frac{\partial X}{\partial x}\right) d\left(\frac{\partial X}{\partial y}\right) d\left(\frac{\partial Y}{\partial x}\right) \left(\frac{\partial Y}{\partial y}\right).$$

If we shift to polar coordinates, writing $\nabla X = R_X e^{i\psi_X}$ and $\nabla Y = R_Y e^{i\psi_Y}$, the integral becomes

$$\frac{1}{2\pi}\frac{1}{(\pi K)^2}\int_{R_X=0}^{R_X=\infty}\int_{R_Y=0}^{R_Y=\infty}\int_{\psi_X=0}^{\psi_X=2\pi}\int_{\psi_Y=0}^{\psi_Y=2\pi}(R_X)^2(R_Y)^2|\sin(\psi_Y-\psi_X)|e^{-\frac{R_X^2+R_Y^2}{K}}dR_X dR_Y d\psi_X d\psi_Y .$$

But the integral on the sine function results 8π and the ones on R_X and R_Y give each one $\frac{1}{4}K^{\frac{3}{2}}\sqrt{\pi}$. From these we finally have:

$$d=\frac{1}{2\pi}\frac{1}{(\pi K)^2}8\pi\frac{1}{16}K^3\pi=\frac{K}{4\pi} .$$

These statistical computations, which are only a particular simple example of the links between statistics and geometry, are really interesting on a theoretical (and not only numerical) plane for the following reason. In their 2005 reference book *Random Fields and Geometry* [2], Robert Adler and Jonathan Taylor thoroughly studied the generalisations of the Rice-Kac formula for random fields $\mathscr{F}_a$ defined on a base space M and with values in $\mathbb{R}^k$. Let's take for example $k=1$. One of the main problems, extremely difficult, is to compute $\mathbb{P}\{\sup_{a\in M}\mathscr{F}_a\geq u\}$ for *large* u. This probability of "excursion" in the interval $[u,\infty)$ is well approximated by $\mathbb{E}\{\chi(A_{[u,\infty)})\}$ where, if D is a domain of $\mathbb{R}^k$, $A_D:=\{a\in M:\mathscr{F}_a\in D\}$ and where χ is the *Euler-Poincaré characteristic*. Under the hypothesis of Gaussian laws, stationarity, isotropy and regularity of the correlation functions C, one arrives to (complex) explicit formulas for the $\mathbb{E}\{\chi(A_D)\}$. What is interesting is that proving these formulas involves many fundamental tools of the "philosophy" of Morse-Whitney-Thom which we presented in our first section 1.2.1. First, in order to process enough cases, we suppose that M is a manifold with boundary endowed with a "good" stratification (which satisfies what are called the Whitney's A and B conditions) $M=\cup_{k=0}^{k=N}\partial_k M$ (where k is the dimension of the strata which compose $\partial_k M$) with $\partial_N M=\mathring{M}$ (the topological interior of M), $\partial_0 M=\{\text{vertices of } M\}$. We suppose that the properties of regularity of C imply that the samples F of the field $\mathscr{F}$ are *Morse's functions* on M (we use the generalization of Morse's theory to stratified manifolds owed to MacPherson). The field $\mathscr{F}$ defines a natural *metric* with distance $d_{\mathscr{F}}(a,b)$ on the base space M through the formula $d_{\mathscr{F}}^2(a,b)=\mathbb{E}\left\{\|\mathscr{F}_a-\mathscr{F}_b\|^2\right\}$. So we can use also the resources of Riemannian geometry: curvature tensor, Levi-Civita connection, covariant derivation, Lipschitz-Killing curvatures, etc.

For $k=1$, we apply then to $A_{[u,\infty)}$ the formulas which connect Morse's theory to Euler-Poincaré characteristic. If the sample F of $\mathscr{F}$ is a Morse function and if u is a regular value (*i.e.* not critical) of F, $A_{[u,\infty)}$ is a sub-manifold with boundary of M which is "well" stratified by the intersection strata $\mathring{A}_{[u,\infty)}\cap\partial_k M$ and $\partial A_{[u,\infty)}\cap\partial_k M$. F is not necessarily a Morse function on $A_{[u,\infty)}$ but we can approximate it by a Morse function $\widetilde{F}$ on $A_{[u,\infty)}$ whose critical points correspond to the critical points of F above u.

The Rice-Kac formula corresponds to the case of a rectangle $M=T$ of $\mathbb{R}^N$ and to an $\mathscr{F}$ with values in $\mathbb{R}^N$. Let J be the Jacobian of a sample F (J is a matrix $N\times N$). Let N_u be the number of points of T for which $F(a)=u\in\mathbb{R}^N$. The formula says that

$$\mathbb{E}\{N_u\} = \int_T \mathbb{E}\{|\det(J)| : F(a) = u\}\, p_a(u)\, dt\ .$$

In our case $N = 2$, $\mathscr{F} = \mathscr{Z}$, $u = 0$ and T is a unit square.

1.4.6 The Fibre Bundle $\mathbb{V} = \mathbb{R}^2 \times \mathbb{P}^1$ as a Blowing Up Model for Pinwheels

The similarity between pinwheels and phase dislocations is particularly remarkable. But we have to underline two points:

1. the model presupposes that selectivity to orientation vanishes at the dislocation points (*i.e.* at the centres of the pinwheels);
2. it loses any meaning at the limit case of an infinite number of pinwheels because it gives a field $Z \equiv 0$.

It is therefore necessary to investigate more precisely the neighbourhood of singularities. There exist key results, *e.g.*, by Pedro Maldonado, Imke Gödecke, Charles Gray and Tobias Bonhöffer [79] or by David McLaughlin Robert Shapley and Michael Shelley [85], [118].

Now, many of these works, in particular the ones using in vivo imaging methods based upon two-photon confocal microscopy (methods which enable one to reach the resolution of individual neurons, see Ohki *et al.* [94]), show how orientation selectivity is still good at the singular points of pinwheels. This is why we proposed to reconsider the pinwheels as *blowups*. A blowup is constituted by a singular point a, a fibre $P_a \simeq \mathbb{P}^1$ and a helicoidal local section, above a small neighbourhood U of a, of the fibre bundle $\pi : U \times \mathbb{P}^1 \to U$. The section is constructed by taking as orientation above any point $b \neq a$ of U the orientation of the segment $\overline{ab}$.

The idea is therefore to lift up the orientation field from $\mathbb{R}^2$ to $\mathbb{V} = \mathbb{R}^2 \times \mathbb{P}^1$ by blowing up the dislocations c_i and by lifting up the field lines as curves in $\mathbb{V}$: if a field line γ starts from a source c_1 with an angle θ_1 to end at the sink c_2 with the angle θ_2, γ is lifted to a curve Γ of $\mathbb{V}$ which starts from the height θ_1 in the fibre above c_1 and arriving at height θ_2 in the fibre above c_2. Only the fibres above the c_i are involved and everything is as if we had blew up the c_i in parallel. The tangent structures to this multi-blowing-up in the infinitesimal neighbourhoods of the fibres $P_{c_i} = \Delta_i \to c_i$ are all isomorphic to the local model.

In order to pass to the limit when the mesh of the lattice L of the c_i tends to 0, a method is to use a "non-standard" model[5] $(\mathbb{R}^*)^2$ of $\mathbb{R}^2$ where, around each standard point $a = (x, y)$, there is a "monad" of infinitesimals $\mu(a) = \{(x + dx, y + dy)\}$. In the blowup, the fibre Δ^* above a standard point is then a $\left(\mathbb{P}^1\right)^*$ and $\frac{dy}{dx}$ a non-standard real number $p^* = p + dp$, equivalent to $p \in \mathbb{P}^1$. To first order, field lines boil down to infinitesimal segments from a to $a + da$. The standard part of the non-standard fibration $\mathbb{V}^* = \left(\mathbb{R}^2\right)^* \times \left(\mathbb{P}^1\right)^*$ gives not only the fibre bundle $\pi : \mathbb{V} = \mathbb{R}^2 \times \mathbb{P}^1 \to \mathbb{R}^2$

---

[5] In a non-standard universe of set-theory, if X is a set, we write X^* for its non-standard enlargement

but also the infinitesimal structure defined on it by the differential 1-form $\omega = dy - pdx$, *i.e.* the contact structure, as we will see in section 1.4.9.

The advantage of a non-standard model is to make intuitive the dimensional collapse which characterises the pinwheel structure. We take the fibre bundle $\pi : \mathbb{R}^2 \times \mathbb{P}^1 \to \mathbb{R}^2$, we "compactify" the fibres[6] until they become infinitesimal and we project them into the monads $\mu(a)$. This concept of blowing up where the fibre Δ is made infinitesimal and projected on the base plane has been introduced for reasons of high pure mathematics of a different order (singularities of analytical functions and Gevrey classes) by Pierre Deligne [46] in 1986 in his correspondence with Bernard Malgrange and Jean-Pierre Ramis. Pierre Deligne introduced the concept of a "fat point", which consists in substituting a point $a \in \mathbb{C}$, let's say $a = 0$, with a little disc D with boundary Δ, and in considering the space $\widetilde{\mathbb{C}} = (\mathbb{C} - \{0\}) \cup D$ (union of $\mathbb{C} - \{0\}$ with D) endowed with the topology of the blowing up of 0 in $\mathbb{C}$ along $(\mathbb{C} - \{0\}) \cup \Delta$. In his last text on Gevrey's classes (edited by Jean-Pierre Ramis), Jean Martinet [83] used this construction with discs D which are infinitesimal in the sense of non-standard analysis. We could say that, at the continuous limit, a lattice of pinwheels becomes a continuum of "fat" points, the standard part of this structure returning the fibre bundle $\pi : \mathbb{V} = \mathbb{R}^2 \times \mathbb{P}^1 \to \mathbb{R}^2$.

1.4.7 Independent "Engrafted" Variables and the Transversality Principle

In the previous section, we focused on a single "engrafted" variable, the orientation one, but other "engrafted" variables exist, for example spatial frequency (the size of the receptive profiles, *i.e.* the scale or the resolution) or the ocular dominance (the correlation between a simple neuron and the ipsilateral or contralateral eye). All these "internal" variables are implemented in 2- *dimensional* cortical layers. In the orientation case, an abstract structure of dimension 3 collapsed in dimension 2; now we are concerned with an abstract structure of dimension 5. This drastic dimensional collapse obviously raises the question on how to know how to represent in dimension 2 the *independence* of the "internal" variables. It is a central problem, emphasised by specialists as for example Nicholas Swindale [126].

Experimental data, *e.g.* the ones of Hübener *et al.* [68], show how the boundaries of the ocular dominance domains (ODD) are strongly transversal, and often quasi-orthogonal, to iso-orientation lines (see figure 1.7). It is so also for the boundaries of the frequency domains. Consequently, it seems that the solution found by the biological evolution has been to maximise a condition of *transversality*. Evidently, if more than 2 internal variables exist, then transversality can't be that strong everywhere; nevertheless we can have an optimisation of two antagonist constraints of transversality. This seems really to be the case.

Hongbo Yu *et al.* [140] analysed the relation between the three maps of orientation, ocular dominance, and spatial frequency, by focusing on the way in which a strong transversality codes the independence of the associated continuous variables.

[6] Just as in physics, in Kaluza-Klein models of field theory.

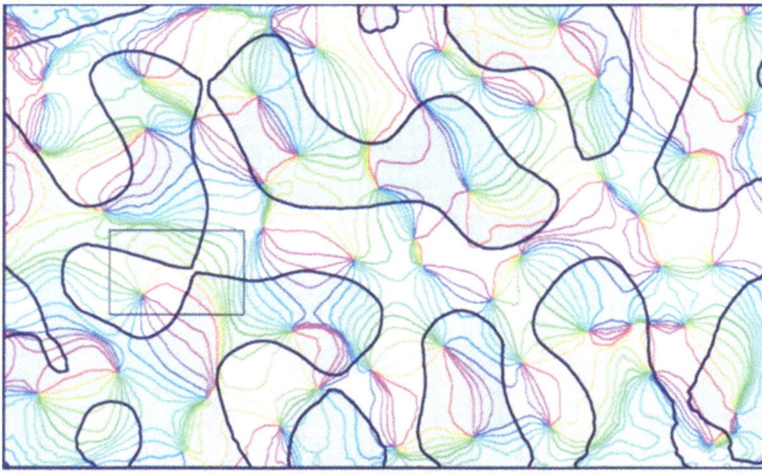

Fig. 1.7 The relations between pinwheels and ocular dominance domains. Many iso-orientation lines cross the boundaries of the ODD quite orthogonally. (From Hübener *et al.* [68]).

They considered the *gradients* of the variables and first showed how they are maximal in *disjoint* zones and how transversality is maximal when the two gradients are jointly sufficiently strong. As they say, (p. 277), "two features are mapped orthogonally in their high-gradient overlap regions". In figure 1.8, we see the field lines of the orientation field (which are the level lines of the "orientation" variable) and the level lines of the ODD. The centres of the pinwheels are the points where the orientation gradient is maximal and the boundaries of the ODD are the level lines where the gradient of ocular dominance is maximal. These singularities *avoid* each other according to many experiments which show how the centres of the pinwheels are placed on the symmetry axes of the ODD. In the grey regions the two gradients are *jointly* high. We find that in these regions transversality is also very high, almost an orthogonality. These two constraints of avoidance and of transversality express the independence.

It is particularly remarkable to see how the structural principle of transversality, to which René Thom gave so much importance, could be used by biological evolution for functional purposes.

1.4.8 Horizontal Connections and Association Field

Hypercolumns correspond to the "vertical" retino-geniculo-cortical connectivity. One of the great experimental findings of the 90s were the lateral *cortico-cortical "horizontal"* connections which are *internal* to cortical layers, long-ranged (up to 6-8mm), excitatory, slow (about $0,2\,\mathrm{m}/\mathrm{s}$) and distributed in a very anisotropic and "patchy" way. This second set of connections is particularly important for neurogeometry, because it implements the functional architecture which enables contour

Fig. 1.8 The iso-orientation lines and the level lines of the ODDs. In the grey zones the two gradients are jointly high. We notice how transversality is then very strong. (From Yu *et al.* [140]).

integration. The fibre bundle structure $\pi : M \times P \rightarrow M$ is not sufficient to this purpose because the visual system needs also to *compare* orientations above different points a and b of M (and so fibres P_a and P_b). It is this process, which geometers call *parallel transport* since Elie Cartan, which is neurally implemented through the lateral "horizontal" connections. Indeed, experimental data show how these connections link neurons with similar preferred orientations in distant hypercolumns.

Figure 1.9, from Bosking *et al.* [25], shows how biocytine locally injected in a zone of about $100\mu m$ of the layer $2/3$ of $V1$ of a tupaya (tree shrew), diffuses along horizontal connections in a selective, "patchy"anisotropic way. Short-ranged diffusion is isotropic and corresponds to *intra*-hypercolumnar inhibitory connections. On the contrary, long-ranged diffusion is highly anisotropic, corresponding to excitatory *inter*-hypercolumnar connections and is restricted to domains sharing the *same* orientation as the injection site. We notice also how the marked synaptic buttons cluster along a great diagonal top-left →bottom-right. The interpretation of this striking phenomenon is that "horizontal" connections preferentially link not only contact elements (a,p) and (b,q) with p and q approximately parallel, but above all

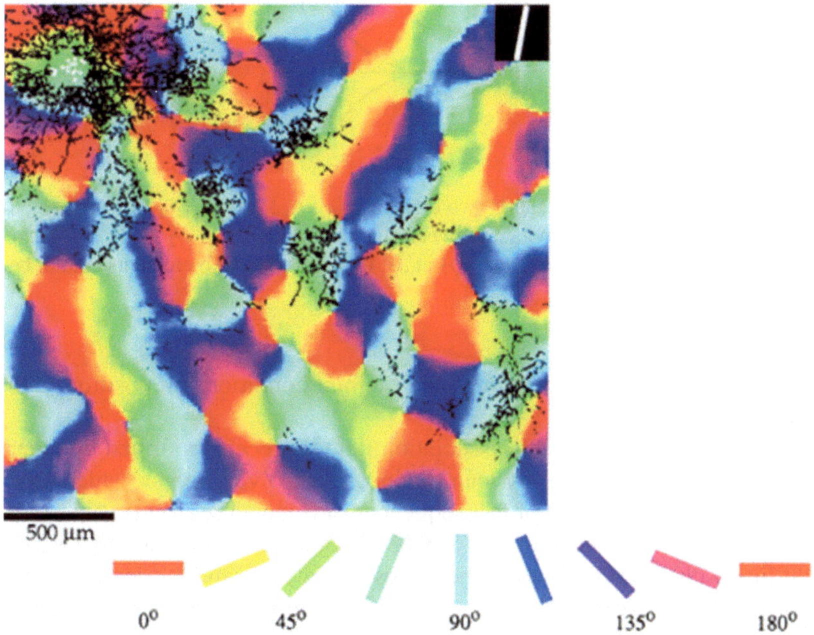

Fig. 1.9 Diffusion of biocytine along the horizontal connections of the 2/3 layer of $V1$ in a tree shrew (tupaya). The injection site is marked by white dots in the upper-left corner. The synaptic buttons reached by diffusion are marked in black. The distribution is anisotropic and "patchy", clustering in domains of the same orientation as the injection site, and globally concentrated along the diagonal top-left → bottom-right. (From Bosking *et al.* [25]).

approximately *coaxial* elements where the common orientation p is the orientation of the axis $\overline{ab}$.

This major result has been confirmed on a psychophysical plane by David Field, Anthony Hayes and Robert Hess' works on perception of the orientation of segments, in their 1993 article Field *et al.* [50]. If we consider a set of randomly distributed little segments, perception does not identify any global structure in them. However, if a curve γ with a weak curvature, composed of aligned segments $v_i = (a_i, p_i)$ is embedded in a background of other randomly-oriented distractors, subjects perceive very well global alignment through a phenomenon of *pop-out* (of perceptual *saliency*). As the authors explain, it is a *low-level integration*: there exist local neurophysiological binding rules which let a global perceptual organization emerge.

The measure of the variations of the detection rate in function of the spatial positions and relative orientations of elements $v_i = (a_i, p_i)$ enabled Field, Hayes and Hess to conclude that the tendency of the elements to be perceived as aligned along a curve comes from the existence of a *specific connectivity* which they called

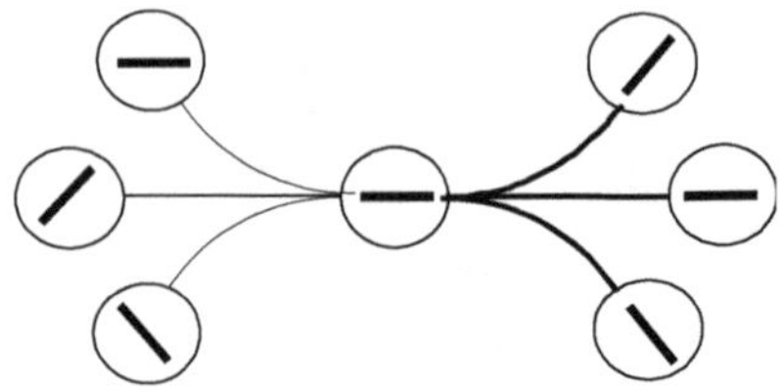

Fig. 1.10 Schema of the association field. Two elements (a_1, p_1) and (a_2, p_2) are connected (thick lines) if it is possible to interpolate between the positions a_1 and a_2 using a curve γ with low curvature, tangent to p_1 at a_1 and to p_2 at a_2. Otherwise the two elements cannot be connected (thin lines).

association field. This connectivity is defined by joint conditions on positions and orientations (see figure 1.10).

The psychophysical reality of the association field has been confirmed by other experiments, in particular the ones by Jean Lorenceau and Yves Frégnac which used the method of the apparent speed of fast sequences of oriented segments ("speedup illusion"). One presents to the subject a series of segments moving along a vertical straight line with a certain speed, where these segments can be oriented in the direction of the motion (vertical colinearity) or in the orthogonal one (horizontal parallelism). One finds that the apparent speed is over-estimated in the colinearity case and under-estimated in the parallelism case. It seems that over-estimation is an effect of the association field and results from the propagation of activity along the horizontal lateral connections. Besides, in the colinearity case, the increase of the apparent speed measured through these psychophysical methods turns to be essentially comparable (about $0.2\ m/s$) to the propagation speed in the horizontal cortico-cortical connections measured by electrophysiological means. The association field seems then to be an effect of the functional architecture.

In a recent article [97], J. Lorenceau and his colleagues confirmed through MEG (magnetoencephalography) this speeding up of vertical apparent motion. The imagery data show how along the horizontal cortico-cortical lateral connections there exists a wave propagation and more precisely a spike train alignment mechanism "that synchronises the neuronal activity tied to a figure contour". Co-alignment shortens the response latency and induces a phase advance along contours. This seems to explain the pop-out and perceptual saliency phenomenon observed by psychophysical means.

1.4.9 *The Contact Structure*

1.4.9.1 Legendrian Lifts and the Integrability Condition

The experimental results provided by the new imaging techniques had an utmost importance for neurogeometry. Firstly, on the plane of the history of geometry, they show from what neuronal processes, inherited from a vast phylogeny, can emerge the *primitive* geometric form of a *line*. They allow to measure the abyss which separates the neuronal materiality from the geometrical ideality. A line is obtained through the integration of oriented segments $v_i = (a_i, p_i)$ (*i.e.* contact elements, see below) whose orientations p_i are tangent to a curve interpolating the positions a_i. This means that, in $V1$, a line is represented as the *envelope of its tangents* and not as a set of points. In some sense, neural darwinism invented what in classical geometry is called "projective duality". Here we reach the *origin* of geometry. Then, as far as functional architecture is concerned, these results show also that $V1$ is mainly dedicated to the extraction of edges through integration. Finally they allow understanding of how $V1$ can "calculate" geometry.

The last point is of a particular importance, and it has been well explained by Jan Koenderink [77]. With the firing rates of spikes along their axons, neurons can only code numerical values of features localised at their receptive field. As Koenderink says, they are "point processors". Retinotopic fields of neurons activated in parallel can only calculate fields of numerical values, that is functions $f(a)$. Thus, how can they perform computations of differential geometry which impose to compute derivatives and therefore limits? "Differential calculus" must come from very specific functional architectures which are structured in such a way that propagation of activity along their connections is *equivalent* to a differential calculus. But this is possible only if we *add supplementary variables* as the orientation p, and a functional architecture *forcing* the interpretation of p as a *tangent* orientation. This is why the subtle notions of *jet* and of *contact structure* must be involved in a natural way imposed by experience. In his 1989 article "The visual cortex is a contact bundle" [64] William Hoffman had already explicitly formulated the idea that contours lift the discontinuities of the retinal stimulus into a retinotopic contact fibre bundle implemented in the cortical hypercolumns.

To explain all these concepts, let us begin with the model $\pi_J : J^1(\mathbb{R},\mathbb{R}) = \mathbb{V}_J = \mathbb{R}^2 \times \mathbb{R} \to \mathbb{R}^2$ with $M = \mathbb{R}^2$, $P = \mathbb{R}$ the line orientations θ measured by $\tan(\theta)$, and $J^1(\mathbb{R},\mathbb{R})$ the fibre bundle of 1-jets of smooth curves in $\mathbb{R}^2$, which is defined in the following way. Consider in $\mathbb{R}^2$ a smooth curve γ which is the graph $\{x, f(x)\}$ of a real function f on $\mathbb{R}$. The first order jet of f at x, $j^1f(x)$, is characterized by 3 slots: the coordinate x, the value $y = f(x)$ of the function f at x, and the value $p = f'(x)$ of the first derivative of f at x. So, a 1-jet is nothing else than a triple $v = (a = (x,y), p)$, what geometers call a *contact element*. Conversely, to every contact element $v = (a, p)$, one can associate the set of smooth functions f whose graph is tangent to v at a.

One can give a more geometric version of the fibration π_J. Consider at every point a of M, not the tangent plane T_aM, but the set C_aM of its lines through the origin 0. C_aM is isomorphic to the projective space $\mathbb{P}^1$. The total space $CM = \mathbb{R}^2 \times \mathbb{P}^1$ gluing these fibres is called the *contact bundle* of M. We will denote it by $\mathbb{V}_P$. $\mathbb{V}_P$ is the compactification of the space of 1-jets $J^1(\mathbb{R},\mathbb{R}) = \mathbb{V}_J$ and its fibre $\mathbb{P}^1$ corresponds to the fibre $\mathbb{R}$ of $\mathbb{V}_J$ via the stereographic projection $\mathbb{P}^1 - \{\frac{\pi}{2}\} \to \mathbb{R}$, $\theta \mapsto \tan(\theta)$.

Now, let γ be a parametrised smooth curve $a(s) = (x(s), y(s))$ in the base plane $\mathbb{R}^2$ (a contour) with $x'(s) \neq 0$. It can be lifted to $\mathbb{V}_J$ using the 1-jet map $j^1\gamma(a(s))$ of γ that associates to $a(s) = (x(s), y(s))$ the contact element $\left(a(s), p_{a(s)}\right)$ where $p_{a(s)} = \frac{y'(s)}{x'(s)}$ is the slope of the tangent to γ at $a(s)$. If there are vertical tangents we have to use the compactification $\mathbb{V}_P$ of $\mathbb{V}_J$. So $\Gamma = \widetilde{\gamma} = v(s) = (a(s), p(s)) = (a(s), p_{a(s)}(s))$. This lift $\Gamma = \widetilde{\gamma}$ – called the *Legendrian* lift of γ – represents γ as the envelope of its tangents. So, to every smooth curve γ in $\mathbb{R}^2$ (without any vertical tangent) is associated a skew curve Γ in $\mathbb{V}_J$ or $\mathbb{V}_P$. But the converse is of course completely false. If $\Gamma = v(s) = (a(s), p(s)) = (x(s), y(s), p(s))$ is a skew curve in $\mathbb{V}_J$ or $\mathbb{V}_P$, the projection $a(s) = (x(s), y(s))$ of Γ is a curve γ in $\mathbb{R}^2$. But Γ is the Legendrian lift $\widetilde{\gamma}$ of γ if and only if $p(s) = p_{a(s)}(s)$. In other words, if Γ is locally defined by equations $y = f(x)$, $p = g(x)$, there exists a curve γ in $\mathbb{R}^2$ such that $\Gamma = \widetilde{\gamma}$ if and only if $g(x) = f'(x)$, that is if and only if $p = y'$. This condition is called an *integrability condition*. It is the *geometric interpretation of the functional architecture of* $V1$ *and of the association field.*

The integrability condition can be formulated in a more interesting way. Let $t = (a, p; \alpha, \pi) = (x, y, p; \xi, \eta, \pi)$ be tangent vectors to $\mathbb{V}_J$ at $v = (a, p) = (x, y, p)$. Along γ (we suppose x is the independent variable) $t = (x, y, p; 1, y', p')$ and the integrability condition $p = y'$ means that we have in fact $t = (x, y, p; 1, p, p')$. It is straightforward to verify that this condition is equivalent to the fact that t is in the kernel of the differential 1-form $\omega_J = dy - p dx$, $\omega_J = 0$ meaning simply that $p = \frac{dy}{dx}$. Indeed, to compute the value of a 1-form ϖ on a tangent vector $t = (\xi, \eta, \pi)$ at (x, y, p), one applies the rules $dx(t) = \xi$, $dy(t) = \eta$, $dp(t) = \pi$. If $\varpi(t) = \sum \varpi_i t_i$ (where t_i and ϖ_i are the components of t and ϖ with respect to the bases of $T\mathbb{V}_J$ and $T^*\mathbb{V}_J$ associated to the coordinates (x, y, p)), one gets $\omega_J(t) = -p.1 + 1.p + 0.p' = -p + p = 0$ since $\omega_J = -pdx + 1.dy + 0.dp$ and dx (resp. dy, dp) applied to $(1, p, p')$ selects the first (resp. second, third) component 1 (resp. p, p'). It must be emphasised that if $p = y'$ the "vertical" component $\pi = p'$ of the tangent vector t in the direction of the p-axis is the *curvature* of the projection γ at a. Indeed, $p = y'$ implies $p' = y''$ and therefore $\pi = p' = y''$.

The 1-form ω_J is called the *contact form* and its kernel is the field $\mathcal{K}$ of *planes* K_v – called the *contact planes* – with equation $-p\xi + \eta = 0$. The tangent vectors $X_1 = \frac{\partial}{\partial x} + p\frac{\partial}{\partial y} = (\xi = 1, \eta = p, \pi = 0)$ and $X_2 = \frac{\partial}{\partial p} = (\xi = 0, \eta = 0, \pi = 1)$ are evident generators. Now we can express purely geometrically the integrability condition: a curve Γ in $\mathbb{V}_J$ is the Legendrian lift $\widetilde{\gamma}$ of its projection γ if and only if it is everywhere tangent to the field $\mathcal{K}$ of contact planes, *i.e.* if and only if it is an *integral curve* of $\mathcal{K}$.

1.4.9.2 The Contact Structure as a Cartan Connexion

The contact structure has several interesting properties. First, if we don't consider any more the projection $\pi_J : \mathbb{V}_J = \mathbb{R}^2 \times \mathbb{R} \to \mathbb{R}^2$ on the plane (x, y) but the projection $\pi'_J : \mathbb{V}_J = \mathbb{R}^2 \times \mathbb{R} \to \mathbb{R}^2$ on the plane (x, p), we can show how the 1-form ω_J defines a *connection* in the sense of Elie Cartan, the contact plane K_v becoming what are called the "horizontal" planes. This representation deepens the projective duality which we already mentioned. Instead of taking as base plane the (x, y) plane and as fibre the axis of the tangents p calculated through the *derivation* $p = \frac{dy}{dx}$, we take as base plane the (x, p) plane and as fibre the y axis, the curves γ being now given as functions $p = g(x)$, *i.e.* as envelopes of their tangents and y being calculated through the *integration* $y = \int y'dx = \int pdx$. The *curvature* $d\omega_J$ of the connection 1-form ω_J has to be a symplectic form on the new base plane. It is evident, since $d\omega_J = dx \wedge dp$ is the standard symplectic form on the (x,p) plane.

1.4.9.3 The Non Integrability of the Contact Structure

The contact structure $\mathscr{K}$ is the field of planes $K_v \subset T_v\mathbb{V}_J$ defined by the equations $\eta = p\xi$ parametrised by p. As the Legendrian lifts are its integral curves, there exists therefore a lot of 1-dimensional integrals. But, nevertheless, there exist *no* 2-dimensional integrals, no *surfaces* S of $\mathbb{V}_J$ which are tangent to K_v at every point $v \in S$, *i.e.* such that $T_vS = K_v$. This is due to the fact that the field K_v spins too rapidly with p to be integrable: K_v is the "vertical" plane above the "horizontal" line of slope p and, when p varies along the fibre $\mathbb{R}_a$ above a, it rotates with p.

More precisely, the non integrability of $\mathscr{K}$ – called *non holonomy* – results from the violation of the *Frobenius integrability condition* saying that a 1-form ϖ admits integral surfaces if and only if $\varpi \wedge d\varpi = 0$ (that is $d\varpi(t,t') = 0$ for all tangent vectors t and t' such that $\varpi(t) = \varpi(t') = 0$). Now, for $\omega_J = dy - pdx$ one gets

$$\begin{aligned} d\omega_J &= -\left(\frac{\partial p}{\partial x}dx \wedge dx + \frac{\partial p}{\partial y}dy \wedge dx + \frac{\partial p}{\partial p}dp \wedge dx\right) + d^2y - pd^2x \\ &= -dp \wedge dx = dx \wedge dp \end{aligned}$$

and therefore

$$\omega_J \wedge d\omega_J = (-pdx + dy) \wedge dx \wedge dp = dy \wedge dx \wedge dp = -dx \wedge dy \wedge dp\,.$$

But this 3-form is a volume form of $\mathbb{V}_J$ and vanishes nowhere. By the way, for the basis $X_1 = \frac{\partial}{\partial x} + p\frac{\partial}{\partial y} = (1,p,0)$, $X_2 = \frac{\partial}{\partial p} = (0,0,1)$ of K_v one has $[X_1,X_2] = -X_3 = -\frac{\partial}{\partial y} = (0,-1,0)$ and $X_3 = (0,1,0) \notin K_v$ since $\omega_J(X_3) = 1 \neq 0$.

1.4.9.4 Scale and Characteristic Vectors

It must be emphasised that the definition of the contact structure using the 1-form ω_J contains more information than the definition using its kernel, *i.e.* the distribution $\mathscr{K}$ of contact planes K_v. Indeed, the 1-forms ω_J and $\alpha\omega_J$ ($\alpha \neq 0 \in \mathbb{R}$) have the same kernel and define the same distribution. The supplementary information encoded in ω_J is the numerical value of ω_J on the "characteristic" tangent vector X_3 transverse to K_v.

1.4.9.5 $SE(2)$-Invariance

The contact structure is invariant under the action of the special Euclidean group $G = SE(2)$ of rigid motions in the plane, which is the semi-direct product $SE(2) = \mathbb{R}^2 \rtimes SO(2)$ of the rotation group $SO(2)$ and the translation group $\mathbb{R}^2$. If (b, r_θ) is an element of $SE(2)$, it acts on a point a of $\mathbb{R}^2$ by $(b,r_\theta)(a) = b + r_\theta(a)$. If (b,r_θ) and

(c, r_φ) are two elements of $SE(2)$, their (non commutative) product is given by the formula:

$$(c, r_\varphi) \circ (b, r_\theta) = (c + r_\varphi(b), r_{\varphi+\theta}).$$

This product is noncommutative since $(b, r_\theta) \circ (c, r_\varphi) = (b + r_\theta(c), r_{\theta+\varphi})$ and $c + r_\varphi(b) \neq b + r_\theta(c)$ in general. The rotation r_θ acts on the fibre bundle $\mathbb{V}_J$ by $r_\theta(a, p(\varphi)) = (r_\theta(a), p(\varphi + \theta))$ (where $p(\varphi)$ is the orientation of angle φ), this very particular form of action expressing the fact that the alignment of preferred orientations is an invariant property.

1.4.9.6 The Polarised Heisenberg Group

A key point concerning the contact structure of $\mathbb{V}_J$ is that it is left-invariant for a noncommutative *Lie group structure* which is isomorphic to the Heisenberg group and called the *polarised Heisenberg group*. The product law is given by the formula:

$$(x, y, p).(x', y', p') = (x + x', y + y' + px', p + p') .$$

It is straightforward to verify that this law is associative, that the origin $(0,0,0)$ of $\mathbb{V}_J$ is its neutral element, and that the inverse of $v = (x, y, p)$ is $v^{-1} = (-x, -y + px, -p)$. Due to the asymmetry of the coupling term px', the product is noncommutative. $\mathbb{V}_J$ is a semi-direct product $\mathbb{V}_J = \mathbb{R}^2 \rtimes \mathbb{R}$. The base plane $\mathbb{R}^2$ of the fibration $\pi_J : \mathbb{V}_J = \mathbb{R}^2 \times \mathbb{R} \to \mathbb{R}^2$ is the commutative subgroup of translations and the centre Z of $\mathbb{V}_J$ is the y-axis. Indeed, $v' = (x', y', p')$ commutes with all $v \in \mathbb{V}_J$ if and only if for every $v = (x, y, p)$ we have $px' = p'x$, which implies $x' = p' = 0$.

If $t = (\xi, \eta, \pi)$ are vectors of the Lie algebra $\mathcal{V}_J = T_0\mathbb{V}_J$ of $\mathbb{V}_J$, $\mathcal{V}_J$ has Lie bracket

$$[t, t'] = [(\xi, \eta, \pi), (\xi', \eta', \pi')] = (0, \xi'\pi - \xi\pi', 0)$$

and is generated as a Lie algebra by the basis of K_v: $X_1 = \frac{\partial}{\partial x} + p\frac{\partial}{\partial x} = (1, p, 0)$ and $X_2 = \frac{\partial}{\partial p} = (0, 0, 1)$ at $v = 0$. Indeed, at 0, $X_1 = (1,0,0)$, $X_2 = (0,0,1)$ and $[X_1, X_2] = (0, -1, 0) = -X_3$ (the other brackets $= 0$). The fundamental fact that the basis $\{X_1, X_2\}$ of the distribution $\mathcal{K}$ is *bracket generating*, *i.e.* Lie-generates the whole tangent bundle $T^*\mathbb{V}_J$ is called the *Hörmander condition*. It is the key property for generalising to higher dimensions and general manifolds our very simple contact structure $\mathbb{V}_J$. Moreover this group is *nilpotent* of step 2, which means that all brackets of the form $[t, [u, v]]$ vanish.

We will measure in the following sections the importance of Hörmander's condition. Lars Hörmander (Fields Medal 1962) was one of the main specialists of hypoelliptic differential operators such as sub-Riemannian Laplacians. His 1983 four volumes *Analysis of Linear Partial Differential Operators* [66] is a fundamental reference. Another classical opus is Elias Stein's 1993 *Harmonic Analysis, Real Variable Methods, Orthogonality, and Oscillatory Integrals* [124]. With Linda Preiss Rothschild, Stein applied the theory to nilpotent groups (see [109]).

Computations in $\mathbb{V}_J$ become very easy if we use the matrix representation

$$v = (x, y, p) = \begin{pmatrix} 1 & p & y \\ 0 & 1 & x \\ 0 & 0 & 1 \end{pmatrix}$$

and

$$t = (\xi, \eta, \pi) = \begin{pmatrix} 0 & \pi & \eta \\ 0 & 0 & \xi \\ 0 & 0 & 0 \end{pmatrix}$$

Indeed, the product in $\mathbb{V}_J$ becomes the matrix product $v.v'$ and the Lie product in $\mathcal{V}_J$ becomes the commutator $[t, t'] = t.t' - t'.t$. Using this trick, it is easy to see that the contact structure is left-invariant. The *left translation* L_v of $\mathbb{V}_J$ is defined by $L_v(v') = v.v'$ and is a diffeomorphism of $\mathbb{V}_J$ whose tangent map at 0 is the linear map

$$\begin{aligned} T_0L_v : \ \mathcal{V}_J = T_0\mathbb{V}_J &\to \qquad T_v\mathbb{V}_J \\ t = (\xi, \eta, \pi) &\mapsto T_0L_v(t) = (\xi, \eta + p\xi, \pi) \end{aligned}$$

The matrix of T_0L_v is

$$T_0L_v = \begin{pmatrix} 1 & 0 & 0 \\ p & 1 & 0 \\ 0 & 0 & 1 \end{pmatrix}.$$

This shows that the basis $\left\{\frac{\partial}{\partial x}, \frac{\partial}{\partial y}, \frac{\partial}{\partial p}\right\}$ of the tangent bundle $T\mathbb{V}_J$ associated to the coordinates $\{x, y, p\}$ is *not* left-invariant. It is the source of non holonomy. To get a left-invariant basis we must translate via L_v the basis $\left\{\frac{\partial}{\partial x}, \frac{\partial}{\partial y}, \frac{\partial}{\partial p}\right\}_0$ at 0. We get the basis $\left\{\frac{\partial}{\partial x} + p\frac{\partial}{\partial y}, \frac{\partial}{\partial y}, \frac{\partial}{\partial p}\right\}$, that is $\{X_1, X_3, X_2\}$.

Let now t be a vector of the contact plane K_0 at 0. Since $\eta = p\xi$ and $p = 0$, we have $\eta = 0$. Its translated $T_0L_v(t)$ is therefore $(\xi, p\xi, \pi)$, and since $\eta = p\xi$, $T_0L_v(t)$ is an element of the contact plane K_v and the contact structure $\mathcal{K} = \{K_v\}$ is nothing else than the left-invariant field of planes left-translated from K_0. In fact, the 1-form ω_J itself is left-invariant and left-translates $\omega_{J,0} = dy$.

Using the matrix form, it is also very easy to analyse other aspects of the Lie group structure of $\mathbb{V}_J$, its adjoint and coadjoint representations and its unitary irreducible representations (unirreps). According to a variant of the Stone-von Neumann theorem concerning the Heisenberg group, every unirrep of the polarised Heisenberg group $\mathbb{V}_J$ which is not trivial on its centre Z (the y-axis) is equivalent to a Schrödinger representation $\pi_\lambda(x, y, p)$ acting on the infinite dimensional Hilbert space $\mathcal{H} = L^2(\mathbb{R}, \mathbb{C})$ via

$$\pi_\lambda(x, y, p)\, u(s) = e^{i\lambda(y+xs)} u(s + p), \text{ with } \lambda \neq 0\,.$$

For $\lambda = 0$ these unirreps degenerate into trivial representations of dimension 1:

$$\pi_{\mu,\nu}(x, y, p)\, u(s) = e^{i(\mu x + \nu p)} u(s)\,.$$

According to a deep theorem due to Kirillov, these unirreps correspond to the orbits of the coadjoint representation of $\mathbb{V}_J$: the planes $(\mathbb{R}, \eta^*, \mathbb{R})$ if $\eta^* \neq 0$ and, if $\eta^* = 0$, every point of the $(\xi^*, 0, \pi^*)$ plane ((ξ^*, η^*, π^*) are vectors of the dual Lie algebra $\mathscr{V}_J^*$ of $\mathbb{V}_J$). We will return on these topics in section 1.5.4.3.

1.4.9.7 Contact Structure and "Simplexity"

The contact structure of the space of 1-jets is a good example of what Alain Berthoz proposed to call "*simplexity*", *i.e.* the original solution found by biological evolution "to decompose complex problems in simpler sub-problems, thanks to specialised modules, and to recompose the whole later" ([17], p.22). In our case, the complex problem is to compute the derivatives of some functions with respect to retinal position variables $a = (x, y)$. The "simplex" solution phylogenetically "invented" by simple neurons of $V1$ consists in adding a new independent variable p to the variables (x, y), in organising in specialised modules (orientation hypercolumns) the values of the three variables, *i.e* the contact elements (a, p), in processing the inputs by measuring the point values of the (a, p) they activate, and finally in recomposing the whole through a functional architecture which guarantees the equivalence between, on the one hand, taking the point values of the three variables (x, y, p) and, on the second hand, deriving the initial variables (x, y). This equivanence is exactly the definition of a 1-jet: $\{x, y, p\} \simeq j^1 y(x)$ (see. [108]).

1.4.10 Illusory Contours as Sub-Riemannian Geodesics

In this initial neuro-geometrical framework, we can easily interpret the variational process giving rise to Kanizsa illusory contours evoked in the Introduction. We presented in section 1.3.3 Mumford's model defined in $\mathbb{R}^2$ by an energy of the type $E = \int_\gamma (\alpha\kappa^2 + \beta) ds$. But we can now use what we know concerning the functional architecture of $V1$. The pacmen define two contact elements (a, p) and (b, q) and an illusory contour interpolating between (a, p) and (b, q) is a skew curve Γ in $\mathbb{V}_J$ from (a, p) to (b, q) which is at the same time:

1. a Legendrian lift $\tilde{\gamma}$ of a curve γ in the base plane $\mathbb{R}^2$, *i.e.* an integral curve of the contact structure (integrability condition $p(x) = y'(x)$);
2. a curve "as straight as possible", as it was already emphasised by Shimon Ullman in 1972 [135] when he introduced the idea of a variational model "minimising total curvature".

The simplest way to satisfy these two requisites is to model illusory contours using *geodesics* for a natural metric because, since the variation of p measures the curvature κ of γ, we minimise at the same time the length and the curvature of the projection γ. But, due to condition (1), the metric has to be defined *only on the distribution $\mathscr{K}$* of contact planes K_v. In our 1999 first synthesis of neurogeometry *Vers une neurogéométrie. Fibrations corticales, structures de contact et contours subjectifs modaux* [101], geodesics were computed solving the Euler-Lagrange equations with the Lagrange multiplier expressing the condition of integrability. And these equations were reformulated in the framework developed by Robert Bryant and Phillip Griffiths for variational models on Lie groups. Later on, the search for constraint minima was naturally interpreted in terms of a suitable metric defined on the contact distribution $\mathcal{K}$. Such a metric is called *sub-Riemannian* and so the modelling of illusory contours is embedded into the mathematical context of left-invariant sub-Riemannian metrics on nilpotent Lie groups.

We will return to sub-Riemannian geodesics in section 1.5.2. We will then explain in what rigorous sense Kanizsa illusory contours are "*geodesics of a sub-Riemannian geometry defined on the contact structure of the fibre bundle of 1-jets of planar curves*". But before explaining this sub-Riemannian geometry of $\mathbb{V}_J$, let us sketch how Bryant-Griffiths' more sophisticated framework enables to work directly in $SE(2)$.

1.4.11 From Maurer-Cartan to Bryant-Griffiths

A powerful tool to solve this kind of variational problem is the method of the *moving frame* introduced by Elie Cartan. It is rather sophisticated but gives a remarkable insight. The (direct) euclidean frames of $\mathbb{R}^2$ constitute the Lie group $G = SE(2) = \mathbb{R}^2 \rtimes SO(2)$ which is a principal fibre bundle of base $\mathbb{R}^2$. An essential structure associated to G and to its Lie algebra $\mathcal{G}$, a structure which explains the problems of non-holonomy, is what is callled the *Maurer-Cartan form*. Let's start from the general expression $g = \begin{pmatrix} 1 & 0 \\ b & e^{i\theta} \end{pmatrix}$ of an element g of G represented through a 3×3 matrix where b is a column vector of translation $\{u, v\}$ and $e^{i\theta}$ the 2×2 matrix of rotation angle θ. We consider the differential of g, $dg = \begin{pmatrix} 0 & 0 \\ db & ie^{i\theta}d\theta \end{pmatrix}$ and we interpret it as a differential 1-form on G with values in $\mathcal{G}$. This means that the components of dg are 1-forms on G, but that dg has the type of an element of $\mathcal{G}$. In other terms, if $\Omega^1(G)$ is the vector space of 1-forms on G, dg is an element of the tensor product $\Omega^1(G) \otimes \mathcal{G}$. If $g = e$, we verify that $dg(e)$ is the *identity* of $\mathcal{G}$. This means that $dg(e)$ is a 1-form on $\mathcal{G}$ and is therefore applied to vectors $\varsigma \in \mathcal{G}$. But as it has values in $\mathcal{G}$, we have $dg(e)(\varsigma) \in \mathcal{G}$. Identity means that $dg(e)(\varsigma) = \varsigma$. More generally, dg can be interpreted as the *identity function* of the tangent fibre bundle TG. The problem of non-holonomy is that dg is *not* invariant for the left-translations L_g. It is the identity function of TG, but not the one of TG globally trivialised by the L_g. Indeed, a 1-form on G with values in $\mathcal{G}$ which is G-invariant by left-translations must have *constant* components in the invariant bases of the T_g^*G dual to the invariant bases of the tangent spaces T_gG. Now, this is not the case for dg.

Then, Cartan's idea is to translate $dg(e)$ in order to get a 1-form on G with values in $\mathcal{G}$ which will be *by construction* invariant under the left-translations L_g. Let $\Lambda_G : TG \to \mathcal{G}$ be this 1-form, called the Maurer-Cartan form. We can easily interpret it in a geometrical way. Indeed, we have by definition $\Lambda_G(g) = \left(T_g L_{g^{-1}}\right)^* dg(e)$.[7] If $\varsigma \in T_g G$ is a tangent vector to G at g, $\Lambda_G(g)(\varsigma) = T_g L_{g^{-1}}(\varsigma)$, in other terms, Λ_G transports ς in $\mathscr{G}$ through the global trivialisation provided by the left-translations L_g. We verify that we have $\Lambda_G(g) = \begin{pmatrix} 0 & 0 \\ e^{-i\theta}db & id\theta \end{pmatrix}$. Traditionally, one writes Λ_G under the compact form $\Lambda_G = g^{-1}dg$ where g^{-1} symbolises $\left(T_g L_{g^{-1}}\right)^*$.

From now on we will write ω and ρ the 1-forms (with scalar values respectively in $\mathbb{C}$ and $\mathbb{R}$) $e^{-i\theta}db$ and $d\theta$. With these notations, the 1-form Λ_G (with values in $\mathscr{G}$) is written:

$$\Lambda_G = \begin{pmatrix} 0 & 0 \\ \omega & \rho \end{pmatrix} = \omega \otimes \xi + \rho \otimes \tau$$

where (ξ, τ) is the natural basis of $\mathscr{G}$ which corresponds to infinitesimal translations and rotations.

The remarkable fact is that the structure of the Lie algebra of $\mathscr{G}$ can be retieved from the Maurer-Cartan form Λ_G and is given by the universal formula $d\Lambda_G = -\frac{1}{2}[\Lambda_G, \Lambda_G]$. The exterior derivative of Λ_G is a G-invariant 2-form with values in $\mathscr{G}$. In order to calculate it, we calculate first the exterior derivatives of the components ω and ρ. We get, since for every scalar differential form σ we have a $d^2\sigma = 0$:

$$\begin{cases} d\omega = -ie^{-i\theta}d\theta \wedge db + e^{-i\theta}d^2 b = ie^{-i\theta}db \wedge d\theta = i\omega \wedge \rho \\ d\rho = d^2\theta = 0 \end{cases}$$

and so

$$d\Lambda_G = d\omega \otimes \xi + d\rho \otimes \tau = i(\omega \wedge \rho) \otimes \xi \,.$$

Now, the space $\Omega^1(G) \otimes \mathscr{G}$ of the 1-forms on G with values in $\mathscr{G}$ is endowed with an exterior product which allows interaction between the exterior product of the 1-forms and the Lie bracket of $\mathscr{G}$. Indeed, let $\varsigma = \mu \otimes \xi + \nu \otimes \tau$ and $\varsigma' = \mu' \otimes \xi + \nu' \otimes \tau$ be any two elements of $\Omega^1(G) \otimes \mathscr{G}$. We have by definition:

$$[\varsigma, \varsigma'] = (\mu \wedge \mu') \otimes [\xi, \xi] + (\mu \wedge \nu') \otimes [\xi, \tau] + (\nu \wedge \mu') \otimes [\tau, \xi] + (\nu \wedge \nu') \otimes [\tau, \tau]$$

which is a vector 2-form with values in $\mathscr{G}$, *i.e*. an element of $\Omega^2(G) \otimes \mathscr{G}$. As $[\xi, \xi] = [\tau, \tau] = 0$, $[\tau, \xi] = -[\xi, \tau]$ and $[\xi, \tau] = -i\xi$, we finally obtain:

$$[\varsigma, \varsigma'] = ((\mu \wedge \nu') - (\nu \wedge \mu')) \otimes [\xi, \tau] = -i((\mu \wedge \nu') - (\nu \wedge \mu')) \otimes \xi \,.$$

When $\varsigma = \varsigma'$, we therefore obtain – unlike what happens for a scalar 1-form – $[\varsigma, \varsigma] = -2i(\mu \wedge \nu) \otimes \xi$. In particular, $[\Lambda_G, \Lambda_G] = -2i(\omega \wedge \rho) \otimes \xi$. When we compare the expressions which give $d\Lambda_G$ et $[\Lambda_G, \Lambda_G]$ we obtain the universal equations of Maurer-Cartan which code the geometry of every Lie group G:

[7] We have $T_g L_{g^{-1}} = (T_e L_g)^{-1} : T_g G \to T_e G$ and therefore by duality $(T_g L_{g^{-1}})^* : T_e^* G = \mathscr{G}^* \to T_g^* G$.

$$d\Lambda_G = -\frac{1}{2}[\Lambda_G, \Lambda_G].$$

After expressing the geometry of G through the Maurer-Cartan equations, let's return to the curves γ in the plane $\mathbb{R}^2$. If we follow the moving frame $F = F_a$ constituted by the unitary tangent and normal vectors at a when the point $a = (x,y)$ runs through γ, we get a curve Γ in G that lifts γ and that is called his *Frénet lift*. As G is a principal fibre bundle on $\mathbb{R}^2$ of fibre $SO_2(\mathbb{R}) \simeq \mathbb{S}^1$, we can reinterpret in this new context the Legendrian lifts we studied before. If γ is parametrised by t, Γ is also parametrised by t. The element of arclength on γ (and not on Γ) is given by $ds = \sigma(t)dt$ with $\sigma(t) = \|a'(t)\|$. Moreover, the Frénet frame F_a is given by the unitary tangent vector $e_1(t) = \frac{a'(t)}{\sigma(t)} = (\cos(\theta), \sin(\theta))$ and by the unitary normal vector $e_2(t) = e_1(t)^{\perp} = (-\sin(\theta), \cos(\theta))$. The differentials of e_1 et e_2 are consequently the vector 1-forms on $\mathbb{R}^2$ with values on $\mathbb{R}^2$:

$$\begin{cases} de_1 = (-\sin(\theta)d\theta, \cos(\theta)d\theta) = d\theta \otimes e_2 = \rho \otimes e_2 \\ de_2 = -(\cos(\theta)d\theta, \sin(\theta)d\theta) = -d\theta \otimes e_1 = -\rho \otimes e_1 . \end{cases}$$

Moreover, as $\omega = e^{-i\theta}da$ by definition, we have $da = e^{i\theta}\omega = \omega_u \otimes e_1 + \omega_v \otimes e_2$ where ω_u and ω_v are the components in the moving frame. But as $da = e^{i\theta}ds$, we have also $\omega = ds$, *i.e.* $\omega_u = ds$ and $\omega_v = 0$. Hence the following reinterpretation of dg in the moving frame F of M :

$$\begin{cases} dp = \omega_u \otimes e_1 + \omega_v \otimes e_2 \\ de_1 = \rho \otimes e_2 \\ de_2 = -\rho \otimes e_1 . \end{cases}$$

This differential system is associated to G and is therefore *independent* from any particular curve γ.

But we also know that, on a particular curve γ, we have, by definition of ds and of the curvature κ[8], the following expression for the infinitesimal variation dF of the moving frame F:

$$\begin{cases} da = ds \otimes e_1 = \sigma(t)dt \otimes e_1 \\ de_1 = ds\kappa(t) \otimes e_2 = \kappa(t)\sigma(t)dt \otimes e_2 \\ de_2 = -ds\kappa(t) \otimes e_1 = -\kappa(t)\sigma(t)dt \otimes e_1 . \end{cases}$$

If we compare the two expressions we see that, on the curves $\gamma(t)$, the following Pfaff system is verified:

$$\Pi = \begin{cases} \omega_u - \sigma(t)dt = 0 \\ \omega_v = 0 \\ \rho - \kappa(t)\sigma(t)dt = 0 . \end{cases}$$

In the same way as the Legendrian lifts in the jet space were the integral curves of the contact structure, the Frénet lifts in G are the *integral curves* of Π. We remark

[8] Remember that $\kappa = d\theta/ds$ and therefore that $\rho = d\theta = \kappa ds = \kappa\sigma dt$.

that for these lifts we have $d\omega = i\omega \wedge \rho = 0$ because $d\omega_u = -\omega_v \wedge \rho = 0$ and $d\omega_v = \omega_u \wedge \rho = \sigma(t)dt \wedge \sigma(t)\kappa(t)dt = \sigma^2(t)\kappa(t)dt \wedge dt = 0$.

The Pfaff system Π is defined on G but depends upon t together with the functions $\sigma(t)$ and $\kappa(t)$. Then we can apply exactly the same strategy we applied for the contact fibre bundle and the 1-jets fibre bundle, namely to introduce *supplementary coordinates* $(\sigma,\kappa,t) \in \mathbb{R}^+_\sigma \times \mathbb{R}_\kappa \times \mathbb{R}_t = Y$[9] and consider Π as actually defined on the direct product $X = G \times Y$[10]. As $(\omega_u, \omega_v, \rho)$ are the components of $\Lambda_G = \omega \otimes \xi + \rho \otimes \tau$, it simply consists in the vanishing of the 1-form :

$$\mu = \Lambda_G - Pdt$$

where $P = (\sigma, 0, \kappa\sigma)$. As Λ_G is now seen as a 1-form on X with values in $\mathcal{G}$, the coherence of the types imposes the same type for $\mu = \Lambda_G - Pdt$. Thus it is natural to consider that P is a vector of $\mathcal{G}$, namely the vector:

$$P = \begin{pmatrix} 0 & 0 \\ \sigma & \kappa\sigma \end{pmatrix} = \sigma \otimes \xi + \kappa\sigma \otimes \tau \,.$$

Then, if we note $\mathcal{A}$ the subspace of $\mathcal{G}$ constituted by the P of this form, we can redefine Y as $\mathcal{A} \times \mathbb{R}_t$ and consider that Π becomes actually defined on $X = G \times Y = G \times \mathcal{A} \times \mathbb{R}_t$.

In this context Bryant and Griffith reformulated the variational problems on Lie groups and in [101] we used their work to reformulate the $\mathbb{V}_J$ model.

1.5 Developments in Neurogeometry after 2000

Starting from the late 1990s and the early 2000s, neurogeometry knew a number of developments in an international context where interactions between mathematics and neurophysiology became increasingly numerous and fertile, somehow as the ones between statistical physics and neuronal nets during the 80s. When I assumed the direction of the CREA and of the Master in Cognitive Sciences, I could easily start at the Ecole Polytechnique a class in Cognitive Neurosciences and a seminar on “Brain and Cognition” organised with my colleague of the ENS Patrick Charnay. The lectures were later published in 2008 under the title *Neurogéométrie de la Vision* [105]. In this context some collaborations had been particularly strong and fruitful. Quickly, they led to a new synthesis [102] published in 2003 in the *Journal of Physiology-Paris*.

1.5.1 Collaborations with Alessandro Sarti and Giovanna Citti

At the end of 1997 Alessandro Sarti, a young specialist from Bologna in models of vision and image processing, contacted me. We discussed on different aspects of neurogeometry and, in 2001, when he returned from Berkeley where he had worked

[9] The index represents the coordinate of the considered exemplar of $\mathbb{R}$.

[10] We trivially extend a 1-form on G to a 1-form on X and we keep the same symbol.

from 1997 to 2000 with Sethian on Kanizsa illusory contours, we started a rich co-operation which involved also his colleague at the University of Bologna, Giovanna Citti, an outstanding specialist in functional analysis, diffusion PDE, and harmonic analysis.

1.5.1.1 The Synchronisation of Oscillators and the Mumford-Shah Model

One of the first works by Sarti and Citti in this field has been linking the models of synchronisation of fields of oscillators with the Mumford-Shah model of segmentation which we analysed in section 1.3.3. The Mumford-Shah variational model is fundamental in image processing and in computational vision, but it lacks of an explicit neurophysiological meaning. However, specialists as Bard Ermentrout and Nancy Kopell showed as continuous nets of coupled oscillators, whose frequencies code the intensity of the input, works the same way as algorithms of segmentation. This is the reason of the interest in a comparison with the Mumford-Shah model.

The main idea is the following (see Sarti *et al.* [115]). We consider a 2*D* field of oscillators where the phase $\theta(x,t)$ is a function of the spatial position x. Let ξ be a variable of distance between neighbouring oscillators on a lattice of mesh ε. We introduce a PDE of the form:

$$\frac{\partial\theta(x,t)}{\partial t} = \omega(x) + \sum \frac{1}{|\xi|^2} \{K(x-\xi)\left[\varphi\left(\theta(x,t) - \theta(x-\xi,t)\right)\right] - K(x)\left[\varphi\left(\theta(x+\xi,t) - \theta(x,t)\right)\right]\}$$

where the function φ is analogous to a sine function, $K(x)$ is a local coupling function, and the sum Σ is taken over the neighbours $x+\xi$ of x. Then, if we *encode* in the coupling function $K(x)$ the anisotropic geometry of the functional architecture and if we let the mesh of the lattice tend to 0, we get a model which Γ-converges (Γ-convergence being a convergence adapted to variational models) to the Mumford-Shah variational model for the sub-Riemannian contact metric defined by $K(x)$ (see [115]).

1.5.1.2 The $SE(2)$ Model

The jet-space $\mathbb{V}_J$ model implies choosing a priviledged x-axis. We saw in section 1.4.11 that the group $SE(2)$ naturally operates on the model, but the asymmetry between x and y in the base space $\mathbb{R}^2$ was reflected in the "polarisation" of the Heisemberg group. As in every case in which a group G operates on a fibre bundle, here $G = SE(2)$ on $\pi_J : \mathbb{V}_J = \mathbb{R}^2 \times \mathbb{R} \to \mathbb{R}^2$, it is relevant to consider the *principal* associated fibre bundle, here

$$\pi_S : G = SE(2) = \mathbb{R}^2 \rtimes SO(2) \simeq \mathbb{V}_S = \mathbb{R}^2 \times \mathbb{S}^1 \to \mathbb{R}^2.$$

We worked with G in section 1.4.11 with Bryant and Griffith when using the form of Maurer-Cartan. But we can do this more directly with Citti, Sarti and Manfredini [115], [37]. In this case, the contact form is $\omega_S = -\sin(\theta)\,dx + \cos(\theta)\,dy$, that is $\cos(\theta)(dy - pdx) = \cos(\theta)\omega_J$. The contact planes are spanned by the tangent vectors $X_1 = \cos(\theta)\frac{\partial}{\partial x} + \sin(\theta)\frac{\partial}{\partial y}$ and $X_2 = \frac{\partial}{\partial \theta}$ with Lie bracket $[X_1, X_2] = \sin(\theta)\frac{\partial}{\partial x} - \cos(\theta)\frac{\partial}{\partial y} = -X_3$. Contrary to the polarised Heisenberg case, the X_j constitute an Euclidean orthonormal basis and are therefore more natural. The distribution $\mathscr{K}$ of contact planes is still bracket generating (Hörmander condition) and maximally non integrable since $d\omega_S = \cos(\theta)dx \wedge d\theta + \sin(\theta)dy \wedge d\theta$, and $\omega_S \wedge d\omega_S = -dx \wedge dy \wedge d\theta$ cannot vanish because it is a volume form. The Frobenius condition $\omega_S \wedge d\omega_S = 0$ is not satisfied, and there exists no integral surface of $\mathscr{K}$ in $\mathbb{V}_S$ (but there exist a lot of integral curves of $\mathscr{K}$: all the Legendrian lifts Γ in $\mathbb{V}_S$ of curves γ in the base plane $\mathbb{R}^2$). As for the characteristic vector field (or Reeb field) X_3, it is orthogonal to K_v for the Euclidean metric and defines a *scale* through

$$\omega_S(X_3) = (-\sin(\theta)dx + \cos(\theta)dy)(X_3) = \sin^2(\theta) + \cos^2(\theta) = 1 .$$

When we work with $\mathbb{V}_S$, the natural *sub-Riemannian* metric is the one making $\{X_1, X_2\}$ an orthonormal basis of the contact plane K_v.

The two contact structures on $\mathbb{V}_J = \mathbb{R}^2 \times \mathbb{R}$ and $\mathbb{V}_S = \mathbb{R}^2 \times \mathbb{S}^1$ seem to be alike but are nevertheless very different. Indeed, let us look at their respective Lie algebras. For $\mathbb{V}_J$ we have the algebra $\mathscr{V}_J$ generated by $\{t_1, t_2, t_3\}$ ($t_1 = \frac{\partial}{\partial x} + p\frac{\partial}{\partial y}$, $t_2 = \frac{\partial}{\partial p}$, $t_3 = \frac{\partial}{\partial y}$) with $[t_1, t_2] = -t_3$ and $[t_1, t_3] = [t_2, t_3] = 0$ (we denote these vectors by t_i and no longer by X_i to avoid any confusion). As we have seen, $\mathscr{V}_J$ is a nilpotent algebra because the coefficients $\{1, p, 1\}$ are polynomials whose derivatives vanish after a certain rank (here 2). On the contrary, for $\mathbb{V}_S$ we have the algebra $\mathscr{V}_S$ generated by $\{X_1, X_2, X_3\}$ (with $X_3 = -\sin(\theta)\frac{\partial}{\partial x} + \cos(\theta)\frac{\partial}{\partial y}$) satisfying $[X_1, X_2] = -X_3$, $[X_1, X_3] = 0$ and $[X_2, X_3] = -X_1$, which is therefore *not nilpotent*. Nevertheless we can notice that for small θ, we have at first order $p \sim \theta$, $\sin(\theta) \sim \theta$ and $\cos(\theta) \sim 1$, and so $\omega_S = -\sin(\theta)dx + \cos(\theta)dy$ can be approximated by $\omega = -\theta dx + dy$ which is nothing else than the 1-form $\omega_J = dy - pdx$. $\mathbb{V}_J$ is in some sense "*tangent*" to $\mathbb{V}_S$. In fact it is called the "*tangent cone*" of $\mathbb{V}_S$ or its "nilpotentisation" (see e.g. Mitchell [86], Rothschild, Stein [109], Margulis, Mostow [81] and Bellaïche [12]).

So $\mathbb{V}_J$ and $\mathbb{V}_S$ becomes two sub-Riemannian models, and we will see in section 1.5.5 how it is possible to construct a continuous family interpolating between them. $\mathbb{V}_J$ is defined on a nilpotent group (Carnot group) and $\mathbb{V}_S$ on a non nilpotent group. Methods for neurogeometry, that is for the modelling of neural functional architectures of vision, become therefore part of sub-Riemannian geometry with their geodesics, Laplacians, heat kernels, harmonic analysis, etc.

Many great geometers have studied these very rich structures since the 1980s-1990s. We already cited Hörmander [66], [65], Stein [124] and Rothschild-Stein [109]. We shall also cite Folland-Stein [52], Nagel-Stein-Wainger [92] for the structure of sub-Riemannian spheres, and papers [58] and [59] of Daryl Geller for the wavelet analysis on the Heisenberg group[11].

[11] We thank the anonymous referee for very interesting bibliographical informations.

We were personally highly interested in works by Misha Gromov, Andrei Agrachev, Richard Beals, Bernard Gaveau, Peter Greiner, Luca Capogna, Vladimir Gershkovich, John Mitchell, Richard Montgomery, Robert Strichartz, Anatoly Vershik, Jean-Pierre Pansu, Jean-Michel Bismut, André Bellaïche, and Jean-Jacques Risler. Reference works have been for us Montgomery's *Tour of Sub-Riemannian Geometries* [88], Gromov's *Carnot-Carathéodory spaces seen from within* [62], Agrachev-Sachkov's *Control Theory from the Geometric Viewpoint* [3], Vershik-Gershkovich's *Non Holonomic Dynamical Systems, Geometry of Distributions and Variational Problems* [137], Capogna *et al. An Introduction to the Heisenberg Group and to the sub-Riemannian Isoperimetric Problem* [31], or Strichartz's *Sub-Riemannian Geometry* [125].

1.5.1.3 Sub-Riemannian Diffusion and Perceptual Completion

A specialist of analysis in Lie groups endowed with sub-Riemannian metrics, Giovanna Citti tackled many difficult problems, and in particular, with Luca Capogna, the mean curvature flow. We evoked in section 1.3.2 the non linear diffusion equation

$$\frac{\partial I}{\partial \sigma} = |\nabla I| \operatorname{div}\left(\frac{\nabla I}{|\nabla I|}\right) = \Delta I - \frac{H(\nabla I, \nabla I)}{|\nabla I|^2} = \sum_{i,j}\left(\delta_{ij} - \frac{\left(\frac{\partial I}{\partial x_i}\right)\left(\frac{\partial I}{\partial x_j}\right)}{|\nabla I|^2}\right)\frac{\partial^2 I}{\partial x_i \partial x_j}$$

(where H is the Hessian of I) which makes the level curves C_σ of $I(x,y;\sigma)$ evolve with a normal velocity equal to their curvature. When generalising this situation to Carnot groups as $\mathbb{V}_J$, one meets several difficulties. First, one works on surfaces S_σ evolving in a 3 dimensional space, and we have seen in section 1.3.2 that for dimension ≥ 3 the mean curvature flow can present singularities. Secondly, if one wants to adapt diffusion techniques to sub-Riemannian geometry, one must substitute normals n to S_σ with their projections n_K onto the contact planes K_v. We have also seen section 1.4.9.3 that, because the non-integrability of the contact structure, no piece of S_σ, as small as it may be, can be everywhere tangent to the contact distribution $\mathscr{K}$. But, of course, there will exist in general points $v \in S_\sigma$ where the tangency $T_v S_\sigma = K_v$ occurs. At these points, called *characteristic points*, the projection n_K is no longer defined, and the diffusion becomes singular. Except at these characteristic points, the diffusion equation becomes

$$\frac{\partial I}{\partial \sigma} = \sum_{i,j}\left(\delta_{ij} - \frac{(X_i I)(X_j I)}{\sum_i (X_i I)^2}\right) X_i X_j I$$

In [32], Luca Capogna and Giovanna Citti, study this equation.

These theoretical works can be applied to image completion, image segmentation and subjective surfaces as A. Sarti did in Berkeley with Sethian (see *e.g.* [117]). Given an image of intensity function $I(x,y)$ which is supposed to be regular on a domain W of $\mathbb{R}^2$ (if this is not the case we can smooth it infinitesimally), we can consider the Legendrian lifts of its level curves in $\mathbb{V}_S$. We get a surface Σ in $\mathbb{V}_S$.

Let's suppose that the image is corrupted and contains a gap Λ. To restore the image and to fill-in Λ, the idea is to process a propagation along the association field, *i.e.* a highly anisotropic diffusion driven by the sub-Riemannian geometry of $\mathbb{V}_S$. The idea is simple in practice but very difficult on the mathematical plane. Indeed, even in the simplest case of the heat equation, diffusion lets evolve the functions $f(x)$ on $\mathbb{V}_S$ through the equation $\frac{\partial}{\partial t} = \Delta$, where $\Delta = X_1^2 + X_2^2$ is the *sub-Riemannian* Laplacian, and it is processed by transport along the geodesics. As the heat equation is a linear parabolic PDE, it is sufficient to compute what is called the *heat kernel*, *i.e.* the way in which a Dirac delta function is diffused. Then we write $I(x,y)$ as an infinite superposition of weighted deltas, and we apply linearity. This is equivalent to performing the convolution of I with the heat kernel.

For links of sub-Riemannian diffusion with the mean curvature flow of section 1.3.2, one can consult Chen-Giga-Goto [33] and their proof of the convergence of the Bence-Merriman-Osher algoritm.

But, as we will see later in section 1.5.2 and 1.5.4.1, sub-Riemannian geodesics are far more complex than Riemannian geodesics, making computation of the heat kernel really difficult. G. Citti dedicated many works to this technical question most studied since the classical works of Lars Hörmander. A good introduction to this topic is the book [20] of Bonfiglioli, Lanconelli, Uguzzoni, *Stratified Lie Groups and Potential Theory for their sub-Laplacians*. For general hypoelliptic sub-Laplacians the reader could consult Jerison, Sánchez-Calle [72] and Sánchez-Calle [112] which present approximation techniques for finding the fundamental solution using series expansions. For more precise results on the Heisenberg group and nilpotent groups, the reader could consult, *e.g.*, Folland, Stein [52], Rothschild, Stein [109], and Citti, Uguzzoni [36]. For methods using harmonic analysis, he/she could consult, *e.g.*, Hulanicki [71] and Cygan [41].

But for the applications we are intrested in, general results are not sufficient, and we need also operational computational techniques.We will return to this point in section 1.5.2.

1.5.1.4 Curvature, 2-Jets and Engel Structure

Specialists of vision such as Steve Zucker put forward some experimental data to support the hypothesis that not only detectors of orientation but also detectors of *curvature* exist in $V1$. If we admit this hypothesis, the problem is to know if we have to process these new detectors as the orientation ones by introducing the 2-jets of curves in $\mathbb{R}^2$, and so, in addition to x, y, p or x, y, θ, to add a *fourth* independent variable κ and a supplementary 1-form which forces its interpretation as a curvature. This is what Sarti, Citti, and I did together.

In the $\{x,y,p\}$ case, we work in $\mathbb{V}_J = \mathbb{R}^2 \times \mathbb{R}$ with the contact 1-form $\omega_J = dy - pdx$ and the nonholonomic basis of the contact planes $\left\{X_1 = \frac{\partial}{\partial x} + p\frac{\partial}{\partial y}, X_2 = \frac{\partial}{\partial p}\right\}$, the third base tangent vector X_3 being given by the Lie bracket $[X_1, X_2] = -X_3 = -\frac{\partial}{\partial y}$. We introduce a fourth variable κ, we work on the space $\widetilde{\mathbb{V}_J} = \mathbb{R}^2 \times \mathbb{R} \times \mathbb{R}$, and we write that the natural interpretation of κ is associated to the second derivative

$f''(x)$ for curves of equation $y = f(x)$. The space $\widetilde{\mathbb{V}_J}$ of the $\{x, y, p, \kappa\}$ is the space of 2-jets $J^2(\mathbb{R}, \mathbb{R})$ and its canonic structure, called Engel structure[12], is the Pfaff system constituted by the two 1-forms ω_J and $\tau_J = dp - \kappa dx$.

The kernel of τ_J in $\widetilde{\mathbb{V}_J}$ is generated by the 3 tangent vectors $X_1^\kappa = \frac{\partial}{\partial x} + p\frac{\partial}{\partial y} + \kappa\frac{\partial}{\partial p} = X_1 + \kappa X_2$, $X_3 = \frac{\partial}{\partial y}$, $X_4 = \frac{\partial}{\partial \kappa}$ whereas the kernel of ω_J extended to $\widetilde{\mathbb{V}_J}$ is generated by X_1^κ, X_2 and X_4. The distribution of planes is now $\mathrm{Span}\{X_1^\kappa, X_4\}$, and it generates the whole Lie algebra because $[X_1^\kappa, X_4] = -X_2 = -\frac{\partial}{\partial p}$ and $[[X_1^\kappa, X_4], X_1^\kappa] = -X_3 = -\frac{\partial}{\partial y}$.

In the $SE(2)$ case, we work in $\mathbb{V}_S = \mathbb{R}^2 \times \mathbb{S}^1$ with the contact 1-form $\omega_S = -\sin(\theta)dx + \cos(\theta)dy$ and the nonholonomic bases of the contact planes $\left\{X_1 = \cos(\theta)\frac{\partial}{\partial x} + \sin(\theta)\frac{\partial}{\partial y}, X_2 = \frac{\partial}{\partial \theta}\right\}$, the third base tangent vector X_3 being given by the Lie bracket $[X_1, X_2] = -X_3 = \sin(\theta)\frac{\partial}{\partial x} - \cos(\theta)\frac{\partial}{\partial y}$. We introduce the curvature K and work in the space $\widetilde{\mathbb{V}_S} = \mathbb{R}^2 \times \mathbb{S}^1 \times \mathbb{R}$. The Pfaff system which defines the Engel structure is now composed of the two 1-forms ω_S and $\tau_S = d\theta - Kds = d\theta - K(\cos(\theta)dx + \sin(\theta)dy)$, where s is the parametrisation of the curves of the (x, y) plane by their arch length. Indeed, we know that the curvature, which is given in Cartesian coordinates for a curve of equation $y = f(x)$ by the formula $K = \frac{f''(x)}{\left(1+f'(x)^2\right)^{3/2}}$, is also given by $K = \frac{d\theta}{ds}$. The kernel of the 1-form τ_S is now generated by the 3 tangent vectors

$$\begin{cases} X_1^K = \cos(\theta)\frac{\partial}{\partial x} + \sin(\theta)\frac{\partial}{\partial y} + K\frac{\partial}{\partial \theta} = X_1 + KX_2 \\ X_3 = -\sin(\theta)\frac{\partial}{\partial x} + \cos(\theta)\frac{\partial}{\partial y} \\ X_4^K = \frac{\partial}{\partial K} \end{cases}$$

while the kernel of ω_S extended to $\widetilde{\mathbb{V}_S}$ is generated by X_1, X_2 and X_4^K. The distribution of planes is now $\mathrm{Span}\{X_1^K, X_4^K\}$. It generates the whole Lie Algebra because $[X_1^K, X_4^K] = -X_2 = -\frac{\partial}{\partial \theta}$ and $[[X_1^K, X_4^K], X_1^K] = X_3 = -\sin(\theta)\frac{\partial}{\partial x} + \cos(\theta)\frac{\partial}{\partial y}$.

1.5.1.5 Scale and Symplectic Structure

We saw in section 1.4.9.4 how the distribution $\mathscr{K}$ of the contact planes K_v is underdetermined with respect to the 1-form ω_J because all the 1-forms $\alpha\omega_J$ ($\alpha \neq 0 \in \mathbb{R}$) have the same kernel. Whence the idea of explicitly taking into account the scale factor α in the model. We developed it with Sarti and Citti in [116].

We work now in the 4-dimensional space $\mathbb{W} = \mathbb{R}^2 \times \mathbb{S}^1 \times \mathbb{R}$ where the coordinates are (x, y, θ, σ). We extend the group $G = SE(2)$ in order to take into account the multiplicative law of scales, and we use the left-invariant basis:

---

[12] Friedrich Engel was one of the principal disciples and collaborators of Sophus Lie.

$$\begin{cases} X_1 = e^{\sigma}\left(\cos(\theta)\frac{\partial}{\partial x} + \sin(\theta)\frac{\partial}{\partial y}\right) \\ X_2 = \frac{\partial}{\partial \theta} \\ X_3 = e^{\sigma}\left(-\sin(\theta)\frac{\partial}{\partial x} + \cos(\theta)\frac{\partial}{\partial y}\right) \\ X_4 = \frac{\partial}{\partial \sigma}\,. \end{cases}$$

For each scale σ, we have now the contact 1-form $\omega = e^{-\sigma}(-\sin(\theta)dx + \cos(\theta)dy)$ defined on the sub-space $\mathbb{V}_{\sigma} = \mathbb{R}^2(x,y) \times \mathbb{S}^1(\theta) \times \{\sigma\}$ of $\mathbb{W}$. The main point is that the 2-form $d\omega$ obtained by differentiating ω with respect to all its variables, including the scale, induces a *symplectic structure* on $\mathbb{W}$. We have:

$$\begin{aligned} d\omega &= \left(e^{-\sigma}\cos(\theta)dx + e^{-\sigma}\sin(\theta)dy\right) \wedge d\theta \\ &+ \left(-e^{-\sigma}\sin(\theta)dx + e^{-\sigma}\cos(\theta)dy\right) \wedge d\sigma \\ &= \omega_1 \wedge \omega_2 + \omega_3 \wedge \omega_4 \end{aligned}$$

where ω_i is the dual 1-form of X_i. This way, the 2-form $d\omega$ can be identified with the left-invariant 2-form deduced by left-translations from the standard symplectic 2-form on $T_0\mathbb{W}$:

$$dx \wedge d\theta + dy \wedge d\sigma\,.$$

We remark that as the orientation is the conjugated variable of x, the scale is the conjugated variable of y.

1.5.1.6 Eikonal Equation and Skeletonisation of Forms

The great interest of the "symplectisation" of the contact structure through the scale consists in the fact that on any symplectic manifold it is possible to define Hamiltonian mechanics analogous to "geometrical optics". In our case this allows a definition of a neural implementation of the "grassfire" models we evoked in section 1.3.2.

In [116], given a simple image constituted by regions E delimited by edges $C = \partial E$, Sarti and Citti associate to every point (x,y) of the extension $W \subset \mathbb{R}^2$ of the image the maximal values $\bar{\theta}$ and $\bar{\sigma}$ of θ and σ of the corresponding hypercolumn (winner-take-all strategy) and interpret them (i) as the direction of the edge C at the point which is the closest to (x,y) and (ii) as the distance (up to a factor) of (x,y) to this point. By lifting all the points (x,y) of W, they generate a surface Σ in the fibre bundle $\mathbb{W}$, $\Sigma = \left\{(x,y,\bar{\theta}(x,y),\bar{\sigma}(x,y))\right\}$. Under natural hypotheses, it is possible to deduce "good" properties of $\bar{\theta}(x,y)$ and $\bar{\sigma}(x,y)$ and "good" properties of Σ with respect to the symplectic structure. Indeed, Σ is a Lagrangian surface of $\mathbb{W}$ (see section 1.2.4). Then, if $\bar{\theta}(x,y)$ is the orientation of the closest edge to (x,y) and $\bar{\sigma}(x,y)$ the distance of (x,y) from this edge, the projections of the level curves of $\bar{\theta}(x,y)$ are orthogonal to $C = \partial E$ and the ones of $\bar{\sigma}(x,y)$ are parallel to $C = \partial E$ (see figure 1.11). They propagate the edge C through wave-fronts parallel to C (Huyghens model) and they are solutions of the *eikonal* equation of geometrical optics we met in sections 1.2.4, 1.3.2 and 1.4.5.2.

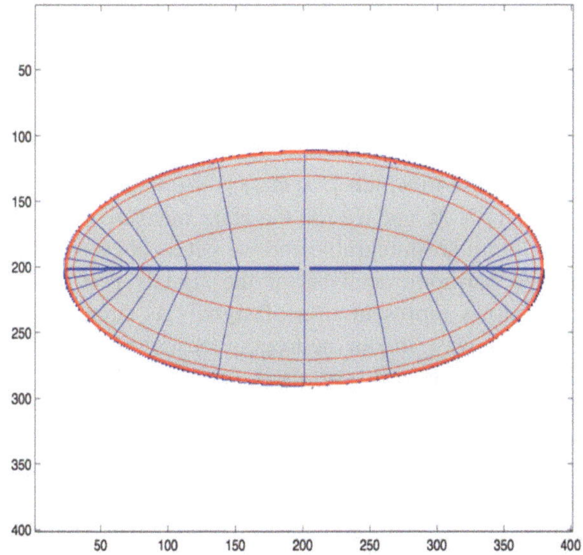

Fig. 1.11 Level curves of $\bar{\theta}(x,y)$ (blue) and $\bar{\sigma}(x,y)$ (red). The first ones are orthogonal to $C = \partial E$ and the second ones parallel to $C = \partial E$. (From Sarti *et al.* [116]).

The singularities of this "optical" propagation define what is called the "generalised axis of symmetry" or "medial axis" S, or also "*skeleton*", of the form E. For a circle it is a point (the centre). For a contour it is generically a graph of dimension 1 constituted by pieces of lines S_i connected by end points and triple points. The fundamental perceptual role of this *virtual* structure associated to the contour C has been underlined, after pioneering works by Harry Blum [19], by many great geometers and specialists in vision as René Thom, David Marr, David Mumford, Steve Zucker or James Damon because S allows canonical decomposition of C in a set of cylinders C_i whose axes are the S_i. Consequently, it is important to notice how, as the illusory contours, the medial axes are constitutive of perceptual geometry and have a neurophysiological reality even if they are not present in the sensorial inputs. This is a consequence of the functional architecture of the primary visual areas, and we can give a neurogeometrical explanation about it.

1.5.1.7 Coherent States and Harmonic Analysis

G. Citti and A. Sarti also examined in depth the idea that the link between neurogeometry and signal analysis (*e.g.* wavelets) is given through the notion of a *coherent state*.

1. We want to analyse signals considered as vectors of a Hilbert space $\mathscr{H}$ (here $L^2(\mathbb{V})$).

2. A locally compact group G is available (here the group $SE(2)$) which acts irreducibly and unitarily on $\mathcal{H}$ through a representation π.

3. A receptive profile $\varphi_0 \in \mathcal{H}$ is also available, which is well localised both in the position space and in the Fourier space.

4. We take the orbit $\{\varphi_g\}_{g\in G}$ of φ_0 under the action of G and we suppose that φ_0 is "admissible" in the sense in which $\int_G \left|\langle\varphi_g, f\rangle\right|^2 d\mu(g) < \infty$ for any $f \in \mathcal{H}$, with $\langle .,. \rangle$ the scalar product on $\mathcal{H}$ and $d\mu(g)$ the Haar measure on G. We then say that the representation π is square-integrable and we get a coherent state.

5. We process the harmonic analysis of the signals f of $\mathcal{H}$ using these coherent states. This allows representation of signals as superpositions of elementary functions and this way, in our neurogeometrical context, to measure them neurophysiologically. The general formula is $f(x) = \int_G T_f(g)\varphi_g(x)d\mu(g)$, where $T_f(g) = \langle f, \varphi_g\rangle \in L^2(G)$ is the transform of f. We can notice the structure of this formula. We want to analyse functions $f(x)$ and we have receptive profiles $\varphi_g(x)$ parametrised by $g \in G$. The function $f(x)$ is constructed as an integral on G considering that x is fixed, where the coefficients $T_f(g)$ are the respective weights of the receptive profiles $\varphi_g(x)$ in the synthesis of $f(x)$. The coherent state $\{\varphi_g\}_{g\in G}$ allows to represent the $f \in \mathcal{H}$ through transforms $T_f(g) = \langle f, \varphi_g\rangle \in L^2(G)$, where the $\langle f, \varphi_g\rangle$ are the "measures" of the signal f given by the receptive profiles φ_g.

These works on coherent states are completed by a research on the optimal forms of receptive profiles φ_g. We saw in section 1.4.2 that the statistics of natural images imposes strong constraints to these profiles. Other converging works showed how functional architectures and association fields, which materialise the Gestalt principle of good continuation, reflect the statistical properties of lines and edges in natural images. For example in [120], Mariano Sigman *et al.* confirmed that "the geometry of the pattern of interactions in primary visual cortex parallels the interactions of oriented segments in natural scenes" (p.1939). Their experimental method consists of measuring the correlation of the orientations of the edges between an origin 0 and a second point a on a corpus of $N = 4.000$ natural images. In his thesis [114], Gonzalo Sanguinetti showed how the results by Sigman converge in a remarkable way with the neurogeometrical models of good continuation.

Moreover, taking inspiration from the fact that Gabor functions have been introduced in quantum mechanics as functions sufficiently well localised both in position and in frequency in order to optimise Heisenberg uncertainty relations, Citti, Sarti, and their PhD students Davide Barbieri and Sanguinetti showed that optimal receptive profiles can be deduced from the structure of the group $SE(2)$ (see [10]).

1.5.1.8 International Conferences, Seminars and Special Issues

These works have been associated with international symposia in neurogeometry organised by G. Citti and A. Sarti. The first was held in Bologna (1-3 July 2004), *Mathematical Models in Visual Perception*. Then came the symposium of the Scuola Normale Superiore of Pisa (4-9 September 2006) *Neuromathematics of Vision*, then, again in Bologna (31 August-4 September 2009) the conference *Sub-Riemannian*

Geometry and Vision. Many scholars as, *e.g.*, Paul Bressloff, Jack Cowan, Guy David, Alain Destexhe, Olivier Faugeras, Yves Frégnac, Walter Gerbino, Jan Koenderink, Jean Lorenceau, Lamberto Maffei, Marc Mézard, Jean-Michel Morel, Scott Pauls, Martin Reimann, James Sethian, Wolf Singer, or Steve Zucker gave their contribution. We also edited together in 2009 a double special issue $(103, 1-2)$ *Neuromathematics of Vision* of the *Journal of Physiology, Paris*. Since 2011, a seminar *Neuromathématiques et modèles de perception* has been held at the Institut Henri Poincaré of Paris.

1.5.2 The Geodesics of the $\mathbb{V}_J$ Model

We have seen in section 1.4.10 how to interpret illusory contours as sub-Riemannian geodesics of the $\mathbb{V}_J$ model. Let us now give an idea of these geodesics, and emphasise that, as the metric is *sub*-Riemannian, they are extremely different from Riemannian ones, even at the infinitesimal level.

As the distribution $\mathcal{K}$ of the contact planes K_v is bracket generating and satisfies Hörmander condition, a celebrated theorem of Chow says that every pair of points (v, v') of $\mathbb{V}_J$ can be connected by an integral curve of $\mathcal{K}$. If $\mathcal{K}$ is endowed with a sub-Riemannian metric, we can compute the length of such integral curves and look for geodesics, which are integral curves of minimal length. The problem of computing geodesics is quite difficult to solve. The 1981 work of Brockett [29] *Control Theory and Singular Riemannian Geometry* is a classic reference. Excellent other references are the already cited books of Montgomery [88] and Strichartz [125] and also Ge [57] and Hammenstädt [63]. One of the main difficulties is that, contrary to the Riemannian case, there can exist "abnormal" geodesics, that is geodesics which do not satisfy the differential equation canonically associated to the geodesic variational problem. Fortunately we will not meet this "abnormality" since our models, even if they are non trivial, remain rather elementary.

Richard Beals, Bernard Gaveau and Peter Greiner who solved with *explicit* formulas the geodesic problem for the (non polarised) Heisenberg group emphasised ([11], p. 634): "how complicated a control problem can become, even in the simplest situation." It was a new mathematical result since in 1977 Bernard Gaveau still said ([56], p. 114) that the variational problem of minimising "the energy of a curve in the base manifold under the Lagrange condition that its lifting is given in the fiber bundle" seemed "not yet (...) studied."

We adapted Beals, Gaveau and Greiner computations to the *polarised* Heisenberg group $\mathbb{V}_J = J^1(\mathbb{R}, \mathbb{R})$ with coordinates $(x, y, p = \tan(\theta))$, product $(x, y, p).(x', y', p') = (x + x', y + y' + px', p + p')$ and contact planes generated by $X_1 = \frac{\partial}{\partial x} + p\frac{\partial}{\partial y} = (1, p, 0)$ and $X_2 = \frac{\partial}{\partial p} = (0, 0, 1)$ with Lie bracket $[X_1, X_2] = -X_3 = (0, -1, 0) = -\frac{\partial}{\partial y}$. Following the approach of Agrachev-Sachkov [3] (see section 1.5.4), let us formulate the geodesic problem as a *control problem*. If $\Gamma = \{v(s)\}$ is a smooth parametrised curve in $\mathbb{V}_J$, to say that it is an integral curve of the contact structure is to say that $\dot{v}(s) = u_1 X_1 + u_2 X_2$ for appropriate controls u_1 and u_2 or, in other words, that $\dot{x} = u_1$, $\dot{y} = pu_1$, $\dot{p} = u_2$, the integrability condition $\frac{\dot{y}}{\dot{x}} = p$ being automatically satisfied. To find the geodesics for the chosen sub-Riemannian metric SR with scalar product $\langle .,. \rangle_{SR}$ and norm $\|.\|_{SR}$, one minimises the Lagrangian given by the kinetic energy $L = \frac{1}{2}\|\dot{v}\|_{SR}^2$ along such curves. L is defined on the tangent bundle $T\mathbb{V}_J$. Using Legendre transform, it can be transformed into the Hamiltonian

$$\begin{aligned} h(v,\varpi) &= \langle \varpi, \dot{v} \rangle - \frac{1}{2}\|\dot{v}\|_{SR}^2 \\ &= \varpi(u_1X_1 + u_2X_2) - \frac{1}{2}\|u_1X_1 + u_2X_2\|_{SR}^2 \end{aligned}$$

defined on the cotangent bundle $T^*\mathbb{V}_J$. If $\varpi = \xi^*dx + \eta^*dy + \pi^*dp = (\xi^*, \eta^*, \pi^*)$ is a 1-form on $\mathbb{V}_J$, then

$$h(v,\varpi) = \xi^*u_1 + \eta^*u_1p + \pi^*u_2 - \frac{1}{2}\left(u_1^2\|X_1\|_{SR}^2 + 2u_1u_2\langle X_1, X_2\rangle_{SR} + u_2^2\|X_2\|_{SR}^2\right) .$$

It is natural to choose a left-invariant metric namely the sub-Riemannian metric SR_J making $\{X_1, X_2\}$ an orthonormal basis of the contact plane K_v since $\{X_1, X_2\}$ is the left-invariant basis translating the standard Euclidean orthonormal basis of K_0. This metric is not the Euclidean metric $\langle .,. \rangle_E$, $\|.\|_E$ since, due to non holonomy, Euclidean metric is not left-invariant. By the way, even if $\|X_2\|_E = 1$ and $\langle X_1, X_2\rangle_E = 0$, we have $\|X_1\|_E = 1 + p^2 \neq 1$ if $p \neq 0$: it is only on the (x,y) plane $p = 0$ that the two metrics are the same. If we choose SR_J, then $\|X_1\|_{SR_J} = \|X_2\|_{SR_J} = 1$, $\langle X_1, X_2\rangle_{SR_J} = 0$, and

$$\begin{aligned} h(v,\varpi) &= \xi^*u_1 + \eta^*u_1p + \pi^*u_2 - \frac{1}{2}\left(u_1^2 + u_2^2\right) \\ &= \varpi(u_1X_1 + u_2X_2) - \frac{1}{2}\left(u_1^2 + u_2^2\right) . \end{aligned}$$

One can then apply a fundamental result of control theory called the *Pontryagin maximum principle*, which generalises the classical method of variational calculus using Euler-Lagrange equations and Lagrange multipliers (see Agrachev, Gamkrelidze [4]) we employed in our first synthesis [101]. It says that geodesics are projections on $\mathbb{V}_J$ of the trajectories of the Hamiltonian H having *maximising* controls $u_{j,\max}$. The maximisation conditions are $\frac{\partial h}{\partial u_1} = \varpi(X_1) - u_1 = 0$ and $\frac{\partial h}{\partial u_2} = \varpi(X_2) - u_2 = 0$ and therefore

$$H(v,\varpi) = u_1\varpi(X_1) + u_2\varpi(X_2) - \frac{1}{2}\left(u_1^2 + u_2^2\right) = \frac{1}{2}\left(u_1^2 + u_2^2\right) = \frac{1}{2}\left(\langle \varpi, X_1\rangle^2 + \langle \varpi, X_2\rangle^2\right)$$

and in terms of coordinates:

$$H(x,y,p,\xi^*,\eta^*,\pi^*) = \frac{1}{2}\left[(\xi^* + p\eta^*)^2 + \pi^{*2}\right] .$$

The structure of geodesics implies that the sub-Riemannian *sphere* S (the ends of geodesics from 0 of sub-Riemannian length $= 1$, which are *global* minimisers) and the *wave front* W (the ends of geodesics from 0 of sub-Riemannian length $= 1$, which are not necessarily global minimisers) are rather strange. In particular, the *cut locus* of 0 (that is the ends of geodesics when they cease to be globally minimising), and the *conjugate locus* or *caustic* of 0 (that is the singular locus of the exponential map $\mathscr{E}$ integrating geodesics) are rather complex.

The Hamilton equations on $T^*\mathbb{V}_J$ derived from the Hamiltonian H are

$$\begin{cases} \dot{x}(s) = \frac{\partial H}{\partial \xi^*} = \xi^* + p\eta^* \\ \dot{y}(s) = \frac{\partial H}{\partial \eta^*} = p\left(\xi^* + p\eta^*\right) = p\dot{x}(s) \text{ (i.e. } p = \frac{\dot{y}}{\dot{x}} = \frac{dy}{dx}\text{, integrability condition)} \\ \dot{p}(s) = \frac{\partial H}{\partial \pi^*} = \pi^* \\ \dot{\xi}^*(s) = -\frac{\partial H}{\partial x} = 0 \\ \dot{\eta}^*(s) = -\frac{\partial H}{\partial y} = 0 \\ \dot{\pi}^*(s) = -\frac{\partial H}{\partial p} = -\eta^*\left(\xi^* + p\eta^*\right) = -\eta^*\dot{x}(s) \end{cases}$$

As H is independent from x and y, the derivatives $\dot{\xi}^*(s) = -\frac{\partial H}{\partial x}$ and $\dot{\eta}^*(s) = -\frac{\partial H}{\partial y}$ vanish, and the momenta ξ^* and η^* are therefore constant along any geodesic: $\xi^* = \xi_0^*$ and $\eta^* = \eta_0^*$. This fact simplifies the equations since $\dot{x}(s) = \xi_0^* + p\eta_0^*$, $\dot{y}(s) = p\left(\xi_0^* + p\eta_0^*\right)$, and $\dot{\pi}^*(s) = -\eta_0^*\left(\xi_0^* + p\eta_0^*\right)$. We emphasise the relations $\ddot{p} = \dot{\pi}^* = -\eta^*\dot{x}$ and $\ddot{x} = \eta^*\dot{p}$, or $(\ddot{x}, \ddot{p}) = \eta^*\left(\dot{p}, -\dot{x}\right)$, which means that in the (x, p) plane the acceleration is orthogonal to the velocity and geodesics are circles whose radius increases when η_0^* decreases (at the limit $\eta_0^* = 0$ the circle becomes a straight line). By the way, $H\left(x, y, p, \xi^*, \eta^*, \pi^*\right) = \frac{1}{2}\left(\dot{x}^2 + \dot{p}^2\right)$ since, by construction, the Hamiltonian H is the kinetic energy of the projection of the trajectories on the (x, p) plane.

Computations show how the $(x, p) = z$ part of the geodesics from 0 to $(x_1 = x(\tau), y_1 = y(\tau), p_1 = p(\tau))$ is given by the formulas:

$$\begin{cases} x(s) = \frac{\sin\left(\frac{s}{2}\eta_0^*\right)}{\sin\left(\frac{\tau}{2}\eta_0^*\right)}\left(\cos\left(\frac{(\tau-s)}{2}\eta_0^*\right)x_1 - \sin\left(\frac{(\tau-s)}{2}\eta_0^*\right)p_1\right) \\ p(s) = \frac{\sin\left(\frac{s}{2}\eta_0^*\right)}{\sin\left(\frac{\tau}{2}\eta_0^*\right)}\left(\sin\left(\frac{(\tau-s)}{2}\eta_0^*\right)x_1 + \cos\left(\frac{(\tau-s)}{2}\eta_0^*\right)p_1\right) \end{cases}$$

which are effectively the equations of a circle

$$x^2 + p^2 - x\left(x_1 + p_1\cot\left(\frac{\eta_0^*\tau}{2}\right)\right) - p\left(p_1 - x_1\cot\left(\frac{\eta_0^*\tau}{2}\right)\right) = 0$$

passing through 0 and (x_1, p_1), with center

$$x_c = \frac{1}{2}\left(x_1 + p_1\cot\left(\frac{\eta_0^*\tau}{2}\right)\right), y_c = \frac{1}{2}\left(p_1 - x_1\cot\left(\frac{\eta_0^*\tau}{2}\right)\right)$$

and radius

$$r^2 = \frac{1}{4}\left(x_1^2 + p_1^2\right)\left(1 + \cot\left(\frac{\eta_0^*\tau}{2}\right)\right) = \frac{1}{4\sin^2\left(\frac{\eta_0^*\tau}{2}\right)}\left|z_1\right|^2 .$$

One verifies that the constant value of the Hamiltonian along a trajectory is:

$$H_0 = \frac{\eta_0^{*2}}{8\sin^2\left(\frac{\eta_0^*\tau}{2}\right)}|z_1|^2 = \frac{\eta_0^{*2}}{2}r^2 .$$

For $y(s)$, computations are more involved. We get

$$\begin{aligned} y(s) = {} & \frac{1}{8\left(\cos\left(\eta_0^*\tau\right)-1\right)}\left[-2\eta_0^* s\left(x_1^2+p_1^2\right) - 4x_1p_1\cos\left(\eta_0^*(s-\tau)\right)+\right. \\ & 2\left(x_1^2-p_1^2\right)\sin\left(\eta_0^*(s-\tau)\right)+ \\ & 2x_1p_1\cos\left(\eta_0^*(2s-\tau)\right) - \left(x_1^2-p_1^2\right)\sin\left(\eta_0^*(2s-\tau)\right)+ \\ & 2x_1p_1\cos\left(\eta_0^*\tau\right) + \left(x_1^2-p_1^2\right)\sin\left(\eta_0^*\tau\right)+ \\ & \left. 2\left(x_1^2+p_1^2\right)\sin\left(\eta_0^* s\right)\right] . \end{aligned}$$

In terms of ξ_0^*, π_0^*, η_0^* and τ the formula writes:

$$\begin{aligned} y(s) = {} & \xi_0^{*2}\frac{2\eta_0^* s+\sin\left(2\eta_0^* s\right)}{4\eta_0^{*2}} - \xi_0^{*2}\frac{\sin\left(\eta_0^* s\right)}{\eta_0^{*2}} + \xi_0^*\pi_0^*\frac{\sin^2\left(\eta_0^* s\right)}{\eta_0^{*2}} - \\ & \xi_0^*\pi_0^*\frac{1-\cos\left(\eta_0^* s\right)}{\eta_0^{*2}} + \pi_0^{*2}\frac{2\eta_0^* s-\sin\left(2\eta_0^* s\right)}{4\eta_0^{*2}} . \end{aligned}$$

The key point is that these equations explain the origin of the striking *multiplicity* of sub-Riemannian geodesics connecting two points. Let us indeed compute $y_1 = y(\tau)$. We find

$$y_1 = \frac{1}{2}x_1p_1 + \frac{x_1^2+p_1^2}{4}\left[\frac{\left(\frac{\eta_0^*\tau}{2}\right)}{\sin^2\left(\frac{\eta_0^*\tau}{2}\right)} - \frac{\cos\left(\frac{\eta_0^*\tau}{2}\right)}{\sin\left(\frac{\eta_0^*\tau}{2}\right)}\right] .$$

If we introduce the new variable $\varphi = \frac{\eta_0^*\tau}{2}$, we see that we must solve the equation

$$4\left(y_1 - \frac{1}{2}x_1p_1\right) = \mu(\varphi)|z_1|^2$$

where $\mu(\varphi)$ is the function

$$\mu(\varphi) = \frac{\varphi}{\sin^2(\varphi)} - \cot(\varphi) .$$

It is the function $\mu(\varphi)$ which is the key of the strange behaviour of sub-Riemannian geodesics. It is an odd function that diverges for $\varphi = k\pi$ ($k \neq 0$) (*i.e.* $\eta_0^*\tau = 2k\pi$) and presents critical points when $\varphi = \tan(\varphi)$. But when $\varphi = \tan(\varphi)$, we have

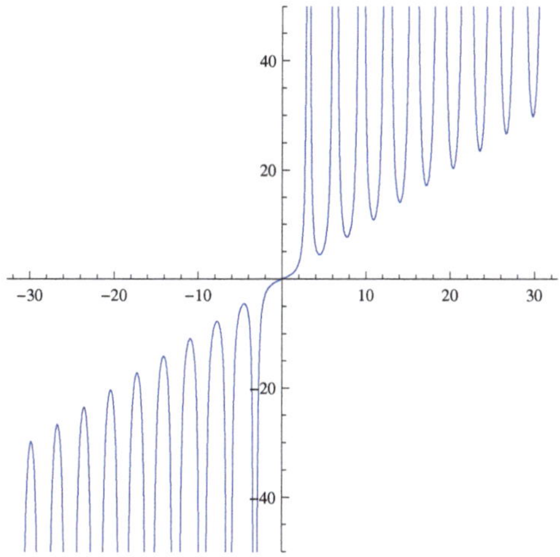

Fig. 1.12 The function $\mu(\varphi)$ occuring in the construction of sub-Riemannian geodesics of the polarised Heisenberg group $\mathbb{V}_J$ (the scale of the two axes are not the same)

$$\begin{aligned}\mu(\varphi) &= \frac{\tan(\varphi)}{\sin^2(\varphi)} - \cot(\varphi) \\ &= \frac{1-\cos^2(\varphi)}{\cos(\varphi)\sin(\varphi)} = \tan(\varphi) = \varphi \ ,\end{aligned}$$

and the minima of $\mu(\varphi)$ are on the diagonal. The graph of $\mu(\varphi)$ is represented in figure 1.12.

Let us compute the length of geodesics. Let γ be a geodesic starting at 0 and ending at time τ at $(x_1, y_1, p_1) = (z_1, p_1)$. If L is its length, we have $L = \int_0^\tau \ell ds$ with $\ell^2 = (\xi^* + p\eta^*)^2 + \pi^{*2}$ the squared norm of $\dot{\gamma}$ in the contact plane endowed with the orthonormal basis $\{X_1 = \partial_x + p\partial_y, X_2 = \partial_p\}$. But $\ell^2 = 2H = 2H_0$ since the Hamiltonian is constant along its trajectories, and we know that

$$H_0 = \frac{\eta_0^{*2}}{8} \frac{1}{\sin^2\left(\frac{\eta_0^* \tau}{2}\right)} |z_1|^2 \ .$$

So, with $\frac{\eta_0^* \tau}{2} = \varphi$,

$$L = \sqrt{2}\left(\frac{\eta_0^* \tau}{2}\right) \frac{1}{\left|\sin\left(\frac{\eta_0^* \tau}{2}\right)\right|} |z_1| = \sqrt{2} \frac{\varphi}{|\sin(\varphi)|} |z_1| \ .$$

In the sub-Riemannian geometry of $\mathbb{V}_J$, the sphere S and the wave front W (with radius $\sqrt{2}$) are given by the fundamental equation

$$|z_1| = \frac{|\sin(\varphi)|}{\varphi} \ .$$

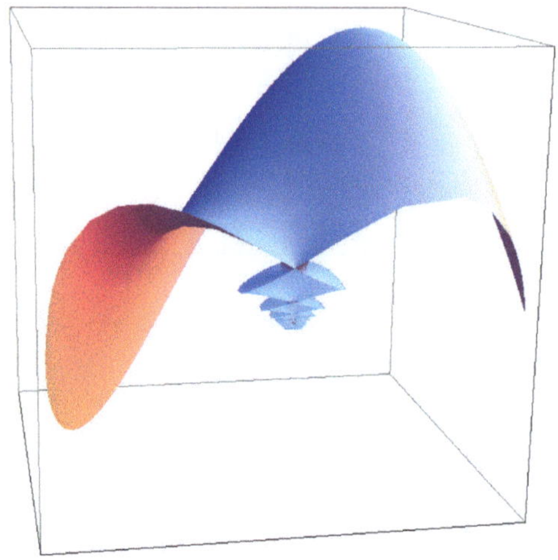

Fig. 1.13 A piece of the sub-Riemannian wave front W. The external surface is the sub-Riemannian sphere S. The internal part is $W - S$. It presents smaller and smaller circles of cusp singularities which converge to 0.

We get therefore

$$\begin{cases} x_1 = \frac{|\sin(\varphi)|}{\varphi} \cos(\theta) \\ p_1 = \frac{|\sin(\varphi)|}{\varphi} \sin(\theta) \\ y_1 = \frac{1}{2} x_1 p_1 + \frac{\varphi - \sin(\varphi)\cos(\varphi)}{4\varphi^2} \\ = \frac{1}{2} \frac{\sin^2(\varphi)}{\varphi^2} \cos(\theta) \sin(\theta) + \frac{\varphi - \cos(\varphi)\sin(\varphi)}{4\varphi^2} \\ = \frac{\varphi + 2\sin^2(\varphi)\cos(\theta)\sin(\theta) - \cos(\varphi)\sin(\varphi)}{4\varphi^2} \end{cases}$$

We present in figure 1.13 pieces of S and W. The external surface is the sub-Riemannian sphere S. It has a saddle form with singularities at the intersections with the y-axis. The internal part is $W - S$. It presents smaller and smaller circles of cusp singularities which converge to 0. Such a complex behaviour is impossible in Riemannian geometry.

1.5.3 New Collaborations with Neurosciences and Psychophysics

In parallel to these mathematical developments, other enrichments of neurogeometry during the 2000s came from the continuation of the dialogue with the colleagues of neurosciences and psychophysics we already cited.

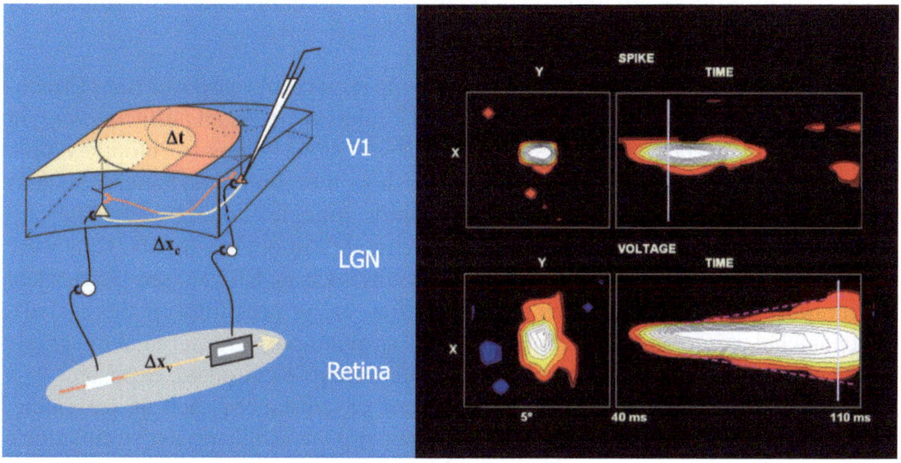

Fig. 1.14 In the left figure, Δx_v represents the eccentricity of the distal retinal stimulus (the white segment to the left) with respect to the stimulus localised in the MDF (the white segment to the right, in the grey rectangle). Δx_c represents the cortical distance between the neurons of $V1$ activated by the two stimuli. Δx_t is the time of latency induced by the propagation along the horizontal connection. In the right figure, above we see the MDF ("minimal discharge field" defined by the spiky responses) in the (x,y) plane and its temporal evolution in the space (x,t). We see below the SIF ("synaptic integration field") defined by the subliminal activity. The SIF is distinctly more extended than the MDF. (From Frégnac *et al.* [55]).

1.5.3.1 The UNIC Lab of Yves Frégnac

The links with the laboratory UNIC of the CNRS (Gif-sur-Yvette) directed by Yves Frégnac (my successor in 2010 at the Ecole Polytechnique) have been particularly fruitful. We cite some results relevant for neurogeometry.

(i) The functional architecture of $V1$ leads to redefinition of the classic concept of receptive field as a domain of the retinal field where localised stimuli elicit spiky responses. This minimal discharge field (MDF) is narrow and does not takes into account the fact that horizontal lateral connections induce strong *contextual* effects which are important for contour integration, surface perception, segmentation, figure-ground distinction, etc. Figure 1.14 by Yves Frégnac *et al.* [55] schematises the way in which the MDF widens, thanks to the propagation of cortical waves along the horizontal connections, into a synaptic integration field (SIF).

(ii) Experimental advances allowed consideration of the response of visual neurons not only to simple stimuli, as bars or gratings, but also to *natural* images whose structure is much more complex. We already mentioned that in section 1.4.2. The results of the UNIC team are really interesting, for example the replicability of the fine structure of the trains of spikes emitted by an axon is far more strong.

1.5.3.2 Entoptic Vision and the Ermentrout-Cowan-Bressloff-Golubitsky Model

Furthermore, Yves Frégnac indicated to me some remarkable works of Bard Ermentrout, Jack Cowan, Paul Bressloff and Martin Golubitsky on visual *hallucinations*, in particular in their 2001 article [28]. Visual hallucinations belong to entoptic vision, in particular the purely geometric ones which show morphological patterns such as tunnels and funnels, spirals, lattices (honeycombs, triangles), or cobwebs. They are very interesting because they are completely virtual (without any input). They were studied during the 1920s by Gestalt theoreticians such as Klüver (see [75]). They are morphologies of phosphenes perceived after strong pressions on the eyeballs (mechanichal stimulation), electro-magnetic stimulations (transcranial magnetic stimulation, electrical stimulation via implanted micro-electrodes), exposures to a violent flickering light, headaches, absorptions of substances such as mescaline, LSD, psilocybin, ketamin, some alkaloids (peyote) (neuropharmaco stimulation), or near death experiences (see Fregnac [54]). They depend upon an increased abnormal excitability of the photoreceptors and of $V1$. In the case of ingestion of a substance, a qualitative explanation is that the substance shifts the balance of activity of the brain away from its ground state, by a vector representing the profile of binding affinities at different receptors. The bifurcation of the brain state explains the hallucination.

The key result of Bressloff *et al.* [28] is that these hallucinations can be deduced from the encoding of the functional architecture of $V1$ into the Hopfield equations of a neural net (see section 1.3.5). They work with the fibre bundle $\pi_S : \mathbb{V}_S = \mathbb{R}^2 \times \mathbb{S}^1 \to \mathbb{R}^2$ with coordinates $v = (a, \theta)$ labelling the "simple" neurons. Let $E(a, \theta, t)$ be the activity of $V1$. They look for the PDE governing the evolution of E using a standard Hopfield equation:

$$\frac{\partial E(a,\theta,t)}{\partial t} = -\alpha E(a,\theta,t) + \frac{\mu}{\pi}\int_0^{\pi}\int_{\mathbb{R}^2} w\langle a,\theta|a',\theta'\rangle\,\sigma\left(E(a',\theta',t)\right) da'd\theta' + h(a,\theta,t)$$

where σ is a non linear gain function (with $\sigma(0) = 0$), h an external input, and $w\langle a,\theta|a',\theta'\rangle$ the weight of the connection between the neuron $v = (a,\theta)$ and the neuron $v' = (a',\theta')$, α a parameter of decay (α can be taken $= 1$) and μ a parameter of excitability of $V1$. The increasing of μ models an increasing of the excitability of $V1$ due to the action of substances on the nuclei which produce specific neurotransmitters (such as serotonin or noradrenalin).

Then the authors specify this general model by encoding the characteristic geometry of coaxial alignments into the synaptic weights, an hypothesis which we interpreted as a fundamental link between neural nets theory and sub-Riemannian neurogeometry. They impose (i) that the local "vertical" connections inside a single hypercolumn must be of the form $w\langle a,\theta|a',\theta'\rangle = w_{loc}(\theta-\theta')\,\delta(a-a')$ where $\delta(a-a')$ is a Dirac distribution prescribing $a = a'$ and $w_{loc}(\theta-\theta')$ is a given even function (in general a Gaussian centered at 0), while (ii) the lateral "horizontal" connections between different hypercolumns must be of the form $w\langle a,\theta|a',\theta'\rangle = w_{lat}(s)\,\delta(a-a'-se_\theta)\,\delta(\theta-\theta')$ where the factor $\delta(\theta-\theta')$

prescribes parallelism $\theta = \theta'$, the factor $\delta(a - a' - se_\theta)$ (where e_θ is the unit vector along the direction θ) prescribes alignment, and $w_{lat}(s)$ is a given function depending only upon the distance. It is straightforward to verify that the synaptic weights w are $E(2)$-invariant under the group $E(2) = \mathbb{R}^2 \rtimes O(2)$ and that the PDE is itself $E(2)$-equivariant if $h = 0$.

If there exists no external input ($h = 0$) and if $\mu = 0$, that is if the subject is in the dark and sees nothing, then his $V1$ activity is in its "ground state" (which can be very complex: endogeneous activity, spontaneous noise, etc.). In the model, the "ground state" is the homogeneous state $E \equiv 0$. It is stable and the activity E measures the shift of the $V1$ state away from the "ground state" when $h = 0$ but $\mu \neq 0$. Now, the analysis of the PDE shows that, as the parameter μ increases, this initial state $E \equiv 0$ can become unstable and *bifurcate*, for critical values μ_c of μ, towards new stable states presenting spatial patterns generated by an $E(2)$ *symmetry breaking*. The bifurcations can be analysed using classical methods: (i) linearisation of the PDE near the solution $E \equiv 0$ and the critical value μ_c; spectral analysis of the linearised equation; computation of its eigenvectors (eigenmodes); hypothesis of periodicity with respect to a lattice Λ of $\mathbb{R}^2$. The last step is to reconstruct from eigenmodes in $V1$ virtual retinal images using the inverse of the retinotopic map between the retina and $V1$. The mathematical model fits extremely well with the empirical data.

This model by Bressloff's *et al.* [28] was later improved, for example in [27]. It is fundamental for (at least) three reasons.

1. First, it is in resonance with the two main streams in modelling we previously evoked: (i) brain dynamics, their attractors and bifurcations (section 1.2.6), (ii) neural nets (section 1.3.5).
2. As much as neural nets are concerned, it is one of the first examples in which the synaptic weights are not considered in a statistical way but as something which encodes a functional architecture. The consequences are spectacular.
3. It offers a perfect example of the enormous gap separating, on one side, a phenomenological lived-experience of perceptual geometry (to experience the hallucinations) and, on the other side, a neural explanation. We can use it to test the contemporary philosophical discussion between phenomenology and neurosciences on "neural correlates of consciousness", developed by colleagues as David Chalmers, Alva Noë, Evan Thompson or Shaun Gallagher. It shows how, in contrast to the beliefs of many philosophers, what Noë and Thompson call the "structural coherence of perceptual experience" can be build up out of receptive fields if we take into account neural functional architectures (see [105]).

(iv) Another important cooperation has been the organisation, with Jean Lorenceau in 2003 in the *Journal of Physiology-Paris*, whose Editor-in-Chief is Yves Frégnac, of a double special issue *Neurogeometry and Visual Perception*. The interested reader will find, in addition to the articles of the three editors, texts by Hess-Hayes-Field, Lee, Angelucci-Bullier, Kimia,.Zucker *et al.*, Ermentrout *et al.*, Bressloff-Cowan, McLaughlin-Shapley-Shelley, Wolf-Geisel, Morel *et al.*, Leaci *et al.*, Tallon Baudry, Van Rullen, Sarti-Citti-Manfredini.

Many other conferences (Valparaiso, Rome, Bolzano, etc.) played an important role in these developments from 2000, as well as many working groups. First the seminar "*Géométrie et Cognition*", organised at the ENS of Paris by Giuseppe Longo and Bernard Teissier. Then, from 2004, the research program NIM of the CNRS ("Nouvelles Interfaces des Mathématiques") on the "*Neurogéométrie de V*1" which grouped colleagues from the LPPA of Alain Berthoz at the Collège de France (Jacques Droulez, specialist in models of vision, Chantal Milleret, specialist of the corpus callosum, Daniel Bennequin, specialist of contact structures and singularities), and of the UNIC of Yves Frégnac. In 2006, the annual day of the SMF (Société Mathématique de France) has been dedicated to the topic *Géométrie et Vision* with talks by Jean-Michel Morel, Stéphane Mallat and myself.

An important development on the "vertical" part of the models $\mathbb{V}_J$ and $\mathbb{V}_S$ has been introduced in 2006 by Olivier Faugeras and Pascal Chossat [34]. Their main idea is that hypercolumns of $V1$ encode (of course at a given scale defined by the size of the receptive fields) not only local features such as orientation or curvature but the whole symmetric definite positive "structure tensor" $\mathcal{T}$ of the (smooth) stimulus $I(x,y)$. $\mathcal{T}$ is given by

$$\mathcal{T} = \begin{pmatrix} \left(\frac{\partial I}{\partial x}\right)^2 & \frac{\partial I}{\partial x}\frac{\partial I}{\partial y} \\ \frac{\partial I}{\partial y}\frac{\partial I}{\partial x} & \left(\frac{\partial I}{\partial y}\right)^2 \end{pmatrix}.$$

As they claim: "a hypercolumn in $V1$ can represent the structure tensor in the receptive field of its neurons as the average membrane potential values of some of its neuronal population". Now, the space $\mathcal{H}$ of 2×2 symmetric definite positive matrices $\mathcal{T}$ is the 3D hyperbolic quotient space $GL(2,\mathbb{R})/O(2)$ foliated in 2D leaves by $\det(\mathcal{T})$ (det = determinant). For $\det(\mathcal{T}) = 1$, the leaf is the quotient $SL(2,\mathbb{R})/SO(2)$ which is isomorphic to the hyperbolic Poincaré disk $\mathbb{D}$. An interesting problem is then to add the spatial variables a and look at the bifurcations of activity functions $E(a,z,t)$ defined on the fibre bundle $\mathbb{R}^2 \times \mathbb{D}$ with base space $\mathbb{R}^2$ (group $E(2)$) and fibre $\mathbb{D}$ (group $SU(1,1)$).

In 2011, Olivier Faugeras, Paul Bressloff, Nicolas Brunel, Wulfram Gersmer and Viktor Jirsa organised the first special trimester of the CIRM (Centre International de Rencontres Mathématiques de Marseille-Luminy) dedicated to *Theoretical, Mathematical and Computational Neuroscience*. The program included a course of neurogeometry as well as four conferences, among which one has been organised by Paul Bressloff and Stephen Coombes and a second by Viktor Jirsa and Gustavo Deco.

1.5.4 Neurogeometry and Control Theory (Agrachev's Group)

We saw in section 1.5.2 how to compute the sub-Riemannian geodesics of the $\mathbb{V}_J$ model. What about the $\mathbb{V}_S$ model?

1.5.4.1 The Geodesics of the $\mathbb{V}_S$ Model

In 2006 a colleague of mine at CREA, Helena Frankowska, specialist in control theory and in the Hamilton-Jacobi-Bellman equation, put me in contact with Andrei Agrachev of the SISSA (International School for Advances Studies) in Trieste. With A. Sarti and G. Citti, we organised a meeting with him at the IHP. He was interested in our discussions on the use of sub-Riemannian geometry in problems related to vision, and some fruitful cooperation quickly started with him, and with some members of his group, in particular Jean-Paul Gauthier, Ugo Boscain and Yuri Sachkov.

Andrei Agrachev rapidly found the formulas of the geodesics for $\mathbb{V}_S = SE(2)$ endowed with the sub-Riemannian metric making $\{X_1, X_2\}$ an orthonormal basis of K_v. The formulation of the problem in terms of control theory yields the differential system $\{\dot{x} = u_1 \cos(\theta), \dot{y} = u_1 \sin(\theta), \dot{\theta} = u_2\}$. Applying Pontryagin maximum principle, one gets the Hamiltonian on $T^*\mathbb{V}_S$

$$\begin{aligned} H(v,\varpi) &= \frac{1}{2}\left(u_1^2 + u_2^2\right) = \frac{1}{2}\left(\langle \varpi, X_1(v)\rangle^2 + \langle \varpi, X_2(v)\rangle^2\right) \\ &= \frac{1}{2}\left(\varpi_1^2 + \varpi_2^2\right) \\ &= \frac{1}{2}\left((\xi^* \cos(\theta) + \eta^* \sin(\theta))^2 + \vartheta^{*2}\right) \end{aligned}$$

where $\{\varpi_1, \varpi_2, \varpi_3\}$ are the components of the covector ϖ in the dual basis of $\{X_1, X_2, X_3\}$. Hence the Hamilton equations:

$$\begin{cases} \dot{x} = \frac{\partial H}{\partial \xi^*} = \xi^* \cos^2(\theta) + \eta^* \cos(\theta)\sin(\theta) \\ \dot{y} = \frac{\partial H}{\partial \eta^*} = \eta^* \sin^2(\theta) + \xi^* \cos(\theta)\sin(\theta) \\ \dot{\theta} = \frac{\partial H}{\partial \vartheta^*} = \vartheta^* \\ \dot{\xi}^* = -\frac{\partial H}{\partial x} = 0 \\ \dot{\eta}^* = -\frac{\partial H}{\partial y} = 0 \\ \dot{\vartheta}^* = -\frac{\partial H}{\partial \theta} = (\xi^* \cos(\theta) + \eta^* \sin(\theta))(-\xi^* \sin(\theta) + \eta^* \cos(\theta)) \end{cases}$$

The sub-Riemannian geodesics are the projections on $\mathbb{V}_S$ of the solutions. As $\xi^* = \xi_0^*$ and $\eta^* = \eta_0^*$ are constant, if one writes them $(\xi_0^*, \eta_0^*) = \rho_0 e^{i\beta_0}$, then $\dot{\vartheta}^* = \frac{1}{2}\rho_0^2 \sin(2(\theta - \beta_0))$ and the constant Hamiltonian $H = \frac{1}{2}\left(\rho_0^2 \cos^2(\theta - \beta_0) + \vartheta^{*2}\right)$ yields the energy first integral $\rho_0^2 \cos^2(\theta - \beta_0) + \vartheta^{*2} = c$ (with $c = 1$ if $H = \frac{1}{2}$) and the ODE for $\dot{\theta}$ (c, ρ_0 and β_0 are constants) $\dot{\theta}^2 = \vartheta^{*2} = c - \rho_0^2 \cos^2(\theta - \beta_0)$. For $\beta_0 = 0$ (which is allowed by rotation invariance), the equations become:

$$\begin{cases} \dot{x} = \rho_0 \cos^2(\theta) \\ \dot{y} = \rho_0 \cos(\theta)\sin(\theta) = \frac{1}{2}\rho_0 \sin(2\theta) \\ \dot{\theta} = \vartheta^* \\ \ddot{\theta} = \dot{\vartheta}^* = \frac{1}{2}\rho_0^2 \sin(2\theta) \end{cases}$$

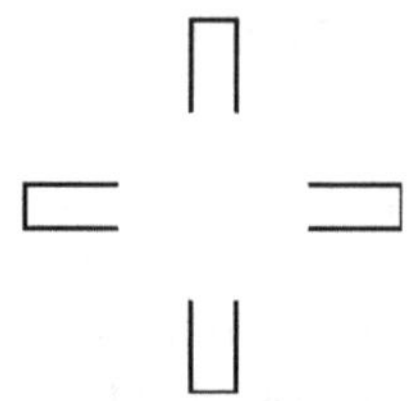

Fig. 1.15 The segments induce an illusory contour which can be perceived as a square or a circle. In general, initial perception consists in a circle, but it can bifurcate after a certain period of sight.

For $\rho_0 = 1$, $2\theta = \pi - \mu$, and $\mu = 2\varphi = \pi - 2\theta$, one gets a *pendulum equation* $\ddot{\mu} = -\sin(\mu)$ with first integral $\dot{\varphi}^2 + \sin^2(\varphi)$. As

$$dt = \pm \frac{1}{\sqrt{c}} \frac{d\varphi}{\sqrt{1 - \frac{1}{c}\sin^2(\varphi)}}$$

the system can be explicitly integrated using *elliptic functions*.

A fundamental property of these geodesics is that, under certain conditions, when the deviation from parallelism between the boundary conditions (a_1, θ_1) and (a_2, θ_2) becomes too wide, they become *singular* and present some cusps. At the level of the underlying pendulum equation, this corresponds to *oscillating* solutions. Yuri Sachkov and Igor Moiseev [87] studied these cusps and explicitly constructed the sphere, the wave-front and the cut locus of the sub-Riemannian geometry of $SE(2)$. Their complexity is remarkable.

These singularities are very interesting and can perhaps explain an intriguing aspect of illusory contours, namely their *bistability*. Let's consider for example the cross in figure 1.15. The segments induce (with the cooperation of $V2$ which induces orientations orthogonal to the segments at their end points) an illusory contour which can be perceived as a square or a circle. If we look at the image for a sufficiently long period the percepts spontaneously and periodically bifurcates from a case to the other one.

In the context of a variational explanation of illusory contours, this shows how two models can compete: a geodesic model where the curvature is maximally spread out and, on the other side, a piecewise linear model which concentrates all the curvature in some angular points (the curvature is null everywhere except at these points where it is infinite). In the case of an illusory contour between (a_1, θ_1) and (a_2, θ_2), experimental data seem to show that, if the difference $|\theta_1 - \theta_2|$ exceeds a certain threshold, the geodesic model is replaced by the piecewise linear model. We can formulate the conjecture that this *bifurcation of variational models* occurs when the geodesics become singular because of the emergence of cusps. In fact a cusp occurs when a geodesic has a "vertical" tangent, *i.e.* a tangent to the fibre of the fibration $\pi_S : \mathbb{V}_S = \mathbb{R}^2 \times \mathbb{S}^1 \to \mathbb{R}^2$. But, in neurophysiological terms this means that some "horizontal" excitatory connections between different hypercolumns must be identified with "vertical" inhibitory connections internal to a single hypercolumn, which is not possible. Bi-modality could then be caused by the fact that the period of fixation "stresses" the selected model and allows its bifurcation.

1.5.4.2 Elastica Revisited

In section 1.3.3 we cited the elastica variational model [90] proposed in 1992 by David Mumford for illusory contours. It consists in minimising an energy $E = \int_\gamma (\alpha\kappa^2 + \beta)ds$ where γ is a curve in $\mathbb{R}^2$ with element of arc length ds. For $\alpha = \beta = 1$, its formulation as a control problem on the group $G = SE(2) = \mathbb{V}_S$ can be written $\{\dot{x} = \cos(\theta), \dot{y} = \sin(\theta), \dot{\theta} = \kappa\}$ where the derivatives are taken with respect to the arc length s and where $\kappa = \frac{d\theta}{ds}$ is the curvature of γ. This model is defined in the base plane $\mathbb{R}^2$ with Euclidean metric, and not with respect to the sub-Riemannian metric in G, because, in G, ds is *not* the element of arc length. The element of arc length in G is $dt = \sqrt{1+\kappa^2(s)}ds$ and the curvature $\kappa_G(s) = \frac{d\theta(t(s))}{dt} = \frac{d\theta}{ds}\frac{ds}{dt} = \frac{\kappa(s)}{\sqrt{1+\kappa^2(s)}}$ satisfies the pendulum equation (with respect to s) $\ddot{\kappa}_G(s) = \kappa_G(s)$. Yuri Sachkov [110] investigated the elastica and compared them to the sub-Riemannian geodesics. The problem is difficult because of the cusps. He later deepened his investigation with Ugo Boscain, Remco Duits and Francesco Rossi [23].

We also saw in section 1.3.3 how David Mumford gave a stochastic interpretation of his elastica model by supposing that the curvature $\kappa(s)$ of γ in $\mathbb{R}^2$ is a white noise and that the angle $\theta(s)$ is therefore a Brownian motion. In terms of control theory, this is equivalent to consider the stochastic process $\{\dot{x} = \cos(\theta), \dot{y} = \sin(\theta), \dot{\theta} \sim N(0,\sigma^2)\}$ where $\dot{\theta}$ is now a normal random Gaussian variable of mean 0 and variance σ^2. This process has been studied by Gonzalo Sanguinetti (in his thesis supervised by G. Citti and A. Sarti) and also by Remco Duits and Markus van Almsick. It is no longer a mere diffusion but an advection-diffusion mechanism described by a Fokker-Planck equation. The advection (the drift) occurs along the X_1 direction and the diffusion of θ occurs along the X_2 direction. The fundamental solution of the Fokker-Planck equation being too complex in the $\mathbb{V}_S$ model, the authors come back to the first order approximation of $G = SE(2)$ (its tangent cone or nilpotentisation), that is to our $\mathbb{V}_J$ model based on the polarised Heisenberg group. Let $v_0 = (x_0, y_0, \theta_0) = (a_0, \theta_0)$ be an initial point in G and let us follow a random walk starting at v_0. Whithout noise, the trajectory is of course deterministic and is a straight line satisfying the principle of strict coaxiality (without any curvature): $\{\theta = \theta_0, x = x_0 + \cos(\theta_0)t, y = y_0 + \sin(\theta_0)t\}$. If $v = (x,y,\theta) = (a,\theta)$ is a generic element of G and if $P(v,t)$ is the probability to find the random walk at v at time t, the evolution equation for P with initial condition $P_0(v) = P(v,0)$ is

$$\frac{\partial P}{\partial t}(v,t) = -\left(\cos(\theta)\frac{\partial P}{\partial x}(v,t) + \sin(\theta)\frac{\partial P}{\partial y}(v,t)\right) + \frac{\sigma^2}{2}\frac{\partial^2 P}{\partial \theta^2}(v,t)$$

$$\frac{\partial P}{\partial t}(v,t) = -X_1(P(v,t)) + \frac{\sigma^2}{2}(X_2)^2(P(v,t)) \ .$$

For the $\mathbb{V}_J$ model where θ is small, $\theta \sim \tan(\theta) = p$, the Fokker-Planck equation is therefore

$$\frac{\partial P}{\partial t}(v,t) = -\left(\frac{\partial P}{\partial x}(v,t) + p\frac{\partial P}{\partial y}(v,t)\right) + \frac{\sigma^2}{2}\frac{\partial^2 P}{\partial p^2}(v,t)$$

The authors solve this equation and, to complete a contour with boundary conditions $v_0 = (a_0, \theta_0)$ and $v_1 = (a_1, \theta_1)$, consider two direction processes, a forward process starting at v_0 and a backward process starting at v_1. They compute the probability of collision of these two random walks.

1.5.4.3 Sub-Riemannian Diffusion, Heat Kernel, and Noncommutative Harmonic Analysis (Gauthier, Boscain)

We underlined in section 1.5.1.3 the importance and the difficulties of the sub-Riemannian diffusion techniques. We evoked the works of some specialists and also the neurogeometrical applications made by G. Citti and A. Sarti in this domain. These techniques belong to the general theory of heat kernels on Riemannian and sub-Riemannian manifolds. The specialised literature on them is enormous. We already cited Rothschild-Stein [109] and Nagel-Stein-Wainger [92]. Let us cite also Davies [44], Varopoulos [136], Saloff-Coste [111], Kusuoka-Stroock [78], Coulhon-Grigor'yan [40], and ter Elst-Robinson [127].

On their side, A. Agrachev, J-P. Gauthier, U. Boscain and their Ph.D. student F. Rossi, following previous results by Hulanicki [71], gave an "intrinsic" formulation of the sub-Riemannian Laplacian and proved in 2009 [5] a general theorem for the unimodular Lie groups (*i.e.* whose Haar measures, invariant to the left and to the right, are identical) of dimension 3 endowed with a left-invariant sub-Riemannian geometry. They use the noncommutative generalised Fourier transform (GFT) defined on the dual space G^* of G (the set of irreducible unitary representations in Hilbert spaces) to compute the heat kernel associated to the hypoelliptic Laplacian $\Delta_{\mathcal{K}} = X_1^2 + X_2^2$, *i.e.* the sum of squares of the generators $\{X_1, X_2\}$ of the distribution $\mathcal{K}$. The Laplacian is *hypoelliptic* due to the fact that $\mathcal{K}$ is bracket generating, *i.e.* satisfies Hörmander condition.

The use of the Fourier transform on groups to compute heat kernels and fundamental solutions of diffusion equations has a long history. The interested reader could consult the already cited pater of Geller [58], Christ *et al.* [35] and more recently the book of Calin-Chang-Furutani-Iwasaki *Heat Kernels for Elliptic and Sub-Elliptic Operators* [30].

In the case of the polarised Heisenberg Lie group $\mathbb{V}_J$, we have seen in section 1.4.9.6 that, according to the Stone-von Neumann theorem, the non trivial unirreps are group morphisms π_λ from $\mathbb{V}_J$ to the group $\mathcal{U}(\mathcal{H})$ of unitary automorphisms of the Hilbert space $\mathscr{H} = L^2(\mathbb{R},\mathbb{C})$, morphisms parametrised by a real scalar $\lambda \neq 0$. They are of the form ($u(s) \in \mathscr{H}$):

$$\begin{aligned} \pi_\lambda : \mathbb{V}_J &\to \mathscr{U}(\mathscr{H}) \\ v &\mapsto \pi_\lambda(v) \quad : \mathscr{H} \to \mathscr{H} \\ &\qquad\qquad u(s) \mapsto e^{i\lambda(y+xs)}u(s+p)\,. \end{aligned}$$

There exists a measure on the dual sace $\mathbb{V}_J^*$, called the *Plancherel measure*, given by $dP(\lambda) = \lambda d\lambda$ which enables making integrations. To compute the Fourier transform of the sub-Riemannian Laplacian $\Delta_{\mathscr{K}}$, one looks at the action of the differential of the unirreps on the left-invariant vector fields X on $\mathbb{V}_J$, which are given by the left translation of vectors $X(0)$ of the Lie algebra $\mathscr{V}_J$ of $\mathbb{V}_J$. By definition,

$$d\pi_\lambda : X \to d\pi_\lambda(X) := \left.\frac{d}{dt}\right|_{t=0} \pi_\lambda\left(e^{tX}\right)$$

and one gets the Fourier transform $\widehat{X_i}^\lambda = d\pi_\lambda(X_i)$. Computations yield $X_1(0) = (1,0,0)$, $e^{tX_1} = (t,0,0)$, $\pi_\lambda\left(e^{tX_1}\right)u(s) = e^{i\lambda ts}u(s)$,

$$\begin{aligned} \widehat{X_1}^\lambda u(s) = d\pi_\lambda(X_1)u(s) &= \left.\frac{d}{dt}\right|_{t=0} \pi_\lambda\left(e^{tX_1}\right)u(s) \\ &= \left.\frac{d}{dt}\right|_{t=0} e^{i\lambda ts}u(s) = i\lambda s u(s) \end{aligned}$$

and $X_2(0) = (0,0,1)$, $e^{tX_2} = (0,0,t)$, $\pi_\lambda\left(e^{tX_2}\right)u(s) = u(s+t)$,

$$\begin{aligned} \widehat{X_2}^\lambda u(s) = d\pi_\lambda(X_2)u(s) &= \left.\frac{d}{dt}\right|_{t=0} \pi_\lambda\left(e^{tX_2}\right)u(s) \\ &= \left.\frac{d}{dt}\right|_{t=0} u(s+t) = \frac{du(s)}{ds}\,. \end{aligned}$$

The GFT of the sub-Riemannian Laplacian is therefore the Hilbert sum (integral on λ with the Plancherel measure $dP(\lambda) = \lambda d\lambda$) of the $\widehat{\Delta_{\mathscr{K}}}^\lambda$ with

$$\widehat{\Delta_{\mathscr{K}}}^\lambda u(s) = \left(\left(\widehat{X_1}^\lambda\right)^2 + \left(\widehat{X_2}^\lambda\right)^2\right)u(s) = \frac{d^2u(s)}{ds^2} - \lambda^2 s^2 u(s)\,.$$

This equation is nothing else than the equation of the *harmonic oscillator*.

The heat kernel is then

$$P(v,t) = \int_{\mathbb{V}_J^*} Tr\left(e^{t\widehat{\Delta_{\mathscr{K}}}^\lambda}\pi_\lambda(v)\right)dP(\lambda), t \geq 0\,.$$

If the $\widehat{\Delta_{\mathscr{K}}}^\lambda$ have discrete spectrum and a complete set of normalised eigenfunctions $\{u_n^\lambda\}$ with eigenvalues $\{\alpha_n^\lambda\}$ then

$$P(v,t) = \int_{\mathbb{V}_J^*} \left(\sum_n e^{\alpha_n^\lambda t} \left\langle u_n^\lambda, \pi_\lambda (v) \left(u_n^\lambda \right) \right\rangle \right) dP(\lambda), t \geq 0 .$$

It is the case here. The eigenfunctions of the harmonic oscillator are well know and satisfy:

$$\frac{d^2 u_n^\lambda (s)}{ds^2} - \lambda^2 s^2 u_n^\lambda (s) = \alpha_n^\lambda u_n^\lambda (s)$$

with $\alpha_n^\lambda = -\frac{2n+1}{\lambda}$. They are essentially the *Hermite functions* scaled by λ:

$$u_n^\lambda (s) = \left(2^n n! \sqrt{\pi}\right)^{-\frac{1}{2}} \lambda^{\frac{1}{4}} e^{-\lambda \frac{s^2}{2}} H_n \left(\sqrt{\lambda} s\right)$$

H_n being the n-th Hermite polynomial.

In the case of $SE(2) = \mathbb{V}_S$, A. Agrachev, J-P. Gauthier, U. Boscain, and F. Rossi found *explicit* formulas for the heat kernel. The dual $\mathbb{V}_S^*$ of $\mathbb{V}_S$ is this time the set of unirreps in the Hilbert space $\mathscr{H} = L^2 \left(\mathbb{S}^1, \mathbb{C}\right)$. These unirreps are parametrised by a positive real λ and are of the form:

$$\begin{aligned}
\mathscr{X}^\lambda : \mathbb{V}_S &\to \mathscr{U}(\mathscr{H}) \\
v &\mapsto \mathscr{X}^\lambda (v) : \mathscr{H} \to \mathscr{H} \\
&\qquad \psi(\theta) \mapsto e^{i\lambda(x\sin(\theta)+y\cos(\theta))} \psi(\theta+\alpha) .
\end{aligned}$$

The Plancherel measure on $\mathbb{V}_S^*$ is still $dP(\lambda) = \lambda d\lambda$. As we have previously explained it, we get $\mathscr{X}^\lambda \left(e^{tX_1}\right) \psi(\theta) = e^{i\lambda t \sin(\theta)} \psi(\theta)$ and

$$\begin{aligned}
\widehat{X_1}^\lambda \psi(\theta) = d\mathscr{X}^\lambda (X_1) \psi(\theta) &= \left.\frac{d}{dt}\right|_{t=0} \mathscr{X}^\lambda \left(e^{tX_1}\right) \psi(\theta) \\
&= \left.\frac{d}{dt}\right|_{t=0} e^{i\lambda t \sin(\theta)} \psi(\theta) = i\lambda \sin(\theta) \psi(\theta)
\end{aligned}$$

and $\mathscr{X}^\lambda \left(e^{tX_2}\right) \psi(\theta) = \psi(\theta + t)$ and

$$\begin{aligned}
\widehat{X_2}^\lambda \psi(\theta) = d\mathscr{X}^\lambda (X_2) \psi(\theta) &= \left.\frac{d}{dt}\right|_{t=0} \mathscr{X}^\lambda \left(e^{tX_2}\right) \psi(\theta) \\
&= \left.\frac{d}{dt}\right|_{t=0} \psi(\theta + t) = \frac{d\psi(\theta)}{d\theta} .
\end{aligned}$$

The GFT of the sub-Riemannian Laplacian is therefore the Hilbert sum of the $\widehat{\Delta_{\mathscr{K}}}^\lambda$ with

$$\widehat{\Delta_{\mathscr{K}}}^\lambda \psi(\theta) = \left(\left(\widehat{X_1}^\lambda\right)^2 + \left(\widehat{X_2}^\lambda\right)^2\right) \psi(\theta) = \frac{d^2\psi(\theta)}{d\theta^2} - \lambda^2 \sin^2(\theta) \psi(\theta)$$

which is nothing else than the *Mathieu equation*. The heat kernel is

$$P(v,t) = \int_{\mathbb{V}_S^*} Tr\left(e^{t\widehat{\Delta_{\mathscr{X}}}^{\lambda}} \mathscr{X}^{\lambda}(v)\right) dP(\lambda), t \geq 0 .$$

The $\widehat{\Delta_{\mathscr{X}}}^{\lambda}$ have discrete spectrum and a complete set of normalised eigenfunctions $\{\psi_n^{\lambda}\}$ with eigenvalues $\{\alpha_n^{\lambda}\}$ and therefore

$$P(v,t) = \int_{\mathbb{V}_S^*} \left(\sum_n e^{\alpha_n^{\lambda} t} \left\langle \psi_n^{\lambda}, \mathscr{X}^{\lambda}(v)\left(\psi_n^{\lambda}\right)\right\rangle\right) dP(\lambda), t \geq 0 .$$

The 2π-periodic eigenfunctions of the Mathieu equation satisfy:

$$\frac{d^2\psi(\theta)}{d\theta^2} - \lambda^2 \sin^2(\theta)\,\psi(\theta) = E\psi(\theta)$$

and, as $\sin^2(\theta) = \frac{1}{2}(1-\cos(2\theta))$, this means:

$$\frac{d^2\psi(\theta)}{d\theta^2} - \frac{\lambda^2}{2}\psi(\theta) - E\psi(\theta) + \frac{\lambda^2}{2}\cos(2\theta)\,\psi(\theta) = 0$$

$$\frac{d^2\psi(\theta)}{d\theta^2} + (a - 2q\cos(2\theta))\,\psi(\theta) = 0, \text{ with } a = -\left(\frac{\lambda^2}{2}+E\right) \text{ and } q = -\frac{\lambda^2}{4} .$$

The normalised 2π-periodic eigenfunctions are known: they are even or odd and denoted $\text{cen}(\theta,q)$ and $\text{sen}(\theta,q)$. The associated $a_n(q)$ and $b_n(q)$ are called characteristic values. There can exist *parametric resonance* phenomena (Arnold tongues) when $a = -\left(\frac{\lambda^2}{2}+E\right) = n^2$.

The authors solved also the problem for $SU(2)$, $SL(2)$ and $SO(3)$.

The sub-Riemannian diffusion is highly *anisotropic* since it is restricted to an angular diffusion of θ and a spatial diffusion only along the X_1 direction. It is strongly constrained by the "good continuation" Gestalt law and its difference with classical (Euclidean) diffusion is spectacular. The figure 1.16, due to Jean-Paul Gauthier, starts with the image of an eye masked by a white grid and applies sub-Riemannian diffusion until the grid has vanished. In spite of this very important diffusion the geometry of the image remains quite excellent.

During Fall 2014, U. Boscain and L. Rifford will organise a special Trimester at the IHP in Paris on *Geometry, Analysis and Dynamics on Sub-Riemannian Manifolds*. Several geometers we have cited will be present: Montgomery, Bryant, Ambrosio, Agrachev, Gauthier, Pansu, Bellaïche, and many others. A workshop on Neurogeometry will be organised with G. Citti and A. Sarti.

1.5.5 Confluence between $\mathbb{V}_J$ and $\mathbb{V}_S$ Models

We analysed two neurogeometrical models of $V1$, $\mathbb{V}_J$ and $\mathbb{V}_S$. It is interesting to notice that one can easily construct an *interpolation* between the two models. Mohammed Brahim Zahaf and Dominique Manchon [143] constructed

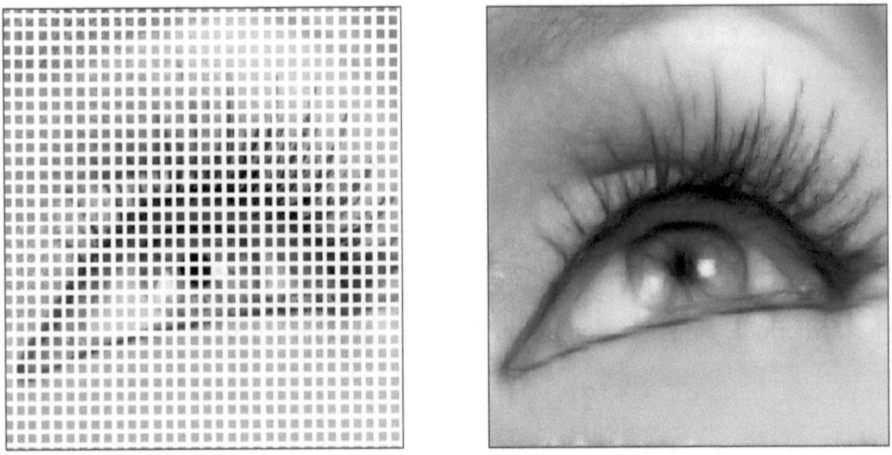

Fig. 1.16 Sub-Riemannian diffusion in $\mathbb{V}_S$ according to Jean-Paul Gauthier. The initial image is an eye masked by a white grid. Sub-Riemannian diffusion is applied until the grid has vanished.

such an interpolation given by a family of models $\mathbb{V}^\alpha$ and studied the confluence of the corresponding differential equations in the Fourier space. The model $\mathbb{V}^\alpha$ can be summarised by the following table: $X_1^\alpha = \cos(\theta)\frac{\partial}{\partial x} + \frac{1}{\alpha}\sin(\alpha\theta)\frac{\partial}{\partial y}$, $X_2^\alpha = \frac{\partial}{\partial\theta}$, $X_3^\alpha = -\alpha\sin(\alpha\theta)\frac{\partial}{\partial x} + \cos(\theta)\frac{\partial}{\partial y}$, $[X_1^\alpha, X_2^\alpha] = -X_3^\alpha$, $[X_2^\alpha, X_3^\alpha] = \alpha^2 X_1^\alpha$, $[X_1^\alpha, X_3^\alpha] = 0$, $\mathbb{V}^\alpha = SE_\alpha(2)$ with $\mathbb{S}_\alpha^1 = \frac{\mathbb{R}}{2\pi\alpha^{-1}\mathbb{Z}}$, $\mathscr{H} = L^2(\mathbb{S}_\alpha^1, \mathbb{C})$, $X_1^\alpha(\psi(\theta)) = i\lambda\alpha^{-1}\sin(\alpha\theta)\psi(\theta)$, $X_2^\alpha(\psi(\theta)) = \psi'(\theta)$, $\hat{\Delta}^\lambda : \psi''(\theta) - \frac{\lambda^2}{\alpha^2}\sin^2(\alpha\theta)\psi(\theta)$, $\psi''(\theta) + \left(\mu - \frac{\lambda^2}{\alpha^2}\sin^2(\alpha\theta)\right)\psi(\theta) = 0$. For $\alpha = 1$, $\mathbb{V}^1$ yields the $\mathbb{V}_S$ model and when $\alpha \to 0$, for small θ denoted p, $\mathbb{V}^0$ yields the $\mathbb{V}_J$ model.

1.6 Conclusion

In this survey, we tried to situate the elements of neurogeometry in their context. We summarised their principles both on the experimental neurophysiological plane and on the mathematical one. We showed how neurogeometry connects two mathematical worlds: (i) the one of perceptual geometry, in reference to Thom, Zeeman, Berry, Koenderink and Mumford, (ii) the one of neurogeometry in its proper sense, modelling the functional architectures in terms of phase-dislocations, Lie groups, Carnot groups, Cartan connections, sub-Riemannian geometry, wavelets, coherent states and noncommutative harmonic analysis in reference to Citti and Sarti, Agrachev, Faugeras and Mallat.

Everything remains to be done in this field. First to "go down" to the underlying microphysical level of the individual neurons and their spikes, governed by equations of Hodgkin-Huxley type. In fact, Neurogeometry works at a *meso*-neuronal level. Then to "go up" to higher visual areas of the extrastriate cortex, from $V2$ to MT (we focused on the striate area $V1$). However, in spite of the very partial character of these results, we hope we showed how we can start to understand the *constitution* of an external and "transcendent" perceptual geometry from internal and "immanent" neurogeometrical algorithms.

Sometimes we proposed a parallel with fundamental physics. It is more than a vague analogy. We think that, as Gestalt theoreticians anticipated in a speculative way, perception depends upon a "field theory" and we tried to present some mathematical structures which enable to conceptualise and compute it.

References

1. Abrahmsen, P.: A Review of Gaussian Random Fields and Correlation Functions (1997), `http://publications.nr.no/917_Rapport.pdf`
2. Adler, R.J., Taylor, J.E.: Random Fields and Geometry. Springer, Berlin (2007)
3. Agrachev, A.A., Sachkov, Y.: Control Theory from the Geometric Viewpoint. Springer, Berlin (2004)
4. Agrachev, A.A., Gamkrelidze, R.V.: The Pontryagin Maximum Principle 50 years later. In: Proceed. Steklov. Math. Inst., Dynamical Systems: Modeling, Optimization, and Control, suppl. 1, pp. S4–S12 (2006)
5. Agrachev, A.A., Boscain, U., Gauthier, J.-P., Rossi, F.: The intrinsic hypoelliptic Laplacian and its heat kernel on unimodular Lie groups. Journal of Functional Analysis 256, 2621–2655 (2009)
6. Alvarez, L., Lions, P.L., Morel, J.M.: Image selective smoothing and edge detection by non linear diffusion. SIAM Journal on Numerical Analysis 29, 845–866 (1992)
7. Amit, D.: Modeling Brain Function. Cambridge University Press, Cambridge (1989)
8. Atick, J.: Could Information theory provide an ecological theory of sensory processing? Network 3, 213–251 (1992)
9. Azais, J.-M., León, J.R., Wschebor, M.: Rice formulae and Gaussian waves. Bernoulli 17(1), 170–193 (2011)
10. Barbieri, D., Citti, G., Sanguinetti, G., Sarti, A.: An uncertainty principle underlying the functional architecture of V1. Journal of Physiology Paris 106(5-6), 183–193 (2012)
11. Beals, R., Gaveau, B., Greiner, P.C.: Hamilton-Jacobi theory and the heat kernel on Heisenberg groups. Journal de Mathématiques Pures et Appliquées 79(7), 633–689 (2000)
12. Bellaïche, A.: The tangent space in sub-Riemannian geometry. In: Bellaïche, A., Risler, J. (eds.) Sub-Riemannian Geometry. Progress in Mathematics, vol. 144, pp. 4–78. Birkhäuser, Basel (1996)
13. Berry, M.V.: Optical Currents. Journal of Optics A: Pure and Applied Optics 11(9), 094001 (2009)
14. Berry, M.V., Dennis, M.R.: Phase singularities in isotropic random waves. Proceedings of the Royal Socety of London A 456, 2059–2079 (2000)
15. Berry, M.V., Dennis, M.R.: Topological events on wave dislocation lines: birth and death of loops, and reconnection. Journal of Physics, A: Mathematical and Theoretical 40, 65–74 (2007)
16. Berry, M.V., Dennis, M.R.: Reconnections of wave vortex lines. European Journal of Physics 33, 723–731 (2012)
17. Berthoz, A.: La simplexité. Odile Jacob, Paris (2009)

18. Berthoz, A., Petit, J.-L.: Physiologie de l'Action et Phénoménologie. Odile Jacob, Paris (2006)
19. Blum, H.: Biological Shape and Visual Science. Journal of Theoretical Biology 38, 205–287 (1973)
20. Bonfiglioli, A., Lanconelli, E., Uguzzoni, F.: Stratified Lie Groups and Potential Theory for their sub-Laplacians. Springer, Berlin (2007)
21. Bonnet, A., David, G.: Cracktip is a global Mumford-Shah minimizer. Asterisque, vol. 274. Société Mathématique de France, Paris (2001)
22. Boscain, U., Duplaix, J., Gauthier, J.-P., Rossi, F.: Anthropomorphic image reconstruction via hypoelliptic diffusion, arXiv:1006.3735v6 (2012)
23. Boscain, U., Duits, R., Rossi, F., Sachkov, Y.: Curve cuspless reconstruction via sub-Riemannian geometry. arXiv: 1203.3089v4 (2013)
24. Boscain, U., Chertovskih, R., Gauthier, J.: Hypoelliptic diffusion and human vision. Accepted on Journal of Mathematical Imaging and Vision (2013)
25. Bosking, W.H., Zhang, Y., Schoenfield, B., Fitzpatrick, D.: Orientation Selectivity and the Arrangement of Horizontal Connections in Tree Shrew Striate Cortex. Journal of Neuroscience 17(6), 2112–2127 (1997)
26. Brakke, K.A.: The motion of a surface by its mean curvature. Princeton University Press, Princeton (1978)
27. Bressloff, P., Cowan, J.: The functional geometry of local and horizontal connections in a model of $V1$. In: Petitot, J., Lorenceau, J. (eds.) Neurogeometry and Visual Perception, Journal of Physiology-Paris 97(2-3), 221–236 (2003)
28. Bressloff, P., Cowan, J., Golubitsky, M., Thomas, P., Wiener, M.: Geometric visual hallucinations, Euclidean symmetry and the functional architecture of striate cortex. Philosophical Transactions of the Royal Society of London B 356, 299–330 (2001)
29. Brockett, R.: Control Theory and Singular Riemannian Geometry. Springer, New York (1981)
30. Calin, O., Chang, D.C., Furutani, K., Iwasaki, C.: Heat Kernelks for Elliptic and Sub-elliptic Operators: Methods and Techniques. Springer, New York (2011)
31. Capogna, L., Danielli, D., Pauls, S., Tyson, J.: An Introduction to the Heisenberg Group and to the sub-Riemannian Isoperimetric Problem. Birkhaüser, Berlin (2007)
32. Capogna, L., Citti, G.: Generalized mean curvature flow in Carnot groups. Communications in PDE 34(79), 937–956 (2009)
33. Chen, Y.-G., Giga, Y., Goto, S.: Uniqueness and existence of viscosity solutions of generalized mean curvature flow equations. Journal of Differential Geometry 33, 749–786 (1991)
34. Chossat, P., Faugeras, O.: Hyperbolic planforms in relation to visual edges and textures perception, arXiv:0907.0963v3
35. Christ, M., Geller, D., Głowacki, P., Polin, L.: Pseudodifferential operators on groups with dilatations. Duke Mathematical Journal 68(1), 31–65 (1992)
36. Citti, G., Uguzzoni, F.: Critical semilinear equations on the Heisenberg group: the effect of the topology of the domain. Nonlinear Analysis 46, 399–417 (2001)
37. Citti, G., Sarti, A.: A cortical based model of perceptual completion in the roto-translation space. Journal of Mathematical Imaging and Vision 24(3), 307–326 (2006)
38. Calin, O., Chang, D.C., Furutani, K., Iwasaki, C.: Heat Kernels for Elliptic and Sub-Elliptic Operators: Methods and Techniques. Springer, New York (2011)
39. Andler, D., Bienenstock, E., Laks, B. (eds.): COMPCOG I-II, Interdisciplinary Workshop on Compositionality in Cognition and Neural Networks. Abbaye de Royaumont (1991-1992)

40. Coulhon, T., Grigor'yan, A.: Heat kernels, volume growth and anti-isoperimetric inequalities. Comptes Rendus de l'Académie des Sciences de Paris 322, 1027–1032 (1996)
41. Cygan, J.: Heat kernels for class 2 nilpotent groups. Studia Mathematica 64(3), 227–238 (1979)
42. Damon, J.: Local Morse Theory for Solutions to the Heat Equation and Gaussian Blurring. Journal of Differential Equations 115(2), 368–401 (1995)
43. David, G., Semmes, S.: On the singular sets of minimizers of the Mumford-Shah functional. Journal de Mathématiques Pures et Appliquées 75, 299–342 (1996)
44. Davies, E.B.: Heat Kernels and Spectral Theory. Cambridge University Press, Cambridge (1989)
45. DeAngelis, G.C., Ozhawa, I., Freeman, R.D.: Receptive-field dynamics in the central visual pathways. Trends in Neuroscience 18(10), 451–458 (1995)
46. Deligne, P., Malgrange, B., Ramis, J.-P.: Singularités irrégulières. Correspondance et documents. Société Mathématique de France (2007)
47. Dennis, M.R.: Topological Singularities in Wave Fields. Thesis, H.H. Wills Laboratory, University of Bristol (2001)
48. Doyon, B., Cessac, B., Quoy, M., Samuelides, M.: Chaos in Neural Networks with Random Connectivity. International Journal of Bifurcation and Chaos 3(2), 279–291 (1993)
49. Duistermaat, J.J.: Oscillatory Integrals, Lagrange Immersions and Unfolding of Singularities. Com. Pure Appl. Math., 207–281 (1974)
50. Field, D.J., Hayes, A., Hess, R.F.: Contour integration by the human visual system: evidence for a local "association field". Vision Research 33(2), 173–193 (1993)
51. Fodor, J., Pylyshyn, Z.: Connectionism and Cognitive architecture: A critical analysis. Cognition 28(1/2), 3–71 (1988)
52. Folland, G.B., Stein, E.M.: Estimates for the $\overline{\partial b}$ complex and analysis on the Heisenberg group. Communications on Pure and Applied Mathematics 27, 429–522 (1974)
53. Folland, G.B., Stein, E.M.: Hardy Spaces on Homogeneous Groups. Princeton University Press, Princeton (1982)
54. Frégnac, Y.: Neurogeometry and entoptic visions of the functional architecture of the brain. In: Petitot, J., Lorenceau, J. (eds.) Neurogeometry and Visual Perception, Journal of Physiology-Paris 97(2-3), 87–92 (2003)
55. Frégnac, Y., Baudot, P., Chavane, F., Lorenceau, J., Marre, O., Monier, C., Pananceau, M., Carelli, P., Sadoc, G.: Multiscale functional imaging in $V1$ and cortical correlates of apparent motion. In: Masson, G.S., Ilg, U.J. (eds.) Dynamics of Visual Motion Processing. Neuronal, Behavioral, and Computational Approaches, pp. 73–94. Springer, New York (2010)
56. Gaveau, B.: Principe de moindre action, propagation de la chaleur et estimées sous elliptiques sur certains groupes nilpotents. Acta Mathematica 139, 96–153 (1977)
57. Ge, Z.: Horizontal Path Spaces and Carnot-Carathéodory metrics. Pacific Journal of Mathematics 161(2), 255–286 (1993)
58. Geller, D.: Fourier Analysis on the Heisenberg Group. I: Schwarz space. Journal of Functional Analysis 36, 205–254 (1980)
59. Geller, D., Mayeli, A.: Continuous Wavelets and Frames on Stratified Lie Groups. I, arXiv:math/0602201v3
60. Grayson, M.: The Heat Equation Shrinks Embedded Plane Curves to Round Points. Journal of Differential Geometry 26, 285–314 (1987)
61. Gregor, K., Le Cun, Y.: Learning Fast Approximations of Sparse Coding, `www.icml2010.org/papers/449.pdf`

62. Gromov, M.: Carnot-Carathéodory spaces seen from within. In: Bellaïche, A., Risler, J. (eds.) Subriemannian Geometry. Progress in Mathematics, vol. 144, pp. 79–323. Birkhäuser, Basel (1996)
63. Hammenstädt, U.: Some regularity in Carnot-Carathéodory metrics. Journal of Differential Geometry 32, 192–201 (1991)
64. Hoffman, W.C.: The visual cortex is a contact bundle. Applied Mathematics and Computation 32, 137–167 (1989)
65. Hörmander, L.: Linear Partial Differential Operators. Springer, Berlin (1963)
66. Hörmander, L.: The Analysis of Linear Partial Differential Operators, I-IV. Springer, Berlin (1983-1985)
67. Hubel, D.H.: Eye, Brain and Vision. Scientific American Library, W.H. Freeman & Co., New York (1988)
68. Hübener, M., Shoham, D., Grinvald, A., Bonhöffer, T.: Spatial relationships among three columnar systems in cat area 17. Journal of Neurosciences 17, 9270–9284 (1997)
69. Huisken, G.: Flow by mean curvature of convex surfaces into spheres. Journal of Differential Geometry 20, 237–266 (1984)
70. Huisken, G.: Asymptotic behavior for singularities of the mean curvature flow. Journal of Differential Geometry 31, 285–299 (1990)
71. Hulanicki, A.: The distribution of energy in the Brownian motion in the Gaussian field and analytic-hypoellipticity of certain subelliptic operators on the Heisenberg group. Studia Mathematica 56(2), 165–173 (1976)
72. Jerison, D.S., Sánchez-Calle, A.: Estimates for the heat kernel for a sum of squares of vector fields. Indiana University Mathematics Journal 35, 835–854 (1986)
73. Kanizsa, G.: Grammatica del Vedere. Il Mulino, Bologna (1980)
74. Keil, W., Wolf, F.: Coverage, continuity, and visual cortical architecture. Neural Systems & Circuits 1(17) (2011)
75. Klüver, H.: Mescal vision and eidetic vision. Am. J. Psychol. 37, 502–515 (1926)
76. Koenderink, J.J., Van Doorn, A.J.: The Singularities of the Visual Mapping. Biological Cybernetics 25, 51–59 (1976)
77. Koenderink, J.J., Van Doorn, A.J.: Representation of local geometry in the visual system. Biological Cybernetics 55, 367–375 (1987)
78. Kusuoku, S., Stroock, D.: Long time estimates for the heat kernel associated with a uniform subelliptic symmetric second order operator. Annals of Mathematics 127, 165–189 (1988)
79. Maldonado, P.E., Gödecke, I., Gray, C.M., Bonhöffer, T.: Orientation Selectivity in Pinwheel Centers in Cat Striate Cortex. Science 276, 1551–1555 (1997)
80. Mallat, S.: A Wavelet Tour of Signal Processing. Academic Press, New York (1998)
81. Margulis, G.A., Mostow, G.D.: Some remarks on the definition of tangent cones in a Carnot-Carathéodory space. Journal d'Analyse Mathématique 80, 299–317 (2000)
82. Marr, D.: Vision. W.H. Freeman, San Francisco (1982)
83. Martinet, J., Ramis, J.-P.: Les derniers travaux de Jean Martinet. Annales de l'Institut Fourier 42(1-2), 15–47 (1992)
84. Maslov, V.P.: Perturbation Theory and Asymptotic Methods. Dunod, Paris (1972)
85. McLaughlin, D., Shapley, R., Shelley, M., Wielaard, D.J.: A Neuronal Network Model of Macaque Primary Visual Cortex (*V1*): Orientation Selectivity and Dynamics in the Input Layer 4Cα. PNAS (Proceedings of the National Academy of Sciences USA) 97(14), 8087–8092 (2000)
86. Mitchell, J.: On Carnot-Carathéodory metrics. J. Diff. Geometry 21, 35–45 (1985)
87. Moiseev, I., Sachkov, Y.: Maxwell Strata in sub-Riemannian problem on the group of motion of a plane. ESAIM: Control, Optimisation and Calculus of Variations 16(2), 380–399 (2010)

88. Montgomery, R.: A Tour of Subriemannian Geometries, Their Geodesics and Applications. American Mathematical Society (2002)
89. Morel, J.-M., Solimini, S.: Variational Methods in Image Segmentation. Birkhäuser, Berlin (1995)
90. Mumford, D.: Elastica and computer vision. In: Bajaj, C. (ed.) Algebraic Geometry and Applications, pp. 491–506. Springer, Heidelberg (1992)
91. Mumford, D., Shah, J.: Optimal Approximations by Piecewise Smooth Functions and Associated Variational Problems. Communications on Pure and Applied Mathematics XLII, 577–685 (1989)
92. Nagel, A., Stein, E.M., Wainger, S.: Balls and metrics defined by vector fields I: Basic Properties. Acta Mathematica 155(1), 103–147 (1985)
93. Petitot, J., Varela, F.J., Roy, J.-M., Pachoud, B. (eds.): Naturalizing Phenomenology: Issues in Contemporary Phenomenology and Cognitive Science. Stanford University Press, Stanford (1999)
94. Ohki, K., Chung, S., Kara, P., Hübener, M., Bonhöffer, T., Reid, R.C.: Highly ordered arrangement of single neurons in orientation pinwheels. Nature 442, 925–928 (2006)
95. Olshausen, B.A., Field, D.J.: Sparse coding of sensory inputs. Current Opinion in Neurobiology 14, 481–487 (2004)
96. Osher, S., Sethian, J.A.: Fronts Propagating with Curvature-Dependent Speed: Algorithms Based on Hamilton-Jacobi Formulations. Journal of Computational Physics 79(1), 12–49 (1988)
97. Paradis, A.-L., Morel, S., Seriès, P., Lorenceau, J.: Speeding up the brain: when spatial facilitation translates into latency shortening. Frontiers in Human Neuroscience 6, 330, doi: 10.3389/fnhum.2012.00330
98. Petitot, J.: Eléments de théorie des singularités, `http://jeanpetitot.com/ArticlesPDF/Petitot_Sing.pdf`, Introduction aux phénomènes critiques, `http://jeanpetitot.com/ArticlesPDF/Petitot_CritPh.pdf`
99. Petitot, J.: Morphogenèse du Sens. Presses Universitaires de France, Paris (1985), English trans. by Manjali, F.: Morphogenesis of Meaning. Peter Lang, Bern (2003)
100. Petitot, J.: Morphodynamics and Attractor Syntax. Dynamical and morphological models for constituency in visual perception and cognitive grammar. In: van Gelder, T., Port, R. (eds.) Mind as Motion, pp. 227–281. MIT Press, Cambridge (1995)
101. Petitot, J. (with a collaboration of Y. Tondut): Vers une Neurogéométrie. Fibrations corticales, structures de contact et contours subjectifs modaux. Mathématiques, Informatique et Sciences Humaines 145, 5–101 (1999)
102. Petitot, J.: The Neurogeometry of Pinwheels as a Sub-Riemannian Contact Structure. In: Petitot, J., Lorenceau, J. (eds.) Neurogeometry and Visual Perception, Journal of Physiology-Paris 97(2-3), 265–309 (2003)
103. Petitot, J.: An introduction to the Mumford-Shah segmentation model. In: Petitot, J., Lorenceau, J. (eds.) Neurogeometry and Visual Perception, Journal of Physiology-Paris 97(2-3), 335–342 (2003)
104. Petitot, J.: Théorie des singularités et équations de diffusion. René Thom, Gazette des Mathématiciens, 89–107 (2004)
105. Petitot, J.: Neurogéométrie de la Vision. Modèles mathématiques et physiques des architectures fonctionnelles. Les Editions e l'Ecole Polytechnique, Palaiseau (2008)
106. Petitot, J., (with Doursat, R.): Cognitive Morphodynamics. Dynamical Morphological Models of Constituency in Perception and Syntax. Peter Lang, Bern (2011)
107. Petitot, J.: Complexity and self-organization in Turing. The Legacy of A.M. Turing. International Academy of Philosophy of Science, Urbino (2012)

108. Petitot, J.: La simplexité de la notion géométrique de jet. Séminaire de Jean-Luc Petit "Complexité-Simplexité", Collège de France (2012)
109. Rothschild, L.P., Stein, E.M.: Hypoelliptic differential operators and nilpotent groups. Acta Mathematica 137, 247–320 (1976)
110. Sachkov, Y.L.: Conjugate points in the Euler elastica problem. Journal of Dynamical and Control Systems 14(3), 409–439 (2008)
111. Saloff-Coste, L.: The heat kernel and its estimates. Advanced Studies in Pure Mathematics, Probabilistic Approach to Geometry 57, 405–436 (2010)
112. Sánchez-Calle, A.: Fundamental solutions and geometry of the sum of squares of vector fields. Inventiones Mathematicae 78, 143–160 (1984)
113. Sanguinetti, G.: Invariant models of vision between phenomenology, image statistics and neurosciences. Thesis, Montevideo University (2011)
114. Sanguinetti, G., Citti, G., Sarti, A.: A model of natural image edge co-occurence in the roto-translation group. Journal of Vision 10(14), 1–16 (2010)
115. Sarti, A., Citti, G., Manfredini, M.: From neural oscillations to variational problems in the visual cortex. In: Petitot, J., Lorenceau, J. (eds.) Neurogeometry and Visual Perception, Journal of Physiology-Paris, 97(2-3), 87–385 (2003)
116. Sarti, A., Citti, G., Petitot, J.: On the Symplectic Structure of the Primary Visual Cortex. Biological Cybernetics 1, 33–48 (2008)
117. Sarti, A., Malladi, R., Sethian, J.A.: Subjective Surfaces: a Geometric Model for Boundary Completion. International Journal of Computer Vision 46(3), 201–221 (2002)
118. Shelley, M., McLaughlin, D.: Coarse-Grained Reduction and Analysis of a Network Model of Cortical Response: I. Drifting Grating Stimuli. Journal of Computational Neuroscience 12, 97–122 (2002)
119. Shmuel, A., Grinvald, A.: Coexistence of linear zones and pinwheels within orientation maps in cat visual cortex. PNAS (Proceedings of the National Academy of Sciences USA) 97(10), 5568–5573 (2000)
120. Sigman, M., Cecchi, G.A., Gilbert, C.D., Magnasco, M.O.: On a common circle: Natural scenes and Gestalt rules. PNAS (Proceedings of the National Academy of Sciences USA) 98(4), 1935–1940 (2001)
121. Simoncelli, E.P., Olshausen, B.A.: Natural Image Statistics and Neural Representation. Annual Review of Neuroscience 24, 1193–1216 (2001)
122. Smolensky, P.: On the Proper Treatment of Connectionism. The Behavioral and Brain Sciences 11, 1–23 (1988)
123. Sompolinsky, H., Crisanti, A., Sommers, H.-J.: Chaos in Random Neural Networks. Physical Review Letters 61, 259–262 (1988)
124. Stein, E.M.: Harmonic Analysis: Real-Variable Methods, Orthogonality, and Oscillatory Integrals. Princeton University Press, Princeton (1993)
125. Strichartz, R.S.: Sub-Riemannian Geometry. Journal of Differential Geometry 24, 221–263 (1986)
126. Swindale, N.V.: How many maps are there in visual cortex? Cerebral Cortex 7, 633–643 (2000)
127. ter Elst, A.F.M., Robinson, D.W.: Reduced Heat Kernels on Nilpotent Lie Groups. Communications in Mathematical Physics 173(3), 475–511 (1995)
128. Thom, R.: Les Singularités des applications différentiables. Annales de l'Institut Fourier 6, 43–87 (1956)
129. Thom, R.: Ensembles et morphismes stratifiés. Bulletin of the American Mathematical Society 75(2), 240–284 (1968)
130. Thom, R.: Stabilité structurelle et morphogenèse, Benjamin, New York, Ediscience, Paris (1972)

131. Thom, R.: Modèles mathématiques de la morphogenèse. Bourgois, Paris (1980)
132. Tirozzi, B., Tsodyks, M.: Chaos in Highly Diluted Neural Networks. Europhysics Letters 14, 727–732 (1991)
133. Turiel, A., Nadal, J.-P., Parga, N.: Orientational minimal redundancy wavelets: from edge detection to perception. Vision Research 43(9), 1061–1079 (2003)
134. Turing, A.: The Chemical Basis of Morphogenesis. Philosophical Transactions of the Royal Society of London, Series B, Biological Sciences 237(641), 37–72 (1952)
135. Ullman, S.: Filling in the gaps: the shape of subjective contours and a model for their generation. Biological Cybernetics 25, 1–6 (1976)
136. Varopoulos, N.T.: The heat kernel on Lie groups. Revista Matemática Iberoamericana 12(1), 147–186 (1996)
137. Vershik, A.M., Gershkovich, V.Y.: Non Holonomic Dynamical Systems, Geometry of Distributions and Variational Problems. Encyclopædia of Mathematical Sciences 16, 1–81 (1994)
138. Visual Cortex, `https://www.lst.inf.ethz.ch/teaching/lectures/ws05/229/LectureSlides/`
139. Xu, X., Bosking, W.H., White, L.E., Fitzpatrick, D., Casagrande, V.A.: Functional Organization of Visual Cortex in the Prosimian Bush Baby Revealed by Optical Imaging of Intrinsic Signals. Journal of Neurophysiology 94, 2748–2762 (2005)
140. Yu, H., Farley, B.J., Jin, D.Z., Sur, M.: The Coordinated Mapping of Visual Space and Response Features in Visual Cortex. Neuron 47, 267–280 (2005)
141. Wolf, F., Geisel, T.: Spontaneous pinwheel annihilation during visual development. Nature 395, 73–78 (1998)
142. Wolf, F., Geisel, T.: Universality in visual cortical pattern formation. In: Petitot, J., Lorenceau, J. (eds.) Neurogeometry and Visual Perception, Journal of Physiology-Paris 97(2-3), 253–264 (2003)
143. Zahaf, M.B., Manchon, D.: Confluence of singularities of differential equations: A Lie algebras contraction approach, hal-00292676 (2008)
144. Zeeman, C.: Catastrophe Theory: Selected Papers 1972–1977. Addison-Wesley, Redwood City (1977)

Chapter 2
Shape, Shading, Brain and Awareness

Jan Koenderink and Andrea van Doorn

Abstract. Shading is one of the generic "monocular depth (and shape) cues". It is of conceptual interest because it apparently implies "causal relations" between the geometry of the scene in front of the observer, the formal description of brain activity, and the visual awareness of the observer. These are three disjunct ontological levels, so the very notion of "causal connections" is problematic. Some silent assumptions in current accounts indeed invoke "magic", we identify internal and external local sign as instances. We attempt an account of the shading cue that avoids at least some of these pitfalls. We conclude that (for the human observer, machine vision has different objectives) the shading cue allows "direct perception" of surface shape.

2.1 Structure of the Scene in Front of an Observer: Radiometry

The radiometric problem of "Shape From Shading" is simple in principle, but frequently intractable in practice [3, 7, 12, 16, 57]. Consider the simplest case.

In the simplest setting, one considers a surface being illuminated with a uniform, unidirectional beam. This constraint may obtain in real life, for instance,

Jan Koenderink
Laboratory of Experimental Psychology, Katholieke Universiteit Leuven, Tiensestraat 102, 3000 Leuven, Belgium
e-mail: jan.koenderink@ppw.kuleuven.be
Universiteit Utrecht, Faculteit Sociale Wetenschappen, Afdeling Psychologische Functieleer, Postbus 80140, 3508 TC Utrecht, The Netherlands
e-mail: J.J.Koenderink@uu.nl

Andrea van Doorn
Universiteit Utrecht, Faculteit Sociale Wetenschappen, Afdeling Psychologische Functieleer, Postbus 80140, 3508 TC Utrecht, The Netherlands
e-mail: A.J.vanDoorn@uu.nl
Department of Industrial Design, Delft University of Technology, Landbergstraat 15, 2628 CE Delft, The Netherlands
e-mail: A.J.vanDoorn@tudelft.nl

G. Citti and A. Sarti (eds.), *Neuromathematics of Vision*, Lecture Notes in Morphogenesis,
DOI: 10.1007/978-3-642-34444-2_2,

direct sunlight is a good approximation [18, 30, 43]. (However, it is the worst setting for effective shading, overcast sky being much more useful, as any professional photographer knows.) The surface is supposed to be uniform. This may obtain in real life, at some reasonable scale. For instance, a blank piece of paper will do at the millimeter scale. The BRDF (Bidirectional Reflectance Distribution Function [47]) is supposed to be constant. This is the so called "Lambertian assumption" [12, 41]. Although such surfaces don't exist, the constant may obtain approximately, if the range of surface spatial attitudes is not too large. For instance, a piece of blotting paper is a good approximation for angles of incidence and observation not too far from normality. Vignetting is supposed to be absent. "Vignetting" indicates occultation of the source by parts of the object [27]. Thus, various parts of the object are illuminated by mutually different "effective" sources. An example is an apple seen under overcast sky illumination. The constraint can be met in many cases, for instance direct sunlight away from the attached shadow boundary. Multiple scattering is supposed to be absent. That constraint can only be met for convex objects, which is very restrictive [27]. However, a surface that is fairly flattish yields a good approximation [27].

Notice that these constraints are quite limiting in their totality. However, the constraints are automatically met if you sufficiently limit the region of interest (ROI). Such a constraint serves to select approximately homogeneous and flattish surface patches. However, the "effective" source might well be quite different from the nominal one. For instance, it could be modulated by vignetting and/or multiple scattering effects. If one considers smaller ROIs the problem becomes simpler, but the effective direction of illumination becomes more variable.

This makes it likely that biological "shape from shading" [8–10, 45, 56] will be limited to rather smallish ROI's, and makes it likely that methods that do not explicitly require knowledge of the effective direction of illumination will be preferred.

2.1.1 Shading Geometry

In the simplest case, the radiance incident upon the eye is independent of the viewing direction. It depends only on the structure of the incident beam. In the simplest case we can summarize the incident beam by a "light vector", which is a suitable average over the directions represented in the beam [36–38, 46]. The radiance scattered to the eye is proportional to the irradiance caused by the incident beam. If the BRDF is approximately constant, the angle of incidence is crucial. The radiance scattered to the eye is proportional to the illuminance, which is proportional to the cosine of the angle of incidence. This implies that the scattered radiance as a function of location is approximately proportional to the change of spatial surface attitude [28, 29, 31, 33, 53]. Notice that there are numerous complicating factors whose influence we have somehow ignored. The degree to which these approximations "work" depend all on the size of the ROI. They also depend upon the form of the BRDF, vignetting, and so forth, something we will ignore from here on, but should not be forgotten.

2.1.2 *Outline of the Paper*

In order to proceed, we need to connect brain activity to scene geometry, to physics, and to visual awareness. We will do this in steps. First we discuss the nature of the relevant brain activity. Then we discuss the relevant nature of the "scene", which involves merely the local surface shape landscape. Then we have to connect these (ontologically distinct) universes in some way. Finally, we need to establish the relation to visual awareness. Needless to say, this is a very ambitious program, and we are unlikely to succeed. The goal of the exercise is mainly to obtain a more focussed conceptual grip on the problem.

2.2 Visual Front-End

Visual awareness is perhaps best understood as a "user interface" [25]. The user interface depends mainly on templates of various sorts. Microgenesis tries templates by running "reality checks" against the front end neural activity. As a result, such "hallucinations" may gain any desired degree of objectivity.

The "reality checks" are against the activity of the "visual ("optical" would have been more apt) front end". It is hard to define the boundaries of the front end. Here we simply talk of V1, the primary visual cortex. However, we are prepared to extend this definition when opportune, neuro-anatomy proper not being our primary interest.

The visual front end is a volatile buffer, that is continually being overwritten by the world (the radiance at the corneas). It is in many respects like the beach, which "represents" footprints. Wet (but not too wet) beaches are better than dry beaches in that respect. Likewise, the front end is optimized to retain useful structure, and discard useless junk. Moreover, it has a number of limiting constraints. For instance, at any location, the front end optimizes dynamic range by adjusting the local set point and gain. These are not forwarded to more remote parts of the brain. The local set point and gain are simply lost. Only in very local regions can one count on the gain to be constant. An example might be a "column", but we can't be sure.

2.2.1 *Brain Activity due the Optical Structure*

One simple principle used by the front end is to prefer derivatives of some useful degree over absolute values [24]. This is useful (at least in principle), because it obviates the need for absolute calibration. The generic example is the Laplacean ("Mexican hat receptive field"), which encodes the difference between a local region and a larger one. Another principle is to prefer ratios over absolute values. The ratio of two values of a non-negative quantity is evidently independent of the absolute calibration. A combination of these principles is to retain the "contrast gradient". The "contrast gradient" may also be defined as a "logarithmic derivative" (same thing). For a retinal illuminant $I(x,y)$ (Cartesian spatial coordinates $\{x,y\}$), the contrast gradient is (defining $I_0 = I(x_0,y_0)$):

$$\mathbf{C}(x_0,y_0) \propto \frac{1}{I_0}\nabla I(x,y) = \nabla \log I(x,y) \tag{2.1}$$

Such a contrast gradient might be available in the visual front end, at least in regions of limited extent. Over larger regions one runs into problems because the location dependent gain factors cannot be assumed to be available.

The contrast gradient is a vector quantity. The front end represents it in terms of an overcomplete basis of first order directional derivatives. The kernel of such a derivative in the x-directions is [22]

$$E(x,y,s) = x\frac{\mathrm{e}^{-\frac{x^2+y^2}{2s^2}}}{2\pi s^4}, \tag{2.2}$$

where the parameter s parameterizes the scale of the derivative operator. In polar coordinates $\{\rho = \sqrt{x^2+y^2}, \varphi = \arctan y/x\}$, the derivative in the direction φ_0 is

$$E(\rho,\varphi,s,\varphi_0) = \rho\cos(\varphi-\varphi_0)\frac{\mathrm{e}^{-\frac{\rho^2}{2s^2}}}{2\pi s^4} \tag{2.3}$$

The overcomplete basis is indexed by φ_0, it is a "cortical column" of "edge detectors". The contrast gradient is represented through the total activity in the column. It is the first Fourier component of the activity. All other Fourier components may be ignored. They represent noise. The overcomplete basis has the advantage of yielding a very robust representation, and not forcing you to decide on a fiducial coordinate frame.

Notice that the edge detector kernel depends upon an additional parameter s. This is a nonnegative quantity, that represents the *scale*. The edge detector yield exact first order directional derivatives *at some finite scale* [22]. We assume that the brain has a range of scales available. Thus, the topic of "scale space" is of basic importance.

Both the size of the ROI, and the scale of computation, will depend upon the task at hand. The square of the ratio of ROI diameter and scale value represents the number of degrees of freedom in the local representation. We expect it to be typically limited, not exceeding a small "icon" (as familiar from computer desktops) in complexity.

2.2.2 *Representation of the Optical Structure*

In the previous subsection we discussed the activity of the front end "at a point". This is obviously not sufficient for our goal. The contrast gradient (modulo gain) at a point is useless. What is needed is the *spatial variation* over a region of (at least approximately) constant gain. We need a contrast gradient *field*.

2.2.2.1 Local Sign

The first problem one meets here is that of *local sign*. How do brain "algorithms" take account of location? This is a fundamental problem, first noticed by Lotze [42], that is conventionally ignored in contemporary accounts. Many people apparently believe that somatotopy renders the problem a non-problem. This is naive. It is enough to consider a thought experiment in which a super-surgeon carefully permutes V1 cells, taking the utmost care to leave all mutual connections intact. Will this produce a local deformation of the visual field? If you don't count on magic, your answer has to be no. The "machine" is still the same, but the somatotopy has been destroyed.

There is one way to deal with local sign locally. One designs a complex receptive field that implements the complete algorithm. This takes care of the problem. Local sign has been "encapsulated". An "edge detector" is the simplest instance [12]. It is really a bilocal entitity, wrapped up into a purely local one.

In general, the local sign problem has never really been solved. The problem is too hard to tackle in this paper. We will simply ignore it, but we will acknowledge the problem, and use it to change our treatment to rather formal, and abstract, whenever the local sign issue arises. There seems to be little use in pretending to suggest "neural implementations" when major basic problems are left open. Of course, our formal treatment will be implementable in principle once ideas concerning local sign take more rigid form.

2.2.2.2 The "Contrast Gradient Image"

The "contrast gradient image" is a formal description of the structure of V1 activity that depends upon the local sign issue. Thus we treat it formally, instead of suggesting some "neural network" implementation. We make the essential role of local sign explicit.

The contrast gradient field is a map of part of the visual field (this is where local sign comes in) to the space of possible gradient values. Both spaces are two-fold extended. The patch of the visual field is a topological disk, say. A convex region will be most convenient. The "gradient space" is a vector space, with a well defined origin. We will mostly consider a disk centered on that origin. The diameter of the region depends upon the dynamic range of the edge detectors.

The map is "from the plane into the plane" [58], thus its generic singularities will be folds and cusps. Non-generic cases will no doubt occur, mostly of a degenerate kind. For instance, a region of uniform illumination will map on the origin, thus the map will collapse on a point. We will mainly be interested in the generic case.

The singular entities are curves (the folds) and points (the cusps) [1]. (See figure 2.1.) The configuration of singular elements alone suffices to characterize the contrast gradient image in a qualitative (or, rather, semi-quantitative) way. We will consider this contrast gradient image as the "representation of the optical structure" in the visual front end. It contains all that is of relevance to the shading cue ("cue" is a notion due to Berkeley [4]), and nothing else.

Folds occur when the Jacobian of the image vanishes, that is to say when $\frac{\partial C_x}{\partial x}\frac{\partial C_y}{\partial y} - \frac{\partial C_x}{\partial y}\frac{\partial C_y}{\partial x} = 0$. This implies $I_{xx}I_{yy} - I_{xy}^2 = 0$. This occurs on curves. Cusps occur when the gradient of the Jacobian is tangent to the curve, they occur as isolated points [1].

The contrast gradient image is a data structure that makes the spatial variation of the local edge detector activities *explicit*. It is necessary, because the raw presence of a distribution of edge detector activities in the brain only contains this structure *implicitly*. Because of the retinotopic structure of the brain it is visible to an external observer of V1. For instance, it might be revealed through some smart imaging technique. However, it is not available to V1 itself, *because V1 cannot "see itself"*. It is a sufficient summary of what might be "seen" (that is to say, made explicit as some data structure) by some "higher" part of the brain.

The very concept of "contrast gradient image" depends upon the existence of a local sign. The more primitive apparatus (the edge detectors) is simple and well understood. The local sign is not. We will not speculate on the form the contrast gradient image might take in terms of higher brain activity.

2.3 Local Shape

The description of "local shape" has nothing to do with the brain *per se*. It is a geometry of certain aspects of the scene in front of the observer. It does not even specify the optical structure impinging on the eyes directly. With local shape we mean the "curvature landscape" of the boundary surfaces of environmental objects. "Local" implies that we study the curvature in the neighborhood of a point. Of course, a certain scale is always implied, since there exist no smooth environmental surfaces if the scale is left unspecified. A local, smooth surface, is an entity that can be described sufficiently well by way of a Taylor expansion up to some reasonable order (say less than ten). This implies both a scale and an extent [23]. One typically has a choice here. For instance, a treetop can be treated as a surface on one scale, but

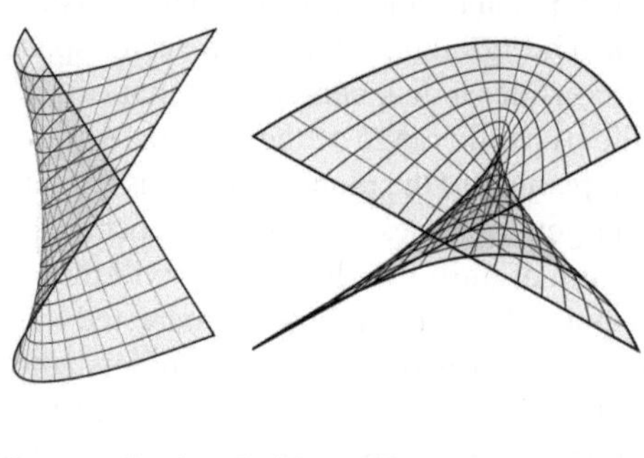

Fig. 2.1 Examples of the generic singularities of the contrast gradient image. At left a fold, at right a cusp. Near a fold gradient space is either covered zero or twice, near a cusp once and thrice.

not on the scale where individual leaves appear. This type of description has been intuitively used in the visual arts for centuries.

A local surface element can be located by its visual direction and distance. It has a spatial attitude that may be specified by its slant and tilt with respect to the visual direction and the vertical. This specifies the surface element as a "planelet" in the sense of Barrow. The deviations from the tangent plane may be denoted "surflets". In the lowest relevant order a surflet is described by its *curvature*, a "shapelet". The curvature varies from point to point, one has a "curvature landscape". The formal description is simply the classical differential geometry of Euclidean space as pioneered by Gauss.

Perhaps unfortunately, the classical theory is not particularly fit to describe the geometry as relevant to a stationary, monocular observer. We develop the necessary geometry in the next subsection.

2.3.1 The Geometry of "Pictorial Space"

When you look *at* a painting you are visually aware of a flat object, embedded in Euclidean space (that is the "space you move in"), covered with pigments in some simultaneous arrangement. When you look *into* the painting (we assume a "realistic" rendering, say a generic late nineteenth century landscape painting), you become aware of a "pictorial space". This pictorial space is fully detached from the space you move in. For instance, both your eye and the picture surface are in the space you move in, but neither of them is to be found in pictorial space. The space you move in and pictorial space don't even meet in the picture surface (as the world and the reflected world do in a mirror surface).

The geometry of pictorial space has been extensively researched, and formal accounts with excellent predictive power exist [32–35]. We use this formalism here, as it applies equally to pictorial space and to the visual space of a stationary, monocular observer.

Pictorial space has the structure of a fiber bundle [17, 21, 33], namely the depth domain over the visual field. For simplicity, we describe the visual field as a Euclidean plane $\mathbb{E}^2$. As a convenience, we fit it with a Cartesian coordinate frame, we denote the coordinates $\mathbf{r} = \{x, y\}$. The origin is arbitrary, for convenience we place it at the center of the ROI. The depth domain (parameterized by the z–coordinate) has the structure of the affine line $\mathbb{A}^1$. There is no origin of the depth domain, since "absolute depth" is a non-entity, and there is no preferred scale. Thus, pictorial space has the structure $\mathbb{E}^2 \times \mathbb{A}^1$. In the simplest cases depths on different fibers are coordinated with a global, linear gauge field. The gauge can be geometrically represented by two parallel, planar cross sections, one defining an arbitrary origin, the other an arbitrary unit point, on each fiber. The gauges are idiosyncratic and often change over time, even for a single observer. The group of similarities ("proper movements" for $\eta = \gamma = 1$)

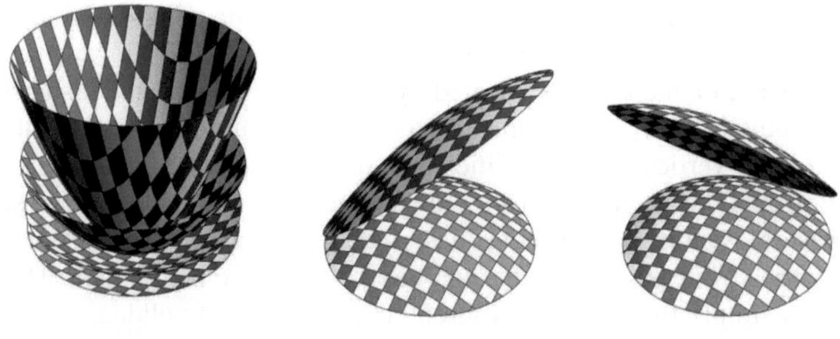

Fig. 2.2 Action of similarities on spheres of the second kind in pictorial space. At left the action of similarities of the second kind (parameter γ in eq. 2.4). At center and right the action of isotropic rotations (parameters ρ_x, ρ_y in eq. 2.4).

$$x' = \eta(x\cos\varphi - y\sin\varphi) + \tau_x, \tag{2.4}$$
$$y' = \eta(x\sin\varphi + y\cos\varphi) + \tau_y, \tag{2.5}$$
$$z' = \rho_x x + \rho_y y + \gamma z + \tau_z, \tag{2.6}$$

describes gauge transformations [33,49,54,59]. It is an 8-parameter group, whereas the corresponding group in Euclidean $\mathbb{E}^3$ is only 7-parameter. We will henceforth set $\eta = 1$, $\varphi = 0$, and $T = \{\tau_x, \tau_y\} = 0$, that is to say, ignore the Euclidean transformations in the visual field. The remaining transformations are a parabolic rotation, parameterized by $R = \{\rho_x, \rho_y\}$, a "similarity of the second kind" parameterized by γ, and a depth translation, parameterized by τ_z. (See figure 2.2.) The translation merely shifts the gauge planes along the fibers, and is generally irrelevant. We ignore it here. The parabolic rotation affects the spatial attitude of the gauge planes, and the similarity affects their spacing.

A "(depth) relief" is a cross section of the fiber bundle. It can be specified as a depth map $\{x, y, z(x,y)\}$. We will consider depth maps *modulo* arbitrary gauge transformations. This describes the nature of pictorial reliefs in considerable quantitative detail.

The geometry of pictorial space is no doubt due to the fact that the optical structure at the eye specifies the scene in front of the observer only partially. The gauge transformations describe the generic ambiguity for many "depth cues". Consider the shading cue for instance. Suppose the ROI is filled with a uniform illuminance. Could it be due to an illuminated surface in the scene? Sure, it could, although this is not necessarily the case. Suppose it is, what may one infer with regard to the shape of the surface? Well, if the illumination is uniform, then (in the generic case), the surface has to be *planar*. Notice that *any plane will do*. Thus the set of possible inferences is simply $z(x,y) = 0$, *modulo* arbitrary gauge transformations.

2.3.2 *Differential Geometry of Pictorial Space*

The differential geometry of pictorial space is similar to, but different from, the familiar differential geometry of Euclidean space $\mathbb{E}^3$ [20, 33, 49, 54, 59]. Consider the metric of $\mathbb{E}^2 \times \mathbb{A}^1$ induced by the gauge transformations. The Euclidean distance in the visual field is conserved, it may be used as the metric of pictorial space. Notice that this renders all points on a single fiber as coincident. The fibers are isotropic (null-)directions. Two points on a single fiber can be assigned a "special distance", which is also conserved. However, the special distance applies only to such "parallel points". The angle measure in the visual field is elliptic, just the familiar (periodic) Euclidean angle. In an isotropic plane, the angle measure is parabolic, thus not periodic. It is measured as the arc length of a "unit circle of the first kind". For instance, in the plane $y=0$, the unit circle with center at the origin consists of the lines $x=\pm 1$. Thus the slope of the line from $\{0,0,0\}$ to $\{x,0,z\}$ is simply z/x.

A regular plane is a planar cross section. Thus, it does not contain an isotropic direction. The fibers meet this plane orthogonally, that is to say, the isotropic angle is infinite. Thus the isotropic direction is the normal of *any* regular plane. The implication for differential geometry is that the concept of "surface normal" cannot play the dominant role as it does in the conventional treatments of Euclidean differential geometry. One uses the tangent planes instead. The tangent planes can be parameterized by their slopes, that is their depth gradient $\{z_x, z_y\}$. The map of a cross section to its gradients, is the "gradient image" of the relief. It can be regarded as the isometric stereographic projection of the "spherical image" of the relief. The spherical image of a surface is a map of the surface on the "unit sphere of the second kind" $\{x, y, (x^2+y^2)/2\}$ by parallel tangent planes. The stereographical projection maps $\{x, y, (x^2+y^2)/2\}$ on $\{x, y, 0\}$ (the "center" of the sphere is $\{0,0,\infty\}$). The stereographical projection is evidently isometric, not just conformal. Notice that this is analogue to the construction of the Gaussian normal spherical image in Euclidean differential geometry. Gauge transformations simply translate and/or magnify the spherical image. Thus, the relative metrical structure of the spherical image defines the curvature landscape.

Near the origin we may transform any relief to the form

$$z(x,y) = \frac{1}{2!}\left(a_{xx}x^2 + 2a_{xy}xy + a_{yy}y^2\right) + \mathrm{O}[x,y]^3, \tag{2.7}$$

by adopting a suitable gauge. This is the shapelet descrion we use in this chapter. The differential invariants $K = a_{xx}a_{yy} - a_{xy}$ and $2H = a_{xx} + a_{yy}$ then define the Gaussian and the mean curvatures [15, 23]. Notice that these expressions are much simpler than the corresponding expressions for the Euclidean case. Reason is the absence of "foreshortening".

2.3.2.1 Shapelet Space

The deviation from planarity near a point is denoted the "surflet' at that point, adapting Barrow's formalism [5, 11]. (See figure 2.3.) The lowest non-trivial description

Fig. 2.3 "Surflets" can be scaled and added. Here the umbilical surflet at left and the saddle-shaped surflet at center are added so as to yield the cylindrical surflet at right.

is in terms of the second order terms in a Taylor expansion about the origin. In the geometry of pictorial space that implies equation 2.7. A shapelet may be parameterized by the coefficients $\{a_{xx}, a_{xy}, a_{yy}\}$, and indicated as a point in "shapelet space". This is useful, because it allows us to regard "curvature landscapes" as surfaces (maps of the visual field, thus immersions) in shapelet space.

The parameterization by $\{a_{xx}, a_{xy}, a_{yy}\}$ is not very useful, because referred to the arbitrary Cartesian frame. One may do better by rewriting the form as

$$r\frac{x^2-y^2}{2}+sxy+t\frac{x^2+y^2}{2}, \tag{2.8}$$

where

$$r = \frac{1}{2}(a_{xx}-a_{yy}), \tag{2.9}$$

$$s = a_{xy}, \tag{2.10}$$

$$t = \frac{1}{2}(a_{xx}+a_{yy}), \tag{2.11}$$

the point being that the shapelet $(x^2+y^2)/2$ is rotationally invariant, whereas the shapelets $(x^2-y^2)/2$ and xy transform as a pair under rotations of the Cartesian frame. Thus, t and $w=\sqrt{r^2+s^2}$ are differential invariants, whereas $\varphi=\frac{1}{2}\arctan s/r$ describes the "orientation of principal curvature" with respect to the Cartesian frame. It may be called the "attitude" of the shapelet, whereas the ratio $w:t$ describes its shape *proper*, and $\sqrt{t^2+w^2}$ its *amount* of curvature.

The "Casorati curvature" [6] $C=\sqrt{r^2+s^2+t^2}$ can be interpreted as the R.M.S. deviation from planarity (simply defined through a suitable limiting process), or also as its R.M.S. sectional curvature, or (again equivalently) $\sqrt{(\kappa_1^2+\kappa_2^2)/2}$, where $\kappa_{1,2}$ are the "principal curvatures": in a frame rotated to set $\varphi=0$ the shapelet is described as $(\kappa_1 x^2+\kappa_2 y^2)/2$. Vanishing planarity implies $C=0$, thus for a proper shapelet one has $C>0$.

The parameter

$$\sigma=\arctan\frac{\kappa_2+\kappa_1}{\kappa_2-\kappa_1}=\arctan\frac{t}{w}, \tag{2.12}$$

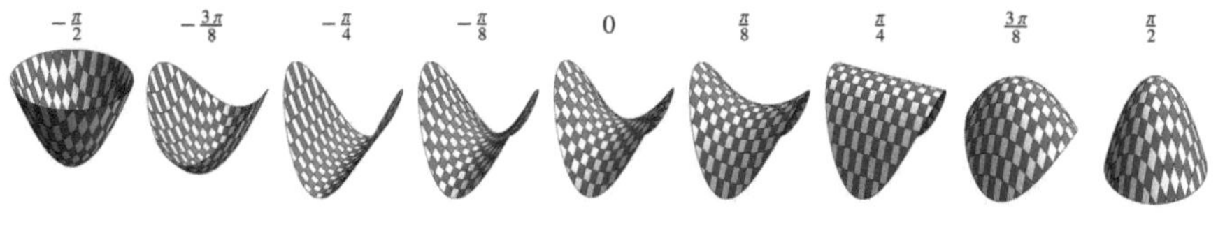

Fig. 2.4 The shape index series of quadric surflets. Notice that the umbilicals are at the endpoints. The symmetrical saddle at zero is congruent to its own mould. Shapes of opposite signs are related as object to mould.

(where we use the convention $\kappa_1 \geq \kappa_2$) is a pure shape descriptor, the "shape index" [23]. It assumes values in the range $[-\pi/2, +\pi/2)$. (See figure 2.4.) At the endpoints of the range the shapelet is "umbilical", that is rotationally symmetric, so the orientation φ is not defined. At $\sigma = 0$ the shapelet is a "symmmetrical saddle", meaning that its inverse (inverted depth) is congruent to itself. For non-zero shape index the shapelet and its inverse are in the relation of a shape to its mold.

One has

$$t = C \sin\sigma, \tag{2.13}$$
$$r = C \cos\sigma \cos 2\varphi, \tag{2.14}$$
$$s = C \cos\sigma \sin 2\varphi, \tag{2.15}$$

thus $\{r, s, t\}$ and $\{C, \sigma, 2\varphi\}$ are natural Cartesian and polar coordinates of "shape space". The space is naturally polarized by the r-direction. The line $r = s = 0$ contains umbilicals, and the principal directions are undefined on it. The right circular cones of semi-top-angle $\pi/4$ with this line as axis are the locus of cylindrical shapelets. Inside the cones one finds hyperbolical (saddle-like) shapelets, outside elliptical ones (either like the outside, or like the inside of egg shells).

Because absolute size is largely irrelevant in vision, it is natural to define a Riemann line element [52]

$$\frac{\mathrm{d}r^2 + \mathrm{d}s^2 + \mathrm{d}t^2}{r^2 + s^2 + t^2} = \mathrm{d}\mu^2 + \mathrm{d}\sigma^2 + 4\cos^2\sigma\,\mathrm{d}\varphi^2, \tag{2.16}$$

(where $\mu = \log C$) as a natural metric for shape space. The geodesics are planar logarithmic spirals in planes through the origin (of course only arcs contained in a half-space are relevant). In this metric the shape index scale (for constant Casorati curvature and orientation) is linear, so is the log-Casorati curvature (for constant shape index and orientation) scale, and so is the orientation (for constant shape index and Casorati curvature) scale. On spheres of constant Casorati curvature the spherical distance scaled by log-Casorati curvature is a geodesic distance.

2.3.2.2 Curvature Landscapes

A "curvature landscape" is a field of shapelets. We can represent it as a map of the ROI in the visual field into shape space. It will generically be an immersed

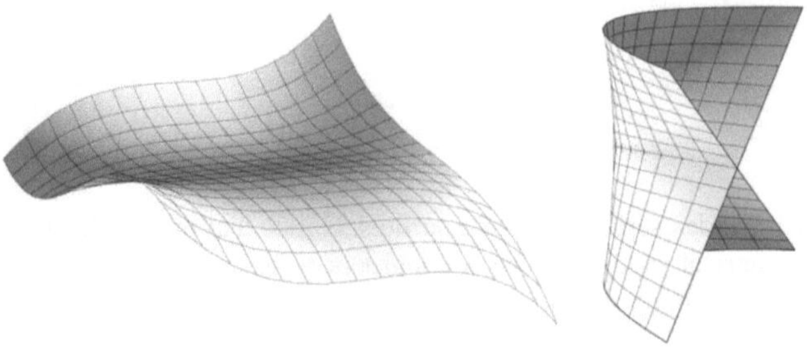

Fig. 2.5 Examples of a "Whitney umbrella". At left the surface, at right the curvature landscape in shape space.

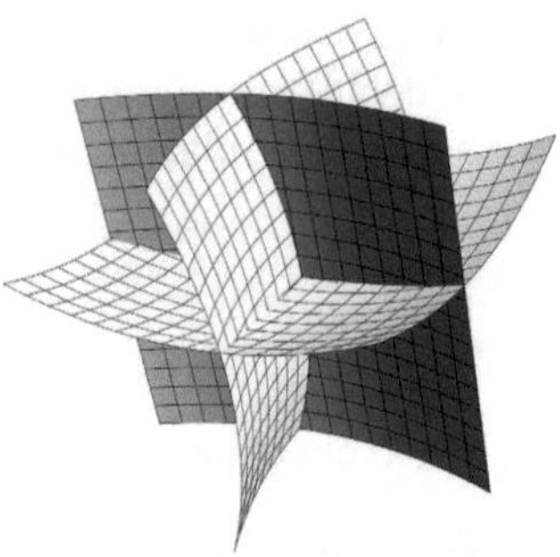

Fig. 2.6 A triple point

surface. Such immersions tend to have only mild singularities (there is lots of room in the space), generically they are Whitney umbrellas [14] (or cross-caps) (figure 2.5) and triple points (figure 2.6). Whitney umbrellas occur when two relations between the cubic terms in the Taylor expansion are simultaneously satisfied, thus at isolated points (the condition is $(a_x)_{xx}(a_x)_{yy} - (a_x)^2_{xy} = 0 \wedge (a_y)_{xx}(a_y)_{yy} - (a_y)^2_{xy} = 0$). Because shading is proportional with the slope of the surface in some direction, it implies $I_{xx}I_{yy} - I^2_{xy} = 0$, thus a point on a fold of the contrast gradient image, when the surface is illuminated.

Notice that not just any immersed surface represents a curvature landscape, in order to classify, the so called "integrability equations" (in the Euclidean case these are the Codazzi-Mainardi equations) have to be satisfied. We are confronted with a Pfaffian problem. For instance, it is evidently required that $\partial a_{xx}/\partial y = \partial a_{xy}/\partial x$, and $\partial a_{yy}/\partial x = \partial a_{xy}/\partial y$. This means that there is a constraint on curvature landscapes, and we cannot simply apply the generic taxonomy of singularities.

In one experiment we generated hundreds of high order, random polynomial surfaces. One finds Whitney umbrellas galore. Triple points are much harder to find, but that is because one has to search over large regions, they are not local phenomena. They certainly occur, because one may always start with three intersecting planes and assemble them into a single surface.

The notion of a "curvature landscape" is necessary if the local description stops at the second order. Of course, similar notions will still be necessary if one includes the cubic and quartic structure in the local description. The curvature landscape yields a global data structure in terms of a map. This is similar to a geographic description that mentions the relative locations of local features like mountains, lakes, and so forth. ("Mountain range" being a simple example.) Without such a global map one has only a bag of mutually unrelated features. In our case the "glue" is what might be called "external local sign", an awareness of the directions in the space external to the eye corresponding to retinal locations.

2.4 The "Shape From Shading" Problem

The "Shape From Shading Problem" can be framed in a large variety of ways [12]. Most of the conventional settings are hardly relevant to human (or animal) vision [2, 4, 26, 30, 40, 43, 48, 50, 51, 53, 55]. Here we impose the following a priori constraints on the matter:

— the "data" are the contrast gradient image, based upon the edge detector activity of the visual front end;
— the desired inference is a curvature landscape, that is the inverse of the image of the ROI in shape space.

Shape space is not a brain activity, or anything like that, it is a formal construction used in microgenesis.

The "microgenesis" of visual awareness is a hypothetical pre-conscious process that generates awareness. From experimental phenomenology we know that microgenesis is a systolic process that regenerates awareness continuously, a single "beat" taking less than a tenth of a second. The process generates hypotheses (or "hallucinates") and runs reality checks against visual front end activity. In a single beat it launches a volley of threads (representing different hypotheses), that may be novel, but typically are diversified threads from the previous beat. Some threads from the previous beats might be terminated. A competition between the threads leads to a final "winner" that enters visual awareness. Thus, momentary awareness is the result of an evolutionary process that runs on a very short time scale. In the generation of each volley current situational awareness ("gist") and goals (input from cognition and emotional states) play a role, though the process is largely autonomous.

"Shape from shading" starts when the microgenetic process selects "shaded surface" as a hypothesis. The same retinal illuminance may be interpreted in many different ways. "Shaded surface" is just one. (For instance, "painted flat picture" is another, "non-uniformly illuminated surface" yet another, and so forth.) It will typically involve a number of mutually related hypotheses, some aimed at material properties, others at the light field, etc. Here we concentrate on the "shaded surface" hypothesis. The act of generating a "shaded surface" hallucination serves to turn the front end structure into (meaningful) "data". The meaning derives from the hypothesis.

2.4.1 *Naive Radiometry*

The microgenetic process may assume very little about the geometrical layout, and physics of the scene. All it can do is assume an "uniformly illuminated surface", which involves a number of associated hypotheses, all of which might become falsified in reality checks at a number of levels, from front end activity to reflective thought. These include:

— the surface is a smooth Lambertian surface (no space-variant pigmentation, no specularities inside the ROI);
— the illumination is oblique (no frontal illumination);
— the surface is illuminated throughout (no attached shadow terminator in the ROI).

Full analysis of these assumptions and their interrelations would take many pages, we skip it here.

Notice that these are no detailed assumptions concerning surface attitude, nor about the light field. The shading is subject to the well known "bas-relief ambiguity". What this all implies is that the magnitude of the contrast gradient cannot be distinguished from the obliquety of the illumination. The structure of the contrast gradient field depends upon the local shapelet and the tangential component of the light vector over the surface ("surface illuminance flow"). Let the shapelet be parameterized as in equation 2.7, and let the direction of surface illuminance flow as projected in the visual field be $\{\cos\vartheta, \sin\vartheta\}$ in the xy-plane.

In order to simplify the formalism we set (arbitrarly, but without loss of generality) $\vartheta = 0$. Then the contrast gradient will (to a good approximation) be proportional with

$$C(x,y) \propto \{a_{xx}, a_{xy}\}. \tag{2.17}$$

The constant of proportionality depends on many things (spatial albedo or BRDF variations, obliquity of the light vector, vignetting, ...). If the surface is significantly corrugated, the contrast gradient will be mainly determined by the curvature landscape though.

2.4.2 *The Shape Cue Inference*

Notice that the "data", that is the contrast gradient image based upon the front end edge detector activity, is a projection of the curvature landscape in $\{a_{xx}, a_{xy}, a_{yy}\}$-space obtained by dropping a_{yy}, that is the second derivative of the depth in the direction orthogonal to the illumination flow direction in the visual field. Since shape space is a simple linear transformation of $\{a_{xx}, a_{xy}, a_{yy}\}$-space, we find that microgenesis finds a projection of the shape image in the front-end data, and is not in need of any further computation. (See figures 2.7 and 2.8.) There is even a check on the viability of the hypothesis: simply find the curl of the gradient field, if it vanishes the hypothesis can be upheld.

Notice that $a_{yy} = 0$ implies $r = t$. Thus, the "view direction" of the view into shape space subtends a $\pi/4$ angle with the axis of umbilicals.

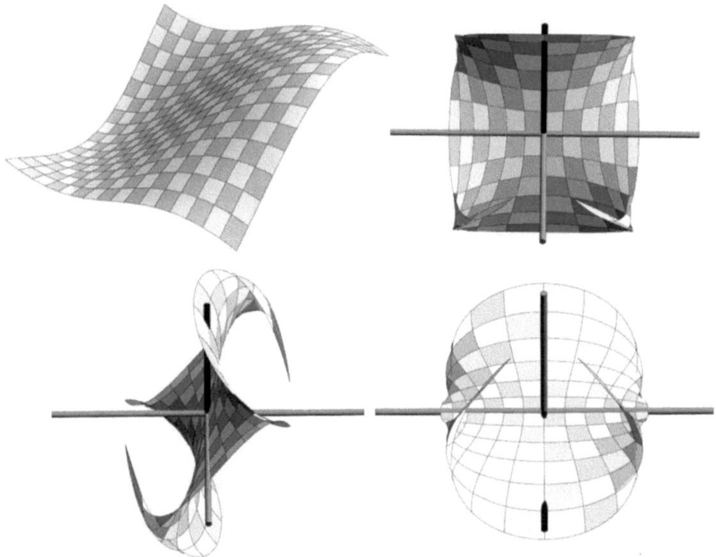

Fig. 2.7 An example of "Shape From Shading". At top left the habitus of a surface. The other subfigures show three views of the shape landscape in shape (that is *rst*-) space. This surface has no singularities.

One might regard this as an example of "direct perception" as propagated by Gibson [13], in his ecological approach to visual perception. What is conceptually interesting is that the shape from shading problem becomes formally identical to the "shape from movement" problem. One observes a two-dimensional projection of a surface immersed in a three-dimensional space, and attempts to make inferences about the immersed surface (e.g., its projections as viewed from other directions). Only the space is different, it is the "space you move in" in the case of shape from movement, and the space of shapes in the case of shape from shading. It is hard to see that this should make much difference to the brain: these spaces are just as "abstract" as seen from the brain's perspective. Both play some role in certain perception-action cycles.

The shape image has generically fold and cusp singularities. The projection generically generates folds and cusps in the contrast gradient image. The latter might be called "spurious", since they depend upon the direction of the projection, whereas singularities in the gradient image that derive from the singularities in the shape image have intrinsic meaning.

Variation of the (relative) illumination direction will clear up such ambiguities. Such variations could be due to movements of the light source (relatively rare), or movements of the surface with respect to the source (common). The latter type of variation can often be brought about by manipulation, thus opening a way to active exploration.

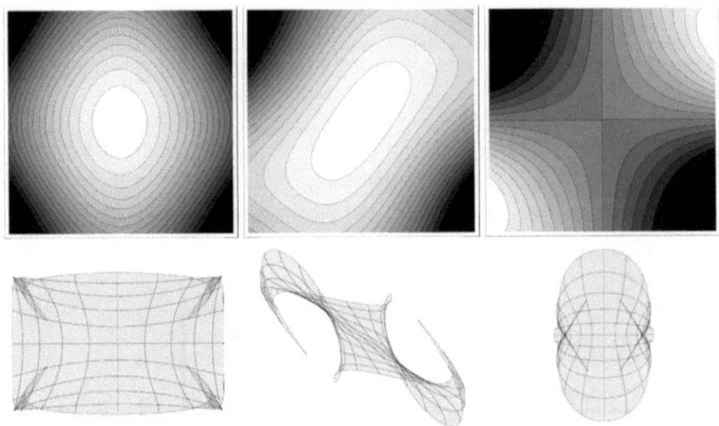

Fig. 2.8 The example from figure 2.7. At top three shadings of the surface for different illumination directions. At bottom the corresponding contrast gradient images. Compare these to the projections of the curvature landscape: they are the same, except for a $\cos\frac{\pi}{4} = \frac{1}{2}\sqrt{2}$ foreshortening.

A final point of some interest is the relation to the "Shape From Shading Problem" as it is usually framed in computer vision with the present setting. One difference is that computer vision frames the problem in Euclidean terms, which introduces some algebraic complications. More interestingly, in computer vision one would not be satisfied with a curvature landscape, but would require an explicit representation of a surface. In the case of the human observer the curvature landscape should most likely be regarded as the end result (see below).

The difference is the need of an integration. One has a Pfaffian system of local tangent quadrics and seeks for an integral surface. There will be a solution if the surface integrability conditions are satisfied (the Codazzi-Mainardi equations), which is never exactly the case if the field of quadrics is due to observations. The computer vision methods differ primarily in the way they handle this problem, which has nothing to do with the shading cue *per se*, it is just a problem of numerical analysis. If one has clean data the integration poses no problem (for an $N \times N$-pixel image one has about $2N^2$ equations for N^2 unknowns, it is mere matter to deal with the ambiguity), things start to become interesting in case the data are "dirty", which they always are in real life. Again, this is not an issue of much biological interest.

2.5 The Shading Cue and Visual Awareness: Phenomenology

The shading cue has been studied extensively in experimental phenomenology. The topic is closely related to that of pictorial space in general. How are pictorial reliefs "represented"? We have been able to show empirically that the representation

Fig. 2.9 *Left*: The Mach book may be the simplest "shading" stimulus ever invented. Many observers see an open book, or two planar facets meeting in a common edge, subtending a roughly right dihedral angle. Ernst Mach [44] interpreted this as a direct causal connection between intensity and the awareness of spatial attitude. *Right*: The circular disk filled with a linear gradient is the "canonical stimulus" that has been used for almost two centuries in psychophysical shape from shading research [19,50,51]. Many observers become aware of a spherical pictorial surface when viewing this stimulus. The contrast gradient image is degenerate (a point). The square contains the same gradient. Many observers see it as a cylinder. *Any* quadric surface could yield this gradient, for instance, a symmetrical saddle is a perfectly valid inference. It is *never* reported.

is not a depth map, but more likely a map of spatial surface attitudes (a depth gradient field), or possibly (this issue is still open) a curvature landscape. The present treatment of shape from shading fits perfectly in this general framework.

Most of the psychophysical work has been concentrated upon very singular cases. The simplest instance is no doubt the "Mach book" (figure 2.9 left), but the most widely used stimulus in shape from shading research is a linear illuminance gradient limited to a circular disk. (See figure 2.9 right.) Thus the contrast gradient image is a point that does not coincide with the origin. Thus, the corresponding curvature landscape would generically be a point. (A line in the direction of projection in the *rst*-space being evidently non-generic.) Hence, the inference would be "any quadric". Observers report convex or concave elliptical shapes, which are indeed "solutions", albeit very specific ones. Hyperbolical shapes are never reported, although they are equally valid inferences. Apparently microgenesis applies additional constraints. In this case it is the shape of the outline (circular). If the outline is square (two edges parallel to the gradient), observers report convex or concave cylindrical shapes (see figure 2.9 right).

One would expect that the next round of empirical research would concentrate on curvature landscapes that are degenerated to curves instead of points. However, we know of no instances. So the next round should involve generic curvature landscapes. Unfortunately, we haven't seen much activity on this topic either (except from some work of our own). The problem of "shape from shading" is pretty much open in the context of experimental phenomenology.

2.6 Conclusion

We have presented a discussion of the shading cue that is decidedly different from its conventional formulation in machine vision. The reason is that the present treatment has been focussed upon the phenomenology of human visual awareness, and upon an understanding of the brain "from the inside" as it were. We have refrained from "representations" that are only available to an external observer. The result is a description that renders the "shape from shading problem" trivial. The inference is essentially identical to the observation. Thus, we end up with a theory of "direct perception". This is not to say that the inference is complete, as it cannot be. The resulting ambiguity is very simple in that one obtains a single perspective of the curvature landscape in shape space. This is much like the visual projection itself: you see only the fronts of objects in the scene in front of you.

The treatment requires the existence of both *local sign* (as defined by Lotze), and *external local sign* [39]. None of these is well understood. There exist a number of theories on the genesis of local sign, and mainly speculations on the origin of external local sign. Recent empirical work has revealed that the structure of external local sign in human observers is surprising, as already intuited by Helmholtz and Kepler.

The connection of our treatment with the experimental phenomenology of human visual awareness is still weak. The reason is mainly the lack of a solid body of quantitative empirical results.

References

1. Banchoff, T., Gaffney, T., McCrory, C.C.: Cusps of Gauss Mappings. Pitman Advanced Pub. Program, Boston (1982)
2. Baxandall, M.: Shadows and enlightenment, 2nd edn. Yale University Press, London (2005)
3. Belhumeur, P.N., Kriegman, D.J., Yuille, A.L.: The bas-relief ambiguity. International Journal of Computer Vision 35, 33–44 (1999)
4. Berkeley, G.: An Essay Towards a New Theory of Vision. Printed by Aaron Rhames, at the Back of DickÃs Coffee-House, for Jeremy Pepyat, Bookseller in Skinner-Rows, Dublin (MDCCIX)
5. Boyd, R.W.: The History of the Calculus and its Conceptual Development. Dover, New York (1949)
6. Casorati, F.: Nuova denizione della curvature della supercie e suo confronto con quella di Gauss. Rend. Inst. Matem. Accad. Lomb. 2(22), 335–346 (1889)
7. van Diggelen, J.: Photometric properties of lunar crater floors. Recherches Astronomiques de lÃObservatoire dÃUtrecht 14, 1–114 (1959)
8. van Doorn, A.J., Koenderink, J.J., Wagemans, J.: Light fields and shape from shading. Journal of Vision 11(3), 1–21 (2011)
9. van Doorn, A.J., Koenderink, J.J., Todd, J.T., Wagemans, J.: Awareness of the light field: the case of deformation. i-Perception 3(7), 467–480 (2012)
10. Erens, R.G.F., Kappers, A.M.L., Koenderink, J.J.: Perception of local shape from shading. Journal of Vision 11, 1–21 (2011)

11. Eves, H.: An Introduction to the History of Mathematics. Brooks/Cole – Thomson Learning, London (1990)
12. Forsyth, D.A., Ponce, J.: Computer vision: A modern approach. Prentice–Hall, Upper Saddle River (2002)
13. Gibson, J.J.: The Perception of the Visual World. Houghton Mifflin, Boston (1950)
14. Gray, A.: The Whitney Umbrella. In: Modern Differential Geometry of Curves and Surfaces with Mathematica, 2nd edn., pp. 311, 401–402. CRC Press, Boca Raton (1997)
15. Hilbert, D., Cohn–Vossen, S.: Geometry and the Imagination, 2nd edn. Chelsea, New York (1952)
16. Horn, B.K.P., Brooks, M.J. (eds.): Shape from shading. MIT Press, Cambridge (1989)
17. Husemoller, D.: Fiber Bundles. Graduate Texts in Mathematics, vol. 20. Springer, London (1993)
18. Jacobs, T.S.: Light for the artist. Watson–Guptill, New York (1988)
19. Kleffner, D.A., Ramachandran, V.S.: On the perception of shape from shading. Perception & Psychophysics 52, 18–36 (1992)
20. Klein, F.: Vorlesungen über nicht–Euklidische Geometrie. Springer, Berlin (1928)
21. Kobayashi, S., Nomizu, K.: Foundations of Differential Geometry. Wiley–Interscience, New York (1963)
22. Koenderink, J.J.: The structure of images. Biological Cybernetics 50, 363–370 (1984)
23. Koenderink, J.J.: Solid Shape. MIT Press, Cambridge (1990)
24. Koenderink, J.J.: The brain a geometry engine. Psychological Research 52, 122–127 (1990)
25. Koenderink, J.J.: Vision & Information. In: Albertazzi, L., Van Tonder, G.J., Vishwanath, D. (eds.) Perception Beyond Inference: The Information Content of Visual Processes, pp. 27–56. MIT Press, Cambridge (2010)
26. Koenderink, J.J., van Doorn, A.J.: Photometric Invariants related to Solid Shape. Optica Acta 27, 981–996 (1980)
27. Koenderink, J.J., van Doorn, A.J.: Geometrical modes as a general method to treat diffuse interreflections in radiometry. Journal of the Optical Society of America 73, 843–850 (1983)
28. Koenderink, J.J., van Doorn, A.J.: Illuminance texture due to surface mesostructure. Journal of the Optical Society of America A 13, 452–463 (1996)
29. Koenderink, J.J., Pont, S.C.: Irradiation direction from texture. Journal of the Optical Society of America A 20, 1875–1882 (2003)
30. Koenderink, J.J., van Doorn, A.J.: Shape and shading. In: Chalupa, L.M., Werner, J.S. (eds.) The Visual Neurosciences, pp. 1090–1105. MIT Press, Cambridge (2003)
31. Koenderink, J.J., van Doorn, A.J.: Local structure of Gaussian texture. IEEE Transactions on Information and Systems 86, 1165–1171 (2003)
32. Koenderink, J.J., van Doorn, A.J.: The Structure of Visual Spaces. Journal of Mathematical Imaging and Vision 31, 117–187 (2008)
33. Koenderink, J.J., van Doorn, A.J.: Gauge Fields in Pictorial Space. SIAM Journal on Imaging Sciences 5(4), 1213–1233 (2012)
34. Koenderink, J.J., van Doorn, A.J., Kappers, A.M.L.: Surface perception in pictures. Perception & Psychophysic 52, 487–496 (1992)
35. Koenderink, J.J., van Doorn, A.J., Kappers, A.M.L.: Ambiguity and the 'mental eye' in pictorial relief. Perception 30, 431–448 (2001)
36. Koenderink, J.J., van Doorn, A.J., Kappers, A.M.L., Pas, S.F., te Pont, S.C.: Illumination direction from texture shading. Journal of the Optical Society of America A 20, 987–995 (2003)

37. Koenderink, J.J., van Doorn, A.J., Pont, S.C.: Light direction from shad(ow)ed random Gaussian surfaces. Perception 33, 1405–1420 (2004)
38. Koenderink, J.J., Pont, S.C., van Doorn, A.J., Kappers, A.M.L., Todd, J.T.: The visual light field. Perception 36, 1595–1610 (2007)
39. Koenderink, J.J., van Doorn, A.J., Todd, J.T.: Wide distribution of external local sign in the normal population. Psychological Research 73, 14–22 (2009)
40. Koenderink, J.J., van Doorn, A.J., Pont, S.C.: The "shading twist," a dynamical shape cue. International Journal of Computer Vision (accepted April 6, 2013)
41. Lambert, J.H.: Photometria, sive, De mensura et gradibus luminis, colorum et umbræ. V.E. Klett, Augsburg (1760)
42. Lotze, H.: Microcosmus: An Essay Concerning Man and His Relation to the World (1856-58, 1858-64), trans. Hamilton, E., Jones, E.E.C.: 4th edn. T. & T. Clark, Edinburgh (1885, 1899)
43. Luckiesh, M.: Light and shade and their applications. Van Nostrand, New York (1916)
44. Mach, E.: Die Analyse der Empfindungen und das verhältniss des physischen zum Psychischen. G. Fischer, Jena (1903)
45. Metzger, W.: Gesetze des Sehens. Verlag Waldemar Kramer, Frankfurt aM (1975)
46. Moon, P., Spencer, D.E.: The photic field. MIT Press, Cambridge (1981)
47. Nicodemus, F.E., Richmond, J.C., Hsia, J.J., Ginsberg, I.W., Limperis, T.: Geometric considerations and nomenclature for reflectance. Technical Report MN-160, U.S. Department of Commerce, National Bureau of Standards (October 1977)
48. Palmer, S.E.: Vision Science: Photons to Phenomenology. MIT Press, Cambridge (1999)
49. Pottmann, H., Optitz, K.: Curvature analysis and visualization for functions defined on Euclidean spaces or surfaces. Computer Aided Geometric Design 11, 655–674 (1994)
50. Ramachandran, V.S.: Perceiving shape from shading. Scientific American 258, 76–83 (1988)
51. Ramachandran, V.S.: Perception of shape from shading. Nature 331, 163–166 (1988)
52. Riemann, G.F.B.: Über die Hypothesen, welche der Geometrie zu Grunde liegen. In: Habilitationsschrift, Lecture Published in Abhandlungen der Königlichen Gesellschaft der Wissenschaften zu Göttingen, vol. 13 (1868)
53. Rittenhouse, D.: Explanation of an optical deception. Transactions of the American Philosophical Society 2, 37–42 (1786)
54. Sachs, H.: Isotrope Geometrie des Raumes. Friedrich Vieweg & Sohn, Braunschweig (1990)
55. Turhan, M.: Über räumliche Wirkungen von Helligkeitsgefällen. Psychologische Forschung 21, 1–49 (1935)
56. Wagemans, J., van Doorn, A.J., Koenderink, J.J.: The shading cue in context. i-Perception 1, 159–177 (2010)
57. Woodham, R.J.: Photometric method for determining surface orientation from multiple images. Optical Engineerings 19, 139–144 (1980)
58. Whitney, H.: On singularities of mappings of Euclidean spaces. I. Mappings of the plane into the plane. Annals of Mathematics 62, 374–410 (1955)
59. Yaglom, I.M.: A simple non–Euclidean geometry and its physical basis. Springer, New York (1997)

Chapter 3
Why Shading Matters along Contours*

Benjamin Kunsberg and Steven W. Zucker

Abstract. Shape from shading is a classical inverse problem in computer vision. It is inherently ill-defined and, in different formulations, it depends on the assumed light source direction. In contrast to these mathematical difficulties, we introduce a novel mathematical formulation for calculating local surface shape based on covariant derivatives of the shading flow field, rather than the customary integral minimization or P.D.E approaches. Working with the shading flow field rather than the original image intensities is important to both neurogeometry and neurophysiology. To make the calculations concrete we assume a Lambertian model for image formation, but we do not make global light source positional assumptions. In particular, for smooth surfaces in generic position, we show that second derivatives of brightness are independent of the light sources and can be directly related to surface properties. We use these measurements to define the matching local family of surfaces that could result from a given shading patch. In total our results change the emphasis from seeking a single, well-define solution to an ill-posed problem to characterizing the ambiguity in possible solutions to this problem. The result is relevant both mathematically and perceptually, because we then show how the equations simplify and the ambiguity reduces are certain critical points of intensity. We conclude with a discussion of image reconstruction at these critical points.

3.1 Introduction

The inference of shape from shading information is so natural for us that the inherent difficulty in the problem seems paradoxical. Perceptually, when shape-from-shading works, we see crisp surfaces and clear boundaries. However, as a classical

Benjamin Kunsberg · Steven W. Zucker
Department of Computer Science, Yale University
e-mail: {benjamin.kunsberg, steven.zucker}@yale.edu

* This work was supported by the National Science Foundation.

G. Citti and A. Sarti (eds.), *Neuromathematics of Vision*,
Lecture Notes in Morphogenesis,
DOI: 10.1007/978-3-642-34444-2_3, © Springer-Verlag Berlin Heidelberg 2014

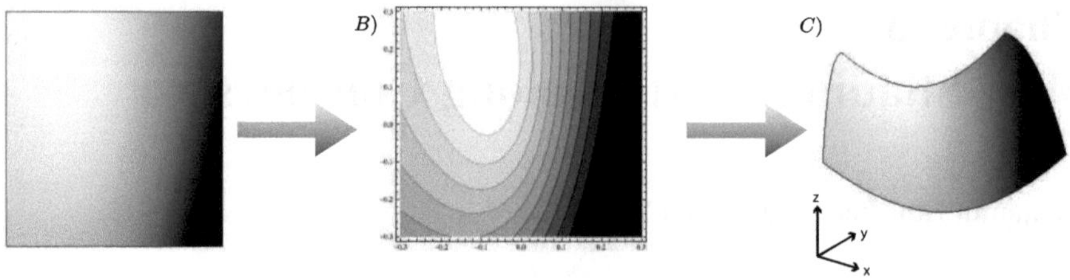

Fig. 3.1 Classical methods attempt to go from pixel values of the image (A) to the surface (C). In this work, we use the intermediate representation of the shading flow field, or the tangent map to the isophote structure (B). This is both supported by biological mechanisms and allows us to use the mathematical machinery of vector fields, transport, and differential goemetry.

inverse problem it is clearly ill-posed, with the two-dimensional image providing much less information than three-dimensional surface calculations require. Egyptian artists took advantage of this inherent ambiguity by carving scenes with deep apparent relief out of shallow stone. But this needs to be reconciled with the ideal for computer vision systems: the exact recovery of unique surface descriptions. How are we to make sense of the apparent contradiction between mathematical uncertainty, perceptual crispness, and computer vision accuracy?

Mach [25], who first formulated the image irradiance equation in 1866, observed that "many curved surfaces may correspond to one light surface even if they are illuminated in the same manner". Instead of working on the general problem, he focused on cylinders. The modern shape-from-shading community seeks to resolve the ambiguity by placing priors on the surface [1], in a sense following along the lines set by Mach, or on the light source direction(s) [1,11] and albedo [1]. But these priors are global and difficult to reconcile with physiology.

We propose an alternative approach: rather than attempting to resolve the ambiguity from the start using global priors, we suggest that the ambiguity can be characterized and exploited. One part of this exploitation derives from neurobiology and the other from mathematics, in particular differential geometry. Thus our goal in this paper is to place the shape-from-shading inference problem in the context of neurogeometry.

The first part of our approach is to build the surface inference process not from the image, but from the manner in which the shading information could be represented in visual cortex; see Fig. 3.1. This is called the shading flow field, and it has the form of a vector field (plus discontinuities); in particular, it is the tangent map to isophote contours. Formulating the shape inference then amounts to working out the relationships between isophotes in the image and isophotes on the surface. The differential geometry arises by formulating the problem in this fashion, and it opens two rich projections back on to neurobiology.

The first of these projections has to do with reducing computations to neural networks. As we show below, the shading flow field can be computed within a neurogeometric architecture of columns of cells tuned to different orientations, plus connections between them (Fig. 3.3). Biologically these connections could be either long-range horizontal connections within an area, or feedforward/feedback loops. Mathematically these connections derive from transport operations, and we show how to derive these transports in this paper. The result is a fibre bundle, and solutions to the inference problem are given by sections through this bundle. In general there are many, which corresponds to the ambiguity in the inverse problem, but we shall have more to say on this momentarily.

The second projection onto neurobiology relates to the question of light source: where, in the brain, might this be represented? How? Since there is no evidence that it is represented in the early visual system, where the shading flow could arise, we seek those image properties that are invariant to the direction of the light source. We prove that ordinary second derivatives of the image irradiance do not depend directly on the direction(s) of the light source(s), but rather only on the local surface derivatives and image gradient properties. The image derivatives can then be used to restrict the potential surfaces corresponding to a local shading patch, regardless of the light source. This effectively "cancels out" the light source from the problem, and is the second major aspect of our approach. It follows, of course, that the light source can be calculated once the surface structure is known, which makes it a kind of *emergent* property in some cases.

Finally, we return to the question of ambiguity. With our characterization we are able to show how other cues, such as the apparent boundary, various highlight lines or cusps, may be sufficent to effectively resolve them. In effect this provides a way to anchor a solution from regions in the image that are clear, and build upon the perceptual effect that the surface is crisp – but only around certain structures! See Fig. 3.2. Understanding these structures and why they are important is the final focus of this paper.

3.1.1 Motivation from Neurobiology

Because V1 is rich in orientationally-selective cells [13], we focus on understanding how sets of orientations could correspond to surfaces. This suggests what at first glance appears to be a small problem change: instead of seeking the map from images to surfaces (and light sources), the image should first be lifted into an orientation-based representation. This lift can be accomplished by considering the image isophotes ([17, 18]); the lift is then the tangent map to these isophotes. This lift has been used earlier in computer vision, and is called the shading flow field [2]. A significant body of evidence is accumulating that such orientation based representations underly the perception of shape([8,9]), but to our knowledge no one has previously formulated the surface inference problem from it.

Since the shading flow field could be computed in V1 (Figure 3.3), we are developing a new approach to shape from shading that is built directly on the information available in visual cortex. Thus it carries the constructions of neurogeometry from contours to shading flows to surfaces. In the end theoretically our computations

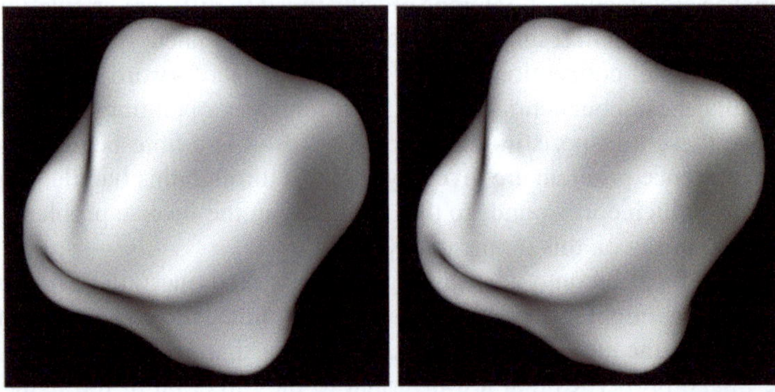

Fig. 3.2 Compare these two images: (Left) a shaded surface; (Right) reconstructed shading values via a linear interpolation algorithm [5] using only the shading values at the highlights and along the contours. Note that the 3D percept is nearly identical, even though the image intensities are not always consistent. Although there are significant image intensity differences between these images, the surface "seen" at a rapid glance is hardly different. We suggest that this is because shape-from-shading is mainly working around certain key image regions. Our goal in this paper is to understand why these regions exist and how they might be characterized.

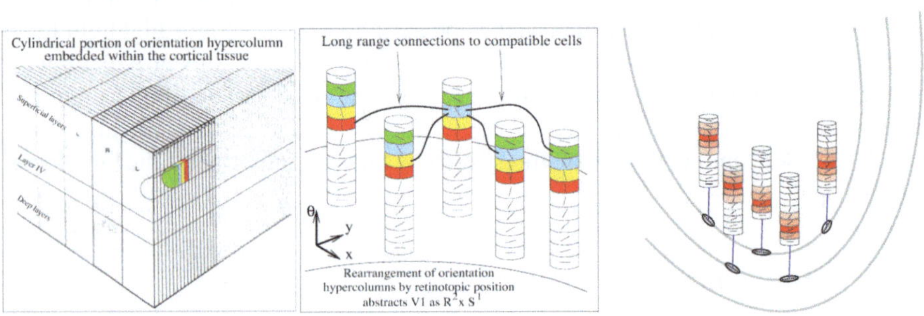

Fig. 3.3 V1 mechanisms applied to the isophote curves result in a shading flow field. (left) Visual cortex contains neurons selective to the local orientation of image patches. In a shading gradient these will respond most strongly to the local isophote orientation; ie, it's tangent. A tangential penetration across V1 yields a column of cells. (middle) Abstracting the column of cells over a given (retinotopic) position spans all orientations; different orientations at nearby positions regularize the tangent map to the isophotes and reduce noise in local measurements. (right) Illustrating the lift of the isophones into a V1 style representation. The mathematical analysis in this paper extends this type of representation to the surface inference prolbem. As such it could be implemented by similar cortical machinery in higher visual areas.

could be implemented by a combination of feedforward and feedback projections, supplemented with the long-range horizontal connections within each visual area, although we do not develop these connections in any detail here. Rather, we concentrate on the calculations and how how a crisp curvature structure emerges from the transport equations. As such it serves as the foundation of a model for understanding feedforward connections to higher levels (surfaces) from lower levels (flows).

3.1.2 Overview

Our approach is summarized in Fig. 3.4. Rather than working globally and attempting to infer a full surface directly from the image and (e.g.) global light source priors, we think of the surface as a composite of local patches (charts). Each patch is described by its (patch of) shading flow, each of which implies a space of (surface patch, light source) pairs. Much of the formal content of this paper is a way to calculate them.

The story, of course, gets most interesting when we consider how to put the patches together. Just as the orientations possible at a position define a fibre bundle, the possible local surface patches define a "fibre" for each patch coordinate; again, over the surface these fibres form a bundle. Conceptually the shape-from-shading problem amounts to finding a section through this bundle. Once a section is obtained, the light source positions emerge; i.e., can be calculated directly.

There are several advantages to this approach.

- Ambiguity now is a measure on these fibres, and it can be reduced by certain (local) conditions, for example curvature at the boundary [27,28]. Thus it is consistent with Marr's Principle of Least Committment.
- The light source positions are essentially an emergent property rather than a prior assumption.
- Mathematically our approach mirrors the composite nature of visual inverse problems: there are those configurations in which solutions are nicely defined, and there are others that remain inherently ambiguous. A powerful illustration of this is provided by artists' drawings: a single new stroke may change the impression completely, such as the indication of a highlight, while others may be lost in the cross-hatching of shading. This lead us to consider the shading flows near critical points and contours as fundamental conditions for the global surface perception.

We now briefly discuss the first stage to our approach – the shading flow field – and follow this with the surface inference calculations.

3.2 The Shading Flow

Consider, from an abstract perspective, how a shaded image would be represented in visual cortex. To start, we note that a smooth surface patch under diffuse Lambertian lighting and orthogonal projection yields smooth image curves of constant brightness. The shading flow derives from these level curves of image intensity $I(x,y)$, or isophotes ([17, 18]). In particular, consider image cells with orientated receptive fields tuned to low spatial frequencies: the strongest responses will be from those

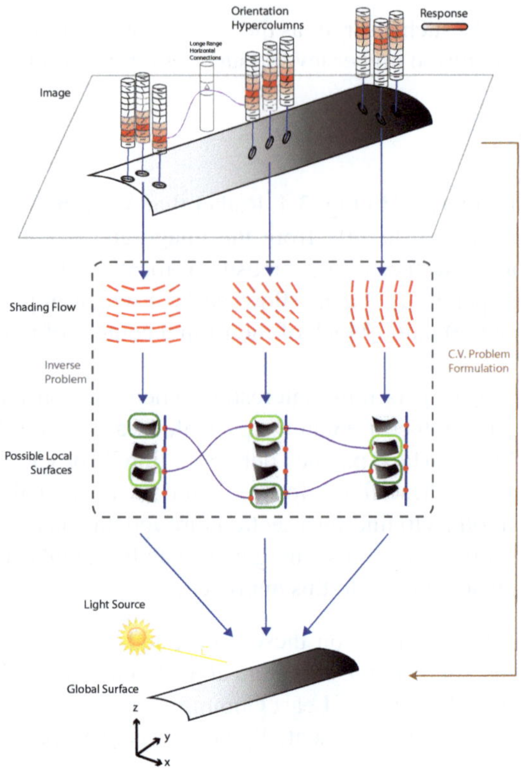

Fig. 3.4 Our approach to shape from shading. Instead of inferring surfaces directly from the image, we impose two fibre bundles between them. The first is the lift of the image into the shading flow, and the second defines the fibre of possible surface patches that are consistent with a given patch of shading flow. We do not assume a light source position, but instead will use assumptions on certain local features to restrict the ambiguity. This amounts to finding a section through the bundle of possible local surfaces. The light source position(s) are then an emergent property.

cells with their excitatory zones on brighter portions of the shading, say, and their inhibitory zones on the darker ones. It is these cells that signal the isophone tangents.

To construct a vector field $V(x,y)$, we quantize these curves by taking their tangents over a predetermined image coordinate grid. We call this vector field of isophote tangents the *shading flow field* and it arose earlier in the literature [2]. In the limit, as the spacing of the grid points goes to zero, integral curves of the shading flow are precisely the intensity level curves. Our goal is to use this 2D vector field to restrict the family of (surface, light source direction) pairs that could have resulted in the image (Fig. 3.7). We shall also require the complementary vector field of brightness gradients. In the limit, these two vector fields of isophote tangents and brightness gradients together encapsulate the same information as the pixel values of the image. For computational work on regularizing and calculating the shading flow, see [2].

Working with the shading flow has a number of advantages over working with the raw image intensities.

- Orientation-selective cells [13] could readily provide the lift for computing it as early as V1 (Figure 3.3).
- The shading flow field is lower dimensional than a pixel representation and is invariant to overall contrast changes.
- There exists psychophysical evidence for the use of oriented flow in surface perception [8, 9] for textures.
- The shading flow is invariant across important transformations in both image space and shape space.
- It allows for the use of machinery from vector calculus on surfaces. Although the same calculus can of course be done point wise, the mathematical intuition and equations are cleaner by working with vector fields.

We now expand on a few of these points.

3.2.1 Psychophysical Evidence

Fleming et al., among others, have focused on psychophysical work on orientation fields [8,9]. He has shown that orientations are stable on specular surfaces and often lie along directions of minimal second derivatives, regardless of the environment map [9]. He then showed that "smearing" white noise via line-integral-convolution can lead to 3D shape perception [8]. Thus, taking noise and adding in an orientation structure yields a percept. Most importantly, it can be shown that 3D shape from shading perception can be preserved even in the presence of certain non-monotonic intensity transformations [10]. Orientation fields are thus a stable and basic structure for shape perception.

3.2.2 Intensity Transformations

There are computational reasons to use the shading flow as a intermediate representation rather than working directly with pixel values. First, it regularizes certain errors due to noise in images [2]. Second, both the isophote tangents and curvatures are invariant under arbitrary monotonic transformations of the intensity [2,21]. These types of transformations include not only simple scalings (albedo changes), but even complex transformations such as the ones in Fig. 3.5.

3.2.3 Surface Scaling

The previous paragraph illustrated invariances with respect to image transformations. But the shading flow is also stable under various surface transformations. Here, we consider the case of a scaling in surface heights. See Fig. 3.6.

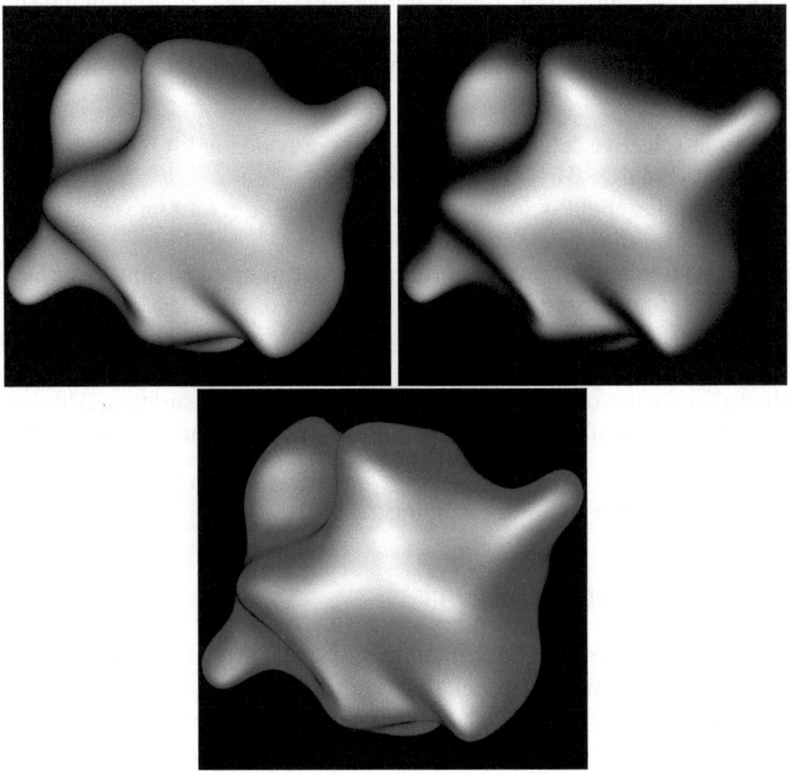

Fig. 3.5 These figures illustrate a shaded surface (left) and two monotonic transformations of intensity (center, right). Although the intensity values change in each of these images, the 3D shading percept remains essentially invariant. This provides some evidence that the surface inference is based on the geometry of isophotes rather than image intensities.

To build intuition, we illustrate the relevant phenomenon mathematically using the simplifying assumption that the light source is behind the viewer. Consider a surface $S(x,y)$ and various scalings of that surface defined by $T : S(x,y) \to cS(x,y), c \in \mathbb{R}^+$. The direction of the isophote tangent at point $\mathbf{p}$ is dependent on the second fundamental form II and the projected light source vector $\mathbf{l_t}$. (We will prove this in the Analysis Section).

Thus, by considering the effects of the scaling on II and $\mathbf{l_t}$, we can understand the scaling effect on the isophote tangents. The projected light source will lie along the gradient of the surface. Thus, its direction will not change by any scaling transformation.

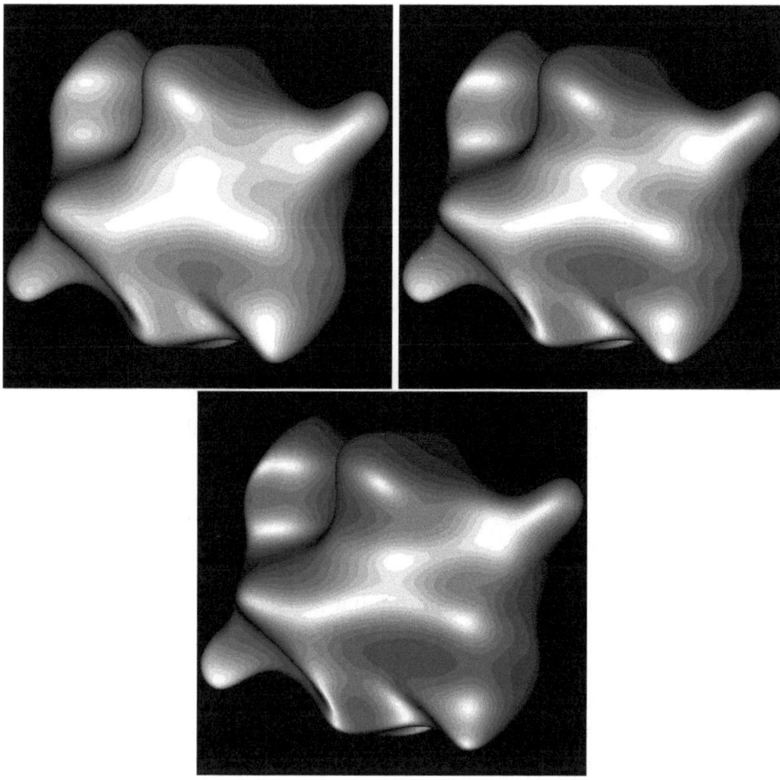

Fig. 3.6 These figures illustrate the isophotes of a shaded surface (left), a 150% scaling of it (center), and a 200 % scaling (right). 15 isophotes have been shown for each surface. Note that the curvature of the isophotes is generally stable throughout the scaling, with the exceptions being due to the singularities of new highlights. Since our percept hardly changes over these images, again this provides evidence that the surface inference is based on the geometry of isophotes rather than image intensities.

Similarly, the second fundamental form – II – is defined by:

$$II = \begin{bmatrix} \mathbf{N}\cdot f_{xx} & \mathbf{N}\cdot f_{xy} \\ \mathbf{N}\cdot f_{xy} & \mathbf{N}\cdot f_{yy} \end{bmatrix} = \begin{bmatrix} \frac{f_{xx}}{\sqrt{1+f_x^2+f_y^2}} & \frac{f_{xy}}{\sqrt{1+f_x^2+f_y^2}} \\ \frac{f_{xy}}{\sqrt{1+f_x^2+f_y^2}} & \frac{f_{yy}}{\sqrt{1+f_x^2+f_y^2}} \end{bmatrix}$$

Any scaling transformation multiplies the derivatives $\{f_x, f_y, f_{xx}, f_{xy}, f_{yy}\}$ by c. Note that T will scale II by a factor of $\frac{c\sqrt{1+f_x^2+f_y^2}}{\sqrt{1+(cf_x)^2+(cf_y)^2}}$. In fact, in the cases where $f_x, f_y \approx 0$ or $f_x, f_y >> 1$, the scaling of II is minimal.

Since both the projected light source and II are only scaled under the transformation T, the shading flow remains unchanged. Note that the magnitude of the

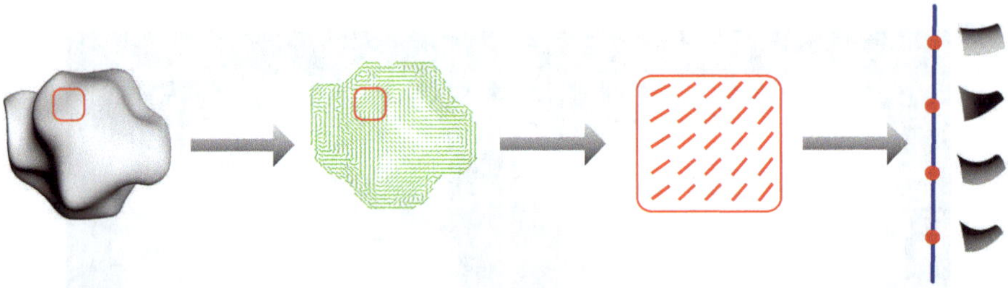

Fig. 3.7 This figure represents the workflow going from image to shading flow to a set of local surfaces. Each surface along the fiber needs a particular light source position to correspond to the given shading flow.

brightness gradients will increase, however. (If we remove the light source behind viewer assumption, it is possible for the shading flow to change, but not generically.)

3.2.4 *Critical Curves*

Regardless of the the image representation that is used to ground the computation, the shape from the shading (flow) problem will remain ill-posed. This is a global statement. However, at certain points on a surface, the ambiguities will collapse in dimension, and we believe understanding which points these are is critical. For example, along the boundary of a smooth object, the view vector lies in the tangent plane [16]. Along a suggestive contour [4], the dot product of the normal vector and the view vector is at a local minimum in the direction of the viewer. At a generic highlight, the dot product of the light source direction and normal vector is at a local maximum. *All of these special types of points are identifiable in the image and provide additional geometric information that reduces the shading inference ambiguity locally.* These points are also important perceptually. This leads us to a novel plan for reconstruction: First, parametrize the shading ambiguity in the general shading case. Then, locate the points where the shading ambiguity vanishes (or reduces greatly) and solve for the local surface shape at those points. Finally, calculate the more complex regions via a compatibility technique [14] or interpolation. Understanding these critical points is one of the most important implications of the analysis that we present next.

3.3 Geometric Analysis of Shape Inference

Our mathematical goal is to translate the problem into the local tangent plane and then use the machinery of covariant derivatives and parallel transport to represent image derivatives (see Fig. 3.9) as a function of the surface vector flows. A similar use of this machinery was developed for shape from texture in [12].

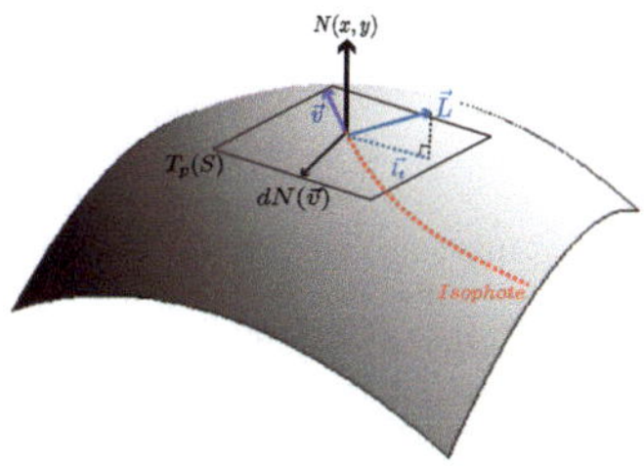

Fig. 3.8 A diagram explaining our defined surface properties

1. We write the brightness gradient and isophote as tangent plane conditions between the projected light source and shape operator.
2. We take the covariant derivative of the projected light source and show it is independent of the direction of the light source.
3. We take the covariant derivative of the isophote condition and separate into the differentiation on the projected light source and the differentiation on the shape operator.
4. We separate the image derivatives into functions of the image and surface properties without reference to the light source direction.

The Lambertian lighting model is defined by:

$$I(x,y) = \rho \mathbf{L} \cdot N(x,y)$$

Consider a small image patch under Lambertian lighting from an unknown light source. This image patch corresponds to a local surface patch. Using Taylor's theorem we represent this as $S = \{x, y, f(x,y)\}$ with $f(x,y) = c_1 x + c_2 y + c_3 x^2 + c_4 xy + c_5 y^2 + c_6 x^3 + c_7 x^2 y + c_8 xy^2 + c_9 y^3$.

Our goal is to understand the derivatives of intensity in terms of the coefficients $\{c_i\}$. It is essential that the order of the Taylor polynomial must be 3 since we shall consider second derivatives of image intensity and intensity is already dependent (via Lambertian lighting) on the first order derivatives of the surface. Other analyses of SFS only consider 2nd order Taylor approximations [26].

For reference, we define our complete notation in the Table 3.1. The symbols will be introduced throughout the analysis. $V(x,y)$ is the shading flow field. We normalize $V(x,y)$ to be of unit length in the image, although the corresponding surface tangent vectors have unknown length. We denote unit length vectors in the image plane with a vector superscript, such as $\mathbf{v}$. The corresponding vectors on the surface tangent plane are defined by the image of $\mathbf{v}$ under the map composition of the differential $df : \mathbb{R}^2 \to \mathbb{R}^3$ and the tangent plane basis change $T : \mathbb{R}^3 \to T_p(S)$. We will use the hat superscript to denote these surface tangent vectors, e.g. $\hat{v}$.

Thus,

$$\hat{v} = T \circ df(\mathbf{v})$$

Because the computation is somewhat involved, we break it up into steps.

Table 3.1 Notation Table

p	a chosen point (x_0, y_0)
$I_p(x,y)$	an image patch centered at p
$\nabla I(x,y)$	the brightness gradient
$S_p(x,y)$	the corresponding (unknown) surface patch
$f(x,y)$	the Taylor approximation at p of S
$\{c_i\}$	the coefficients of the Taylor approximation $f(x,y)$
$T_p(S)$	the tangent plane of S at p
$\mathbf{L}$	the light source direction
$\mathbf{l_t}(p)$	the projection of the L onto the tangent plane
$\mathbf{e_i}$	unit length standard basis vector in direction of coordinate axis i
$N(x,y)$	the unit normal vector field of S
$V(x,y)$	the vector field of isophote directions at each point (x, y)
$\mathbf{v} \in T_p(S)$	the image unit length tangent vector in the direction of the isophote at p
$\mathbf{u} \in T_p(S)$	the image unit length tangent vector in the direction of the brightness gradient at p
$\hat{w} \in T_p(S)$	the tangent vector in direction $\mathbf{w}$ of unit length in the image, expressed in the surface tangent basis
$\mathbf{u}[V]$	the directional derivative of the vector field V in the direction $\mathbf{u}$
$\nabla_{\mathbf{u}} V$	the covariant derivative (on the surface) of the vector field V in the direction $\mathbf{u}$
$D_{\mathbf{u}} V$	the directional derivative (on the image) of the vector field V in the direction $\mathbf{u}$
G	the first fundamental form (also called the metric tensor)
II	the second fundamental form
H	the Hessian
dN	the differential of the Gauss Map, also called the Shape Operator

3.3.1 Calculating the Brightness Gradient

We derive the equations for the brightness gradient as a function of the light source and the second fundamental form. Similar derivations (with different notation) appear in [17].

The brightness gradient ∇I can be defined as a linear 1-form having as input unit length image vectors $\mathbf{w}$ and having as output a real number. The output is the change in brightness along a step on the surface using $\hat{w}$. We write:

$$\nabla I \cdot \mathbf{w} = \hat{w}\left[\langle \mathbf{L}, \mathbf{N} \rangle\right] \tag{3.1a}$$
$$= \langle (\nabla_{\hat{w}} \mathbf{L}), \mathbf{N} \rangle + \langle \mathbf{L}, (\nabla_{\hat{w}} \mathbf{N}) \rangle \tag{3.1b}$$
$$= 0 + \langle \mathbf{L}, dN(\hat{w}) \rangle \tag{3.1c}$$
$$= \langle \mathbf{l_t}, dN(\mathbf{w}) \rangle \tag{3.1d}$$
$$= \mathbf{l}^T II \hat{w} \tag{3.1e}$$

where the first term in equation (3.1b) is zero because the light source is fixed.

Proposition 3.1. *The brightness gradient ∇I can be expressed as the vector $\mathbf{l}_t^T II$.*

Along an isophote surface curve $\alpha(t)$, the brightness is constant. Writing $\mathbf{v} = \alpha'(0)$, we have $\langle \mathbf{l_t}, dN(\mathbf{v}) \rangle = \mathbf{l}^T II \hat{v} = \nabla I \cdot \mathbf{v} = 0$.

Thus, we conclude:

Proposition 3.2. *Each isophote tangent vector $\mathbf{v}$ on S is a function of the normal curvatures and light source and is defined by the equation*

$$\langle \mathbf{l_t}, dN(\mathbf{v}) \rangle = 0.$$

In addition, we calculate each component of the brightness gradient via dot product with $\mathbf{e_i}$.

$$I_x = \langle \mathbf{l_t}, dN(\mathbf{e_1}) \rangle \tag{3.2}$$

$$I_y = \langle \mathbf{l_t}, dN(\mathbf{e_2}) \rangle \tag{3.3}$$

3.3.2 Calculating the Covariant Derivative of Projected Light Source

One of the major advantages of our approach is that we do not need to assume a known light source direction. In fact, using the covariant derivative described below, we can calculate the change in the projected light source vector without knowing where it is!

We briefly remark on the use of covariant derivatives for surfaces in $\mathbb{R}^2$. We consider "movements" in the image plane and syncronize them with "movements" through the tangent bundle of the surface. The difficulty is that the image plane vectors lie on a flat surface, whereas the vectors on the surface tangent planes "live" in different tangent spaces: the surface tangent planes are all different orientations of $\mathbb{R}^2$ in $\mathbb{R}^3$. Thus, to calculate derivatives via limits of differences, we need to parallel transport nearby vectors to a common tangent plane, and this is done with the covariant derivative. We think of the covariant derivative in two ways. The first definition, which we use in this section, is the expression as the composition of a derivative operator in $\mathbb{R}^3$ and a projection operator onto a tangent plane. This is an *extrinsic definition* – it is a definition that requires use of the ambient space. The second definition, which we will use in the following section, will be in terms of parallel transport.

We exploit the structure in $\mathbf{l}_t$: it is the result of a projection from a fixed vector L down into the tangent plane $T_p(S)$. Thus, the change in $\mathbf{l}_t$ just results from changes in the tangent plane, which is dependent only on the surface curvatures and not on L. Importantly, we avoid having to represent L in our calculations by only considering its projected changes. We now show this rigorously.

Lemma 3.1. *The covariant derivative of the projected light source is only dependent on the position of the light source through the observed intensity. Thus,*

$$\nabla_{\mathbf{u}} \mathbf{l_t} = -(\mathbf{L} \cdot \mathbf{N}) dN(\mathbf{u}).$$

Proof. Let Π_{p_0} be the projection operator taking a vector in 3-space onto the tangent plane of S at p_0. Recall that the covariant derivative of a tangent vector can be expressed as the composition of a derivative operator and Π.

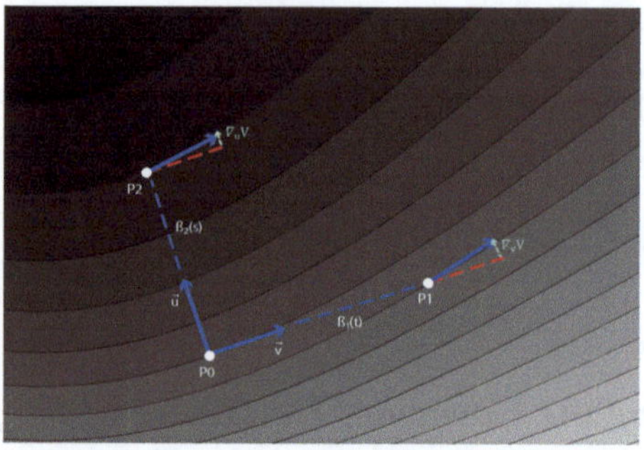

Fig. 3.9 A diagram explaining our use of the shading flow field. As we move on $\beta_1(t)$ in the direction of the isophote $\mathbf{v}$ from $P0$ to $P1$, the flow field $V(x,y)$ changes by $\nabla_{\mathbf{v}}V$. Similarly, we may move in direction $\mathbf{u}$ along $\beta_2(s)$, which is perpendicular (in the image) to the isophote. Then, our flow field changes by $\nabla_{\mathbf{u}}V$. In Proposition 2, we relate these changes in closed form to the curvatures of the surface and the light source direction.

$$
\begin{aligned}
\nabla_{\mathbf{u}}\mathbf{l_t} &= \Pi_{p0}\left(\frac{d\mathbf{l_t}}{dt}\right) && \text{(3.4a)}\\
&= \Pi_{p0}\left(\frac{d}{dt}(\mathbf{L}-(\mathbf{L}\cdot\mathbf{N})\mathbf{N})\right) && \text{(3.4b)}\\
&= \Pi_{p0}\left(\frac{d\mathbf{L}}{dt}-\frac{d}{dt}[(\mathbf{L}\cdot\mathbf{N})\mathbf{N}]\right) && \text{(3.4c)}\\
&= \Pi_{p0}\left(0-\frac{d}{dt}[\mathbf{L}\cdot\mathbf{N}]\mathbf{N}-(\mathbf{L}\cdot\mathbf{N})\frac{d\mathbf{N}}{dt}\right) && \text{(3.4d)}\\
&= \Pi_{p0}\left(-\left[\frac{d\mathbf{L}}{dt}\cdot\mathbf{N}+\mathbf{L}\cdot\frac{d\mathbf{N}}{dt}\right]\mathbf{N}-(\mathbf{L}\cdot\mathbf{N})dN(\mathbf{u})\right) && \text{(3.4e)}\\
&= -(\mathbf{L}\cdot\mathbf{N})dN(\mathbf{u}) && \text{(3.4f)}
\end{aligned}
$$

The fact that this change in the projected light source only depends on surface properties (along with the measurable image intensity) allows us to remove the light source dependence from the second derivatives of intensity as defined by $\{D_{\mathbf{v}}\mathbf{v}, D_{\mathbf{u}}\mathbf{v}, D_{\mathbf{u}}\mathbf{u}\}$.

3.3.3 Covariant Derivative of the Isophote Condition

We now use the changes in the brightness gradient and the isophote directions to restrict our surface parameters. Let $\mathbf{v}$ be the unit length image vector in the direction of the isophote at an arbitrary point p. Let $\mathbf{u}$ be the unit length image vector in the direction of the brightness gradient at p. In the image, $\mathbf{v} \perp \mathbf{u}$ but the projected vectors $\hat{v}$ and $\hat{u}$ may not be orthogonal on the tangent plane at p.

In fact, considering these particular changes in $\mathbf{u}$ and $\mathbf{v}$ is equivalent to choosing a basis. This will result in solving for equations of the three second derivatives $\{D_\mathbf{v}\mathbf{v}, D_\mathbf{u}\mathbf{v}, D_\mathbf{u}\mathbf{u}\}$, although we could have considered the changes in $\{I_x, I_y\}$ and instead solved for $\{I_{xx}, I_{xy}, I_{yy}\}$. However, the equations simplify when choosing the basis defined by the isophote and brightness gradient.

To emphasize the conceptual picture, we will derive the $D_\mathbf{v}\mathbf{v}$ equation here. The remaining equations can be found in [24].

We start by first calculating I_v and then taking the directional derivative of I_v in the direction $\mathbf{v}$. From Section 3.3.1, we can write:

$$0 = I_v = \nabla I \cdot \mathbf{v} = \langle \mathbf{l_t}, dN(\mathbf{v}) \rangle \tag{3.5}$$

Applying the directional derivative with respect to $\mathbf{v}$ on both sides and using the result from Equation 3.4f:

$$\begin{aligned}
0 &= v[\langle \mathbf{l_t}, dN(\mathbf{v}) \rangle] && (3.6a)\\
&= \langle (\nabla_\mathbf{v} \mathbf{l_t}), dN(\mathbf{v}) \rangle + \langle \mathbf{l_t}, \nabla_\mathbf{v} dN(\mathbf{v}) \rangle && (3.6b)\\
&= \langle -(\mathbf{L} \cdot \mathbf{N}) dN(\mathbf{v}), dN(\mathbf{v}) \rangle + \langle \mathbf{l_t}, \nabla_\mathbf{v} dN(\mathbf{v}) \rangle && (3.6c)\\
&= -I \langle dN(\mathbf{v}), dN(\mathbf{v}) \rangle + \langle \mathbf{l_t}, \nabla_\mathbf{v}(dN(\mathbf{v})) \rangle && (3.6d)
\end{aligned}$$

We now unpack $\langle \mathbf{l_t}, \nabla_\mathbf{v} dN(\mathbf{v}) \rangle$ which requires a technical computation using parallel transport and tensor algebra. Due to space constraints, we will state the simplification rather than derive it. However, the derivation can be found in [24].

We recall the second definition of covariant differentiation here. We define it intrinsically, that is, independent of the ambient space $\mathbb{R}^3$. We will not go into the derivations regarding *connections* or Christoffel symbols, which can be found in [7] and [6]. We just summarize that *parallel transport* is a way to "equate" nearby vectors in nearby tangent planes along a curve $\beta(s)$. Using notation as in [7], we will write the parallel transport in the forward direction of the vector field $\mathbf{w}(\beta(s))$ as $\tau_s^{\rightarrow}(\mathbf{w}(\beta(s)))$. Conversely, the parallel transports backwards along the curve is written $\tau_s^{\leftarrow}(\mathbf{w}(\beta(s)))$. Then, the covariant derivative can be defined intrinsically as:

$$\nabla_{\beta'(0)} \mathbf{w} = \lim_{s \to 0} \left(\frac{\tau_s^{\leftarrow}(\mathbf{w}(\beta(s)) - \mathbf{w}(\beta(0))}{s} \right)$$

Thus, covariant differentiation resolves the tangent plane orientation problem by first transporting the vector $\mathbf{w}(\beta(s)) \in T_{\beta(s)}(S)$ back to a "parallel" vector in $T_{\beta(0)}(S)$ before doing the standard derivative subtraction.

Now, we will need to parallel transport the operator dN in addition to the parallel transport of $\mathbf{v}$. Luckily, we can use the fact that dN can be represented as a $(1,1)$ tensor. Thus, the parallel transport of dN can be represented as a sum of tensor products of parallel transports on vectors and 1-forms. We use this in [24] in order to obtain the following simplification.

Simplification of (3.6d) using parallel transport yields the following equation:

$$\nabla I \cdot D_\mathbf{v}\mathbf{v} = -(\mathbf{L} \cdot \mathbf{N}) ||dN(v)||^2 + \nabla I H^{-1}(\mathbf{v}[H]\mathbf{v})$$

3.4 The Second-Order Shading Equations

We have now computed the covariant derivative of the vector **v** in the direction **v**. For an arbitrary point p, let **u** be the image vector of unit length in the direction of the brightness gradient. Then, we can repeat this calculation for the covariant differentiation of **v** in the direction **u**. In addition, we can calculate the covariant derivative of the vector **u** in the direction **u**. Both of these proofs are similar to the one above. This gives us a total of three equations equating the second order intensity information (as represented in vector derivative form) directly to surface properties.

Theorem 3.1. *For any point p in the image plane, let $\{\mathbf{u},\mathbf{v}\}$ be the local image basis defined by the brightness gradient and isophote. Let I be the intensity, ∇I be the brightness gradient, $f(x,y)$ be the height function, H be the Hessian, and dN be the shape operator. Then, the following equations hold regardless of the light source direction:*

$$\nabla I \cdot D_{\mathbf{v}}\mathbf{v} = -I||dN(\mathbf{v})||^2 + (\nabla I)\cdot H^{-1}(\mathbf{v}[H]\mathbf{v}) \tag{3.7}$$

$$\nabla I \cdot D_{\mathbf{u}}\mathbf{u} = -I||dN(\mathbf{u})||^2 - 2\frac{||\nabla I||}{\sqrt{1+||\nabla f||^2}}\langle \nabla f, dN(\mathbf{u})\rangle + (\nabla I)\cdot H^{-1}\cdot(\mathbf{u}[H]\mathbf{u}) \tag{3.8}$$

$$\nabla I \cdot D_{\mathbf{u}}\mathbf{v} = -I\langle dN(\mathbf{v}), dN(\mathbf{u})\rangle - \frac{||\nabla I||}{\sqrt{1+||\nabla f||^2}}\langle \nabla f, dN(\mathbf{v})\rangle + (\nabla I)\cdot H^{-1}\cdot(\mathbf{u}[H]\mathbf{v}) \tag{3.9}$$

These equations are novel; we call them *the 2nd-order shading equations*. Note that there is no dependence on the light source (except through measurable image properties); thus, these equations directly restrict the derivatives of our local surface patch.

3.5 Simplifications of the Shading Equations

In a generic patch, the second-order shading equations are highly complex and non-linear. In addition, we have only three equations on the third order Monge patch, which consists of 9 free parameters. Thus, there is a six dimensional local ambiguity. For analysis in the generic case, see [24].

For this reason, we believe the shape from shading problem can – and should – either be solved at certain points in the image (considered next) or should be combined with other means for obtaining tangent plane information, such as a texture flow. In the following section, we consider the above equations at critical points of the intensity.

3.5.1 Ambiguity Reduction at Critical Points

Much work has focused on the question: "Where should one draw lines on a surface in order to give the best impression of the surface shape?" Recently, Decarlo et al. [4] have considered "suggestive contours" and Judd et al. have suggested apparent ridges [15]. How do we decide which feature lines are "better" [3]? Why are certain curves so helpful in psychophysics? We believe that understanding shading ambiguity can provide a useful metric for deciding between different definitions of "shape representing contours." We can also go the other way: we can use the dimension of the shading ambiguity to define contours (or sets of points) where the surface information is mathematically more restricted by the shading than at a generic point, key among these points are highlights and ridges [23], which we will define below.

Consider the points where the intensity is a local maximum or minimum, i.e. $\nabla I = 0$. Koenderink et al. classified the local surface at these critical points [20] in the case where the tangent plane was frontal-parallel and the principal curvatures were known. By using our shading equations, we can generalize significantly. There are two cases: the Gaussian curvature $K \neq 0$ or $K = 0$. In the first case, we get generic highlight points. The second case is even more interesting; we will use the word "ridge" to denote contours consisting of these points.

3.5.2 Reconstruction from Critical Points

Let M be the map (under constant albedo) taking a surface to its respective image. The shading equations define the inverse image M^{-1}. Now, we are analyzing M^{-1} at specific points in image space. In general, the image under M^{-1} will be high dimensional and complex; however near ridges and highlights, we will see that the inverse image is simply structured. It is possible the human visual system "understands" these relationships at such special points and uses them to anchor the shape from shading inference. For example, given a solution at the critical points interpolation could be used to infer (or otherwise fill-in) the surface between them.

A necessary condition behind this hypothesis is that the image information at critical points is (essentially) complete; that is, is sufficient to fill-in the remainder of the image information. To test this we devised a simple reconstruction experiment. Given a shaded image, we considered the shading flows only in the neighborhoods of the highlights, ridges, and occluding contour. We then linearly interpolated [5] between these critical points to gain the shading values on the rest of the surface. See Figs. 3.10, 3.2.

By and large, the reconstructed image is very close to the original image in both shape percept and intensities. This is evidence that shape from shading could be done by just understanding the shading at critical points because (at least) the remaining image structure could be filled in. We conjecture that the same result holds at the surface level. That is, we conjecture that the visual system may be able to build a robust shape percept using just the simple relationships described below at the highlight and ridge cases and then "filling in" between them in a consistent fashion. As an aside, we find it hard to believe the visual system implements any of the

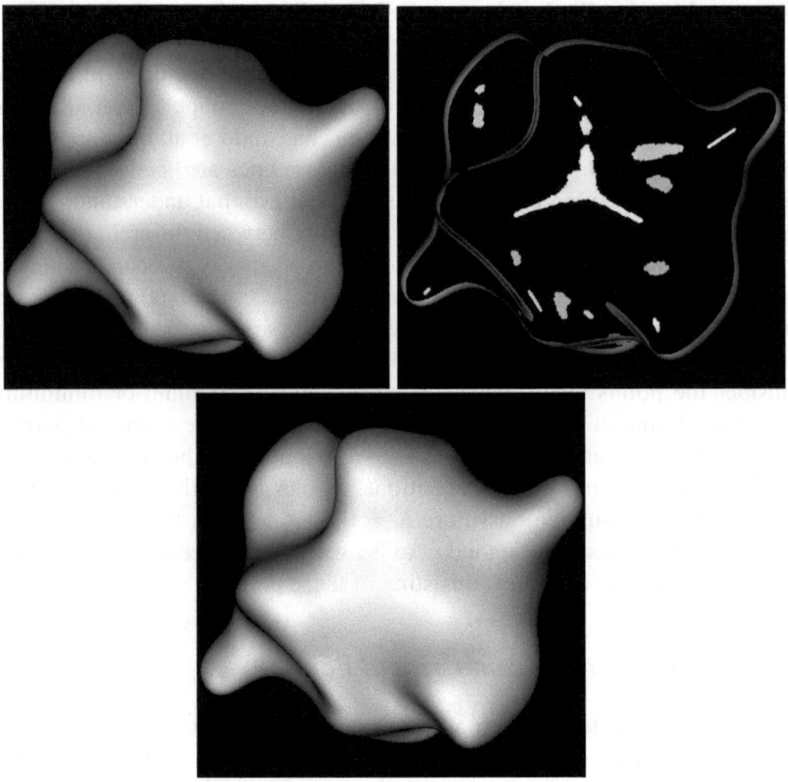

Fig. 3.10 Left – a shaded surface. Center – shading values in 4x4 pixel neighborhoods surrounding critical points. Right – reconstructed shading values via a linear interpolation algorithm [5] using only the shading values in the center figure. Note that the 3D percept for both Left and Right is nearly identical.

current computer vision shape from shading algorithms, due to the extreme complexity and arbitrary priors used in many of them. Note that a linear interpolation of intensity values can be achieved by a quadratic interpolation of normal vectors.

3.5.3 Highlights

We now analyze the two cases of critical points. First, consider the generic case when the Gaussian curvature is not zero, so H^{-1} is well-defined. Since $l_t^T II = \mathbf{0}$, and II is not singular, $l_t = \mathbf{0}$. This is the case of normal incidence of the light source. Then our equations 3.7, 3.8 , 3.9 simplify:

$$\nabla I \cdot D_{\mathbf{v}}\mathbf{v} = -(I)||dN(\mathbf{v})||^2 \tag{3.10a}$$

$$\nabla I \cdot D_{\mathbf{u}}\mathbf{u} = -(I)||dN(\mathbf{u})||^2 \tag{3.10b}$$

$$\nabla I \cdot D_{\mathbf{u}}\mathbf{v} = -(I)\langle dN(\mathbf{v}), dN(\mathbf{u})\rangle \tag{3.10c}$$

At these points, the second derivatives of intensity relate directly to the total change in normal for a unit step in the image. This cursory analysis may explain why highlight lines are so effective at revealing surface shape psychophysically: A simple rule relates the image derivatives to the surface derivatives. Essentially, the widths of the highlights are proportional to the appropriate surface curvatures. See Fig 3.11. Although one does not gain information about the tangent plane, due to the unknown light source(s), one has information about the curvatures as seen from the surface. Once the tangent plane has been found, these equations tell us the second order shape properties.

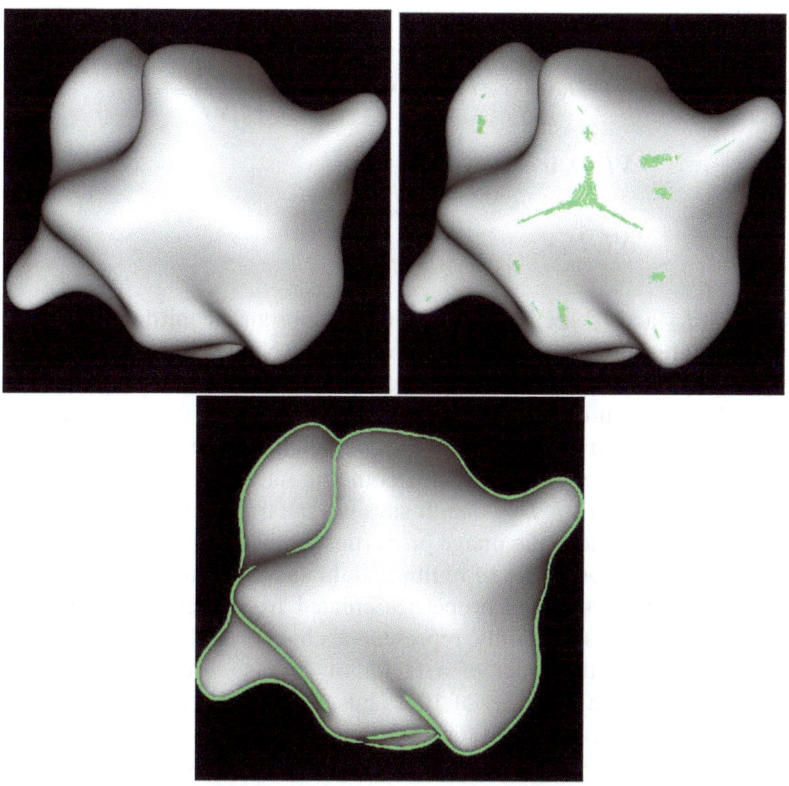

Fig. 3.11 Left – a shaded surface. Center – the green pixels represent the highlight points, determined via the gradient of the image ($\nabla I < \varepsilon$). Right – green pixels represent points on either an occluding boundary or a ridge, calculated via a Laplacian filter.

3.5.4 Ridges

The next class of critical points are ridges. In this work, we use the term "ridge" as an image contour of connected points that are all either local minima or maxima of intensity. However, this time, we assume $K = 0$. Ridges are very salient points of the image; they can often be isolated via a Laplacian filter. See the interior contours in Fig 3.11. Along these contours, we get a simplification of the shading equations.

Let $\mathbf{w} \in T_pS$ denote the unknown nonzero principal direction; $\mathbf{w}$ corresponds to the major axis for the locally cylindrical Taylor approximation. For simplicity, we can define a Frenet basis for the image contour and so express the surface derivatives in that $\{\mathbf{u}, \mathbf{v}\}$ basis. Let $\mathbf{l_t}$ be expressed as $\{l_1, l_2\}$ in this basis. Although $\mathbf{w}$ is unknown, at highly foreshortened tangent planes (as is the case on the ridges in this example), tangent vectors project to either $\{\mathbf{u}, \mathbf{v}\}$, up to the visual system's resolution. Without loss of generality, suppose $\mathbf{w} \approx \mathbf{u}$. (The other possible approximation $\mathbf{w} \approx \mathbf{u}$ leads to analogous equations.) Then, the shading equations again reduce after some algebra:

$$\nabla I \cdot D_{\mathbf{v}}\mathbf{v} = -(I)||dN(\mathbf{v})||^2 + \frac{l_2 f_{vvu}}{\sqrt{1+||\nabla f||^2}} \tag{3.11a}$$

$$\nabla I \cdot D_{\mathbf{u}}\mathbf{u} = \frac{l_2 f_{uuu}}{\sqrt{1+||\nabla f||^2}} \tag{3.11b}$$

$$\nabla I \cdot D_{\mathbf{u}}\mathbf{v} = \frac{l_2 f_{vuu}}{\sqrt{1+||\nabla f||^2}} \tag{3.11c}$$

Thus, both I_{uu} and I_{uv} are proportional to the appropriate third derivatives of the surface, once foreshortening has been accounted for. Note that the foreshortening of the tangent plane and the light source make up the coefficient of proportionality. In particular, the light source only affects I_{uu} and I_{uv} by scaling the necessary f_{uuu} and f_{vuu}. Thus, the ambiguity for several of the coefficients of the Taylor approximation is only a matter of scale. To gain some geometric intuition, see Fig 3.12. The interesting equation is the $\nabla I \cdot D_{\mathbf{v}}\mathbf{v}$ one. It says we can trade off weighted versions of $||dN(\mathbf{v})||$ and f_{vvu} and keep the image properties the same.

We remark on the possibility of multiple light sources. Due to linearity, one can simply sum the various image properties created from each light source. Thus, additional light sources can only represent a change in the value l_2 and often these changes are on the order of a factor of 2 or so. However, the values of $\{D_{\mathbf{v}}\mathbf{v}, D_{\mathbf{u}}\mathbf{v}, D_{\mathbf{u}}\mathbf{u}\}$ are on the order of 10^3. Thus, these ridges tend to be remarkably stable under addition of new light sources.

Although we do not claim that the shading outside the critical points is completely irrelevant, there is significant literature on how contours can "trump" shading information [22, 27, 28]. In addition, it is clear that contours alone can lead to a rich 3D percept: simply view famous artists' line drawings, etchings, etc! Finally, we note that shape perception will also be ambiguous; the exact nature of the surface is dependent on the task and the "beholder's share" [19]. It is not possible and maybe not even be desirable to calculate the precise depth values of the surface. Rather, a

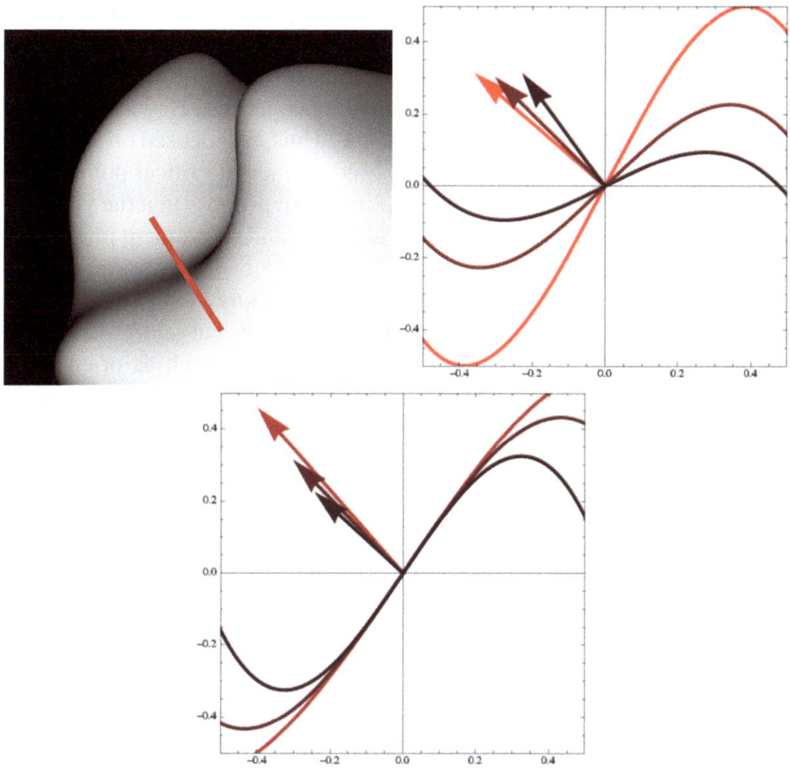

Fig. 3.12 Left – a portion of our shaded surface. The red line defines a normal plane, which indicates a cross section (x-axis) in Center and Right. Center – The curves represent various possible cross sections resulting in the same I_{uu} value. The arrows represent the necessary light source. Note the tangent plane changes. Right – Various cross sections and associated light sources. Here, the tangent plane stays fixed, but the projected light source changes. These two types of transformations generate the possible cross sections.

preferred output representation may be the local quadric class [17, 28], which is a more flexible and less precise representation. We believe studying the critical shading points will lead to mathematically natural ways to segment the surface according to these quadric classes.

3.6 Conclusion

The differential invariants of surfaces are curvatures. Thus a natural framework for formulating surface inferences is in terms of differential geometry. We here propose such a framework, by lifting the image information to a vector field (the shading flow field) and formulating the shape-from-shading problem on it. Our goal is to find those (surface, light source) pairs that are consistent with a given shading flow.

Working with simplifying assumptions, we develop the basic transport machinery in closed form and calculate the full family of solutions.

On any image patch of a smooth surface, the curvatures of the shading flow restrict the possible local surfaces via these shading equations. However, the ambiguous families are, in general, not easy to describe. Thus, we focused on understanding the relationships between the shading flow and surface at critical points. On these points, the geometric relationships are simplified. We propose that the visual system may use the shading flows near these critical points and subsequent interpolation to gain its "first pass" shape percept.

Finally, we close with a neurobiological point. It is known that the higher visual areas are selective for surface properties, including their curvatures [29]. It is also known that many different forms of orientation images, such as oriented texture noise and glossy patterns (see references in [8]) are perceived as surfaces. To our knowledge the calculations here are the first example of how this inference might take place from the shading flow to surfaces. It thus serves as a common "language" for formulating feedback, but also underlines the need for additional information.

Shape inferences are not done in a vacuum, and the rich interconnectivity of the visual system should reflect the rich mathematical connectivity between surfaces inferences from contour, shading, texture, stereo and motion. In our view thinking of these in neurogeometric terms – vector and tensor fields, transport equations, and differential geometry – may well be the research path to understand them.

References

1. Barron, J., Malik, J.: Shape, illumnation, and reflectance from shading. Technical Report (2013)
2. Breton, P., Zucker, S.W.: Shadows and Shading Flow Fields. In: Proc. IEEE Conf. on Computer Vision and Pattern Recognition, CVPR 1996, pp. 782–789 (1996)
3. Cole, F., Sanik, K., DeCarlo, D., Finkelstein, A., Funkhouser, T., Rusinkiewicz, S., Singh, M.: How well do line drawings depict shape? ACM Trans. Graph. 28 (2009)
4. DeCarlo, D., Finkelstein, A., Rusinkiewicz, S., Santella, A.: Suggestive contours for conveying shape. SIGGRAPH, 848–855 (2003)
5. D'Errico, J.: Interpolating elements in a 2d array. Matlab Central File Exchange (2004)
6. Docarmo, M.P.: Differential Geometry of Curves and Surfaces. Prentice-Hall Inc., Upper Saddle River (1976)
7. Dodson, C.T.J., Potson, T.: Tensor Geometry. Springer, Heidelberg (1991)
8. Fleming, R., Holtmann-Rice, D., Bulthoff, H.: Estimation of 3D Shape from Image Orientations. Proceedings of the National Academy of Sciences 108 (2011)
9. Fleming, R., Torralba, A., Edelson, E.: Specular reflections and the perception of shape. Journal of Vision 4 (2004)
10. Fleming, R., Vergne, R., Zucker, S.W.: Predicting the effects of illumination in shape from shading. Journal of Vision 13 (2013)
11. Freeman, W.T.: The generic viewpoint assumption in a framework for visual perception. Nature 368, 542–545 (1994)
12. Garding, J.: Surface orientation and curvature from differential texture distortion. In: Proc. 5th International Conference on Computer Vision (1995)
13. Hubel, D.: Eye, Brain, and Vision. Scientific American Library (1988)

14. Hummel, R., Zucker, S.: On the Foundations of Relaxation Labeling Precesses. IEEE Transactions on Pattern Analysis and Machine Intelligence, PAMI-5 (1983)
15. Judd, T., Durand, F., Adelson, E.H.: Apparent ridges for line drawing. ACM Trans. Graph. 26, 19 (2007)
16. Koenderink, J.J.: What does the Occluding Contour tell us about Solid Shape? Perception 13, 321–330 (1984)
17. Koenderink, J.J.: Solid Shape. The MIT Press, Cambridge (1990)
18. Koenderink, J.J., Van Doorn, A.J.: Photometric invariants related to solid shape. Optica Acta 27, 981–996 (1980)
19. Koenderink, J.J., Van Doorn, A.J., Kappers, A.M.L.: Ambiguity and the 'mental eye' in pictorial relief. Optica Acta 30, 431–448 (2001)
20. Koenderink, J.J., van Doorn, A.J.: Illuminance critical points on generic smooth surfaces. J. Opt. Soc. Am. A 10 (1993)
21. Koenderink, J.J., van Doorn, A.J.: Two-plus-one-dimensional differential geometry. Pattern Recognition Letters 15, 439–443 (1994)
22. Koenderink, J.J., van Doorn, A.J., Wagemans, J.: Sfs? not likely! i-Perception 4, 299–302 (2013)
23. Kunsberg, B., Zucker, S.W.: Shape-from-shading and cortical computation: a new formulation. Journal of Vision 12 (2012)
24. Kunsberg, B., Zucker, S.W.: Characterizing ambiguity in light source invariant shape from shading. SIAM Journal of Imaging Sciences (2013) (submitted)
25. Mach, E.: On the Physiological Effect of Spatially Distributed Light Stimuli (Transl, F. Ratliff). In: Mach Bands: Quantitative Studies on Neural Networks in the Retina. Holden Day, San Francisco (1965)
26. Pentland, A.: Local Shading Analysis. IEEE Transactions on Pattern Analysis and Machine Intelligence, PAMI-6 (1984)
27. Ramachandran, V.S.: Perceiving shape from shading. Scientific American 259, 76–83 (1988)
28. Wagemans, J., Van Doorn, A.J., Koenderink, J.J.: The shading cue in context. i-Perception 1, 159–178 (2010)
29. Yamane, Y., Carlson, E.T., Bowman, K.C., Wang, Z., Connor, C.E.: A Neural Code for Three-Dimensional Object Shape in Macaque Inferotemporal Cortex. Nature Neuroscience: Published Online (2008)

Chapter 4
From Functional Architectures to Percepts: A Neuromathematical Approach

Giovanna Citti and Alessandro Sarti

Abstract. In this paper we will consider mathematical models of the functional architecture of the primary visual cortex based on Lie groups equipped with sub-Riemannian metrics. We will critically review and clarify our line of work, joining together within an integrative point of view geometric, statistical, and harmonic models. The neurogeometry of the cortex in the SE(2) groups introduced recalling the original paper [12]. Amodal perceptual completion is reconsidered in terms of constitution of minimal surfaces in the geometric space of the functional architecture, and a new Lagrangian field model is introduced to afford the problem of modal perceptual completion [14] of the Kanizsa triangle. The neurogeometric structure is considered also from the a probabilistic point of view and compared with the statistics of co-occurence of edges in natural images following [48]. Finally the problem of perceptual units constitution is introduced by means of a neurally based non linear PCA technique able to perform a spectral decomposition of the neurogeometrical operator and produce the perceptual gestalten [52, 53].

4.1 Introduction

The pioneeristic work of Hubel and Wiesel in the seventies [32, 33] allowed the discovery of the modular structure of the mammalian visual cortex. Every module is composed by many families of cells, every one sensible to a specific feature of the image, either position, orientation, scale, color, curvature, velocity or stereo. Modules are spatially arranged in suitable maps always respecting retinotopy in such a way that for every point (x,y) of the retinal plane there is a hypercolumn containing

Giovanna Citti
Dipartimento di Matematica, University of Bologna
e-mail: giovanna.citti@unibo.it

Alessandro Sarti
CAMS Center of Mathematics, CNRS-EHESS, Paris
e-mail: alessandro.sarti@ehess.fr

G. Citti and A. Sarti (eds.), *Neuromathematics of Vision*, Lecture Notes in Morphogenesis,
DOI: 10.1007/978-3-642-34444-2_4,

an entire set of cells sensitive to all feature instances. The functional architecture of the visual cortex is the geometric machine underlying the processing of visual stimuli, and it is defined in terms of hypercolumnar organization and its neural connectivity. A particularly interesting framework to mathematically model the functional architecture of the primary visual cortex has been introduced first by W.C. Hoffmann in [30]. A differential geometry setting has been proposed to model the hypercolumnar organization in in terms of a manifold equipped with a fiber bundle structure. Steven Zucker in [61] followed a similar approach to model the action of simple cells for orientation detection, introducing Frenet frames in a fiber bundle structure to represent hypercolumns. Jean Petitot and Yannique Tondut in [46] introduced in the fiber bundle structure a supplementary constraint, giving rise to the so called contact structure. They propose for the first time the notion of neurogeometry. The contact constraint is fundamental to take into account the anysotropic pattern of connectivity between hypercolumns. Citti and Sarti proposed to represent the functional architecture in terms of Lie group structures equipped with sub-Riemannian metrics, which better describe the symmetry of the cortex [12]. In their model, the hypercolumnar structure is described in terms of the Lie symmetries of the Euclidean group, equipped with the suitable sub-Riemannian metric for modelling anysotropic connectivity. The integral curves of its generating vector fields can also be considered as a mathematical representation of the association fields of Field, Hayes and Hess [23]. The propagation in the sub-Riemannian setting allows to perform amodal contour completion [12]. With similar instruments, both contour and image completion have been achieved in the Lie group of affine transformation in [49] and improved by introducing the hyperbolic plane in [50]. Analytical properties of the model [12] was further studied by R. Hladky and Pauls [29]. Finally we recall the works of Duits, van Almsick, Franken, ter Haar Romeny [19] [20] [21] who proposed new models in different Lie groups, with many applications to image processing. The connectivity kernels resulting from the sub-Riemannian model of the cortex are compared with the statistics of co-occurence of edges in natural images in [48], giving a possible explanation of their emergence. In [14] a Lagrangian field model is introduced to couple the functional architecture of the Lateral Geniculate Nucleus with the one of the visual cortex. The resulting model is able to afford the problem of modal perceptual completion of the classical triangle of Kanizsa. The problem of the constitution of perceptual units in the geometry of the functional architecture is the very question to fill the gap between neurophysiology and phenomenology of perception. In [52] and [53] a mechanism of grouping has been proposed as extension of the model of visual hallucination proposed by Bressloff and Cowan [7]. The model performs a non-linear Principle Component Analysis by using the neurogeometrical kernels of the functional architecture.

In this paper we will critically review and clarify our line of work, joining together within an integrative point of view geometric, statistical and harmonic models previously presented. The paper is organized in six main parts. In section 4.2, the functional architecture of the Lateral Geniculate Nucleus is presented, while in section 4.3, the neurogeometry of the cortex is recalled following [12]. In section 4.5, a model of amodal perceptual completion is proposed in terms of constitution of minimal surfaces in the geometric space of the functional architecture. In section 4.6, a Lagrangian field model is introduced to afford the problem of modal perceptual completion [14]. In section 4.7, we will reinterpret the neurogeometric structure from a probabilistic point of view and compare it with the statistics of co-occurence of edges in natural images, following [48]. In section 4.8, we will afford the problem of perceptual units constitution. By means of a sinlge neural population model [8], we will construct the gestalten performing a spectral decomposition of the neurogeometrical operator.

4.2 Functional Architecture of Lateral Geniculate Nucleus

4.2.1 Receptive Fields and Profiles of Thalamic Cells

A number of cells of visual areas elicit a spike response to a luminance signal applied on the retinal plane $M \subset \mathbb{R}^2$. Every visual neuron is characterized by its *receptive field* (RF) that is classically defined as the domain of the retina to which the neuron is sensitive. Note that in general neurons are not directly connected to the retina. For example, cortical neurons are projecting from the retina to the cortex through the lateral geniculate nucleus along the thalamic way.

The *receptive profile* (RP) of a visual neuron is defined on the domain marked by the RF, and corresponds to the impulse response of the cell as a filter kernel. It is a function $\Psi(x,y)$ (where x,y are retinal coordinates) $\Psi : M \to \mathbb{R}$ which is defined on the retinal plane M and measures the response (ON / OFF) of the neuron to stimulation at the point (x,y). When a visual stimulus $I(x,y) : M \subset \mathbb{R}^2 \to \mathbb{R}^+$ activates the retinal layer, the cells centered at every point (x,y) of M process in parallel the retinal stimulus with their receptive profile. Reverse correlation techniques enable the recording of the RPs in terms of the correlations of the inputs (generally flashes of light and dark spot) with the outputs (spikes) [18]. The correlation of the inputs with the outputs yields the transfer function of the neuron, namely the RP.

RPs of the retinal ganglion cells are usually modelled by Laplacians of Gaussians [39],

$$\Psi_0(\xi,\eta) = \Delta G(\xi,\eta),$$

where $G(\xi,\eta) = e^{-(\xi^2+\eta^2)}$ is the Gaussian bell and Δ is the standard Laplacian. The same receptive profiles are found also in the Lateral Geniculate Nucleus (LGN), that is a copy of the retina but strictly in contact with the visual cortex. The size of the retinal and LGN RPs vary with a scale parameter allowing an entire multiscale analysis of the signal. For the purpose of this study, we will consider a fixed scale, since the phenomena described are invariant with respect to size changes, up to a rescaling.

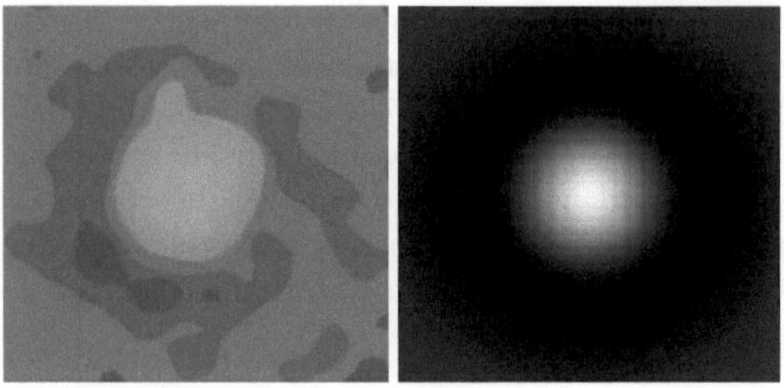

Fig. 4.1 On the left: a measured LGN cell RP (thanks to De Angelis [18]), on the right: a LGN cell RP modelled as Laplacian of a Gaussian

4.2.2 Cell Response and Lateral Connectivity of LGN

The output of the cells in LGN in response to the visual signal is highly nonlinear, and it rescales the visual input in a logarithmic way:

$$O(x,y)_{LGN} = \Delta G(x,y) * \log I(x,y).$$

The output of LGN cells is propagated via the lateral connectivity in LGN itself. Since this connectivity is isotropic [57], it can be modelled by the fundamental solution $\Gamma(x,y)$ of the 2D Laplacian operator

$$\Gamma(x,y) = -\log\sqrt{x^2+y^2}.$$

LGN lateral connectivity with strength $\Gamma(x,y)$ acts linearly on the output $O(x,y)_{LGN}$, giving a total contribution

$$\phi(x,y) = \frac{1}{2}\Big(\Gamma(x,y) * \Delta \log I(x,y)\Big). \tag{4.1}$$

Note that the action of receptive profiles $\Delta \log I(x,y)$ and the one of LGN lateral connectivity $\Gamma(x,y)*$ is dual in a differential sense.

4.2.3 The Retinex Algorithm as a Model of LGN Action

Eq. (4.1) corresponds to the Retinex algorithm in the version proposed by Horn in [31], where the authors proposed a physically based algorithm, which recovers the reflectance f of an image I as

$$\Delta \log f(x,y) = \Delta \log I(x,y), \tag{4.2}$$

where $I(x,y)$ is given. This is a Poisson equation with solution

$$\log f(x,y) = \frac{1}{2}\Big(\Gamma(x,y) * \Delta \log I(x,y)\Big),$$

formally equivalent to eq. (4.1) provided that $\phi(x,y) = logf(x,y)$.

Let us recall that the Retinex algorithm has been introduced in [37,38] to explain lightness perception, i.e. the phenomenon causing a gray patch to appear brighter when viewed against a dark background and darker when viewed against a bright background. Improvements and new models have been proposed by [35,43] among others.

In particular in [43], it has been proved that the original Retinex algorithm can be equivalently espressed by the Poisson equation (4.2), and in [26,27] a new interpretation was given in terms of covariant derivatives and fiber bundles. Indeed setting

$$\mathbf{A}(x,y) = \nabla I(x,y)/I(x,y) \tag{4.3}$$

equation (4.2) can be considered the Euler Lagrange equation of the functional

$$\tilde{F}(x,y) = \int \frac{|\nabla f(x,y) - \mathbf{A}(x,y) f(x,y)|^2}{f(x,y)^2} dxdy. \tag{4.4}$$

This functional is invariant with respect to the transformation

$$f \to fI, \quad \mathbf{A} \to \mathbf{A} + \frac{\nabla I}{I}$$

so that the choice $\mathbf{A}(x,y) = \frac{\nabla I(x,y)}{I(x,y)}$ is compatible with the transformations which leaves the functional invariant. The quantity $\nabla f - \mathbf{A} f$ can be interpreted as a covariant derivative.

Here we can further notice that, setting

$$\phi(x,y) = \log f(x,y), \quad h(x,y) = \log I(x,y), \tag{4.5}$$

equation (4.2) simplifies as

$$\Delta \phi(x,y) = \Delta h(x,y) \tag{4.6}$$

and setting as before: $\mathbf{A}(x,y) = \nabla I(x,y)/I(x,y) = \nabla h(x,y)$, the functional becomes

$$F(x,y) = \int |\nabla \phi(x,y) - \mathbf{A}(x,y)|^2 dxdy = \int |\nabla \phi(x,y) - \nabla h(x,y)|^2 dxdy, \tag{4.7}$$

while the transformations which leave the operator invariant become

$$\phi \to \phi + h, \quad \mathbf{A} \to \mathbf{A} + \nabla h.$$

The functional (4.7) has to be considered as the invariant energy underlying the Retinex Poisson equation (4.2) solved by the action of the LGN connectivity represented by eq. (4.1).

4.3 The Neurogeometrical Structure of the Primary Visual Cortex

4.3.1 The Set of Simple Cells Receptive Profiles as a Lie Group

While LGN cells are fairly isotropic, simple cells of the primary visual cortex V1 are strongly oriented, and their RPs are interpreted as Gabor patches [17, 36] or directional derivatives of Gaussians.

A first derivative of Gaussian models odd simple cell RPs (Fig. 4.2 right):

$$\Psi_0(\xi,\eta) = \partial_\eta G(\xi,\eta). \tag{4.8}$$

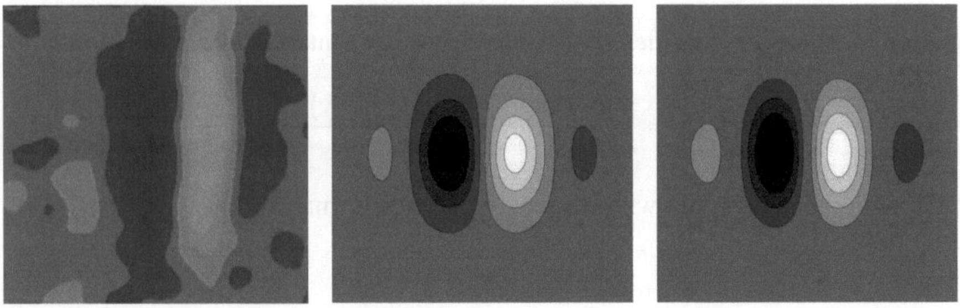

Fig. 4.2 Left: an odd simple cell RPs measured by reverse correlation techniques (thanks to De Angelis [18]). Odd receptive profile modelled as Gabor filter (middle) and first directional derivative of Gaussian (right).

The set of observed profiles can be obtained from the mother profile $\Psi_0(\xi,\eta)$ (see [36]) under the action of a Lie group.

The entire set of RPs of the same point (x,y) is then obtained by rotating the mother profile, explicitly:

$$\Psi_\theta = \Psi_0\Big(\xi\cos\theta + \eta\sin\theta, -\xi\sin\theta + \eta\cos\theta\Big).$$

This structure enlightens the modular structure of the cortex. Each family of cells acts on the same retinal basis but depends on different engrafted variables, and it is described by different groups of symmetry.

We will describe the action of the affine group of rotation and translation on vectors of R^2, since the other two groups are subset of this one. The action $M_{x,y,\theta}$ transforms every vector (ξ,η) in a new vector $(\tilde{\xi},\tilde{\eta})$ as:

$$(\tilde{\xi},\tilde{\eta}) = M_{x,y,\theta}(\xi,\eta) = \begin{pmatrix} x \\ y \end{pmatrix} + \begin{pmatrix} \cos(\theta) & -\sin(\theta) \\ \sin(\theta) & \cos(\theta) \end{pmatrix} \begin{pmatrix} \xi \\ \eta \end{pmatrix}. \tag{4.9}$$

The action of the group on the set of profiles will be:

$$\mathscr{L}_{(x_0,y_0,\theta)}\Psi(x,y) = \Psi(M^{-1}_{(x_0,y_0,\theta)}(x,y)). \tag{4.10}$$

Then the whole set of receptive profile will be $\{\Psi_{x,y,\theta}(\tilde{\xi},\tilde{\eta}) = \mathscr{L}_{(x_0,y_0,\theta)}\psi(x,y)\}$.

After rotation of the axis of an angle θ, the derivative ∂_η becomes

$$X_3 = -\sin(\theta)\partial_\xi + \cos(\theta)\partial_\eta. \tag{4.11}$$

Then the odd cell RP oriented in the direction θ is in general represented by

$$\Psi_\theta = X_3 G(\xi,\eta). \tag{4.12}$$

Finally the expression of filters on different points is obtained by translation:

$$\Psi_{x,y,\theta}(\tilde{x},\tilde{y}) = \Psi_\theta(x-\tilde{x}, y-\tilde{y}).$$

In Fig. 4.3 we visualize the set of odd simple cells.

Note that for simplicity we take Euclidean translations on the cortical plane neglecting the conformal log-polar retino-cortical mapping, that can be easily taken into account by introducing a Riemannian metric on the cortical layer. This feature will not be implemented in the present study.

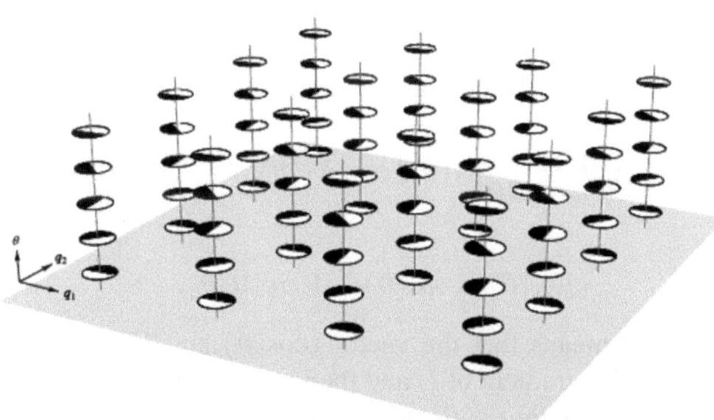

Fig. 4.3 The set of simple cells odd receptive profiles $\Psi_{x,y,\theta}(\tilde{x},\tilde{y})$ represented in the space (x,y,θ)

4.3.2 Simple Cells Response

If classical RPs are considered, the output u is given by linear filtering of the stimulus $h(\xi,\eta) = log(I(\xi,\eta))$ by the set of RPs:

$$u(x,y,\theta) = \int_M h(\tilde{\xi},\tilde{\eta})\Psi_{(x,y,\theta)}(\tilde{\xi},\tilde{\eta})d\tilde{\xi}d\tilde{\eta} \tag{4.13}$$

and by eq. (4.12)

$$u(x,y,\theta) = \int X_3 G(\xi,\eta)h(\tilde{\xi},\tilde{\eta})d\tilde{\xi}d\tilde{\eta} = -X_3 G(\xi,\eta) * h(\tilde{\xi},\tilde{\eta}) = -X_3(\theta)h_s \tag{4.14}$$

where h_s is a smoothed version of h,

$$h_s = h * G(\xi,\eta).$$

4.3.3 Non Maximal Suppression

For a fixed point (x,y) the output is a function just of the variable θ that attains its maximum at the point

$$\max_\theta ||u(x,y,\theta)|| = ||u(x,y,\bar{\theta})||. \tag{4.15}$$

This maximality condition can be mathematically expressed requiring that the derivative of $||u||$ with respect to the variables θ vanishes at the point $(x,y,\bar{\theta})$:

$$\partial_\theta u(x,y,\bar{\theta}) = 0. \tag{4.16}$$

At the maximum point $\bar{\theta}$ the derivative with respect to θ vanishes, and we have

$$0 = \frac{\partial}{\partial\theta}u(x,y,\bar{\theta}) = \frac{\partial}{\partial\theta}X_3(\bar{\theta})I = -X_1(\bar{\theta})I = -\langle X_1(\bar{\theta}), \nabla I\rangle \tag{4.17}$$

where

$$X_1(\theta) = cos(\theta)\partial_\xi + sin(\theta)\partial_\eta.$$

The condition (4.17) means that the vector $(\cos(\bar{\theta}),\sin(\bar{\theta}))$ is orthogonal to the gradient of h (and to the gradient of I) and then is tangent to its level lines. *Then the angle $\bar{\theta}$ maximizing the output indicates the direction of level lines of the stimulus image.*

Calling

$$\bar{\rho}(x,y) = ||u(x,y,\bar{\theta}(x,y))||, \tag{4.18}$$

we can say that the point (x,y) is lifted to the point $(x,y,\bar{\theta}(x,y),\bar{\rho}(x,y))$ in the cortical space (x,y,θ,ρ). With the same procedure it is possible to lift all the level lines of the 2D image I into 4D curves.

It is easy to check (see for example [12]) that the first 3 components of the lifted curves are tangent to the plane generated by the vector fields

$$\mathbf{X}_1 = (\cos(\theta),\sin(\theta),0) \quad \mathbf{X}_2 = (0,0,1). \tag{4.19}$$

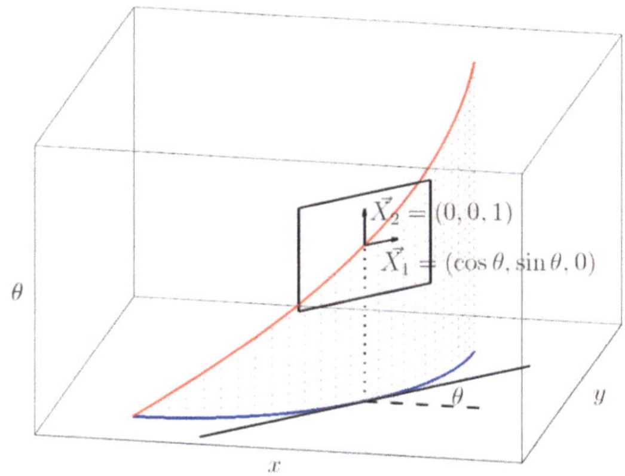

Fig. 4.4 A level line of the image (in blue) and its 3D cortical lifting (in red). The tangent vector to the blue curve is $(\cos(\theta), \sin(\theta))$, so that the tangent vector to its lifting lies in the plane generated by $\mathbf{X}_1 = (\cos(\theta), \sin(\theta), 0)$ and $\mathbf{X}_2 = (0, 0, 1)$.

and visualized in Fig. 4.5. In Fig. 4.4 we depicted the 3D section with the plane (x, y, θ).

Let us notice that all admissible curves of the cortical space are tangent at every point to the subspace generated by $\mathbf{X}_1$ and $\mathbf{X}_2$, which has dimension 2. In the standard Riemannian setting, the number of generators of the tangent space equal the dimension of the space, while in the geometrical model of cortical space the dimension of the manifold is 3, and the admissible tangent space has dimension 2. This endows the 3D space $R^2 \times S^1$ of the variables (x, y, θ) with a sub-Riemannian geometry. While no constraint is imposed on the last variable ρ, the vector field in the ρ direction will be

$$\mathbf{X}_4 = (0, 0, 0, 1).$$

The full 4D space will consequently be

$$R^2 \times S^1 \times R^+,$$

with a sub-Riemnnian metric in the first 3 variables, and a standard metric in the last variable. Since the geometry on the last component ρ is standard, we will describe the geometry of the 3D space generated by the first 3 variables (x, y, θ).

4.3.4 Association Fields and Integral Curves of the Structure

Field, Heyes and Hess in [23] have shown the existence of a perceptual field connecting patches of position and orientation, see Fig. 4.6 left. This connectivity pattern, called association field, is considered at the base of the constitution of

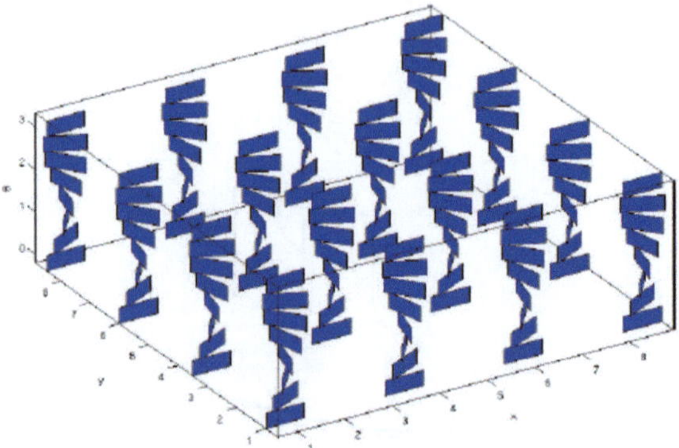

Fig. 4.5 The bundle of subspaces of the tangent manifold generated by the fields $\mathbf{X}_1(\theta)$ and $\mathbf{X}_2(\theta)$

boundaries in visual perception, implementing the gestalt law of good continuation. We have shown in [13] that the association field is well modelled by the integral curves of the vector fields $\mathbf{X}_1$ and $\mathbf{X}_2$, starting from a fixed point (x_0, y_0, θ_0):

$$\gamma'(t) = (x'(t), y'(t), \theta'(t)) = \mathbf{X}_1(x(t), y(t), \theta(t)) + k\mathbf{X}_2(x(t), y(t), \theta(t)) \quad (4.20)$$
$$\gamma(0) = (x_0, y_0, \theta_0),$$

A fan of such integral curves by varying the parameter k is visualized in Fig. 4.6 right.

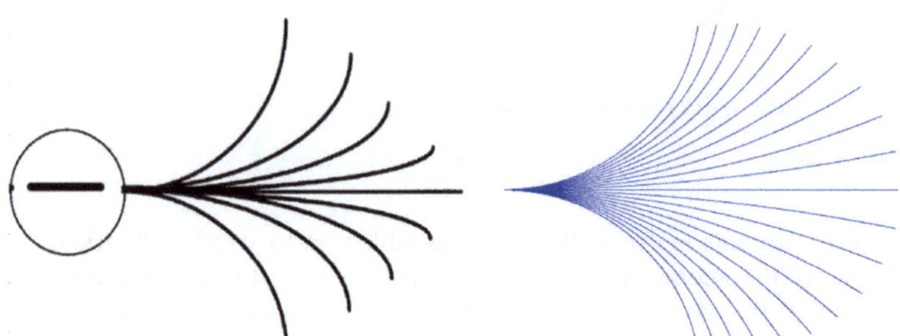

Fig. 4.6 The association fields of Fields, Heyes and Hess [23] (left) and the projected integral curves with constant coefficients (right)

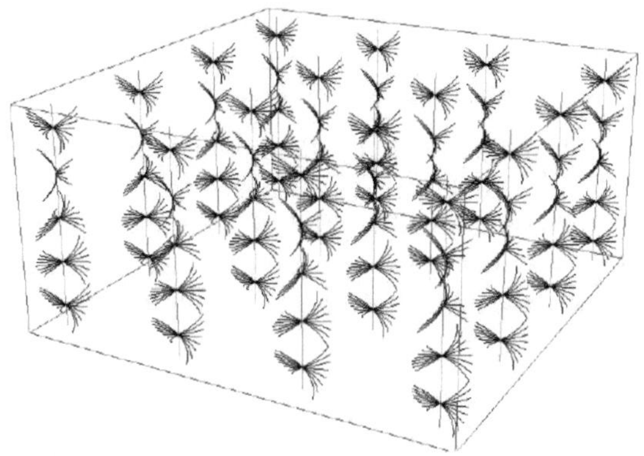

Fig. 4.7 The constant coefficient integral curves of the fields $\mathbf{X}_1(\theta)$ and $\mathbf{X}_2(\theta)$, modeling the local connectivity between points of the cortical space

4.3.5 Length of Lifted Curves and Geodesics

A 2D level line

$$\tilde{\gamma} = (x(t), y(t))$$

parametrized by arc length t, has tangent vectors

$$(x'(t), y'(t)) = (\cos(\theta(t)), \sin(\theta(t)))$$

at every point, where θ denotes the direction of the curve at the point $(x(t), y(t))$. We have seen that the action of simple cells lifts the level line into a 3D cortical curve $\gamma(t) = (x(t), y(t), \theta(t))$. Differentiating x and y we get

$$(x''(t), y''(t)) = (-\sin(\theta(t))\theta'(t), \cos(\theta(t))\theta'(t)) = (-y'(t), x'(t))\theta'(t)$$

so that the euclidean curvature can be computed as

$$k = \frac{y''x' - x'y''}{((x')^2 + (y')^2)^{3/2}} = \theta'$$

The length of the lifted curve is

$$L(\gamma(t)) = \int \sqrt{x'(t)^2 + y'(t)^2 + \theta'(t)^2} dt = \int \sqrt{1 + k(t)^2} dt,$$

considering that $x'(t)^2 + y'(t)^2 = 1$. Notice that the length of lifted curves depends on the length of the 2D level line and on its curvature, analogous to the case of the elastica functional $\int (1 + k(t)^2) dt$, introduced by David Mumford in [44]. The distance between two cortical points (x, y, θ) and $(\bar{x}, \bar{y}, \bar{\theta})$ is defined in term of the length functional as

$$d((x,y,\theta),(\bar{x},\bar{y},\bar{\theta})) = \inf\{L(\gamma) : \gamma \text{ is an horizontal curve} \quad (4.21)$$
$$\text{connecting } (x,y,\theta) \text{ and } (\bar{x},\bar{y},\bar{\theta})\},$$

see [13]. In other words the horizontal path on which the infimum is achieved is a geodesic curve of the cortical space. As suggested by Petitot and Tondout in [46], subjective contours can be computed as geodesics curves in a contact structure. Here we can show (see Fig. 4.8) that subjective contours can be well represented by sub-Riemannian geodesics in the $SE(2)$ group, i.e. as minimizers of (4.21).

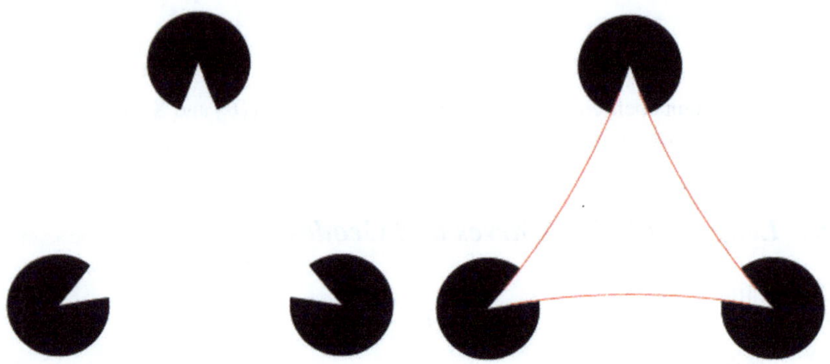

Fig. 4.8 A Kanizsa triangle with curved boundaries (left) and the subjective contours modelled as geodesics of the cortical space (right). The geodesics are not rectilinear, since they minimize the distance (4.21), which is a function of the curvature k.

4.4 The Cortical Implementation of the Neurogeometrical Structure

4.4.1 Response of Simple Cells and Bargmann Transform

Daugmann in [17] first proved that the shape of the simple cells is intimately related to their functionality. His crucial remark is the fact that simple cells try to localize at the same time position (x,y) and frequency ω. Hence the set $SE(2)$ can be interpreted as a real manifold of the phase space, with the frequency variables expressed in polar coordinates:

$$SE(2) \subset \mathbb{C}^2 = \{(x,y,|p|\cos(\theta),|p|\sin(\theta))\}.$$

The classical uncertainty principle in the Heisenberg space asserts that it is not possible to detect with arbitrary precision both position and momentum (see [15]). The principle also provides an explicit condition for functions that minimize uncertainty with respect to the position and momentum operators. These minimizers are called coherent states, and in the Heisenberg setting they are the Gabor filters. This is why these filters have been proposed in [17] as models for the shape of receptive profiles (see Fig. 4.2 Middle). The complete set of coherent states can be obtained via the action of the group (4.10) on a fixed mother state Ψ . Recall that, if Ψ is a coherent state, the Bargmann transform of the function h is

$$B_\Psi(h)(x,y,\theta) = \int \mathscr{L}_{(x,y,\theta)} \Psi(\tilde{x},\tilde{y}) h(\tilde{x},\tilde{y}) d\tilde{x} d\tilde{y}. \tag{4.22}$$

Hence the response of the filters on an image defined in (4.13) can be interpreted as a Bargmann transform (see [4] for the definition and [3] for this interpretation of the cell response).

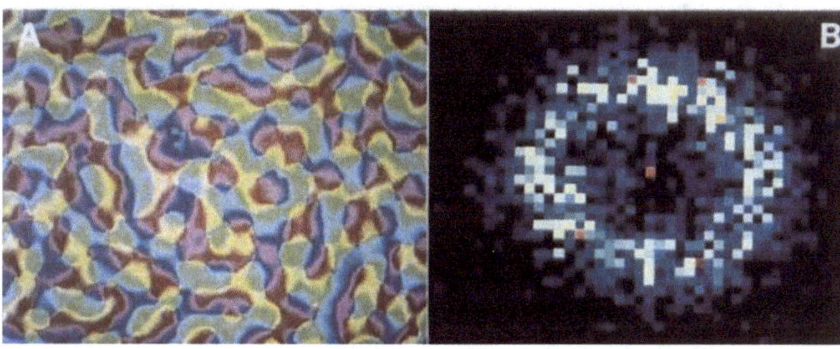

Fig. 4.9 A pinwheel map (left) and its spectral behavior (right). Figure extracted from [41].

4.4.2 The Activity Maps and the Pinwheel Structure

Even if the primary visual cortex shows the symmetries of SE(2), that is a 3-dimensional group, its physical implementation is realized on the 2-dimensional layer provided by the cortex. Orientation columns are radially arranged around singular points like the spokes of a wheel, that are called pinwheels (see Fig. 4.9). This structure has been observed first with optical imaging techniques [6, 5] and more recently by in vivo two-photon imaging proving their organization with single cell precision. In [41] an empirical method was also introduced that is able to reproduce orientation map-like structures as a superposition of plane waves with random phases. More recently, in [16], the activity in V1 has been reconstructed by measuring cell responses to so called gratings. Activated regions depend on the orientation at which they are presented so that if a family of gratings is presented to an observer, the result is a family of real maps $\{u_\theta(x,y)\}$ (see Fig 4.10). The pinwheel image has been reproduced in Fig.4.10 (center) by performing a vector sum of the orientations

$$\mathscr{P}(x,y) = \frac{1}{2} arg \int_0^\pi e^{i2\theta} u_\theta(x,y) d\theta. \tag{4.23}$$

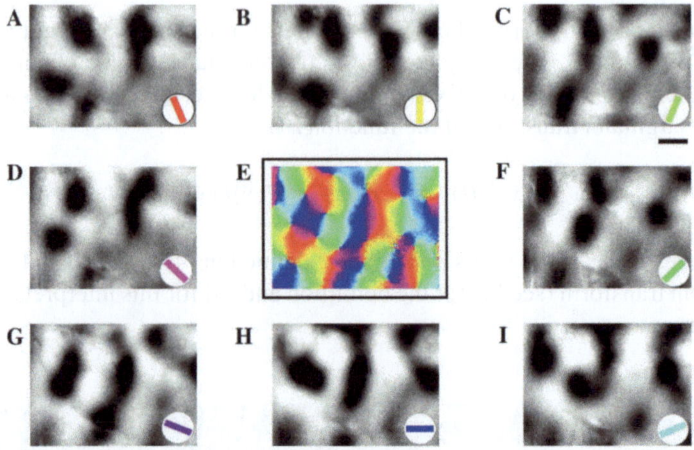

Fig. 4.10 The original results obtained by [16] showing a set of gray level cortical maps u_θ acquired by optical imaging and the reconstructed color coded image at the center with the well known pinwheel structure

4.4.3 An Uncertainty Principle on the Functional Geometry

In a recent paper [2] Barbieri, Sanguinetti and the authors of the present paper proposed a model of the cortical activity and pinwheel structure based on an uncertainty principle in the structure $SE(2)$ of the visual cortex, which reproduces the experiment in [16]. The image of the left-invariant vector fields (4.19) under the differential of the action defined in (4.10) is defined as $Y_i = d\mathscr{L}(X_i)$ and provides the differential operators

$$Y_1 = \partial_{\tilde{y}}, \quad Y_2 = \tilde{y}\partial_{\tilde{x}} - \tilde{x}\partial_{\tilde{y}}. \tag{4.24}$$

Under the action of the Fourier transform, these vector fields (4.24) become respectively

$$\mathscr{F}(Y_1 f) = i\xi_2 \hat{f}, \quad \mathscr{F}(Y_2 f) = (\xi_2 \partial_{\xi_2} - \xi_1 \partial_{\xi_2})\hat{f}.$$

Experimentally we see that the Fourier spectrum of the orientation maps is approximately concentrated on a circle (see Fig.4.9, right). On the other hand no action is performed by the fields in the radial direction, allowing to restrict the study to functions defined on the circle of radius ρ (see [55]). Since it is not possible to further reduce the set where to study these vector fields, the representation of these vector fields on the circle is called irreducible. In polar coordinates $(\xi_1, \xi_2) = \rho(\cos(\tilde{\theta}), \sin(\tilde{\theta}))$ they reduce to an even simpler representation:

$$\hat{Y}_1 \hat{f} = i\rho \sin(\tilde{\theta})\hat{f}, \quad \hat{Y}_2 \hat{f} = \partial_{\tilde{\theta}} \hat{f}. \tag{4.25}$$

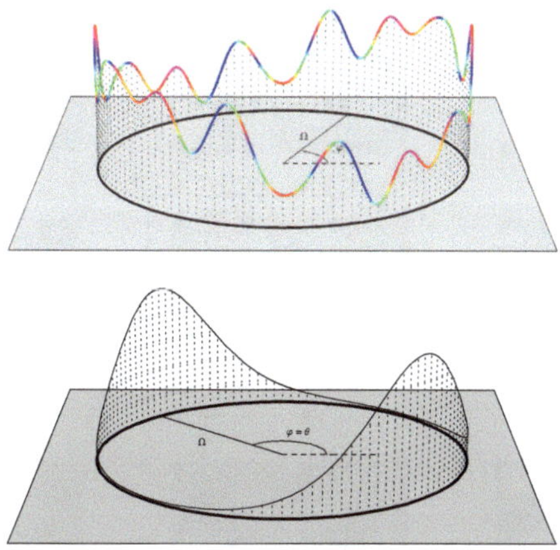

Fig. 4.11 Top: A general function in the reduced Fourier plane, where only a circle is considered. The height of the graph represents the modulus, and the color represents the argument of the function. Bottom: A coherent state in the reduced Fourier plane.

The uncertainty principle can be stated for any couples of noncommuting self-adjoint operators on Hilbert spaces [24]. In terms of operators $\hat{Y}_i$ it reads:

$$|\langle\hat{\Phi},\rho\cos(\tilde{\theta})\hat{\Phi}\rangle| \leq 2\|\rho\sin(\tilde{\theta})\hat{\Phi}\| \; \|\partial_{\tilde{\theta}}\hat{\Phi}\|. \tag{4.26}$$

Minimal uncertainty states can be computed by making use of an equation associated to inequality (4.26), that in this case reads (see also [9])

$$(\hat{Y}_2 - i\lambda\hat{Y}_1)\hat{\Phi}(\tilde{\theta}) = 0 \tag{4.27}$$

where the scaling parameter λ represents frequency. Solutions to equation (4.27) read, up to a normalization constant $c = c_\rho(\lambda)$

$$\hat{\Phi}(\tilde{\theta}) = ce^{\lambda\rho\cos(\tilde{\theta})}. \tag{4.28}$$

The states (4.28) are the most concentrated functions in angular position and momentum, hence allowing optimal localization (from now on we will omit to write the constant c for simplicity). Note that the bigger λ is, the sharper its localization is: for large values of λ, that means $\lambda \gg 1/\rho$, it is maximally concentrated. Due to the uncertainty inequality, the variance of $\frac{d}{d\varphi}$ grows to infinity, i.e. it is maximally undetermined. On the other hand, for small values of λ, i.e. $\lambda \ll 1/\rho$, then $\hat{\Phi}(\varphi)$ approaches a constant, and the angular momentum is maximally concentrated.

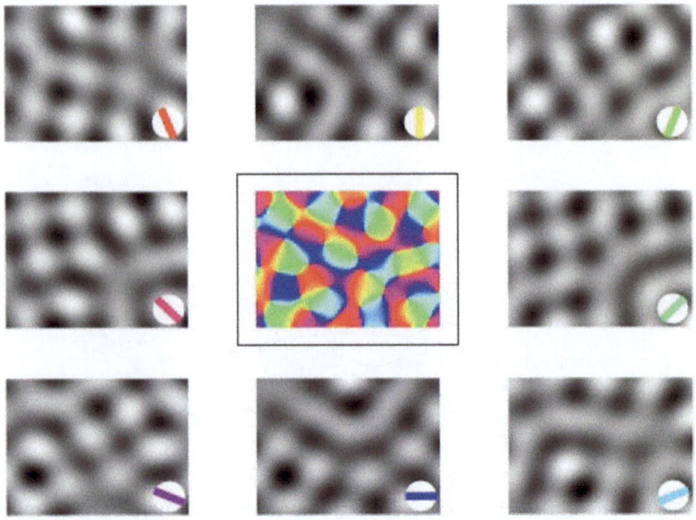

Fig. 4.12 The gray valued maps $u_{\Phi,\theta}(x,y)$ for different values of the orientation and in center the color image obtained as a vector sum, to be compared with the experimental result of [16], reproduced in Fig.4.10

4.4.4 Irreducible Bargman Transform, the Activity Maps and Pinwheels

The notion of Bargmann transform in $SE(2)$ has been introduced by Barbieri et al. in [3]. In analogy with the classical expression in (4.22) it is expressed as the operator associated to the coherent states. The action of the group on the coherent states Φ is the Fourier transform of the action defined in (4.10):

$$\hat{\mathscr{L}}_{(x,y,\theta)}\hat{\Phi}(\tilde{\theta}) = e^{-i\rho_0(x\cos(\tilde{\theta})+y\sin(\tilde{\theta}))}\hat{\Phi}(\tilde{\theta}-\theta).$$

Consequently, the irreducible Bargman transform reads:

$$B_\Phi(h)(x,y,\theta) = \int_{S^1}\hat{\mathscr{L}}_{(x,y,\theta)}\hat{\Phi}(\tilde{\theta})h(\tilde{\theta})d\tilde{\theta}. \tag{4.29}$$

Given a white noise $W = W(\tilde{\theta})$ with values in $[0,2\pi]$, defined on $[0,2\pi]$, and such that $W(\tilde{\theta}+\pi) = -W(\tilde{\theta})$ we can consider the function $B_\Phi(W)(x,y,\theta)$. We impose the symmetries of the cortex to represent orientations. Namely we require that it is $\pi-$periodic in θ and provides opposite response at orthogonal angles as is the case for V1 cells:

$$u_{\Phi,\theta}(x,y) = Re\Big(B_\Phi(W)(x,y,\theta) - B_\Phi(W)(x,y,\theta+\pi/2)\Big). \tag{4.30}$$

By definition this irreducible Bargmann transform is simply the transform in (4.22) restricted to functions with constant frequency. Hence it expresses a response of the cortex, which has been interpreted in [2] as a model of the activity maps measured in [16](see Fig. 4.12). As in (4.23) we perform the sum

$$\mathscr{P}_{\Phi}(x,y) = arg\Big(\int_0^{\pi} u_{\Phi,\theta}(x,y)e^{i2\theta}d\theta\Big) \tag{4.31}$$

reproducing the original experiment. An important property of (4.31) is that it can be represented as a sum of plane waves, with frequency ρ, and random phases in accordance with the mechanism introduced in [41]. This model is indeed able to interpret both the orientation activity maps and the pinwheel-shaped orientation maps as interference figures.

4.4.5 Propagation in the Pinwheel Structure

Techniques of optical imaging associated to tracers allow a large-scale observation of neural signal propagation via cortico-cortical connectivity. These tests have shown that the propagation is highly anisotropic and almost collinear to the preferred orientation of the cell (see Fig. 4.13), confirming at neural level the phenomenological results of Field, Heyes and Hess in [23] and the model of propagation curves developed in section 3. This experiment suggests to introduce not only a propagation along curves but with a differential operator, which can propagate in a neighborhood of a point. Even though the propagation is performed in the cortex at the pinwheel level, we will describe the neural propagation in the 3D model of $R^2 \times S^1$, which is formally simpler to handle, and from which it is possible to project on the pinwheel structure.

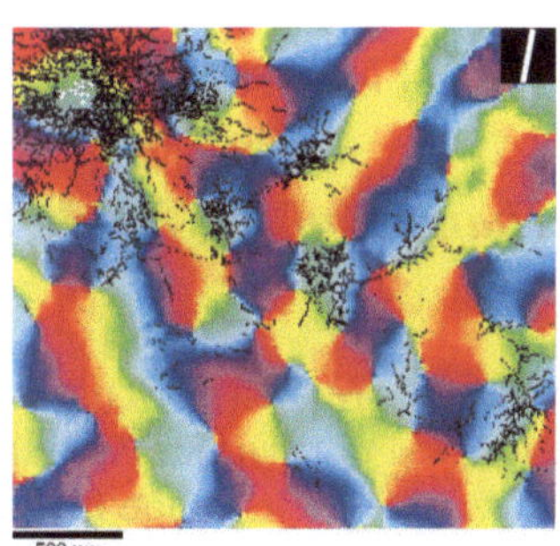

Fig. 4.13 A marker is injected in the cortex, in a specific point, and it diffuses mainly in regions with the same orientation as the point of injection

4.5 Propagation in the Family of Simple Cells

4.5.1 Propagation along the Association Fields

Neural activity develops and propagates only along the integral curves of the structure. If γ is one of the curves in (4.20), the Lie derivative of a function f along the curve γ is defined as

$$X_i u(\xi_0) = \frac{d}{ds}(u \circ \gamma)_{|s=0},$$

and on regular functions it coincides with the directional derivative. In analogy with the Euclidean gradient, the sub-Riemannian gradient is defined by

$$\nabla_R u = (X_1 u, X_2 u).$$

Note that it does not contain derivatives in direction X_3, which can be recovered by commuting:

$$X_3 = [X_2, X_1] = X_2 X_1 - X_1 X_2,$$

see [12] for details. Hence X_3 will play a role similar to a second derivative. In this setting the divergence of a horizontal vector field $v = (v_1, v_2)$ is defined as

$$div_R v = X_1 v_1 + X_2 v_2,$$

so that the sub-Laplacian operator becomes

$$\Delta_R u = div_R(\nabla_R u) = X_{11} u + X_{22} u. \tag{4.32}$$

The time dependent counterpart is the sub-Riemnnian diffusion operator:

$$\partial_t u = \Delta_R u, \tag{4.33}$$

modelling a diffusive mechanism of propagation in the cortical space.

4.5.2 The Lifting Mechanism of the Whole Image: Regular Graphs in $R^2 \times S^1$

The mechanism of non maxima suppression does not lift each level lines independently, but the whole image is lifted to a surface. Condition (4.16) ensures that the lifted surface is identified by a zero level set of the function

$$H(x, y, \theta) = \partial_\theta u(x, y, \theta),$$

meaning that if we consider only strict maxima, the lifted surface becomes

$$\Sigma = \{(x, y, \theta) : H(x, y, \theta) = 0, \partial_\theta H(x, y, \theta) > 0\}. \tag{4.34}$$

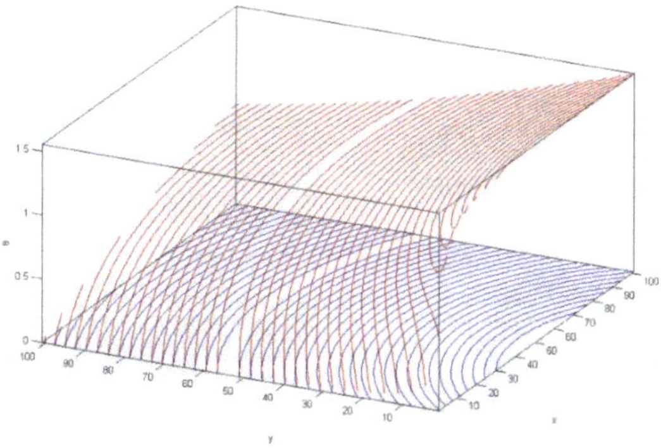

Fig. 4.14 The level lines of a 2D image (in blue), and their 3D lifting (in red)

Since the vector ∂_θ is a horizontal vector, then Σ is a regular surface in the sub-Riemannian metric.

In particular, we can define a horizontal normal to Σ, and denote with ν_R its projection on the horizontal plane of the Euclidean normal:

$$\nu_R = \frac{(X_1 u, X_2 u)}{\sqrt{(X_1 u)^2 + (X_2 u)^2}}.$$

In analogy with the notion of Euclidean curvature, we define the sub-Riemannian curvature as the $R-$divergence of the $R-$normal vector:

$$H_R(\Sigma) = div_R(\nu_R),$$

4.5.3 Completion Model and Minimal Surfaces in the Roto-translation Space

The two cortical mechanisms of sub-Riemannian diffusion (4.33) and non maximal suppression (4.34) will be iteratively applied. The action of these two mechanism can be formalized as a two step algorithm.

- The lifted surface $\Sigma_0 = \Sigma$ defined in (4.34) and the function $\rho_0(x,y) = \rho(x,y)$ defined on it in (4.18) allow to define a measure $\rho_0 \delta_{\Sigma_0}$ concentrated on the surface. Diffusing this measure, we define a function $u_1(x,y,\theta,t)$ concentrated in a neighborhood of the surface.
- The second step is a non maximal suppression, which performs a concentration and allows recovering a surface Σ_1 via the condition $\partial_\theta u = 0$ and a new function $\rho_1 = u_{|\Sigma_1}$ .The surface Σ_1 will be the graph of a function θ_1 .

To the new surface we iteratively apply the same two step procedure.

Fig. 4.15 The two step algorithm: the lifted surface generated by a concentration mechanism (left), and its diffusion (right)

If we fix a time T and a discretization step $h = T/m$, we repeat m times the algorithm. At a general step n we will have a surface Σ_n and a function ρ_n, defined on Σ_n. We diffuse for an interval of time of length h:

$$\begin{cases} \partial_t u = \Delta_R u \; in(\mathbb{R}^2 \times S^1)\backslash\Sigma_0 & t \in [nh, (n+1)h] \\ u(\cdot, nh) = \rho_n \delta_{\Sigma_n}. \end{cases} \tag{4.35}$$

At time $t = (n+1)h$ we have a new function ρ_{n+1} and a new surface defined as

$$\rho_{n+1}(\cdot, (n+1)h) = u(\cdot, (n+1)h) \quad \text{and} \quad \Sigma_{n+1}((n+1)h) = \{\partial_\theta u = 0, \partial_\theta^2 u < 0\}.$$

After m steps we have two sequences: $\rho_{m+1}(\cdot, T)$ and $\Sigma_{m+1}(T)$. Letting m go to $+\infty$ we have:

$$\rho(T) = lim_{m \to +\infty} \rho_{m+1}(T), \quad \Sigma(T) = lim_{m \to +\infty} \Sigma_{m+1}(T)$$

The diffusion followed by a concentration in the normal direction leads to a purely tangential diffusion of the surface giving rise to the surface $\Sigma(T)$ moving by curvature, with initial surface Σ_0. The function $\rho(T)$ coincides with the Laplace Beltrami flow, with initial condition ρ_0.

This mechanism is a generalization of the well known algorithm of Merriman, Osher and Sethian in [42]. The convergence of the Euclidean version of this scheme has been proved by Evans [22] and Barles, Georgelin [5]. The convergence of the analogous flow in the sub-Riemannian setting has been first presented in [13] and then formalized in [10].

For $T \to +\infty$ the surface $\Sigma(T)$ converges to a minimal surface Σ, and $\rho(T)$ tends to the solution $\rho(x,y)$ of the (time independent) Laplace Beltrami equation on the surface.

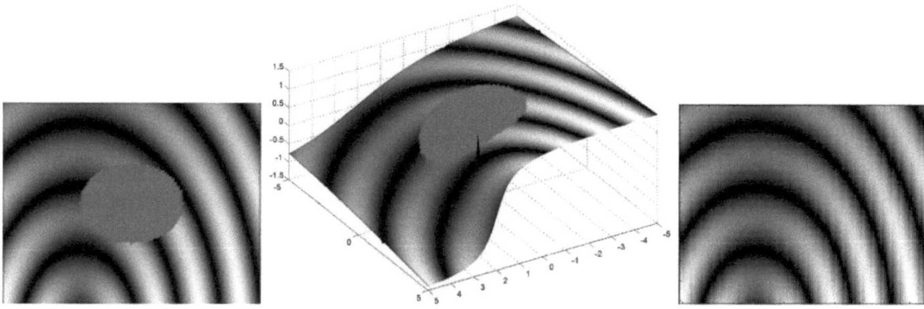

Fig. 4.16 An image with a missing part (left) is lifted to the 3D space (middle) and completed with the previous described algorithm (right)

The algorithm gives rise to an amodal completion of the image. Indeed the initial image (see Fig. 4.16), left) is lifted in the rotranslation space. The lifted surface is completed by iteratively applying the algorithm until a minimal surface is generated.

4.5.4 *Minimal Surfaces as Minima of the Area Functional*

The minimal surface Σ can be expressed as a graph of a function θ. In addition, if we project the vector fields defined in (4.19) on the x,y plane, we end up with a unique vector field

$$X_{1\theta} = cos(\theta(x,y))\partial_x + \sin(\theta(x,y))\partial_y = \langle \nabla, (\cos(\theta(x,y)), \sin(\theta(x,y))) \rangle \quad (4.36)$$

since the projection of the vector X_2 on the same plane is 0. We explicitly note that the vector $X_{1\theta}$ here is only formally similar to the vector X_1 in (4.19). Indeed $\theta(x,y)$ in (4.36) is a function while in (4.19) θ was simply an independent variable of the 3D space.

The minimal surfaces equation can be expressed in terms of the function $\theta(x,y)$ as follows:

$$X_{1\theta}\Big(\frac{\rho^2 X_{1\theta}\theta(x,y)}{\sqrt{|\rho^2 X_{1\theta}\theta(x,y)|^2+1}}\Big) = 0,$$

where θ coincides with $\bar{\theta}$ on the existing boundaries. Taking explicitly the derivative, the equation becomes

$$X_{1\theta}(\rho^2 X_{1\theta}\theta(x,y)) = 0. \quad (4.37)$$

This equation can be interpreted as a second order directional derivative, in the direction $(\cos(\theta), \sin(\theta))$. Hence it is the Euler-Langrange equation of the Dirichlet functional

$$\int |X_{1,\theta}\theta(x,y)|^2 \rho^2(x,y)dxdy.$$

The function ρ satisfies the Laplace Beltrami equation, which is the Euler-Langrange equation of the functional

$$\int |X_{1,\theta}\rho(x,y)|^2 dxdy.$$

Summing these two terms we obtain

$$\int |X_{1,\theta}\theta(x,y)|^2 \rho^2(x,y)dxdy + \int |X_{1,\theta}\rho(x,y)|^2 dxdy. \tag{4.38}$$

In order to take into account the graphs of the two functions θ and ρ we can define a function

$$\mathbf{A}(x,y) = \rho(x,y)(\cos(\theta(x,y)), \sin(\theta(x,y)).$$

In (4.18) we defined ρ as the modulus of the gradient and θ as its orientation. Then, if we denote $\mathbf{A} = (A_1, A_2)$ the derivative X_1 is expressed in terms of $\mathbf{A}$ as

$$X_{1,A} = \frac{A_1}{\sqrt{A_1^2 + A_2^2}}\partial_x + \frac{A_2}{\sqrt{A_1^2 + A_2^2}}\partial_y,$$

and the functional in (4.38) can be expressed in terms of the function $\mathbf{A}$ as

$$\int |X_1\mathbf{A}|^2 dxdy. \tag{4.39}$$

Indeed

$$\int |X_1A_1|^2 = \int |X_1(\rho cos(\theta))|^2 = \int (X_1\rho cos(\theta) + \rho sin(\theta)X_1\theta)^2 =$$

$$= \int (X_1\rho)^2 cos^2(\theta) + \rho^2 sin^2(\theta)(X_1\theta)^2 + 2\rho X_1\rho cos(\theta)sin(\theta)X_1\theta$$

$$\int |X_1A_2|^2 = \int (X_1\rho)^2 sin^2(\theta) + \rho^2 cos^2(\theta)(X_1\theta)^2 - 2\rho X_1\rho cos(\theta)sin(\theta)X_1\theta.$$

Summing up we get

$$\int |X_1\mathbf{A}|^2 = \int (X_1\rho)^2 + \rho^2(X_1\theta)^2.$$

4.6 A Field Lagrangian for Perceptual Completion

4.6.1 The Full Lagrangian

In this section we will study the joint action of LGN cells and cells in V1, taking into account feed-forward, horizontal and feedback connectivity. We propose a complete Lagrangian, sum of three terms: a particle term corresponding to functional (4.7), a field term modelled by functional (4.39), and an interaction term coupling the two terms as in the usual Lagrangian field theories.

The first term is

$$\mathscr{L}_1 = \int |\nabla\phi(x,y) - \nabla h(x,y)|^2 dxdy \tag{4.40}$$

and is directly inspired by the Retinex model, describing the reconstruction of the image from image boundaries. It is considered here as the particle term of the Lagrangian, where $\phi(x,y)$ is the particle and $h(x,y)$ is the stimulus forcing term.

The field term is the functional (4.38), in analogy with the classical fields theories:

$$\tilde{\mathscr{L}}_2 = \int |X_1\mathbf{A}|^2 dxdy.$$

It expresses the spatial propagation of the field $\mathbf{A}(x,y)$ where $\mathbf{A}$ denotes the presence of subjective contours. As we have seen in chapter 4.5, the propagation of level lines (and then of subjective contours) is a strongly anisotropic process, and it is performed with respect to a sub-Riemannian metric, hence the functional is expressed in terms of the vector field X_1.

The last term describes the interaction between the particle $\phi(x,y)$ and the field $\mathbf{A}(x,y)$. In classical physics it is called "minimal coupling" because it tries to minimize the difference between the two terms:

$$\mathscr{L}_3 = \int |\nabla\phi(x,y) - \mathbf{A}(x,y)|^2 dxdy. \tag{4.41}$$

The field $\mathbf{A}(x,y)$ codifies illusory contours. When $\mathbf{A}(x,y)$ is viewed as a forcing term for $\phi(x,y)$, this second term is similar to the Retinex term (4.40) but driven by subjective contour instead of existing ones, while when $\Phi(x,y)$ is viewed as a forcing term for $\mathbf{A}(x,y)$, it drives the constitution of subjective contours.

The resulting functional $\mathscr{L} = \mathscr{L}_1 + \mathscr{L}_2 + \mathscr{L}_3$ is then

$$\mathscr{L} = \int |\nabla\phi - \nabla h|^2 dxdy + \int |\nabla\phi - \mathbf{A}|^2 dxdy + \int |X_1\mathbf{A}|^2 dxdy \tag{4.42}$$

where all the terms are functions of the coordinates (x,y).

4.6.2 The Euler Lagrange Equations

The Euler Lagrange Equations of the functional (4.42) are obtained by variational calculus:

$$\begin{cases} \Delta\phi = \frac{1}{2}(\Delta h + div(\mathbf{A})) \\ \Delta_{\mathbf{A}}\mathbf{A} = -\nabla\phi + \mathbf{A}. \end{cases} \tag{4.43}$$

The first equation (particle equation) is clearly a generalized Retinex equation, forced by the boundaries present in the stimulus Δh and by the subjective boundaries $div(\mathbf{A})$. It performs a contrast invariant reconstruction of the image. Note that the two terms $\mathscr{L}_1$ and $\mathscr{L}_2$ which generalize the Retinex functional give rise to this unique particle equation.

The second equation propagates boundaries in the sub-Riemmanian metric, and allows to recover subjective boundaries. Note that $\mathbf{A} = (A_x, A_y)$ is a vector, hence the first equality is indeed a system, and the sub-Riemannian Laplacian $\Delta_{\mathbf{A}}$ is the directional Laplacian associated to the considered metric.

We explicitly remark that the equation of $\mathbf{A}$ is a non-linear sub-Riemannian equation, and can be solved by iterative linearization.

This means that we need to find an initial approximated solution $\mathbf{A}_0$. A natural choice is the solution of the vector Laplace equation

$$\Delta \mathbf{A}_0 = \nabla\phi.$$

Of course this is only an approximated solution $\mathbf{A}_0$, but we can recover a better one $\mathbf{A}_1$ as a solution of

$$\Delta_{\mathbf{A}_0}\mathbf{A}_1 = \nabla\phi,$$

using the sub-Riemmannian operator associated to $\mathbf{A}_0$. From here we start an iteration:

$$\Delta_{\mathbf{A}_1}\mathbf{A}_2 = \nabla\phi, \cdots \quad \Delta_{\mathbf{A}_{j-1}}\mathbf{A}_j = \nabla\phi.$$

At each step we get a better approximation of the solution, moreover the sequence has a limit $\mathbf{A} = \lim_{j\to+\infty}\mathbf{A}_j$. Passing to the limit in the previous expression, we will get:

$$\Delta_{\mathbf{A}}\mathbf{A} = \nabla\phi,$$

so that the limit provides a solution of the nonlinear equation.

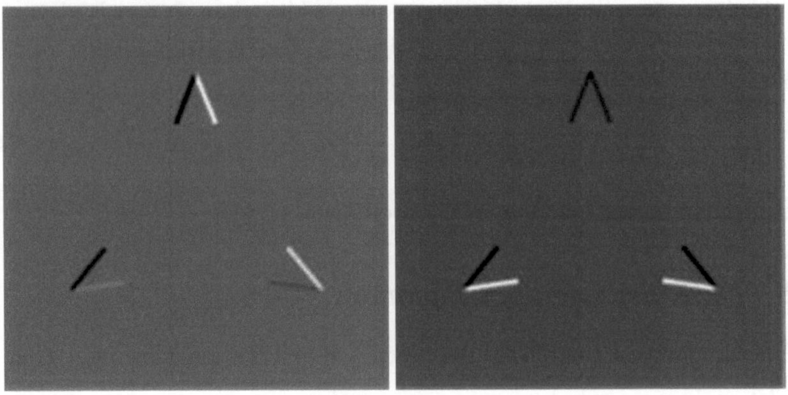

Fig. 4.17 The x and y components of ∇h related to the Kanizsa triangle inducers

4.6.3 Solution of Euler Lagrange Equations

We solve sequentially the coupled system of differential equations (4.43). We first apply the Retinex equation to the initial image:

$$\Delta\phi = \frac{1}{2}\Delta h \tag{4.44}$$

and solve it by linear convolution

$$\phi = \frac{1}{2}\Big(\Gamma(x,y) * \Delta h\Big)$$

with the fundamental solution of the 2D Laplacian:

$$\Gamma(x,y) = -\log|(x,y)|.$$

Then we solve the equation for boundaries propagation. In this first step, we choose $\mathbf{A} = 0$ in the right hand side, and the nonlinear equation reduces to

$$\Delta_{\mathbf{A}}\mathbf{A} = \nabla\phi.$$

As we explained in the previous section, this equation will be solved by linearization, stopped after the first two steps:

$$\begin{cases} \Delta\mathbf{A}_0 = \nabla\phi \\ \Delta_{\mathbf{A}_0}\mathbf{A}_1 = \nabla\phi. \end{cases} \tag{4.45}$$

The first term is computed by convolution

$$\mathbf{A}_0 = \overrightarrow{\Gamma} * \nabla\phi,$$

where $\overrightarrow{\Gamma}$ is the fundamental solution of the vector Laplacian

$$\overrightarrow{\Gamma}(x,y) = (-\log|(x,y)|, -\log|(x,y)|).$$

The second one is computed by approximating functions with centered differences and by means of a standard linear solutor.

The first equation in (4.45) propagates $\nabla\phi$ isotropically and generates a first approximated vector field $\mathbf{A}_0$. In fig. 4.18 the vector field $\mathbf{A}_0$ related to the Kanizsa inducers is visualized. The second equation generates a better approximation $\mathbf{A}_1$ by propagating $\nabla\phi$ in the direction $\mathbf{A}_0$. In fig. 4.19 the components of the vector field $\mathbf{A}_1$ related to the same inducers is shown. Inducers have been manually selected.

Since particle and field equations are coupled, we can now solve the complete particle equation

$$\phi = \frac{1}{2}\Gamma(x,y) * \Big(\Delta h + \frac{1}{2}(div(\mathbf{A}_1))\Big)$$

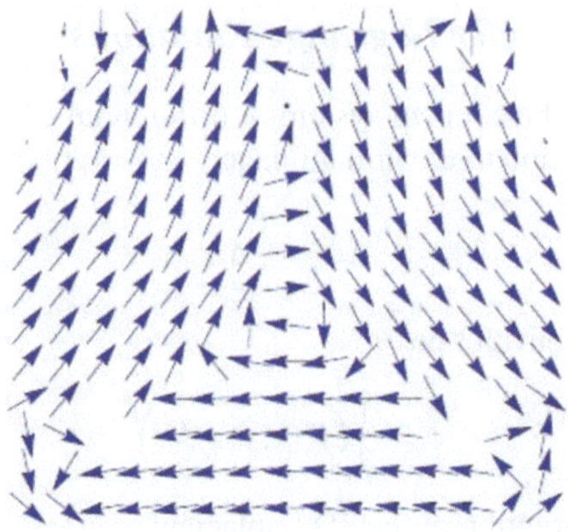

Fig. 4.18 The field $\mathbf{A}_0$ generated by the Kanizsa triangle inducers

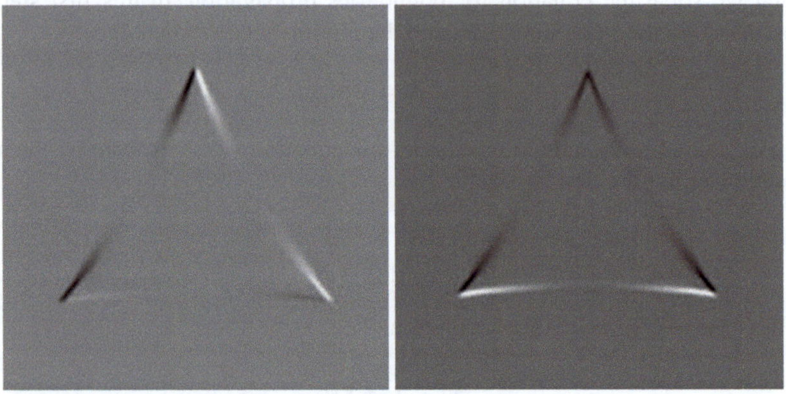

Fig. 4.19 The x and y components of the gauge field $\mathbf{A}_1$ related to the Kanizsa triangle inducers. $\mathbf{A}_1$ is an approximation of the field $\mathbf{A}$, solution of the gauge field equation.

again by convolution with the fundamental solution Γ. This is a version of the Retinex equation able to reconstruct the original image together with the subjective surface. In Fig. 4.20 (left) the forcing term $\frac{1}{2}(\Delta h + div(\mathbf{A}_1))$ of the particle equation is visualized, while in Fig. 4.20 (right) the solution ϕ is shown.

4.7 Stochastic Neurogeometry

The phenomenological experiment conduced by Field, Heyes and Hess [23] shows that the association field allows to perceive configurations of oriented patches preferably fulfilling a co-circularity condition. This result is well modelled by the integral curves of equation (4.20). The experiment shows also that even if the co-circularity

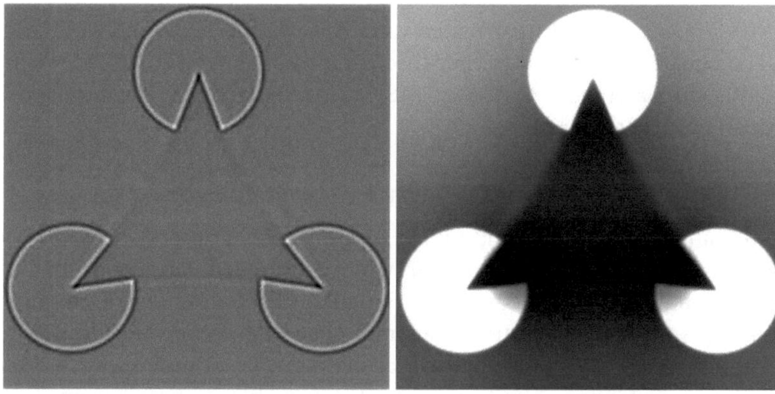

Fig. 4.20 Left: The forcing term $\frac{1}{2}(\Delta h + div(\mathbf{A}_1))$ of the particle equation. Right: The reconstructed Kanizsa triangle as the solution ϕ of the particle equation.

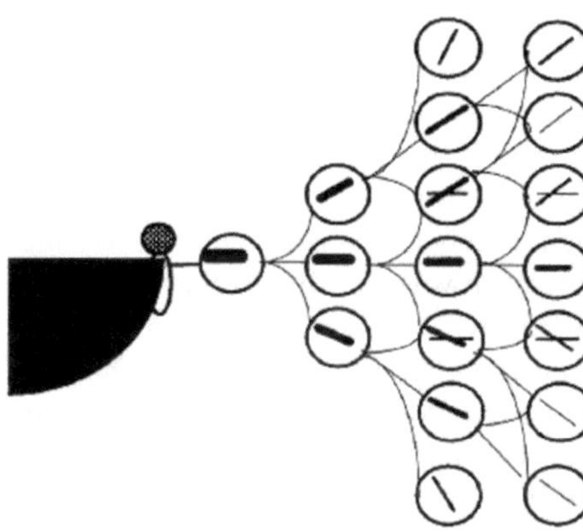

Fig. 4.21 The association field of Field, Heyes and Hess showing that the co-circular configuration of patches is just the most salient among other possible configurations (double bars in a circle).

condition induces a maximum perceptual saliency, all the configurations of patches are perceived with a certain saliency.

To model the saliency of the entire set of perceptual configurations it is necessary to leave the deterministic framework and introduce a stochastic setting that constitutes the probabilistic counter part of the deterministic equation (4.20). David Mumford first introduced a stochastic differential equation in [44] to model partially occluded edges:

$$(x'(t), y'(t), \theta'(t)) = (\cos(\theta(t)), \sin(\theta(t)), N(0, \sigma^2)),$$

where $N(0,\sigma^2)$ is a normally distributed variable with zero mean and variance equal to σ^2. The equation has been further discussed by August-Zucker [1], Williams-Jacobs [59], and Sanguinetti-Citti-Sarti [48] who outlined that it is naturally defined in the SE(2) group structure

$$(x'(t),y'(t),\theta'(t)) = \mathbf{X}_1(\theta(t)) + N(0,\sigma^2)\mathbf{X}_2. \tag{4.46}$$

Indeed both systems (4.20) and (4.46) are represented in terms of left invariant operators of the Lie group, the first one with deterministic curvature, the second with normal random variable curvature. Equation (4.46) describes a random walk with constant speed in a direction randomly changing. Let us denote $p(x,y,\theta,t)$ the probability density to find a particle at the point (x,y) moving with direction $\mathbf{X}_1$ at the instant of time t conditioned by the fact that it started from the point $(x(0)=0, y(0)=0, \theta(0)=0)$. After Ito integration, this probability density satisfies a Kolmogorov Forward Equation or Fokker-Planck equation (FP):

$$\begin{aligned}\partial_t p(x,y,\theta,t) &= X_1 p(x,y,\theta,t) + \sigma^2 X_{22} p(x,y,\theta,t) \qquad (4.47)\\ &= \cos(\theta)\partial_x p(x,y,\theta,t) + \sin(\theta)\partial_y p(x,y,\theta,t) + \sigma^2 \partial_{\theta\theta} p(x,y,\theta,t).\end{aligned}$$

Let us notice that the equation (4.47) consists of an advection term in the direction $\mathbf{X}_1$ and a diffusion term in the direction $\mathbf{X}_2$, where the doubling of the index expresses a second order derivation.

This equation has been largely used in computer vision and applied to perceptual completion related problems. It was used by Williams and Jacobs in [59] to compute stochastic completion field, by S. Zucker and his collaborators in [1] to define the curve indicator random field, and more recently by R. Duits et Al. in [25] applying it to perform contour completion, denoising and contour enhancement. In [48] it was proposed to consider the stationary counterpart of (4.47) to model both the Fields, Heyes and Hess association field and the probability of co-occurence of contours in natural images. Both the phenomenon are indeed stationary.

For this purpose, we integrate eq. (4.47) in time, obtaining the stationary equation

$$X_1 p(x,y,\theta) + \sigma^2 X_{22} p(x,y,\theta) = \delta(x,y,\theta). \tag{4.48}$$

Equation (4.48) is strongly biased in direction $\mathbf{X}_1$, and to take into account the symmetry of both association fields and edge cooccurences, the model for the probability density propagation has been symmetrized considering the backward FP equation in the opposite direction

$$-X_1 p(x,y,\theta) + \sigma^2 X_{22} p(x,y,\theta) = \delta(x,y,\theta). \tag{4.49}$$

The desired fundamental solution is then obtained by summing the Green functions corresponding to forward and backward FP equations.

The fundamental solution has been computed numerically with standard Markov Chain Monte Carlo methods. This is done by generating random paths obtained from numerical solutions of the stochastic system and averaging their passages over

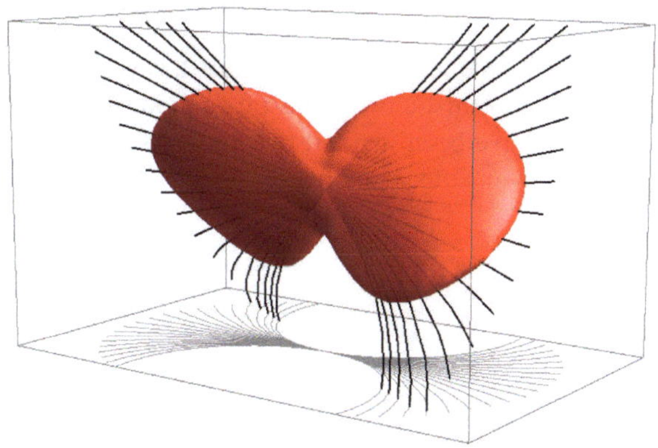

Fig. 4.22 The sum of fundamental solutions of the stationary Fokker Planck equations (4.48) and (4.49). An isosurface of intensity is visualized in red together with the integral curves of the group eq. (4.20).

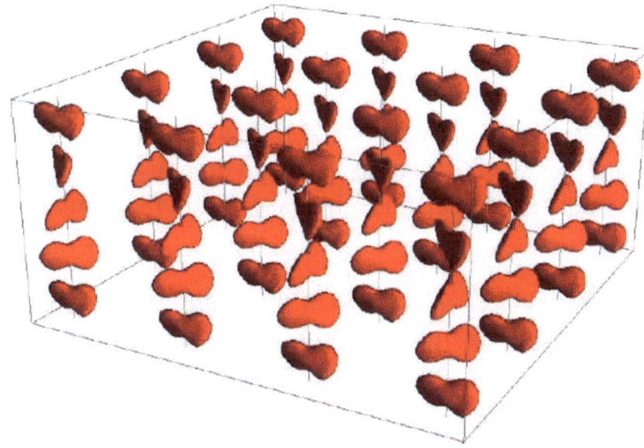

Fig. 4.23 The stochastic neurogeometry as fundamental solution of stationary Fokker Planck equation. It is a "fat" version of the deterministic structure visualized in Fig. 4.7

discrete volume elements. See [40] for a classical and complete description of the technique. In Fig.4.22 the sum of fundamental solutions of the stationary Fokker Planck equations (4.48) and (4.49) is visualized together with the integral curves of the group eq. (4.20). Note that the kernel seems to be a thick version of the fan (4.20). The value of the probability decays slowly along the integral curves and quickly in the direction normal to the surface ruled by the integral curves. In Fig. 4.23 the entire set of fundamental solutions is visualized in the space (x,y,θ). Notice that it is a "fat" version of the deterministic structure visualized in Fig.4.7.

4.7.1 The Statistics of Edges in Natural Images

It can be of some interest to question the origin of the very specific shape of the association field in its deterministic or stochastic version. An intriguing hypothesis is that association fields have been learned by the visual stimuli and then that the particular pattern comes from some statistical property of natural images.

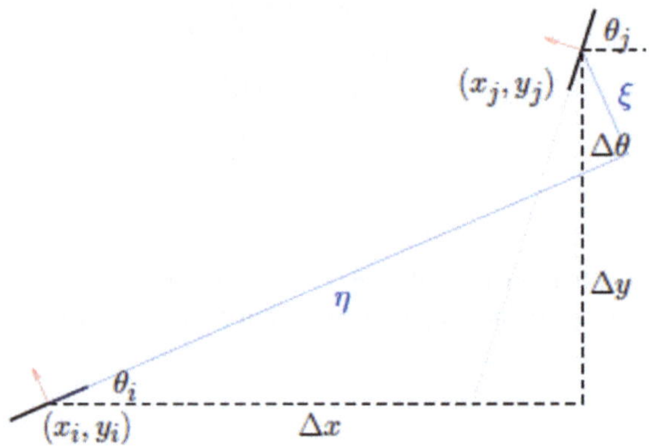

Fig. 4.24 Geometric schemata of co-occurences of edges: a co-occurrence takes place when two edges (x_i, y_i, θ_i) and (x_j, y_j, θ_j) occur at the same time in the image. Only relative positions $(\Delta x, \Delta y)$ and orientations $\Delta\theta$ are taken into account in the computation of histograms.

A specific study to assess the existence of this relation has been proposed in [48]. Research has been focused on the statistics of edges in natural images and particularly in the statistics of co-occurrence of couples of edges taking into account its relative position and orientation. The statistics have been estimated analyzing a number of natural images from which a multidimensional histogram of relative position and orientation of edges has been constructed. Images have been preprocessed by linear filtering with a set of oriented edge detection kernels (Gabor filters) and performing non maximal suppression. A list of pixels corresponding to edges with their respective orientations has been obtained by thresholding and binarization. A four dimensional histogram $(\Delta x, \Delta y, \theta_c, \theta_p)$ has been computed by counting how many times two detected edges with relative positions $(\Delta x, \Delta y)$ have orientations (θ_c, θ_p). Finally a 3D histogram $(\Delta x, \Delta y, \Delta\theta)$ is obtained where the third coordinate is the relative orientation $\Delta\theta = \theta_p - \theta_c$.

4.7.2 Comparison between the Statistics of Edges and the Stochastic Structure

A comparison between the estimated statistical distribution of edges and the computed stochastic neurogeometrical model has been performed in [48]. In Fig. 4.27

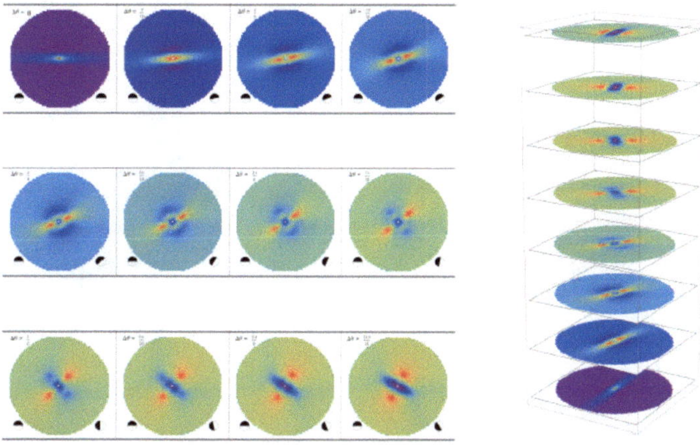

Fig. 4.25 Sections of the co-occurence maps $(\Delta x, \Delta y)$ for different pairs of orientations $\Delta\theta$ (left). Sections are piled up to build the 3D distributions $(\Delta x, \Delta y, \Delta\theta)$ (right).

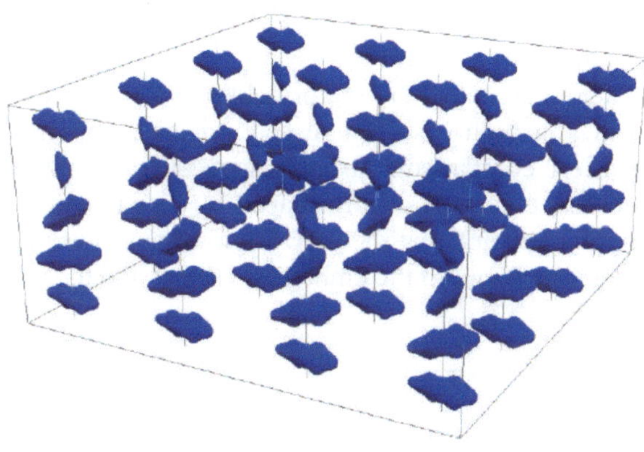

Fig. 4.26 The whole distribution of co-occurrence of edges in natural images in the space (x, y, θ). It can be interesting compare this distribution with the deterministic structure visualized in Fig. 4.7 and the stochastic structure of Fig. 4.23.

the two distributions are visualized and compared. There are two degrees of freedom in the model to match: the variance σ of the Fokker Plank operator and a scale factor. After performing the parameter identification, the relative difference between both functions is less than 0.02, and the reported value for σ is 1.71, a very interesting value since it estimates the variance of the co-occurrence random process.

We found that it is possible to recover the co-circularity pattern from both the distributions with the following procedure. First we compute the surface of maximal probability $\bar{\theta}(x,y)$, where

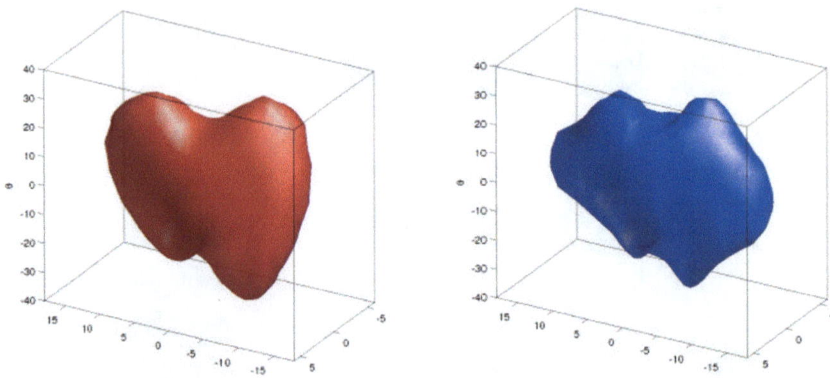

Fig. 4.27 Comparison between the computed stochastic model (left) and the distribution of edge cooccurrences in natural images (right). The percentual error between the two distributions is less than 2.

$$\max_{\theta} p(x,y,\theta) = p(x,y,\bar{\theta}).$$

Secondly, the integral curves of the vector field $(\cos\bar{\theta}(x,y), \sin\bar{\theta}(x,y))$ are computed. We found that in both cases curves present co-circularity. See Fig. 4.28 for a comparison. These comparisons strongly suggest that horizontal connectivity modeled by the neurogeometry is deeply shaped by the statistical distributions of features in the environment and that the very origin of neurogeometry as to be discovered in the interaction between the embodied subject and the world.

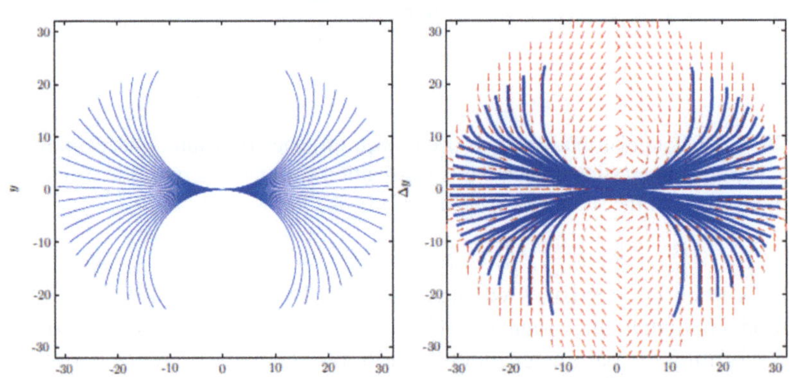

Fig. 4.28 The co-circularity constraint in the neurogeometrical model, as the projection of integral curves (left), and in the statistics of natural images as integral curves of the projected vector field (right)

4.8 A Harmonic Neurogeometrical Model

4.8.1 The Binding Problem and the Constitution of Perceptual Units

The question of constitution of perceptual units is central in the entire body of study of visual organization. The issue deals with the mechanism allowing the information distributed in the visual areas to get bound together into coherent object representations. Electromagnetic radiations (photons) going from the physical world to the eye are completely independent one to the other and do not contain any explicit information about the unitarity of objects. The visual areas are constituted in its turn by million of cells. Along this path the unity of physical objects is completely lost, but at the end of the chain the object shows up again at the perceptual level as a unit. In which way is this reconstruction possible?

This process is known as "binding" or "perceptual grouping", and it has been extensively studied both from a phenomenological point of view by the experimental psychology of Gestalt [34] and from a neuro-physiological point of view aiming to clarify its biological functionalities. Even if the mechanism of grouping is still unknown there is a wide experimental evidence that the constitution of a perceptual unit involves the response of a large number of neurons distributed over a large spatial region. Different answers have been proposed to explain how these distributed neural activities are integrated together to constitute a unit.

If excitatory-inhibitory neural populations are considered, the most accreditated hypothesis is that binding is implemented with a temporal coding, meaning that object segmentation is performed by the synchronization of oscillatory neural responses [28]. Many models have been proposed to explain phase locking of populations of neural oscillators, see for example [56] for a review. In [11, 51] it has been shown that a neurophysiological model of coupled neural oscillators in phase locking tends in its continuous spatial limit to the phenomenological model of segmentation of Mumford-Shah.

On the other hand, if single population models are considered, different mechanisms for binding can arise. For example in [52] and [53] a mechanism of grouping has been proposed as an extension of the model of visual hallucination proposed by Bressloff and Cowan [7]. We will reconsider extensively this model in the following. Particularly we want to investigate what is the role of the neurogeometry in the constitution of perceptual units, considering that a global integration process is needed. As stated by the studies of Gestalt theory, perception is a global process. Moreover visual perception acts as a differentiation process of the entire field of view, performing first a differentiation between figure and ground, and continuing in the segmentation of single objects. We will try to keep in mind these fundamental points and to formalize them in a coherent model.

4.8.2 The Population Model

As shown in [53], the Ermentraut-Cowan mean field population model in presence of a cortical input stimulus $h(x,y,\theta)$ provided by simple cells is

$$\frac{da(x,y,\theta,t)}{dt} = -\alpha a(x,y,\theta,t) + \sigma_\gamma\Big(\omega * a(x,y,\theta,t) + \frac{h(x,y,\theta)}{\alpha}\Big) \tag{4.50}$$

where $a(x,y,\theta,t)$ is the neural activity of the population and α is the decay constant.

The kernel $\omega\Big((x,y,\theta),(x',y',\theta')\Big)$ is the weight of connectivity between the population at the position (x,y) tuned at the orientation θ and the one at the position (x',y') tuned at θ', and it can be modelled by the fundamental solution of the stationary Fokker-Planck equation (4.48). It takes into account the contribution of cortico-cortical connectivity.

The function σ_γ is the transfer function of the population and it has the classical sigmoidal behavior:

$$\sigma_\gamma(\xi) = \frac{1}{1+e^{-\gamma(\xi-\chi)}} - 1/2,$$

where χ is the threshold and γ is the gain. Notice that $\sigma(\chi) = 0$ and this happens when the input term $h = \alpha\chi$. For simplicity we will consider here a piecewise constant input $h = \alpha\chi$ in a domain $\Omega : \{(x,y,\theta) : h(x,y,\theta) = \alpha\chi\}$ and $h = 0$ elsewhere. Note that in case of weak connectivity kernel, neurons outside Ω will stay constant and under threshold in the dynamic of the system.

Then the activity equation becomes

$$\frac{da(x,y,\theta,t)}{dt} = -\alpha a(x,y,\theta,t) + \tilde{\sigma}_\gamma(\omega * a(x,y,\theta,t)) \tag{4.51}$$

for $(x,y,\theta) \in \Omega$, where

$$\tilde{\sigma}_\gamma(\xi) = \sigma_\gamma(\xi - h/\alpha)$$

with the property $\tilde{\sigma}_\gamma(0) = 0$. Let us outline that eq. (4.51) is formally equivalent to the population equation studied by Bressloff and Cowan in [7] but defined in a domain Ω instead of in the entire $\mathbb{R}^2$. We will study it with the same instruments here.

4.8.3 Solutions of the Activity Equation

To compute stationary states of the activity equation (4.51), it is trivial to check that since $\tilde{\sigma}_\gamma(0) = 0$ then the homogeneous state $a(x,y,\theta,t) = 0$ on the domain Ω is a stationary solution.

The stability of the solution can be studied in terms of eigenvalues of the linearized functional

$$La = -\alpha a + \gamma\omega * a$$

in the domain Ω where $\gamma = \sigma'(0)$. Eigenvalues λ verify the equation

$$\lambda a = -\alpha a + \gamma \omega * a,$$

that is equivalent to the eigenvalue problem

$$\tilde{\lambda} a = \omega * a$$

provided that $\tilde{\lambda} = \frac{(\alpha+\lambda)}{\gamma}$. Then the stability problem is reduced to the study of eigenvectors of the connectivity kernel ω in the domain Ω, meaning the operator $K(a) = \omega * a$ in Ω. Since ω is the fundamental solution of the stationary Fokker Planck equation (4.48), the convolution operator $K(a)$ is compact, and its largest positive eigenvalue is real. Moreover it has infinitely many eigenvalues, and the sequence of eigenvalues tends to 0.

This means that, when γ is small, zero is a stable solution, and it remains stable until $\frac{\alpha a_0}{\gamma}$ takes the value of the largest eigenvalue of $K(a)$. While γ further increases, the solution remains unstable.

(The eigenvectors of the operator K become marginally stable. Also note that the corresponding eigen-space is of finite dimension since the operator K is compact.)

4.8.4 The Discrete Case

If the activity $a(x,y,\theta)$ is approximated by a discrete distribution in terms of Dirac masses

$$a = \sum_{i=1}^{h} a_i \delta_{(x_i,y_i,\theta_i)},$$

then the convolution operator becomes

$$\omega * a = \sum_{j=1}^{n} \omega((x_i,y_i,\theta_i)(x_j,y_j,\theta_j))a_j = \tilde{\lambda} a_i.$$

Note that in this case the eigenvalue problem on the activity $a(x,y,\theta)$ reduces to the spectral analysis of the matrix

$$A_{ij} = \omega((x_i,y_i,\theta_i)(x_j,y_j,\theta_j))$$

with usual linear agebra instruments.

This matrix can be considered as the equivalent of the *affinity matrix* introduced by Perona in [45] to perform perceptual grouping.

Perona proposed to model the affinity matrix in term of a euristic distance $d(x,y,\theta)$, facilitating collinear and cocircular couple of elements

$$A_{ij} = e^{-d^2((x_i,y_i,\theta_i)(x_j,y_j,\theta_j))}.$$

In our case it is possible to prove that the distance d corresponds to the formal Carnot Carathéodory distance d_c, meaning the geometric sub-Riemannian distance induced by the horizontal connectivity, since the fundamental solution ω can be approximated by:

$$\omega((x_i,y_i,\theta_i),(x_j,y_j,\theta_j)) \simeq e^{-d_c{}^2((x_i,y_i,\theta_i),(x_j,y_j,\theta_j))}.$$

See [6] for a formal proof.

4.8.5 Dimensionality Reduction

In [45] the problem of perceptual grouping has been faced in terms of reduction of the complexity in the description of a scene. The visual scene is described in terms of the affinity matrix A_{ij} with a complexity of order $O(N^2)$ if N discrete elements are present in the scene. The idea of Perona is to describe the scene by a point-wise function p_i of objects i taken one by one, rather than by a function A_{ij} of object pairs (i,j). This way the complexity of the description will drop to O(N). A way to reduce by one order of magnitude the complexity of the description is to consider the best approximation of the affinity matrix

$$p = argmin_{\hat{p}} \sum_{i,j=1}^{N} (A_{i,j} - \hat{p}_i\hat{p}_j)^2$$

where the term $\mathbf{pp}^T$ is a rank one matrix with complexity order $O(N)$.

Perona proved that the minimizer p of the Frobenius norm is the first eigenvector v_1 of the matrix A with largest eigenvalue λ: $p = \lambda_1^{1/2} v_1$.

Then the problem of grouping is reduced to the spectral analysis of the affinity matrix $A_{i,i}$, where the salient objects in the scene correspond to the first eigenvectors with largest eigenvalues. We just showed in the previous paragraphs that this eigenvalue analysis can be accomplished by the activity of neural populations connected by means of cortico-cortical horizontal connectivity.

4.8.6 Constitution of Perceptual Units

The affinity matrix can be represented as a graph whose nodes are the active cells and the links are the neural connectivities with a weight that is defined in terms of the connectivity kernel $\omega((x_i,y_i,\theta_i)(x_j,y_j,\theta_j))$ (see Fig. 4.29).

Figure-ground articulation and the segmentation of different objects will correspond to different eigenvectors a_i. In Fig. 4.30 the three principle eigenvectors of the neurogeometrical matrix associated to Fig. 4.29 are visualized in gray values, showing the global emergence of perceptual units.

Let us note that this approach interprets the emergent figures in the image as eigenstates of the neurogeometrical matrix A_{ij}. Mathematically it corresponds to the singular value decomposition of the neurogeometrical graph previously defined. Co-occurring features are mapped to the same eigenvector; features that do not co-occur are mapped to different eigenvectors. The eigenvector linked to the highest

Fig. 4.29 Some perceptual units are perceived as coherent structures out of a random distribution of segments. In the image on the top the most salient features are the semicircle and a vertical segment. Neurogeometry lifts every segment in the E(2) group and induces connectivity weights between every couple of segments represented in colors (bottom). The affinity matrix A_{ij} is constructed with the corresponding weights between couples of segments $i-j$.

singular value represents the most important vector in the data (i.e. the vector that explains the most variance of the matrix); the singular vectors linked to the second highest value represent the second most important vector (orthogonal to the first one), and so on.

Models of image segmentation based on singular value decomposition and in general on dimensionality reduction are largely used in contemporary computer vision (see for example [45], [54]), and we refer to [58] for a review of methods. We have shown in this chapter how these sophisticated models of image segmentation can be implemented at a neural level in terms of harmonic analysis of the neurogeometry.

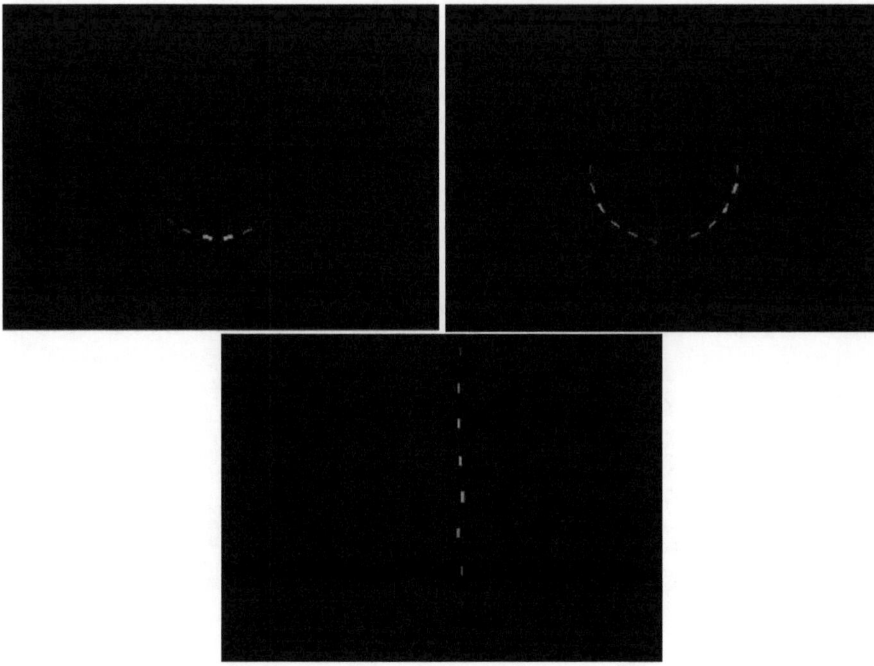

Fig. 4.30 Eigenvectors of the affinity matrix are visualized mapping the intensity in gray levels. The first a_i eigenvectors correspond to the most salient perceptual units.

Acknowledgement. The authors are grateful to Giacomo Cocci and Gonzalo Sanguinetti, who realized some of the figures present in the text and to Davide Barbieri for his patience and dedication in the formatting of the book.

References

1. August, J., Zucker, S.W.: Sketches with curvature: The curve indicator random field and markov processes. IEEE Transactions on Pattern Analysis and Machine Intelligence 25(4), 387–400 (2003)
2. Barbieri, D., Citti, G., Sanguinetti, G., Sarti, A.: An uncertainty principle underlying the functional architecture of V1. Journal of Physiology Paris 106(5-6), 183–193 (2012)
3. Barbieri, D., Citti, G.: Reproducing kernel Hilbert spaces of CR functions for the Euclidean Motion group (submitted)
4. Bargmann, V.: On a Hilbert space of analytic functions and an associated integral transform. Part II. Application to distribution theory. Comm. Pure Appl. Math. 20, 1–101 (1967)
5. Barles, G., Georgelin, C.: A simple proof of convergence for an approximation scheme for computing motions by mean curvature. SIAM Journal on Numerical Analysis 32(2), 484–500 (1995)
6. Bonfiglioli, A., Uguzzoni, F., Lanconelli, E.:

7. Bressloff, P.C., Cowan, J.D.: The functional geometry of local and long-range connections in a model of V1. J. Physiol. Paris 97(2-3), 221–236 (2003)
8. Bressloff, P.C., Cowan, J.D., Golubitsky, M., Thomas, P.J., Wiener, M.: Geometric visual hallucinations, Euclidean symmetry and the functional architecture of striate cortex Phil. Trans. Roy. Soc., B 40, 299–330 (2001)
9. Carruthers, P., Nieto, M.M.: Phase and Angle Variables in Quantum Mechanics. Reviews of Modern Physics 40(2), 411–440 (1968)
10. Capogna, L., Citti, G., Senni, C., Magnani, G.: Sub-Riemannian heat kernel and mean curvature flow of graphsJFA (2013)
11. Citti, G., Manfredini, M., Sarti, A.: Neuronal Oscillations in the Visual Cortex: Gamma-convergence to the Riemannian Mumford-Shah Functional. SIAM Journal Mathematical Analysis 35(6), 1394–1419 (2004)
12. Citti, G., Sarti, A.: A cortical based model of perceptual completion in the Roto-Translation space. Journal of Mathematical Imaging and Vision 24(3), 307–326 (2006)
13. Citti, G., Sarti, A.: A cortical based model of perceptual completion in the Roto-Translation space. Journal of Mathematical Imaging and Vision 24(3), 307–326 (2006)
14. Citti, G., Sarti, A.: A Gauge Field model of modal completion (submitted)
15. Cohen-Tannoudji, C., Dui, B., Laloe, F.: Quantum Mechanics. John Wiley & Sons (1978)
16. Crair, M., Ruthazer, E., Gillespie, D., Stryker, M.: Ocular Dominance Peaks at Pinwheel Center Singularities of the Orientation Map in Cat Visual Cortex. J. Neurophysiol. 77(6), 3381–3385 (1997)
17. Daugman, J.G.: Uncertainty-relation for resolution in space, spatial frequency and orientation optimized by two dimensional visual cortical filters. J. Opt. Soc. Amer. 2(7), 1160–1169 (1985)
18. De Angelis, G.C., Ozhawa, I., Freeman, R.D.: Receptive-field dynamics in the central visual pathways. Trends Neurosci. 18(10), 451–458 (1995)
19. Duits, R., Franken, E.M.: Left invariant parabolic evolution equations on SE(2) and contour enhancement via invertible orientation scores, part I: Linear left-invariant diffusion equations on SE(2). Q. Appl. Math. 68, 255–292 (2010)
20. Duits, R., Franken, E.M.: Left invariant parabolic evolution equations on SE(2) and contour enhancement via invertible orientation scores, part II: Nonlinear left-invariant diffusion equations on invertible orientation scores. Q. Appl. Math. 68, 293–331 (2010)
21. Duits, R., Führ, H., Janssen, B., Bruurmijn, M., Florack, L., Van Assen, H.: Evolution equations on Gabor transforms and their applications. arXiv:1110.6087 (2011)
22. Evans, L.C.: Convergence of an algorithm for mean curvature motion. Indiana Univ. Math. J. 42(2), 533–557 (1993)
23. Field, D.J., Hayes, A., Hess, R.F.: Contour integration by the human visual system: evidence for a local Association Field. Vision Research 33, 173–193 (1993)
24. Folland, G.: Harmonic analysis on phase space. Princeton University Press (1989)
25. Franken, E., Duits, R., ter Haar Romeny, B.M.: Nonlinear Diffusion on the 2D Euclidean Motion Group. In: Sgallari, F., Murli, A., Paragios, N. (eds.) SSVM 2007. LNCS, vol. 4485, pp. 461–472. Springer, Heidelberg (2007)
26. Georgiev, T.: Covariant derivatives and vision. In: Leonardis, A., Bischof, H., Pinz, A. (eds.) ECCV 2006. LNCS, vol. 3954, pp. 56–69. Springer, Heidelberg (2006)
27. Georgiev, T.: Vision, healing brush, and fibre bundles. In: Proceedings of SPIE - The International Society for Optical Engineering, vol. 5666, p. 33, pp. 293–305 (2005)
28. Gray, C.M., König, P., Engel, A.K., Singer, W.: Oscillatory responses in cat visual cortex exhibit inter-columnar synchronization which reflects global stimulus properties. Nature 338, 334–337 (1989)

29. Hladky, R., Pauls, S.: Minimal Surfaces in the Roto-Translation Group with Applications to a Neuro-Biological Image Completion Model. Journal of Mathematical Image and Vision 36, 1–27 (2010)
30. Hoffman, W.C.: The visual cortex is a contact bundle. Applied Mathematics and Computation 32, 137–167 (1989)
31. Horn, B.K.P.: Determining lightness from an image. Comput. Graph. Image Process., 277–299 (1974)
32. Hubel, D.H.: Eye, Brain and Vision. Scientific American Library, New York (1988)
33. Hubel, D.H., Wiesel, T.N.: Ferrier lecture: Functional architecture of macaque monkey visual cortex. Proceedings of the Royal Society of London Series B 198(1130), 1–59 (1977)
34. Kanizsa, G.: Grammatica del vedere. Il Mulino, Bologna (1980)
35. Kimmell, R., Elad, M., Shaked, D., Keshet, R., Sobel, I.: A Variational Framework for Retinex. International Journal of Computer Vision 52(1), 7–23 (2003)
36. Lee, T.S.: Image Representation Using 2D Gabor Wavelets. IEEE Transection of Pattern Analysis and Machine Intelligence 18(10), 959–971 (1996)
37. Land, E.H.: The Retinex theory of color vision. Scientific American 237(6), 108–128 (1977)
38. Land, E., McCann, J.: Lightness and Retinex theory. J. Opt. Soc. Amer. 61(1), 1–11 (1971)
39. Marr, D.: Vision. Freeman (1982)
40. Robert, C.P., Casella, G.: Monte Carlo statistical methods, 2nd edn. Springer (2004)
41. Niebur, E., Wörgötter, F.: Design Principles of Columnar Organization in Visual Cortex. Neural Computation 4(6), 602–614 (1994)
42. Merriman, B., Bence, J., Osher, S.J.: Diffusion generated motion by mean curvature. In: Taylor, J. (ed.) Proceedings of the Computational Crystal Growers Workshop, pp. 73–83. AMS, Providence (1992)
43. Morel, J.-M.: A PDE Formalization of Retinex Theory. IEEE Transactions on Image Processing 19(11) (2010)
44. Mumford, D.: Elastica and computer vision. In: Algebraic Geometry and its Applications, West Lafayette, IN, pp. 491–506. Springer, New York (1990, 1994)
45. Perona, P., Freeman, W.T.: A factorization approach to grouping. In: Burkhardt, H.-J., Neumann, B. (eds.) ECCV 1998. LNCS, vol. 1406, pp. 655–670. Springer, Heidelberg (1998)
46. Petitot, J., Tondut, Y.: Tondut, Vers une Neurogeometrie. Fibrations corticales, structures de contact et contours subjectifs modaux, Mathematiques. Informatique et Sciences Humaines, vol. 145, pp. 5–101. EHESS, CAMS, Paris (1999)
47. Petitot, J., Varela, F., Pachoud, B., Roy, J.-M.: Naturalizing Phenomenology: Issues in Contemporary Phenomenology and Cognitive Science. Stanford University Press (1999)
48. Sanguinetti, G., Citti, G., Sarti, A.: A Stochastic Model for Edges in Natural Images using Lie Groups. To appear in Journal of Vision (2010)
49. Sarti, A., Citti, G., Petitot, J.: The Symplectic Structure of the Primary Visual Cortex. Biol. Cybern. 98, 33–48 (2008)
50. Sarti, A., Citti, G., Petitot, J.: Functional geometry of the horizontal connectivity in the primary visual cortex. Journal of Physiology-Paris 103(1-2), 37–45 (2009)
51. Sarti, A., Citti, G., Manfredini, M.: From Neural Oscillations to Variational Problems in the Visual Cortex. Journal of Physiology Paris 97(2-3), 379–385 (2003)
52. Sarti, A., Citti, G.: On the origin and nature of neurogeometry. La Nuova Critica (2011)
53. Sarti, A., Citti, G.: Visual perceptual units as constrained hallucinations (submitted)

54. Shi, J., Malik, J.: Normalized cuts and image segmentation. In: Proc. IEEE Conf. Comp. Vis. Pat. Rec., pp. 731–737 (1997)
55. Sugiura, M.: Unitary representations and harmonic analysis, an introduction. Elsevier (1990)
56. Wang, D.: Emergent synchrony in locally coupled neural oscillators. IEEE Trans. Neural Networks 6, 941–948 (1995)
57. Webb, B., Tinsley, C.J., Vincent, C.J., Derrington, A.M.: Spatial Distribution of Suppressive Signals Outside the Classical Receptive Field in Lateral Geniculate Nucleus. J. Neurophysiol. 94, 1789–1797 (2005)
58. Weiss, Y.: Segmentation using eigenvectors: a unifying view. In: Proc. IEEE Int. Conf. on Comp. Vis., pp. 975–982 (1999)
59. Williams, L.R., Jacobs, D.W.: Local Parallel Computation of Stochastic Completion Fields. Neural Comp. 9(4), 859–881 (1997)
60. Wilson, H.R., Cowan, J.D.: Excitatory and inhibitory interactions in localized populations of model neurons. Biophys. J. 12, 1–24 (1972)
61. Zucker, S.W.: Differential geometry from the Frenet point of view: boundary detection, stereo, texture and color. In: Paragios, N., Chen, Y., Faugeras, O. (eds.) Handbook of Mathematical Models in Computer Vision, pp. 357–373. Springer, US (2006)

Chapter 5
Cuspless Sub-Riemannian Geodesics within the Euclidean Motion Group $SE(d)$

Remco Duits, Arpan Ghosh, Tom Dela Haije, and Yuri Sachkov

Abstract. We consider the problem $\mathbf{P}_{curve}$ of minimizing $\int_0^{\ell} \sqrt{\beta^2 + |\kappa(s)|^2} \mathrm{d}s$ for a planar curve having fixed initial and final positions and directions. Here κ is the curvature of the curve with free total length ℓ. This problem comes from a 2D model of geometry of vision due to Petitot, Citti and Sarti. Here we will provide a general theory on cuspless sub-Riemannian geodesics within a sub-Riemannian manifold in $SE(d)$, with $d \geq 2$, where we solve for their momentum in the general d-dimensional case. We will explicitly solve the curve optimization problem $\mathbf{P}_{\mathbf{curve}}$ in 2D (i.e. $d = 2$) with a corresponding cuspless sub-Riemannian geodesic lifted problem defined on a sub-Riemannian manifold within $SE(2)$. We also derive the solutions of $\mathbf{P}_{\mathbf{curve}}$ in 3D (i.e. $d = 3$) with a corresponding cuspless sub-Riemannian geodesic problem defined on a sub-Riemannian manifold within $SE(3)$. Besides exact formulas for cuspless sub-Riemannian geodesics, we derive their geometric properties, and we provide a full analysis of the range of admissible end-conditions. Furthermore, we apply this analysis to the modeling of association fields in neurophysiology.

5.1 Introduction

Curve optimization plays a major role both in imaging and visual perception. In imaging there exist many works on snakes and active contour modeling, whereas in visual perception illusionary contours arise in various optical illusions [38, 42]. Mostly, such snake and active contour models involve curve optimization in $\mathbb{R}^d$,

Remco Duits · Arpan Ghosh · Tom Dela Haije
IST/e, Eindhoven University of Technology, IST/e, Den Dolech 2,
NL-5600 MB, The Netherlands
e-mail: {R.Duits,A.Ghosh,T.C.J.Dela.Haije}@tue.nl

Yuri Sachkov
Program Systems Institute, Russian Academy of Sciences
e-mail: sachkov@sys.botik.ru

G. Citti and A. Sarti (eds.), *Neuromathematics of Vision*,
Lecture Notes in Morphogenesis,
DOI: 10.1007/978-3-642-34444-2_5,

$d \geq 2$, that rely on Euler's elastica curves [24] (minimizing $\int(|\kappa|^2 + \beta^2)\mathrm{d}s$) in order to obtain extensions where typically external forces to the data are included, cf. [9, 12, 50–52] .

The elastica problem suffers from the well-known fact that not every stationary curve is a global minimizer, e.g. many local minimizers exist, cf. Figure 5.1. Stationarity of a curve can be reasonably checked by the visual system using local perturbations, whereas checking for (global) optimality [45, 46] is much more difficult. Some visual illusions (e.g. the Kanisza triangle) involve corners requiring abrupt resetting of initial and ending conditions, which are difficult to explain in the elastica model. Another problem with elastica is that it is very hard to solve the boundary value problem analytically [3, 4], and typically require $(2d-1)$-dim. shooting schemes. On top of that elastica curves relate to modes of the direction process (for contour-completion [21, 23, 38, 50]) where the direction of an oriented random walker is deterministic and its orientation is random. Such deterministic propagation only makes sense when the initial orientation is sharply defined. Instead Brownian motion with random behavior both in spatial propagation direction and in orientation direction ([2, 13, 19, 22]), relates to hypo-elliptic diffusion on the planar roto-translation group. Such a Brownian motion models contour enhancement [19] rather than contour completion [21], see [17] for a short overview. The corresponding Brownian bridge measures [22, 57] (relating to so-called completion fields in imaging [3, 21, 53]) tend to concentrate towards optimal sub-Riemannian geodesics [7, 13, 20, 35, 37, 45]. So both elastica curves and sub-Riemannian geodesics relate to two different fundamental left-invariant stochastic processes on sub-Riemannian manifolds on the 2D-Euclidean motion group [21] (respectively to the direction process and to hypo-elliptic Brownian motion).

In short, advantages of the sub-Riemannian geodesic model over the elastica model are:

- If $d = 2$, every cuspless sub-Riemannian geodesic is a global minimizer [8, 15].
- The Euler-Lagrange ODE for momentum (including normalized curvature vector $\underline{\kappa}/\sqrt{|\underline{\kappa}|^2 + \beta^2}$) can be reduced to a linear one,
- The boundary value problem can be tackled via effective analytic techniques,
- If $d = 2$, the locations where global optimality is lost can be derived explicitly.
- Sub-Riemannian geodesics (in contrast to lifted elastica) are parametrization independent in the roto-translation group $SE(d)$. Here we note in case $d = 2$, the sub-Riemannian manifold $(SE(2), \Delta_2, G_\beta)$ is encoded via a pinwheel structure of cortical columns in the primary visual cortex [41].

However, the practical drawback of sub-Riemannian geodesics compared to elastica is that their spatial projections may exhibit cusps and it is hard to analyze when such a cusp occurs. Therefore, in this article we provide a complete analysis of such sub-Riemannian geodesics, their parametrization, solving the boundary value problem, and we show precisely when a cusp occurs. See Figure 5.3 and see Figure 5.2.

A variant of the sub-Riemannian problem that ensures avoiding cusps is the following variational problem, which we will explain next.

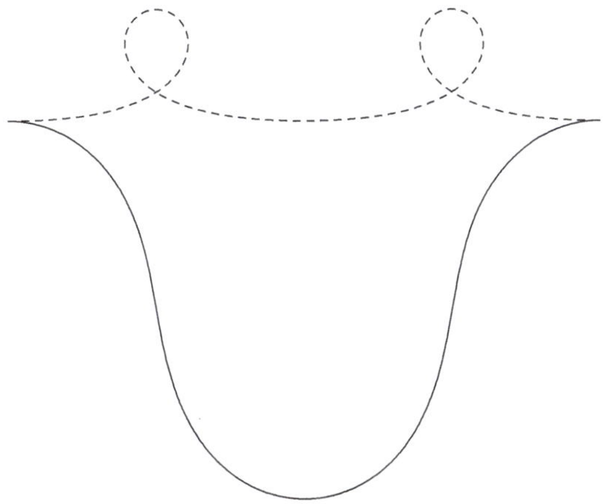

Fig. 5.1 Stationary curves of the elastica problem ($\int_0^\ell \kappa^2(s) + \beta^2 \mathrm{d}s \to \min$) do not need to be global minimizers, cf. [44]. E.g. the non-dashed elastica is a global minimum (for $\beta = 1$), whereas in dashed lines we have depicted a local minimum. This is in contrast to cuspless sub-Riemannian geodesics in $(SE(2), \Delta_2, G_\beta)$ where every stationary curve is globally optimal.

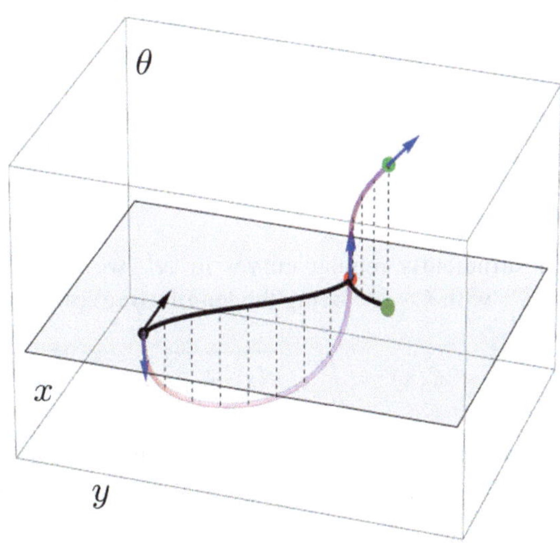

Fig. 5.2 An example of a smooth sub-Riemannian geodesic in $(SE(2), \Delta_2, G_\beta)$ whose spatial projection shows a cusp

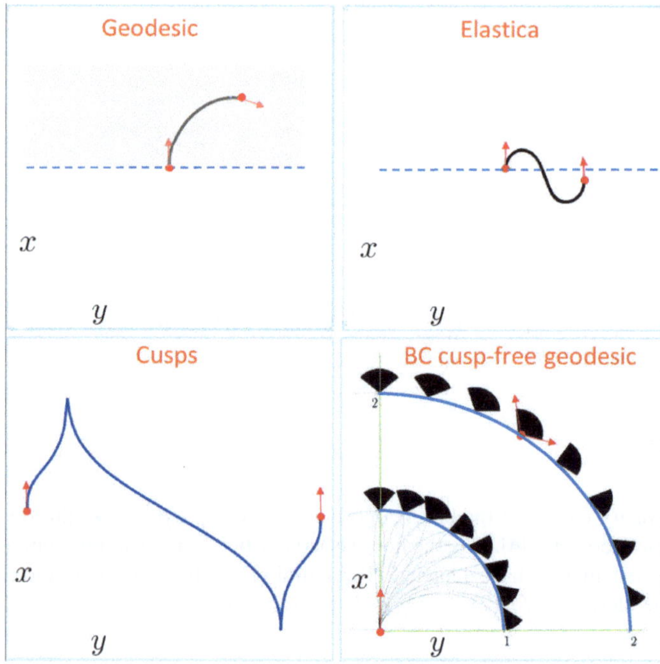

Fig. 5.3 Top left: example of a spatially projected sub-Riemannian geodesic without cusp (i.e. a solution of $\mathbf{P_{curve}}$). Top right: example of an elastica curve reaching points $x < 0$. Such a (weak) connection is not possible with cuspless sub-Riemannian geodesics. Instead we see in the bottom left figure a comparable example of a spatially projected sub-Riemannian geodesic connecting the $g_{in} = (0,0,0)$ with $g_{fin} = (0, y_{fin}, 0)$ via two cusps. Bottom right: not all points in $x \geq 0$ can be reached via a globally minimizing geodesic, here we have depicted the set $\mathfrak{R}$ of admissible end-conditions $g_{fin} = (x_{fin}, y_{fin}, \theta_{fin})$ via black cones on half circles with radius 1 and 2.

On the space of sufficiently regular curves in $\mathbb{R}^d$, we define a functional $\mathscr{E}$: $W^{2,1}([0,\ell],\mathbb{R}^d) \to \mathbb{R}^+$, with $\ell \in \mathbb{R}^+$ being the length (free) of the curves, by

$$\mathscr{E}(\mathbf{x}) := \int_0^\ell \sqrt{\kappa(s)^2 + \beta^2}\, \mathrm{d}s. \tag{5.1}$$

Here, s denotes the arc-length parameter of curve $\mathbf{x}$ and $\kappa : [0,\ell] \to \mathbb{R}^+ \cup \{\infty\}$ denotes the absolute curvature $\|\ddot{\mathbf{x}}(\cdot)\|$ of the curve $\mathbf{x}$ at each arc-length, and $\beta > 0$ is a constant.

The two dimensional case (i.e. $d = 2$) of this variational problem was studied as a possible model of the mechanism used by the primary visual cortex V1 of the human brain to reconstruct curves which are partially hidden or corrupted. The two dimensional model was initially due to Petitot (see [40, 41] and references therein). Subsequently, the sub-Riemannian structure was introduced in the problem by Petitot [42] for the contact geometry of the fiber bundle of the 1-jets of curves in the plane (the polarized Heisenberg group) and by Citti and Sarti [13, 47] for the principal bundle on $SE(2)$ also considered in this article. The stationary curves of the problem were derived and studied by Boscain, Charlot and Rossi in [7], by Duits in [20], by Sachkov in [35] and their global optimality is shown by Boscain, Duits, Rossi and Sachkov in [8, 15]. The two dimensional problem relates to a mechanical problem completely solved by Sachkov [37, 45, 46]. It was also studied by Hladky and Pauls in [32] and by Ben-Yosef and Ben-Shahar in [6]. Within Section 3 of this article we will summarize only the main results from our previous works [8, 15]. For detailed proofs of these results, we refer the reader to [8, 15].

Many imaging applications such as DW-MRI require an extension to higher dimensions, see e.g. [14, 23, 27, 50], which motivates us to study the higher dimensional curve optimization of the functional given by Eq. (5.1).

Let $\mathbf{x}_0, \mathbf{x}_1 \in \mathbb{R}^d$ and $\mathbf{n}_0, \mathbf{n}_1 \in S^{d-1} = \{\mathbf{v} \in \mathbb{R}^d | \|\mathbf{v}\| = 1\}$. The goal is to find an arc-length parameterized curve $s \mapsto \mathbf{x}(s)$ such that

$$\mathbf{x} = \arg \inf_{\substack{\mathbf{y} \in W^{2,1}([0,\ell],\mathbb{R}^d),\ell, \\ \mathbf{y}(0)=\mathbf{x}_0,\ \dot{\mathbf{y}}(0)=\mathbf{n}_0, \\ \mathbf{y}(\ell)=\mathbf{x}_1,\ \dot{\mathbf{y}}(\ell)=\mathbf{n}_1}} \mathscr{E}(\mathbf{y}). \tag{5.2}$$

We shall refer to this curve optimization problem as problem **P**. We assume that the boundary conditions $(\mathbf{x}_0, \mathbf{n}_0)$ and $(\mathbf{x}_1, \mathbf{n}_1)$ are chosen such that a minimizer exists.

Remark 5.1. Due to rotation and translational invariance of the problem **P**, it is equivalent to the problem with the same functional, but with boundary conditions $(\mathbf{0}, \mathbf{a})$ and $(R_{\mathbf{n}_0}^T(\mathbf{x}_1 - \mathbf{x}_0), R^T\mathbf{n}_1)$, where $\mathbf{a} \in S^{d-1}$ is a fixed axis, and with $R_{\mathbf{n}_0}$ is any rotation that maps a fixed reference axis $\mathbf{a}$ to $\mathbf{n}_0 \in S^{d-1}$.

Remark 5.2. The physical dimension of parameter β is $[\text{Length}]^{-1}$. From a physical point of view it is crucial to make the energy integrand dimensionally consistent. However, the problem with $\beta > 0$ is equivalent up to a scaling to the problem with $\beta = 1$: The minimizer $\mathbf{x}$ of **P** with $\beta > 0$ and boundary conditions $(\mathbf{x}_1, \mathbf{n}_1)$ relates to the minimizer $\overline{\mathbf{x}}$ of **P** with $\beta = 1$ and boundary condition $(\beta\mathbf{x}_1, \mathbf{n}_1)$ by spatial re-scaling, $\mathbf{x}(s) = \beta^{-1}\overline{\mathbf{x}}(s)$.

Therefore, without loss of generality, we set (unless explicitly stated otherwise)

$$\beta = 1,\ \mathbf{x}_0 = \mathbf{0},\ \text{and}\ \mathbf{n}_0 = \mathbf{e}_d$$

for the remainder of the article. Hence, the problem now is to find a sufficiently smooth arc-length parameterized curve $s \mapsto \mathbf{x}(s)$ such that

$$\mathbf{x} = \arg \inf_{\substack{\mathbf{y} \in W^{2,1}([0,\ell],\mathbb{R}^d),\ell \geq 0, \\ \mathbf{y}(0)=\mathbf{0},\ \dot{\mathbf{y}}(0)=\mathbf{e}_d, \\ \mathbf{y}(\ell)=\mathbf{x}_1,\ \dot{\mathbf{y}}(\ell)=\mathbf{n}_1}} \mathscr{E}(\mathbf{y}). \tag{5.3}$$

We refer to the above problem as $\mathbf{P_{curve}}$.

We stress that there are restrictions on the boundary conditions for problem **P** and problem $\mathbf{P_{curve}}$ to be well-posed [7]. For instance in case $d=2$, one must have

$$\left(R_{\mathbf{n}_{in}}^{-1}(\mathbf{x}_{in}-\mathbf{x}_{fin}), R_{\mathbf{n}_{in}}^{-1}\mathbf{n}_{fin}\right) \in \mathfrak{R}, \tag{5.4}$$

where in $d=2$ we have $\mathbf{n}_{fin} = (\cos(\theta_{fin}), \sin(\theta_{fin}))^T$ and where $\mathfrak{R}$ denotes the range of the exponential map [8, 15]. Roughly speaking this means that the endpoint in $\mathbf{P_{curve}}$ must be chosen such that it can be connected with a stationary curve.

In the d-dimensional setting criterium (5.4) is necessary, but so far it is still an open problem whether it is sufficient (for $d>2$). Therefore, in this article we will lift and extend problem $\mathbf{P_{curve}}$ to a problem $\mathbf{P_{mec}}$ of finding sub-Riemannian geodesics within $SE(d)$, which is well-posed regardless the end-condition. Then subsequently, we assume the end-condition in problem $\mathbf{P_{curve}}$ is chosen such that this end-condition gives rise to sub-Riemannian geodesic without cusps (i.e. $\dot{x}(t) \neq 0$ at the interior of the curve). Criterium (5.4) is then satisfied. Before we can formally introduce this problem $\mathbf{P_{mec}}$ we need some preliminaries.

5.1.1 Preliminaries and Notations

- The group of rotations in $\mathbb{R}^d$ equals $SO(d) = \{R \in \mathbb{R}^{d\times d} \,|\, R^T = R^{-1}, \det(R) = 1\}$
- The special Euclidean motion group on $\mathbb{R}^d$ is given by the semi-direct product $SE(d) = \mathbb{R}^d \rtimes SO(d)$. Its group elements are denoted by $g = (\mathbf{x}, R)$ and it is endowed with group product $(\mathbf{x}_1, R_1)(\mathbf{x}_2, R_2) = (R_1\mathbf{x}_2 + \mathbf{x}_1, R_1R_2)$. Its unity element equals $e = (\mathbf{0}, I)$ with I denoting the $d\times d$ identity matrix. The group $SE(d)$ acts on the set $\mathbb{R}^d \times S^{d-1}$ via

$$(\mathbf{x}, R)(\mathbf{y}, \mathbf{n}) = (R\mathbf{y} + \mathbf{x}, R\mathbf{n}). \tag{5.5}$$

- Let $\mathbf{a} \in S^d$ be a fixed element on the d-dimensional Euclidean sphere $S^d := \{\mathbf{x} \in \mathbb{R}^d \,|\, \|\mathbf{x}\| = 1\}$. We set $\mathbf{a} = \mathbf{e}_d$, e.g. if $d=3$ we set $\mathbf{a} = \mathbf{e}_z := (0,0,1)^T$, if $d=2$ we set $\mathbf{a} = \mathbf{e}_y := (0,1)^T$.
- Let $d \geq 2$. The coupled space of positions and directions is defined as the Lie group quotient

$$\mathbb{R}^d \rtimes S^{d-1} := SE(d)/(\{\mathbf{0}\} \times SO(d-1)) \tag{5.6}$$

  where we identify $SO(d-1)$ with $\{R \in SO(d) \,|\, R\mathbf{a} = \mathbf{a}\}$. For simplicity elements of $\mathbb{R}^d \rtimes S^{d-1}$ are denoted by $(\mathbf{y}, \mathbf{n})$ with $\mathbf{y} \in \mathbb{R}^d$ and $\mathbf{n} \in S^{d-1}$, where we keep in mind that each element represents a left coset within $SE(d)$. Such a left-coset contains equivalent rigid body motions that map $(\mathbf{0}, \mathbf{a})$ to $(\mathbf{y}, \mathbf{n})$ via the rigid body motion action (5.5):

$$(\mathbf{y}, \mathbf{n}) = (\mathbf{y}, R_{\mathbf{n}})(\mathbf{0}, \mathbf{a})$$

  where $R_{\mathbf{n}} \in SO(d)$ denotes *any* rotation that maps $\mathbf{a}$ onto $\mathbf{n} \in S^{d-1}$.
- The left-invariant vector fields considered as differential operators acting on smooth functions $\phi : SE(d) \to \mathbb{R}$ are given by the push-forward of the left multiplication $L_g : SE(d) \to SE(d)$ given by $L_g h = gh$:

$$\mathscr{A}_g = (L_g)_* \mathscr{A}_e, \text{ i.e. } \mathscr{A}_g \phi = \mathscr{A}_e(\phi \circ L_g), \tag{5.7}$$

where $\mathscr{A}_e \in T_e(SE(d))$, with $T_e(SE(d))$ the tangent space at the unity element e.

- We choose[1] a basis in $T_e(SE(d))$, say $\{A_1,\ldots,A_d\} \cup \{A_{d+1},\ldots,A_{d(d+1)/2}\}$ with vector fields $A_j = \frac{\partial}{\partial x^j}$, $j = 1,\ldots,d$ acting only on the spatial part[2], and $\{A_{d+1},\ldots,A_{d(d+1)/2}\}$ acting only on the $SO(d)$ part. The matrix representations of the spatial generators are given by

$$A_k = \begin{pmatrix} 0 & \mathbf{e}_k \\ 0 & 0 \end{pmatrix} \in \mathbb{R}^{(d+1)\times(d+1)}, k = 1\ldots d.$$

The matrix representations of these angular generators in $T_e(SE(d))$ are given by

$$A_{d(d+1)/2-i+1} = \begin{cases} \begin{pmatrix} -E^n_{n+1} + E^{n+1}_n & \mathbf{0} \\ 0 & 0 \end{pmatrix}, & \text{if } i = n(n+1)/2 \\ & \text{for some } n \in \{1,\ldots,d-1\}, \\ \begin{pmatrix} E^{i_0}_{n+2} - E^{n+2}_{i_0} & \mathbf{0} \\ 0 & 0 \end{pmatrix}, & \text{if } i = n(n+1)/2 + i_0 \\ & \text{with } i_0 \in \{1,\ldots,n\} \\ & \text{for some } n \in \{1,\ldots,d-1\}. \end{cases} \tag{5.8}$$

with $E^i_j \in \mathbb{R}^{d\times d}$ a matrix with all zero elements except for a unity 1 in row i and column j. In this way we have

$$\text{span}\{\mathscr{A}_{d+1},\ldots,\mathscr{A}_{d(d+1)/2}\} \equiv T_e(SO(d)) = so(d) = \text{span}\{E^i_j - E^j_i \mid 1 \le i < j \le d\}.$$

Furthermore, we observe that the angular generators are ordered such that

$$\text{span}\{A_{2d},\ldots,A_{d(d+1)/2}\} \equiv T_e(\{0\} \times SO(d-1)). \tag{5.9}$$

So in view of $\mathbb{R}^d \rtimes S^{d-1}$, Eq. (5.6), the redundant directions in $SE(d)$ (i.e. the angular generators of the stabilizing sub-group of $\mathbf{a}$, which is isomorphic to $SO(d-1)$) come at the end. To this end we note that

$$\begin{aligned} d(d+1)/2 = \dim(SE(d)) &= \dim(\mathbb{R}^d \rtimes S^{d-1}) + \dim(SO(d-1)) \\ &= (2d-1) + (d-2)(d-1)/2. \end{aligned} \tag{5.10}$$

Via the push-forward $(L_g)_*$ of the left-multiplication, Eq. (5.7), this basis $\{A_1,\ldots,A_{d(d+1)/2}\}$ provides us a moving frame of reference in the group $SE(d)$. This basis will be denoted by $\{\mathscr{A}_1,\ldots,\mathscr{A}_{d(d+1)/2}\}$ with

$$\mathscr{A}_i\big|_g = (L_g)_* A_i \text{ , for all } i = 1\ldots,d(d+1)/2, g \in SE(d), \tag{5.11}$$

[1] The main results (in contrast to the structure constants $c^k_{i,j}$) in this article do not depend on this choice of basis, one may choose a different basis with an ordering such that (5.9) holds.

[2] In previous works [19,20] on $SE(2)$, different ordering conventions are used in the Lie-algebra, and we set $\mathbf{a} = \mathbf{e}_x$ instead of $\mathbf{a} = \mathbf{e}_y$. In subsections 3.2–3.5 we will also adhere to that convention.

with $A_i = \mathscr{A}_i|_{g=e}$. The corresponding dual frame $\{\omega^k\}_{k=1,\ldots,d(d+1)/2}$ is given by

$$\langle \omega^k\Big|_g, \mathscr{A}_j|_g\rangle = \delta^k_j, \text{ for all } k,j = 1,\ldots,d(d+1)/2 \text{ and all } g \in SE(d), \quad (5.12)$$

where δ^i_j are the usual components of the Kronecker tensor. Explicit formulas for the frame of left-invariant vector fields and corresponding dual frame are derived in [18, 20]. Note that the vector space of left-invariant vector fields forms a Lie-algebra, with structure constants $c^k_{i,j}$ given by

$$[\mathscr{A}_i, \mathscr{A}_j] = \mathscr{A}_i\mathscr{A}_j - \mathscr{A}_j\mathscr{A}_i = \sum_{k=1}^{d(d+1)/2} c^k_{i,j}\mathscr{A}_k. \quad (5.13)$$

- Within this article we consider the sub-Riemannian manifold $(SE(d), \Delta_d, G_\beta)$, with base manifold $SE(d)$, and with distribution Δ_d, and metric tensor G_β : $SE(d) \times \Delta_d \times \Delta_d \to \mathbb{R}$ given by

$$\begin{aligned} &\Delta_d = \text{span}\{\mathscr{A}_d, \ldots, \mathscr{A}_{2d-1}\}, \\ &G_\beta\Big|_g \left(\sum_{i=d}^{2d-1} b^i \, \mathscr{A}_i|_g, \sum_{j=d}^{2d-1} c^j \, \mathscr{A}_j\Big|_g \right) = \beta^2 b^d c^d + \sum_{i=d+1}^{2d-1} b^i c^i, \end{aligned} \quad (5.14)$$

for all $\mathbf{b} = (b^i)_{i=d}^{2d-1}, \mathbf{c} = (c^i)_{i=d}^{2d-1} \in \mathbb{R}^d$. So by construction the horizontal left-invariant vector fields $\{\mathscr{A}_i\}_{i=d}^{2d-1}$ form an orthonormal basis in Δ_d w.r.t. metric tensor G_1.

5.1.2 *Lifting Problem* $\mathbf{P_{curve}}$ *to Problem* $\mathbf{P_{mec}}$ *on* $(SE(d), \Delta_d, G_\beta)$

Now we relate the problem $\mathbf{P_{curve}}$ to a sub-Riemannian problem $\mathbf{P_{mec}}$ on the Lie group quotient $\mathbb{R}^d \rtimes S^{d-1}$ given by Eq. (5.6). We define this sub-Riemannian problem by means of the left-invariant frame $\{\mathscr{A}_i\}_{i=d}^{d(d+1)/2}$, recall Eq. (5.11), and its left-invariant co-frame $\{\omega^i\}_{i=1}^{d(d+1)/2}$ given by Eq. (5.12). Within this frame, we will consider the horizontal part only, where we recall that the d-dimensional distribution Δ_d is given by Eq. (5.14) where indices run from d to $2d-1$. See Figure 5.4 for a visualization of this left-invariant frame in case $d = 3$. We will define $\mathbf{P_{mec}}$ on the sub-Riemannian manifold $(SE(d), \Delta_d, G_\beta)$, with distribution Δ_d and metric tensor G_β given by Eq. (5.14), i.e.

$$G_\beta = \beta^2 \omega^d \otimes \omega^d + \sum_{i=d+1}^{2d-1} \omega^i \otimes \omega^i.$$

In the geometric control problem $\mathbf{P_{mec}}$ on $SE(d)$, we use the sub-Riemannian arc-length parameter t. In $\mathbf{P_{mec}}$, we aim for curves $\gamma\colon [0,T] \to SE(d)$, with prescribed boundary conditions $\gamma(0) = (\mathbf{0}, I)$ and $\gamma(T) = (\mathbf{x}_1, R_{\mathbf{n}_1})$, such that

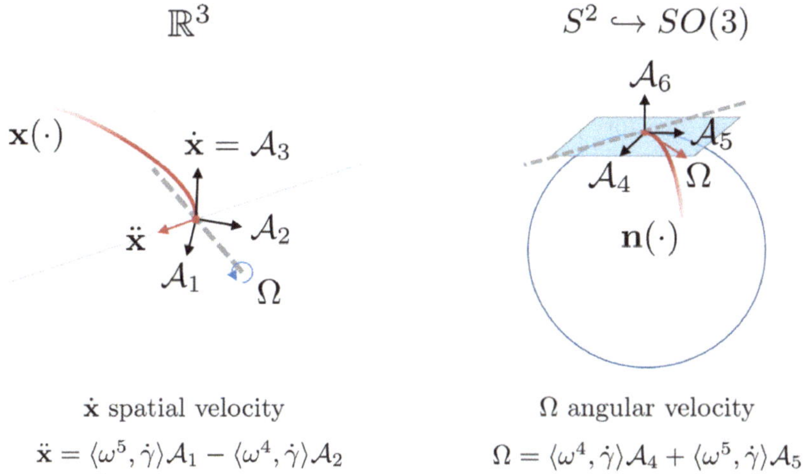

Fig. 5.4 Illustrations of the left invariant frame representing a moving frame of reference along a curve on $\mathbb{R}^3 \rtimes S^2$, i.e. $d = 3$. The spatial velocity and the angular velocity are depicted in the frame to highlight the constraints between the spatial and angular frame.

$$\int_0^T \sqrt{\mathcal{G}_\beta|_{\gamma(t)}(\dot{\gamma}(t), \dot{\gamma}(t))}\, dt = \int_0^T \sqrt{\beta^2 \left(u^d(t)\right)^2 + \sum_{i=d+1}^{2d-1} (u^i(t))^2}\, dt \tag{5.15}$$
$$\to \text{minimize (with free } T)$$

with

$$\dot{\gamma}(t) = \sum_{i=d}^{2d-1} u^i(t) \mathscr{A}_i|_{\gamma(t)} = \sum_{i=d}^{2d-1} \langle \omega^i|_{\gamma(t)}, \dot{\gamma}(t) \rangle \mathscr{A}_i|_{\gamma(t)}$$

where, $u^i \in \mathbb{L}_1([0,T])$ for $i = d, \ldots, 2d-1$ and $R_{\mathbf{n}_1} \in SO(d)$ is any rotation such that $R_{\mathbf{n}_1}\mathbf{a} = \mathbf{n}_1$. In particular, we only consider the stationary curves for which the absolute curvature is $\mathbb{L}_1$ rather than $\mathbb{L}_\infty$.

The existence of minimizers for the problem $\mathbf{P_{mec}}$ is guaranteed by the theorems by Chow-Rashevskii and Filippov on sub-Riemannian structures [1]. We consider those boundary conditions, for which a minimizer of $\mathbf{P_{mec}}$ does not admit an internal cusp (i.e. an interior point with infinite curvature). Clearly, such minimizers are also geodesics. We have the following important remarks about these minimizers.

Remark 5.3.
- We have that for $\mathbf{P_{mec}}$, there are no abnormal extremals. It follows from the fact that any sub-Riemannian manifold with a 2-generating distribution does not allow abnormal extremals (see Chapter 20.5.1 in [1]). This is the case, since for $\Delta_d := \{\mathscr{A}_d, \ldots, \mathscr{A}_{2d-1}\}$ we have $\dim(\Delta_d + [\Delta_d, \Delta_d]) = d(d+1)/2 = dim(SE(d))$.
- Due to the non-existence of abnormal extremals, the minimizers are always analytic [1].

Remark 5.4. Geodesics of $\mathbf{P_{mec}}$ may lose local and/or global optimality after the end condition at a conjugate point, or global optimality at a Maxwell point. A Maxwell point is a point $\gamma(t)$ on a sub-Riemannian geodesic γ such that $\gamma(t) = \tilde{\gamma}(t)$ for another extremal trajectory $\tilde{\gamma}$ with initial condition satisfying $\tilde{\gamma}(0) = \gamma(0)$. A conjugate point on the other hand is a critical point of the exponential map underlying the geometric control problem (cf. Theorem 21.11 in [1]). Here the exponential map maps each allowable pair $(\lambda(0), \ell)$, with initial momentum $\lambda(0)$ and length ℓ, to the endpoint $\gamma(\ell)$ of the corresponding cuspless sub-Riemannian geodesic $s \mapsto \gamma(s)$ that arises from integrating the canonical Euler-Lagrange or Hamiltonian equations (e.g. obtained via the Pontryagin Maximum Principle). We will provide explicit tangible formulas for this exponential map in case $d \in \{2,3\}$.

Remark 5.5. Throughout this article we will associate to a curve γ in $(SE(d), \Delta_d, G_\beta)$ a corresponding curve $\overline{\gamma}$ in $\mathbb{R}^d \rtimes S^{d-1}$ by setting

$$\begin{aligned}&(SE(d), \Delta_d, G_\beta) \ni \gamma(s) = (\mathbf{x}(s), R(s)) \to \\ &\overline{\gamma}(s) := (\mathbf{x}(s), \mathbf{n}(s)) \in \mathbb{R}^d \rtimes S^{d-1}, \text{ with } \mathbf{n}(s) := R(s)\mathbf{a}.\end{aligned} \tag{5.16}$$

In the remainder of this article, we will write $\gamma(s)$ both for curves in $(SE(d), \Delta_d, G_\beta)$ and for its associated curve in $\mathbb{R}^d \rtimes S^{d-1}$ as it is clear from the context what is meant.

The energy functional in Problem $\mathbf{P_{curve}}$ and Problem $\mathbf{P_{mec}}$ coincide for arc-length parameterizable curves $\gamma(\cdot) = (\mathbf{x}(\cdot), R(\cdot))$ in $(SE(d), \Delta_d, G_\beta)$, as we have

$$\begin{aligned}&\sum_{i=d+1}^{2d-1} |\langle \omega^i|_{\gamma(s)}, \dot{\gamma}(s)\rangle|^2 = \|\underline{\kappa}(s)\|^2 = \kappa^2(s), \\ &\langle \omega^d|_{\gamma(s)}, \dot{\gamma}(s)\rangle = \|\dot{\mathbf{x}}(s)\| = 1,\end{aligned} \tag{5.17}$$

where $\underline{\kappa}(s) = \ddot{\mathbf{x}}(s)$ denotes the curvature vector and $\kappa(s)$ the curvature magnitude at $\mathbf{x}(s)$ along the spatial part $\mathbf{x}$ of the curve γ.

Remark 5.6. Stationary curves of Problem $\mathbf{P_{curve}}$ and the spatial part of stationary curves of Problem $\mathbf{P_{mec}}$ coincide if the end-condition $(\mathbf{x}_1, \mathbf{n}_1) \in \mathbb{R}^d \rtimes S^{d-1}$ is chosen such that it can be connected by a stationary curve of $\mathbf{P_{curve}}$ (i.e. if the end condition is contained within the range $\mathfrak{R}$ of the exponential map of Problem $\mathbf{P_{curve}}$).
From now on, such end conditions will be called *admissible end conditions*. E.g. for $d = 2$, we have shown [8] that for each admissible end condition, Problem $\mathbf{P_{curve}}$ is well-posed and there exists a unique stationary curve connecting the origin $(\mathbf{0}, \mathbf{a})$ with $(\mathbf{x}_1, \mathbf{n}_1)$ that is the global minimum of Problem $\mathbf{P_{curve}}$. Furthermore, in [15] we have explicitly derived the set of admissible conditions $\mathfrak{R}$. We will summarize these results in Section 5.3.

5.1.3 Structure of the Article

In Section 5.2, we will derive general results for (cuspless) sub-Riemannian geodesics in $(\mathbb{R}^d \rtimes S^{d-1}, \Delta_d, G_\beta)$. We apply the Pontryagin maximum principle and we will show that for cuspless sub-Riemannian geodesics, the phase portrait of momentum reduces to a d-fold planar hyperbolic phase portrait. We express their momentum in terms of the initial momentum accordingly. We show that momentum is parallel transported along the geodesics w.r.t. a Cartan connection, and we derive a theorem allowing explicit integration to the sub-Riemannian geodesics from their momentum. Finally, we show that cuspless sub-Riemannian geodesics are a good model for association fields obtained in neurophysiology and neuropsychology.

In Section 3, we consider the special case $d = 2$, where we derive the unique globally optimal cuspless sub-Riemannian geodesics and their properties. We also carefully analyze the set $\mathfrak{R}$ of admissible end conditions, which is contained in $x \geq 0$, and for which we solve the boundary value problem associated with Problem $\mathbf{P_{curve}}$ via a semi-analytic method, allowing for a *1D* numerical shooting algorithm to solve the boundary value problem. We also obtain a description and computation of the piecewise smooth boundary $\partial\mathfrak{R}$. From this description we deduce that the extreme orientations per positions are given by endpoints of geodesics ending in a cusp and/or departing from a cusp.

In Section 4, we consider the special case $d = 3$, where we explicitly derive the stationary curves for admissible end conditions (allowing a connection via cuspless sub-Riemannian geodesics). We express their torsion and curvature in terms of momentum, from which we deduce a wide range of geometrical properties. E.g. we show that if the boundary-conditions are co-planar, we obtain the sub-Riemannian geodesics with $d = 2$. Numerical computations show that the sub-Riemannian geodesics are again contained within cones determined by endpoints of those geodesics that end and/or depart from a cusp, supporting (together with the co-planarity results) our *conjecture* that the exponential map of the geometric control problem has similar homeomorphic and diffeomorphic properties as in the case $d = 2$, leaving a challenging open problem for future research. Furthermore, we show that the extreme sub-Riemannian geodesics departing from a cusp will be contained entirely in the half-plane $z \geq 0$.

In Section 5 we consider the special case $d = 4$, where we explicitly derive momentum of the stationary curves.

5.2 Sub-Riemannian Geodesics in $(\mathbb{R}^d \rtimes S^{d-1}, \Delta_d, G_1)$

A general well-established tool to deal with geometric control problems, following a Hamiltonian approach, is the Pontryagin Maximum Principle (PMP) [1, 43, 55]. In Appendix A, we formally apply the Pontryagin maximum principle to problem $\mathbf{P_{mec}}$ of finding sub-Riemannian geodesics in the sub-Riemannian manifold $(SE(d), \Delta_d, G_1)$. There we also include techniques from theoretical mechanics, following a Lagrangian optimization approach as proposed by [10, 11] which produces the same canonical equations, to simplify the canonical equations considerably. The resulting equations are surprisingly simple and structured as we will show next. Let

$$\lambda(t) = \sum_{i=1}^{d(d+1)/2} \lambda_i(t)\, \omega^i\big|_{\gamma(t)} = \sum_{i=1}^{2d-1} \lambda_i(t)\, \omega^i\big|_{\gamma(t)}$$

denote the momentum along the sub-Riemannian geodesics (stationary curves $t \mapsto \gamma(t) = (\mathbf{x}(t), R(t))$) expressed in sub-Riemannian arc-length t. Here,

$$\lambda_i = 0 \text{ for all } 2d \leq i \leq d(d+1)/2 = \dim(SE(d)),$$

since momentum does not contain components in the redundant directions $\omega^{2d} = \omega^{2d+1} = \ldots = \omega^{d(d+1)/2} = 0$, recall Eq. (5.10) and Eq. (5.6). To this end we recall that we are interested in connecting points in the Lie group quotient

$$\mathbb{R}^d \rtimes S^{d-1} = SE(d)/(\{0\} \times SO(d-1)),$$

and the $d(d-1)/2$-dimensional Lie-algebra spanned by $\{\mathscr{A}_{2d}, \ldots, \mathscr{A}_{d(d+1)/2}\}$ is precisely the Lie algebra of the $SO(d-1)$ subgroup (i.e. the Lie-algebra of the stabilizer subgroup of our arbitrarily fixed $\mathbf{a} \in S^{d-1}$). Then the canonical equations are given by

$$\boxed{\begin{aligned} \dot{\gamma}(t) &= \sum_{i=d}^{2d-1} \lambda_i(t)\, \mathscr{A}_i\big|_{\gamma(t)}, \\ \dot{\lambda}_i(t) &= -\sum_{j=d}^{2d-1} \sum_{k=1}^{2d-1} c_{i,j}^k \lambda_k(t) \lambda_j(t), \end{aligned}} \tag{5.18}$$

where the first equation relates to the horizontal part of PMP and the second equation to the vertical part of PMP. These equations can be combined in a single equation using a Cartan connection $\overline{\nabla}$ on a cotangent bundle of the sub-Riemannian manifold $(SE(d), \Delta_d, G_\beta)$ that is derived from a Cartan-Maurer form on the underlying principal fiber bundle (akin to [15, App.C] for $d = 2$ and [30, App.A] for $d = 3$). More explicitly, it turns out that (as we shall prove in Theorem 5.2):

$$\boxed{\overline{\nabla}_{\dot{\gamma}} \lambda = 0 \Leftrightarrow \sum_{i=d}^{2d-1} \left(\dot{\lambda}_i + \sum_{k=1}^{2d-1} \sum_{j=d}^{2d-1} c_{i,j}^k \dot{\gamma}^j \lambda_k \right) \omega^i = 0.}$$

with $\dot{\gamma}^k := \langle \omega^k\big|_\gamma, \dot{\gamma} \rangle$, which according to the first equality in (5.18) (i.e. the horizontal part of PMP) is equal to λ_k. Computation of (5.18), where we omit the vanishing structure constants, see (5.13), yields

$$
\begin{aligned}
&\dot{\lambda}_k(t) = -\lambda_d(t)\lambda_{2d-k}(t)c^d_{k,2d-k} \text{ for } k = 1\dots,d-1,\\
&\dot{\lambda}_d(t) = -\sum_{k=1}^{d-1} \lambda_k(t)\lambda_{2d-k}(t)c^k_{d,2d-k},\\
&\dot{\lambda}_k(t) = -\lambda_d(t)\lambda_{2d-k}(t)c^{2d-k}_{k,d} \text{ for } k = d+1,\dots,2d-1,\\
&\dot{\lambda}_k(t) = 0 \text{ for } k = 2d,\dots,d(d+1)/2.
\end{aligned}
\tag{5.19}
$$

Now, only for cusp-less sub-Riemannian geodesics γ of problem $\mathbf{P_{mec}}$, we switch to spatial arc-length parameter s, where we note that along such curves we have

$$
t'(s) = \sqrt{|\kappa(s)|^2 + 1} \text{ and } s'(t) = \lambda_d(t) \tag{5.20}
$$

which follows from $\|\dot{\mathbf{x}}(s)\| = 1$ and Eq. (5.18). On top of that, we use the cyclic property on the structure constants that holds for structure constants of $SE(d)$:

$$
c^d_{k,2d-k} = c^k_{2d-k,d}, \text{ for } k = 1\dots,d-1,
$$

and we obtain the following remarkably simple ODE system

$$
\begin{aligned}
&\ddot{\overline{\lambda}}_k(s) = \overline{\lambda}_k(s), \text{ for } k \neq d,\\
&(\overline{\lambda}_d(s))\dot{\overline{\lambda}}_d(s) = -\sum_{k=d+1}^{2d-1} (\overline{\lambda}_k(s))\dot{\overline{\lambda}}_k(s),
\end{aligned}
\tag{5.21}
$$

where $\overline{\lambda}_k(s) := \lambda_k(t(s))$ for $k = 1\dots,2d-1$. Then we use the fact (akin to [15,30]) that orbits in the augmented space of position and momentum are contained in the co-adjoint orbits[3] of $SE(d)$, i.e.

$$
\sum_{k=1}^{d} |\overline{\lambda_k}(s)|^2 = \sum_{k=1}^{d} |\overline{\lambda_k}(0)|^2 =: \mathfrak{c}^2, \tag{5.22}
$$

for all $0 \leq s \leq s_{max}$, where s_{max} will be computed later, and the fact that $\overline{\lambda}_d$ is positive (by Eq. (5.20)) to solve for momentum of cuspless sub-Riemannian geodesics:

$$
\begin{aligned}
&\overline{\lambda}_k(s) = \overline{\lambda}_k(0)\cosh(s) + \dot{\overline{\lambda}}_k(0)\sinh(s), \text{ for } k \neq d,\\
&\overline{\lambda}_d(s) = \sqrt{\mathfrak{c}^2 - \sum_{k=1}^{d-1} |\overline{\lambda}(s)|^2}.
\end{aligned}
\tag{5.23}
$$

Remark 5.7. In the remainder of this article, we will just write $\lambda(s)$ instead of $\overline{\lambda}(s)$. When writing $\dot{\lambda}(s)$, we mean $\frac{d}{ds}\lambda(s)$.

Remark 5.8. Besides preservation law (5.22), we deduce the preservation laws

$$
\sum_{i=d}^{2d-1} |\lambda_i|^2 = 1 \text{ and } W(\lambda_i, \lambda_j) := \lambda_i\dot{\lambda}_j - \lambda_j\dot{\lambda}_i = \lambda_i(0)\dot{\lambda}_j(0) - \lambda_j(0)\dot{\lambda}_i(0), \tag{5.24}
$$

[3] Conservation law (5.22) can also be deduced from the second part of Eq.(5.21).

where W denotes the (constant) Wronskian of λ_i and λ_j for each pair $i, j \in \{1, \ldots, d-1\}$ with $i \neq j$.

We represent the momentum co-vector $\lambda(s) = \sum_{i=1}^{2d-1} \lambda_i(s)\, \omega^i\big|_{\gamma(s)}$ by storing its components in a row-vector where we split spatial and angular part

$$\underline{\lambda} = (\underline{\lambda}^{(1)}, \lambda_d; \underline{\lambda}^{(2)}),$$

with $\underline{\lambda}^{(1)} = (\lambda_1, \ldots, \lambda_{d-1})$ (where the component index increases), $\underline{\lambda}^{(2)} = (\lambda_{2d-1}, \ldots, \lambda_{d+1})$ (where the component index decreases). Then Eq. (5.21) becomes

$$\begin{aligned}
&\dot{\underline{\lambda}}^{(1)}(s) = \Lambda \underline{\lambda}^{(2)}(s),\\
&\dot{\lambda}_d(s) = -(\lambda_d(s))^{-1} \sum_{k=1}^{d-1} \lambda_k(s)\lambda_{2d-k}(s) c^k_{d,2d-k},\\
&\dot{\underline{\lambda}}^{(2)}(s) = \Lambda \underline{\lambda}^{(1)}(s),
\end{aligned}$$

where $\Lambda = \mathrm{diag}\left(\{c^d_{k,2d-k}\}_{k=1}^{d-1}\right) \in \mathbb{R}^{(d-1)\times(d-1)}$ is a diagonal matrix whose diagonal elements are determined by the vector $\{c^d_{k,2d-k}\}_{k=1}^{d-1}$ whose elements are within $\{-1, 1\}$. Since $\Lambda^2 = I$, it produces the solutions

$$\begin{aligned}
&\underline{\lambda}^{(1)}(s) = \underline{\lambda}^{(1)}(0)\cosh(s) + \Lambda \underline{\lambda}^{(2)}(0)\sinh(s),\\
&\lambda_d(s) = \sqrt{1 - \|\underline{\lambda}^{(2)}(s)\|^2} = \sqrt{\mathfrak{c}^2 - \|\underline{\lambda}^{(1)}(s)\|^2},\\
&\underline{\lambda}^{(2)}(s) = \underline{\lambda}^{(2)}(0)\cosh(s) + \Lambda \underline{\lambda}^{(1)}(0)\sinh(s).
\end{aligned} \tag{5.25}$$

In turn, these formulas allows us to compute the arc-length towards a cusp

$$s_{max}(\lambda(0)) = \log\left(\frac{\sqrt{1 + \mathfrak{c}^2 + \sqrt{|1+\mathfrak{c}^2|^2 - \|\underline{\lambda}^{(1)}(0) + \Lambda\underline{\lambda}^{(2)}(0)\|\|\underline{\lambda}^{(1)}(0) - \Lambda\underline{\lambda}^{(2)}(0)\|}}}{\|\underline{\lambda}^{(1)}(0) + \Lambda\underline{\lambda}^{(2)}(0)\|}\right), \tag{5.26}$$

since at a cusp, we have $\lambda_d(t_{cusp}) = 0$ and we have

$$\lim_{s\uparrow s_{max}} \|\underline{\lambda}^{(2)}(s)\| = 1 \Leftrightarrow \lim_{s\uparrow s_{max}} |\lambda_d(s)| = 0 \Leftrightarrow \lim_{s\uparrow s_{max}} \kappa(s) \to \infty,$$

recall Eq. (5.20), with $t_{cusp} = t(s_{max})$.

Let us summarize these results on sub-Riemannian geodesics in $(SE(d), \Delta_d, G_1)$ in the following theorems.

Theorem 5.1. *Along the sub-Riemannian geodesics in $(SE(d), \Delta_d, G_1)$, the following canonical equations hold*

$$\begin{aligned} \dot{\gamma}(t) &= \sum_{i=d}^{2d-1} \lambda_i(t)\, \mathscr{A}_i|_{\gamma(t)}, \\ \dot{\lambda}_i(t) &= -\sum_{j=d}^{2d-1}\sum_{k=1}^{2d-1} c_{i,j}^k \lambda_k(t)\, \lambda_j(t), \end{aligned} \tag{5.27}$$

along cuspless sub-Riemannian geodesics (i.e. the sub-Riemannian geodesics that allow parametrization by spatial arc-length s) in $(SE(d), \Delta_d, G_\beta)$. Momentum $\lambda = \sum_{i=1}^{2d-1} \lambda_i \omega^i$ satisfies the simple ODE system Eq. (5.21), whose explicit solution is given by Eq. (5.25). The spatial arc-length towards a cups is given by Eq. (5.26).

Proof. For Eq. (5.27), see Appendix A. The remainder of the proof follows by the earlier derivations in between Eq. (5.19) and Eq. (5.26). □

Corollary 5.1. *The momentum orbit $s \mapsto \lambda(s)$ of a sub-Riemannian geodesic $s \mapsto \gamma(s)$ is determined by a d-fold hyperbolic phase portrait, see Figure 5.5,*

$$\frac{d}{ds}\begin{pmatrix} -c^i_{d,2d-i}\lambda_{2d-i}(s) \\ \lambda_i(s) \end{pmatrix} = \begin{pmatrix} 0 & -1 \\ 1 & 0 \end{pmatrix}\begin{pmatrix} -c^i_{d,2d-i}\lambda_{2d-i}(s) \\ \lambda_i(s) \end{pmatrix} \text{ for } i = 1,\dots,d,$$

and preservation law $\lambda_d(s) = \sqrt{1 - \sum_{i=d+1}^{2d-1} |\lambda_i(s)|^2}$ for all $s \le s_{max}(\lambda(0))$.

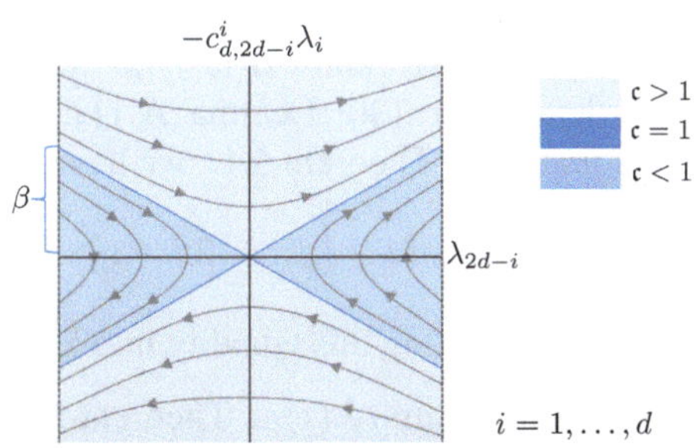

Fig. 5.5 In momentum space, sub-Riemannian geodesics reduce to a d-fold hyperbolic phase portrait, see Corollary 5.1

Theorem 5.2. • *Horizontal exponential curves given by $s \mapsto g_0\, Exp(s \sum_{i=d}^{2d-1} c^i A_i)$, with $c^d = 1$, $g_0 \in SE(d)$, are the auto parallel curves (i.e. $\nabla_{\dot{\gamma}}\dot{\gamma} = 0$) w.r.t. connection $\bar{\nabla}$ on the sub-Riemannian manifold $(SE(d), \Delta_d, G_\beta)$ given by*

$$\bar{\nabla}_X \mathscr{A} = \sum_{k=d}^{2d-1} \left(\dot{a}^k - \sum_{i,j=d}^{2d-1} c_{i,j}^k \dot{\gamma}^i a^j \right) \mathscr{A}_k \tag{5.28}$$

with $X = \sum_{i=d}^{2d-1} \dot{\gamma}^i \mathscr{A}_i$, $\mathscr{A} = \sum_{k=d}^{2d-1} a^k \mathscr{A}_k$.

- *Along an exponential curve, the tangent vectors are covariantly constant, whereas, along a stationary curve, one has covariantly constant momentum, i.e.*

$$\bar{\nabla}_{\dot{\gamma}} \lambda = \sum_{i=1}^{2d-1} \left(\dot{\lambda}_i + \sum_{j=d}^{2d-1} \sum_{k=1}^{2d-1} c_{i,j}^k \lambda_k \dot{\gamma}^j \right) \omega^i = 0.$$

Proof. Define $\dot{\gamma}^i := \langle \omega^i\big|_{\gamma}, \dot{\gamma} \rangle$. Then following the same approach as done in [20], [15, App.C] (for the case $d = 2$) and done in [30, App.A] (for the case $d = 3$), the Cartan connection on the tangent bundle is given by Eq. (5.28). From Eq. (5.28), it is directly clear that the auto-parallel curves are horizontal exponential curves, since

$$\begin{aligned} &\bar{\nabla}_{\dot{\gamma}} \dot{\gamma} = 0 \Leftrightarrow \forall_{i \in \{d,\ldots,2d-1\}} \ddot{\gamma}^i = 0 \Leftrightarrow \forall_{i \in \{d,\ldots,2d-1\}} \dot{\gamma}^i = c^i \text{ for some constants } c_i \\ &\Leftrightarrow \gamma(s) = \gamma(0) Exp \left(s \sum_{i=d}^{2d-1} c^i A_i \right). \end{aligned}$$

and in order to ensure s to be the spatial arclength parameter we must have $c^d = 1$.

Now the Cartan connection on the tangent bundle naturally imposes the following Cartan connection formula on the co-tangent bundle given by

$$\nabla_{\dot{\gamma}} \sum_{i=1}^{2d-1} \lambda_i \omega^i|_{\gamma} = \sum_{i=1}^{2d-1} \left(\dot{\lambda}_i + \sum_{j=1}^{2d-1} \sum_{k=1}^{2d-1} c_{i,j}^k \lambda_k \dot{\gamma}^j \right) \omega^i\big|_{\gamma}, \tag{5.29}$$

which follows from Eq. (5.28) and $\mathrm{d}\langle \omega^k|_{\gamma}, \mathscr{A}_j|_{\gamma} \rangle = \langle \nabla_{\dot{\gamma}} \omega^k|_{\gamma}, \mathscr{A}_j|_{\gamma} \rangle + \langle \omega^k|_{\gamma}, \nabla_{\dot{\gamma}} \mathscr{A}_j|_{\gamma} \rangle = 0$. For details see [30, Lemma A.11]. Now, by the horizontal part of PMP (i.e. first equation in Theorem 5.1), we have

$$\dot{\gamma}^i = \lambda_i \text{ for all } i \in \{d, \ldots, 2d-1\},$$

so that the result follows by substituting this equality into Eq. (5.29). □

Now that we have computed momentum $\lambda(s)$ in Theorem 5.1, we can integrate the equation in Theorem 5.2 to find the sub-Riemannian geodesics.

Theorem 5.3. *Let* $m : G \to Aut(\mathbb{R}^{2d})$ *denote the matrix group representation (see Remark 5.9) such that*

$$\mathrm{d}\lambda\big|_{\gamma} = \lambda\big|_{\gamma} m(\gamma^{-1}) \mathrm{d}m(\gamma),$$

where we represent the covector field $\lambda\big|_{\gamma} = \sum_{i=1}^{2d-1} \lambda_i\, \omega^i\big|_{\gamma}$ *along the geodesic* $\gamma(\cdot) = (\mathbf{x}(\cdot), R(\cdot))$, *by a row-vector* $\lambda\big|_{\gamma} = (\lambda_1, \ldots, \lambda_{2d})\big|_{\gamma}$, *where we note* $\lambda_{2d} = 0$. *Then along the sub-Riemannian geodesics in* $(SE(d), \Delta_d, G_\beta)$ *the following relation with momentum applies*

$$\lambda(s)m(\gamma(s))^{-1} = \lambda(0)m(\gamma(0))^{-1}.$$

Proof. Note that $\nabla_{\dot{\gamma}(s)} \lambda|_{\gamma(s)} = 0$ iff

$$\frac{d}{ds}\lambda(s)|_{\gamma(s)} + \lambda(s)|_{\gamma(s)}\, m((\gamma(s))^{-1})\frac{d}{ds}m(\gamma(s)) = 0$$

for all $0 \leq s \leq s_{max}(\lambda(0))$. The rest follows by

$$\frac{d}{ds}(\lambda(s)(g(s))^{-1}) = -\lambda(s)(g(s))^{-1}\dot{\gamma}(s)(g(s))^{-1} + \dot{\lambda}(s)(g(s))^{-1} = 0\,,$$

with $g(s) = m(\gamma(s))$. The last equation must be multiplied with $g(s)$ from the right to obtain the result. □

Remark 5.9. For $d = 2$, this group representation m is given by Eq. (5.33). For $d > 2$ this group representation m is given by

$$m(\mathbf{x}, R) = \begin{pmatrix} R & \sigma_{\mathbf{x}} R \\ 0 & R \end{pmatrix}, \tag{5.30}$$

where $\sigma_{\mathbf{x}} = \sum_{i=1}^{d} x^i A_{d+i} \in \mathrm{so}(d)$, with $\mathbf{x} = \sum_{i=1}^{d} x^i \mathbf{e}_i$ and $A_{d+i} \in \mathrm{so}(d)$. Here, we have $\sigma_{R\mathbf{x}} = R\sigma_{\mathbf{x}}R^{-1}$ and thereby $m(g_1 g_2) = m(g_1)m(g_2)$ for all $g_1, g_2 \in SE(d)$. Then

$$\begin{aligned} \mathrm{d}\lambda &= \lambda m(\gamma^{-1})\mathrm{d}m(\gamma) = \lambda \begin{pmatrix} R^{-1}\mathrm{d}R & \sigma_{R^{-1}\mathrm{d}\mathbf{x}} \\ 0 & R^{-1}\mathrm{d}R \end{pmatrix} \\ &= \lambda \begin{pmatrix} \sigma_{(\omega^{d+1},\ldots,\omega^{2d})^T} & \sigma_{(\omega^1,\ldots,\omega^d)^T} \\ 0 & \sigma_{(\omega^{d+1},\ldots,\omega^{2d})^T} \end{pmatrix}, \end{aligned}$$

with short notation $\omega^j = \omega^j\big|_\gamma$, $\lambda = \lambda\big|_\gamma$, $\mathrm{d}\lambda = \mathrm{d}\lambda\big|_\gamma$, and where we represent covector $\lambda = \sum_{i=1}^{2d-1} \lambda_i\, \omega^i\big|_\gamma$ by a row-vector $\lambda = (\lambda_1, \ldots, \lambda_{2d})$. Note that $\omega^j\big|_\gamma = 0$ and $\lambda_j = 0$ for all $j \geq 2d$ along sub-Riemannian geodesics $\gamma(\cdot) = (\mathbf{x}(\cdot), R(\cdot))$.

5.2.1 *Summary: The Exponential Map of Control Problem* $\mathbf{P}_{\mathbf{curve}}$

Now let us combine the results of Theorems 5.1, 5.2 and 5.3. Theorem 5.1 provides us momentum $\lambda(s)$ which is entirely determined by $\lambda(0)$. This is not surprising as by Theorem 5.2, one has covariantly constant momentum, as follows from the canonical equations of the Pontryagin maximum principle. The structure of the Cartan connection can be employed to explicitly derive an admissible endpoint $\gamma(\ell)$ of a *cuspless* sub-Riemannian geodesic γ from a pair

$$\begin{aligned} &(\lambda_0, \ell) \in \mathfrak{D} := \{(\lambda_0, \ell) \in \mathscr{C} \times \mathbb{R}^+ \mid \ell \leq s_{max}(\lambda(0)) \neq 0\}, \\ &\text{with } \mathscr{C} := \{\lambda(0) \in T_e^*(SE(d)) \mid \sum_{i=d}^{2d-1} |\lambda_i(0)|^2 = 1\} \end{aligned} \tag{5.31}$$

consisting of momentum λ_0 and length ℓ, with the preservation law of Theorem 5.3. The associated mapping

$$(\lambda_0,\ell) \mapsto \gamma(\ell) =: \widetilde{\mathrm{Exp}}(\lambda_0,\ell)$$

is called exponential map[4] $\widetilde{\mathrm{Exp}}$ of $\mathbf{P_{curve}}$. It coincides with the exponential map for $\mathbf{P_{mec}}$ when restricting to admissible end-conditions.

In the subsequent sections we apply this procedure to get an explicit formula for the exponential map $\widetilde{\mathrm{Exp}}$ for the special cases of interest, respectively $d=2$ and $d=3$. We will also provide analysis and visualization of the range $\mathfrak{R} := \widetilde{\mathrm{Exp}}(\mathfrak{D})$ of the exponential map and show that it provides a reasonable grouping criterium to connect two points, say $(\mathbf{0},\mathbf{a})$ and $(\mathbf{x}_1,\mathbf{n}_1)$ within $\mathbb{R}^d \rtimes S^{d-1} := SE(d)/(\{0\}\times SO(d-1))$. This analysis of problem $\mathbf{P_{curve}}$ (and $\mathbf{P_{mec}}$ for admissible end-conditions) is related to earlier neuro-psychological models of association fields [26, 41].

5.3 The Case $d=2$: Sub-Riemannian Geodesics in $(\mathbb{R}^2 \rtimes S^1, \Delta_2, G_1)$

Let us first apply the results regarding sub-Riemannian geodesics within $(\mathbb{R}^d \rtimes S^{d-1}, \Delta_d, G_\beta)$ to the special case $d=2$. Following our standard conventions, we get

$\mathbf{a} := \mathbf{e}_y$ and furthermore
$A_1 = \partial_x$, $A_2 = \partial_y$, $A_3 = \partial_\theta$,
$\mathscr{A}_1 = \cos\theta\,\partial_x + \sin\theta\,\partial_y$, $\mathscr{A}_2 = -\sin\theta\,\partial_x + \cos\theta\,\partial_y$, $\mathscr{A}_3 = \partial_\theta$,
$\omega^1 = \mathrm{d}\theta, \omega^2 = \cos\theta\,\mathrm{d}x + \sin\theta\,\mathrm{d}y, \omega^3 = -\sin\theta\,\mathrm{d}x + \cos\theta\,\mathrm{d}y$,
$\Delta_2 = \mathrm{span}\{\mathscr{A}_2, \mathscr{A}_3\}$,
$G_\beta = \omega^3 \otimes \omega^3 + \beta^2 \omega^2 \otimes \omega^2$ where we set $\beta = 1$.

This produces the following canonical ODE-system for momentum $\lambda = \sum_{i=1}^{3} \lambda_i \omega^i$ along the sub-Riemannian geodesics:

$$\dot{\lambda}_1(t) = \lambda_2(t)\lambda_3(t),\ \dot{\lambda}_2(t) = -\lambda_1(t)\lambda_3(t),\ \dot{\lambda}_3(t) = \lambda_1(t)\lambda_2(t),$$

expressed in sub-Riemannian arclength parameter t. Along *cuspless* sub-Riemannian geodesics, this ODE-system simplifies to

$$\dot{\lambda}_1(s) = -\lambda_3(s),\ \dot{\lambda}_2(s) = -\tfrac{\lambda_1(s)\lambda_3(s)}{\lambda_2(s)},\ \dot{\lambda}_3(s) = -\lambda_1(s),$$

using the spatial arc-length parameter s, and we find preservation laws

$$\lambda_1^2 + \lambda_2^2 = \mathfrak{c}^2 := \lambda_1^2(0) + \lambda_2^2(0),\ \lambda_2^2 + \lambda_3^2 = 1,$$

[4] In our notation of the exponential map, we include a tilde to avoid possible confusion with the exponential map $Exp: T_e(G) \to G$ from Lie algebra to Lie group.

and solutions

$$\begin{aligned}
\lambda_1(s) &= \lambda_1(0)\cosh(s) - \lambda_3(0)\sinh(s),\\
\lambda_2(s) &= \sqrt{1-|\lambda_3(s)|},\\
\lambda_3(s) &= \tfrac{d\theta}{dt}(t(s)) = \tfrac{\kappa(s)}{\sqrt{\kappa^2(s)+1}} = \lambda_3(0)\cosh s - \lambda_1(0)\sinh s,
\end{aligned}$$

where $\kappa(s)$ denotes the curvature of the spatially projected curve $s \mapsto (x(s), y(s))$. The maximum length towards a cusp is given by

$$s_{max}(\lambda(0)) = \log\frac{1+\mathfrak{c}}{|\lambda_1(0) - \lambda_3(0)|}, \tag{5.32}$$

with $\mathfrak{c} = \sqrt{|\lambda_1(0)|^2 + |\lambda_2(0)|^2}$. Let $m : SE(2) \to \mathbb{R}^{3\times 3}$ be given by

$$m(x,y,\theta) = \begin{pmatrix} \cos\theta & \sin\theta & -x \\ -\sin\theta & \cos\theta & y \\ 0 & 0 & 1 \end{pmatrix}. \tag{5.33}$$

Then along the geodesics we have

$$\lambda(s)\, m(\gamma(s)) = \lambda(0)\, m(\gamma(0)) = \lambda(0) \text{ for all } s \in [0, s_{max}(\lambda(0))),$$

which allows us to compute the endpoint $\gamma(s) = (x(s), y(s), \theta(s)) \in SE(2)$ of a *cuspless* sub-Riemannian geodesic from a pair $(s, \lambda(0))$ with $s \leq s_{max}(\lambda(0))$.

5.3.1 Switching to the Case $\mathbf{a} = \mathbf{e}_x$ and Re-labeling of the Lie-Algebra

So far we have applied the general formula for sub-Riemannian geodesics within $(\mathbb{R}^d \rtimes S^{d-1}, \Delta_d, G_1)$ to the special case $d = 2$, where we kept track of consistency with the case $d \geq 3$.

However, in order to directly relate to previous works by the authors on sub-Riemannian geodesics within the 2D-Euclidean motion group and orientation scores [19, 20], we will in the remainder of this section switch to the case $\mathbf{a} = \mathbf{e}_x$ (instead of $\mathbf{a} = \mathbf{e}_y$), and we will re-label the Lie-algebra as follows:

$$\mathscr{A}_1 := \partial_\theta,\ \mathscr{A}_2 := \cos\theta\,\partial_x + \sin\theta\,\partial_y,\ \mathscr{A}_3 := -\sin\theta\,\partial_x + \cos\theta\,\partial_y. \tag{5.34}$$

The corresponding dual vectors are given by

$$\omega^1 := \mathrm{d}\theta,\ \omega^2 := \cos\theta\,\mathrm{d}x + \sin\theta\,\mathrm{d}y,\ \omega^3 := -\sin\theta\,\mathrm{d}x + \cos\theta\,\mathrm{d}y, \tag{5.35}$$

that we will use to represent the momentum covector $\lambda = \sum_{i=1}^{3} \lambda_i \omega^i$ accordingly.

5.3.2 Explicit Parameterizations of the Cuspless Sub-Riemannian Geodesics and Their Properties

Let us explicitly compute the exponential map for the case $d = 2$ using spatial arc-length parametrization which provides us an explicit formula for the sub-Riemannian geodesics in $(SE(2), \Delta_2, G_1)$ (recall Remark 5.2). We will use the labeling/ordering conventions (5.34) and (5.35).

Theorem 5.1 directly provides us the linear ODE

$$\ddot{\lambda}_1(s) = \lambda_1(s) \Leftrightarrow \frac{d^2}{ds^2}\left(\frac{\kappa(s)}{\sqrt{\kappa^2(s)+1}}\right) = \frac{\kappa(s)}{\sqrt{\kappa^2(s)+1}},$$

which directly provides us with the curvature $\kappa(s)$ of the cuspless sub-Riemannian geodesics in terms of $\lambda(0) = \lambda_1(0)\mathrm{d}\theta + \lambda_2(0)\mathrm{d}x + \lambda_3(0)\mathrm{d}y$ and spatial arc-length s, with $s \leq s_{max}(\lambda(0))$, recall Eq. (5.32). Now instead of integrating a Frenet ODE system, we apply an effective integration procedure via Theorem 5.3. We have

$$\mathrm{d}\hat{\lambda} = \hat{\lambda}\,(m(\gamma))^{-1}\mathrm{d}m(\gamma) \Leftrightarrow \hat{\lambda}(s)\,m(\gamma(s)) = \hat{\lambda}(0)\,m(\gamma(0)) = \hat{\lambda}(0), \tag{5.36}$$

where we use short notation for the row-vector

$$\hat{\lambda} := (-\lambda_3(s), \lambda_2(s), \lambda_1(s)) = (\dot{\lambda}_1(s), \frac{1}{\sqrt{\kappa^2(s)+1}}, \frac{\kappa(s)}{\sqrt{\kappa^2(s)+1}}) \tag{5.37}$$

with $m(\gamma) = \begin{pmatrix} \cos\theta & -\sin\theta & x \\ \sin\theta & \cos\theta & y \\ 0 & 0 & 1 \end{pmatrix}$ the most common group representation of $SE(2)$.

Lemma 5.1. *Let* $\mathfrak{c} := \sqrt{|\lambda_2(0)|^2 + |\lambda_3(0)|^2}$. *There exists a unique* $h_0 \in SE(2)$ *such that* $\hat{\lambda}(0)m(h_0^{-1}) = (\mathfrak{c}, 0, 0)$. *Consequently, we have for* $\tilde{\gamma}(s) := h_0\gamma(s)$ *that*

$$(-\lambda_3(s), \lambda_2(s), \lambda_1(s)) = \hat{\lambda}(s) = (\mathfrak{c}\ 0\ 0)\ m(\tilde{\gamma}(s)). \tag{5.38}$$

Proof. Follows by Theorem 5.3 and the fact that m is a group representation. □

Application of this lemma provides the following explicit formula for the sub-Riemannian geodesics in $(SE(2), \Delta_2, G_1)$.

Theorem 5.4. *The exponential map of* $\mathbf{P_{curve}}$ *expressed in spatial arc-length parametrization is given by*

$$\begin{aligned} &\widetilde{Exp}\left(\sum_{i=1}^{3} \lambda_i(0)\ \omega^i\big|_{\gamma(0)=e}\ , s\right), \\ &= \gamma(s) = (x(s), y(s), \theta(s)), \end{aligned} \tag{5.39}$$

for all $s \in [0, \ell]$ *with total spatial length* $\ell \leq s_{max}(\lambda(0))$. *The cuspless geodesics in* $(SE(2), \Delta_2, G_1)$ *are given by* $\gamma(s) = h_0^{-1}\tilde{\gamma}(s)$, *i.e.*

$$
\begin{aligned}
&\theta(s) = \tilde{\theta}(s) - \overline{\theta}_0 \in [-\pi, \pi], \\
&\textit{with } \cos(\overline{\theta}_0) = \tfrac{-\lambda_3(0)}{\mathfrak{c}} \textit{ and } \overline{\theta}_0 \in [-\pi, 0], \\
&\mathbf{x}(s) = \overline{R}_0^T(\tilde{\mathbf{x}}(s) - \overline{\mathbf{x}}_0), \\
&\textit{with } \overline{R}_0^T = \begin{pmatrix} \cos\overline{\theta}_0 & \sin\overline{\theta}_0 \\ -\sin\overline{\theta}_0 & \cos\overline{\theta}_0 \end{pmatrix} = \tfrac{1}{\mathfrak{c}} \begin{pmatrix} -\lambda_3(0) & -\lambda_2(0) \\ \lambda_2(0) & -\lambda_3(0) \end{pmatrix}
\end{aligned}
\tag{5.40}
$$

with $h_0 = (\overline{\mathbf{x}}_0, \overline{R}_0) \in SE(2)$, $\overline{\mathbf{x}}_0 = (\frac{-\lambda_3(0)}{\mathfrak{c}}, 0)^T$. *Here curve* $\tilde{\gamma} = (\tilde{x}, \tilde{y}, \tilde{\theta})$ *is given by*

$$
\begin{aligned}
&\tilde{x}(s) = \tfrac{\lambda_1(s)}{\mathfrak{c}} = \tfrac{\lambda_1(0)\cosh(s) - \lambda_3(0)\sinh(s)}{\mathfrak{c}}, \\
&\tilde{y}(s) = -\tfrac{1}{\mathfrak{c}} \int_0^s \sqrt{1 - |\lambda_1(\tau)|^2}\, d\tau, \\
&\tilde{\theta}(s) = \arg(\dot{\tilde{x}}(s) + i\dot{\tilde{y}}(s)) \\
&= \arg(-\lambda_3(s) - i\lambda_2(s)) \in [-\pi, 0],
\end{aligned}
\tag{5.41}
$$

with $\lambda_1(s) = \lambda_1(0)\cosh s - \lambda_3(0)\sinh s$, $\lambda_3(s) = \lambda_3(0)\cosh s - \lambda_1(0)\sinh s$ *and where* $\mathfrak{c} = \sqrt{|\lambda_2(0)|^2 + |\lambda_3(0)|^2} \geq 0$.

From these formulas one can directly deduce the following properties:

- If $\mathfrak{c} \leq 1$, the curvature does not switch sign and we obtain U-shaped curves, unless $\lambda_1(0) = \lambda_3(0) = 0$ in which case we get a straight line.
- If $\mathfrak{c} > 1$ and $\lambda_1(0)\lambda_3(0) > 0$, then the curve is an S-shaped curve with bending-point at $s_B = \log \frac{\|\lambda_1(0) + \lambda_3(0)\|}{\|\lambda_1(0) - \lambda_3(0)\|}$.
- If $\mathfrak{c} = 1$ and $\lambda_3(0) = \lambda_1(0)$ we have $s_{max} = \infty$.
- The cuspless sub-Riemannian geodesics are monotonically increasing along the $\lambda_2(0)\mathbf{e}_x + \lambda_3(0)\mathbf{e}_y$-axis:

  $$-\dot{\tilde{y}}(s) \geq 0 \Leftrightarrow \lambda_2(0)\dot{x}(s) + \lambda_3(0)\dot{y}(s) \geq 0,$$

  and even if they tend towards a cusp where curvature tends to infinity, they do not roll up and their sub-Riemannian length stays finite.
- The cuspless sub-Riemannian geodesics are contained within the half-space $x \geq 0$ and the boundary $x = 0$ can only be reached with an angle (w.r.t. the positive x-axis) of π as formally proven in [15, Thm.7 and Thm.8].

5.3.3 The Set $\mathfrak{R}$ and Its Boundary $\partial\mathfrak{R}$

Now that we have computed the exponential map, let us have a look at the range $\mathfrak{R} = \widetilde{\mathrm{Exp}}(\mathfrak{D})$, which according to the results in [8] coincides precisely with the points for which $\mathbf{P_{curve}}$ admits a global minimum. In fact, we have

Theorem 5.5. *In* $\mathbf{P_{curve}}$ *with* $d = 2$, *we set initial condition* $(x_{in}, y_{in}, \theta_{in}) = e = (0,0,0)$ *and consider* $(x_{fin}, y_{fin}, \theta_{fin}) \in \mathbb{R}^2 \rtimes S^1$. *Then*

- $(x_{fin}, y_{fin}, \theta_{fin}) \in \mathfrak{R}$ *if and only if* $\mathbf{P_{curve}}$ *has a unique minimizing geodesic which exactly coincides with the unique minimizer of* $\mathbf{P_{mec}}$.

- $(x_{fin}, y_{fin}, \theta_{fin}) \notin \mathfrak{R}$ *if and only if problem* $\mathbf{P_{curve}}$ *is ill-defined (i.e.* $\mathbf{P_{curve}}$ *does not have a minimizer).*

Corollary 5.2. *Set* $g_{in} = e$. *Then* g_{fin} *is an admissible end-condition for* $\mathbf{P_{curve}}$ *if* $g_{fin} \in \mathfrak{R}$.

According to the next theorem the exponential map has nice smoothness and bijection properties and properly maps analytic trajectories in the hyperbolic phase portrait in momentum space onto analytic sub-Riemannian geodesics in $(SE(2), \Delta_2, G_1)$. For a visualization on how this is achieved physically, see Figure 5.6.

Theorem 5.6. *Let* $\mathfrak{D}$ *and* $\mathfrak{R}$ *denote respectively the domain and range of the exponential map of* $\mathbf{P_{curve}}$ *defined on* $(SE(2), \Delta_2, G_1)$. *Then,*

- $\widetilde{Exp} : \mathfrak{D} \to \mathfrak{R}$ *is a homeomorphism if we equip* $\mathfrak{D}$ *and* $\mathfrak{R}$ *with the subspace topology*[5].
- $\widetilde{Exp} : \mathring{\mathfrak{D}} \to \mathring{\mathfrak{R}}$ *is a diffeomorphism.*

Finally, the boundary $\partial\mathfrak{R}$ *is given by*

$$\partial\mathfrak{R} = \mathscr{S}_B \cup \mathfrak{l} \cup \mathscr{S}_R, \tag{5.42}$$

with $\mathfrak{l} := \{(0,0,\theta) \mid -\pi \leq \theta \leq \pi\}$ *the sphere above the spatial origin,*

$$\mathscr{S}_B := \{\widetilde{Exp}(\lambda_0, s_{max}(\lambda_0)) \mid \lambda_0 \in \mathscr{C}\} \tag{5.43}$$

the set of endpoints of geodesics ending at a cusp (the blue surfaces in Fig.5.6), and

$$\mathscr{S}_R := \left\{\widetilde{Exp}(\lambda_0, s) \mid \lambda_0 \in \mathscr{C} \text{ with } \lambda_3(0) = \pm 1 \text{ and } s \in (0, s_{max}(\lambda(0))\right\} \tag{5.44}$$

the set of endpoints of geodesics departing from a cusp (the red surfaces in Fig.5.6).

Proof. See [15, App.F].

As a result the set $\mathfrak{R}$ is a connected set with a piecewise smooth boundary $\partial\mathfrak{R}$ given by Eq. (5.42). In fact, when taking the intersection with $\{(\mathbf{x}_1, \mathbf{n}) \mid \mathbf{n} \in S^1\}$ with $\mathbf{x}_1 = (x_{fin}, y_{fin}) \in \mathbb{R}^2$ fixed we get a cone in S^1. Sometimes this cone is bounded by a red and a blue surface and sometimes it is bounded by the blue surfaces in Figure 5.6. Also see Figure 5.7.

Let us underpin this observation on the cone of reachable angles with a formal theorem. To this end let $\theta_{\text{endcusp}}(x_{fin}, y_{fin})$ denote the final angle (w.r.t. the positive x-axis) of the geodesic ending in $(x_{fin}, y_{fin}, \cdot)$ with a cusp and where

[5] As $\mathfrak{D}$ and $\mathfrak{R}$ are not open w.r.t. standard topologies on the embedding spaces $T_e(SE(2)) \times \mathbb{R}^+$ and $\mathbb{R}^2 \times S^1$, these subspace topologies do not coincide with the induced topology imposed by the embedding via the identity map. W.r.t. the subspace topologies the set $\mathfrak{D}$, respectively $\mathfrak{R}$, are open sets and the homeomorphism $\widetilde{\text{Exp}} : \mathfrak{D} \to \mathfrak{R}$ is well-defined.

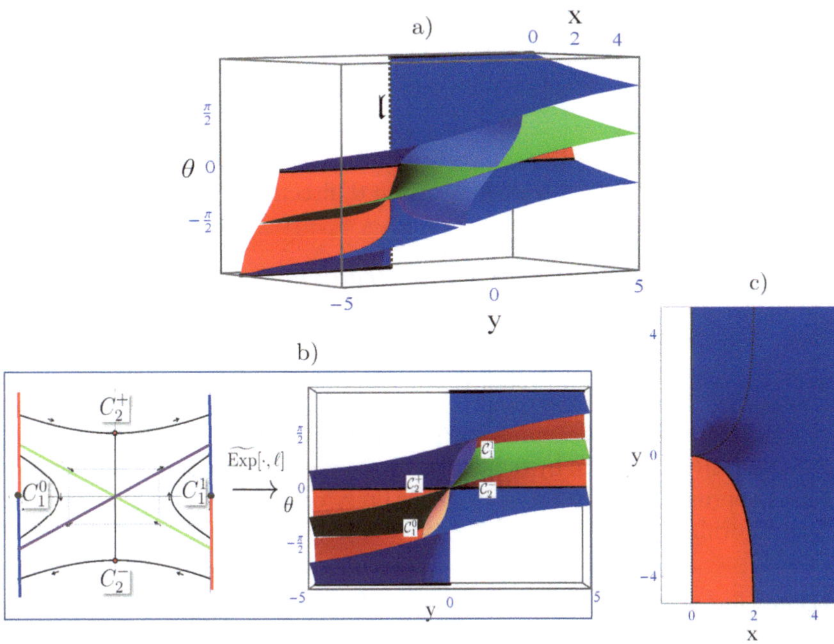

Fig. 5.6 Plots (from 3 perspectives a), b) and c)) of the range $\mathfrak{R}$ of the exponential map of $\mathbf{P_{curve}}$. Red surface: endpoints of geodesics starting from cusps. Blue surface: endpoints of geodesics ending in cusps. The black lines are the intersections of the blue plane with the red plane. Green surface: critical surface ($\mathfrak{c} = 1$) with $\dot{z}_0 = -z_0$. Purple surface: critical surface ($\mathfrak{c} = 1$) with $\dot{z}_0 = z_0$. The critical surface splits the range of the exponential map into four disjoint parts $\mathscr{C}_1^1$, $\mathscr{C}_1^0$, $\mathscr{C}_2^+$ and $\mathscr{C}_2^-$ that relate to the splitting of the phase space into C_1^1, C_1^0, C_2^+ and C_2^- in b) where we have depicted $\mathfrak{R}$ viewed from the x-axis. In c) we have depicted $\mathfrak{R}$ viewed from the θ-axis.

$\theta_{\text{begincusp}}(x_{fin}, y_{fin})$ denotes the final angle of a geodesic ending in $(x_{fin}, y_{fin}, \cdot)$ starting with a cusp. In case there exist two geodesics ending with a cusp at (x_{fin}, y_{fin}), we order them by writing

$$\theta^1_{\text{endcusp}}(x_{fin}, y_{fin}) \leq \theta^2_{\text{endcusp}}(x_{fin}, y_{fin}).$$

Theorem 5.7. *Let* $(x_{fin}, y_{fin}, \theta_{fin}) \in \mathfrak{R}$. *If*

$$|y_{fin}| \leq -x_{fin}\, i E\left(i \operatorname{arcsinh} \frac{x_{fin}}{\sqrt{4 - x_{fin}^2}}, \frac{x_{fin}^2 - 4}{x_{fin}^2} \right), \text{ and } 0 \leq x_{fin} < 2. \tag{5.45}$$

then we have

$$y_{fin} > 0 \Rightarrow \theta_{fin} \in \left[\theta_{begincusp}(x_{fin}, y_{fin}), \theta_{endcusp}(x_{fin}, y_{fin})\right],$$

$$y_{fin} < 0 \Rightarrow \theta_{fin} \in \left[\theta_{endcusp}(x_{fin}, y_{fin}), \theta_{begincusp}(x_{fin}, y_{fin})\right],$$

otherwise (so in particular if $x_{fin} \geq 2$) we have

$$\theta_{fin} \in [\theta^1_{endcusp}(x_{fin}, y_{fin}), \theta^2_{endcusp}(x_{fin}, y_{fin})].$$

Proof. See [15, App.E]. For a direct graphical validation of Theorem 5.7 see Figure 5.6 (in particular the top view along the θ direction).

5.3.4 Solving the Boundary Value Problem

The inverse of the exponential map $(\lambda_0, \ell) \mapsto \widetilde{\text{Exp}}(\lambda(0), \ell) = \gamma(\ell) = g_{fin}$ in Theorem (5.4) can be computed analytically to a large extent. That is, ℓ, $\lambda_2(0)$, $\lambda_3(0)$ can all be analytically expressed in terms of $-1 \leq \lambda_1(0) \leq 1$, which leaves an accurate and efficient *one dimensional* numerical shooting algorithm to find the final remaining unknown $\lambda_1(0)$, as given by the following theorem. Note that recently proposed numerical approaches in the literature [6, 35] rely on three dimensional numerical shooting algorithms.

Theorem 5.8. *Let $g_{fin} \in \mathfrak{R}$. The inverse of the exponential map $(\lambda_0, \ell) \mapsto \widetilde{Exp}(\lambda(0), \ell) = \gamma(\ell) = g_{fin}$ in Theorem 5.4 is given by*

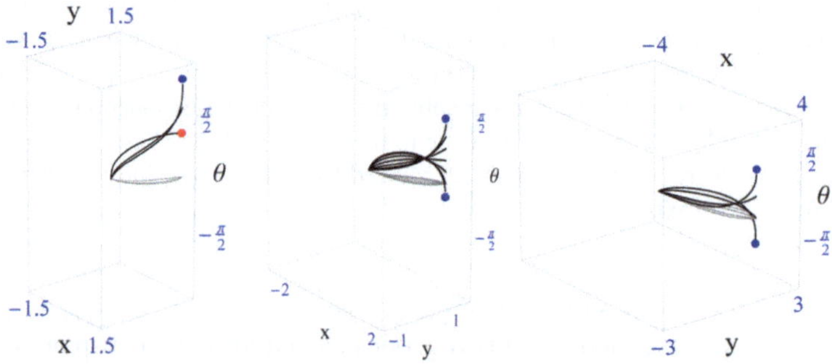

Fig. 5.7 Sub-Riemannian geodesics (and their spatial projections in grey) obtained by our analytical approach to the boundary value problem. We have kept (x_{fin}, y_{fin}) fixed and we have varied θ_{fin} to full range such that our algorithm finds solutions (with relative errors less than 10^{-8}). Left: $(x_{fin}, y_{fin}) = (1, 1.5)$, middle: $(x_{fin}, y_{fin}) = (2, 1)$, right: $(x_{fin}, y_{fin}) = (4, 1)$. At the boundary of cones of reachable angles, the endpoints of the sub-Riemannian geodesics are located on the cusp-surface $\partial\mathfrak{R}$. End-points of geodesics departing from cusps are indicated in red and endpoints of geodesics ending at cusps are indicated in red (as in Figure 5.6).

$$\lambda(0) = \sum_{i=1}^{2} \lambda_i(0)\omega^i,$$

$$\ell(\lambda(0), g_{fin}) = \begin{cases} \log \frac{\lambda_1(0)}{v}, & \mathfrak{c} = 1 \textit{ and } \lambda_3(0) = \lambda_1(0), \\ \log \frac{v}{\lambda_1(0)}, & \mathfrak{c} = 1 \textit{ and } \lambda_3(0) = -\lambda_1(0), \\ \log \frac{v+w}{\lambda_1(0)-\lambda_3(0)}, & \textit{else} \end{cases} \tag{5.46}$$

where $v, w, \mathfrak{c}$ *are given by*

$$v = \lambda_1(\ell) = \lambda_1(0) - x_{fin}\lambda_3(0) + y_{fin}\lambda_2(0),$$
$$w = -\lambda_3(\ell) = -\lambda_3(0)\cos\theta_{fin} + \lambda_2(0)\sin\theta_{fin},$$
$$\mathfrak{c} = \sqrt{|\lambda_2(0)|^2 + |\lambda_3(0)|^2}.$$

Here $\lambda_2(0), \lambda_3(0)$ *are expressed as follows:*

$$\lambda_2(0) = \chi_2(\lambda_1(0)) := \sqrt{1 - |\lambda_1(0)|^2},$$
$$-\lambda_3(0) = \chi_3(\lambda_1(0), g_{fin}) := \frac{-b + sign(g_{fin})\sqrt{D}}{2a},$$
$$\textit{with } a = x_{fin}^2 + \sin^2(\theta_{fin}),$$
$$b = 2x_{fin}(\lambda_1(0) + y_{fin}\lambda_2(0)) - \lambda_2(0)\sin(2\theta_{fin}),$$
$$c = |\lambda_2(0)|^2(y_{fin}^2 - \sin^2(\theta_{fin})) + 2y_{fin}\lambda_1(0)\lambda_2(0),$$
$$D = b^2 - 4ac =: \mathscr{D}(\lambda_1(0), g_{fin}),$$

and with sign function given by

$$\mathrm{sign}(g_{\mathrm{fin}}) = \begin{cases} 1 & \textit{if } g_{fin} \in \mathscr{C}_2^+, \\ 1 & \textit{if } g_{fin} \in \mathscr{C}_1^1 \cup \mathscr{C}_1^0 \textit{ is above } \mathscr{V}, \\ -1 & \textit{if } g_{fin} \in \mathscr{C}_1^1 \cup \mathscr{C}_1^0 \textit{ is below } \mathscr{V}, \\ -1 & \textit{if } g_{fin} \in \mathscr{C}_2^-, \end{cases} \tag{5.47}$$

with surface $\mathscr{V} \in SE(2)$ *(corresponding to the solutions with* $\lambda_3(0) = 0$*)*

$$\mathscr{V} = \left\{ \widetilde{Exp}(z_0\omega^1 + \chi_2(z_0)\omega^2, \ell) \mid z_0 \in [-1,1] \textit{ and } 0 \leq \ell \leq s_{max}(z_0\omega^1 + \chi_2(z_0)\omega^2) \right\}.$$

Finally, $\lambda_1(0)$ *denotes the unique root* $F(\lambda_1(0)) = 0$ *of* $F : I \to \mathbb{R}^+$ *defined on*

$$I = \{z_0 \in [-1,1] \mid \mathscr{D}(z_0, g_{fin}) \geq 0\}$$

given by $F(z_0) = \|\widetilde{Exp}(z_0\omega^1 + \chi_2(z_0)\omega^2 + \chi_3(z_0, g_{fin})\omega^3, \ell(z_0, g_{fin})\ ) - g_{fin}\|$, *where* $\|\cdot\|$ *denotes the Euclidean norm on* $\mathbb{R}^2 \times S^1$.

Proof. By Theorem 5.5, there is a unique stationary curve connecting e and $g_{fin} \in \mathfrak{R}$. The exponential map of $\mathbf{P_{curve}}$ is a homeomorphism by Theorem 5.6 and thereby the continuous function F has a unique zero, since ℓ and $\lambda_3(0)$ are already determined by $\lambda_1(0)$ and g_{fin}. W.r.t the formula for ℓ, Theorem 5.1 (for $d = 2$) implies that:

$$\lambda_1(\ell) = \lambda_1(0)\cosh\ell - \lambda_3(0)\sinh\ell,\ \lambda_3(\ell) = \lambda_3(0)\cosh\ell - \lambda_1(0)\sinh\ell$$

from which e^ℓ can readily be obtained, noting that (by Theorem 5.4 for $s = \ell$):

$$\lambda_1(\ell) = \lambda_1(0) - x_{fin}\lambda_3(0) + y_{fin}\lambda_2(0), \; -\lambda_3(\ell) = -\lambda_3(0)\cos\theta_{fin} + \lambda_2(0)\sin\theta_{fin}.$$

Finally, applying the preservation laws (recall Remark 5.8) $|\lambda_1(s)|^2 + |\lambda_2(s)|^2 = 1$ and $|\lambda_2(s)|^2 + |\lambda_3(s)|^2 = |\lambda_2(0)|^2 + |\lambda_3(0)|^2$ to the case $s = \ell$ provides a quadratic equation for $\lambda_3(0)$ from which the result follows. For details on the choice of the sign in the solution of this quadratic equation and surface $\mathscr{V}$, we refer to [15, Lem.8]. □

Remark 5.10. Theorem 5.8 allows for fast and accurate computations of sub-Riemannian geodesics, see Figure 5.7 where the computed geodesics are instantly computed with an accuracy of relative $\mathbb{L}_2$-errors in the order of 10^{-8}. For an example of the application of Theorem 5.8 see Figure 5.8. Finally, we note that Theorem 5.6 implies that (our approach to) solving the boundary-value problem is well-posed, i.e. the solutions are both unique and stable.

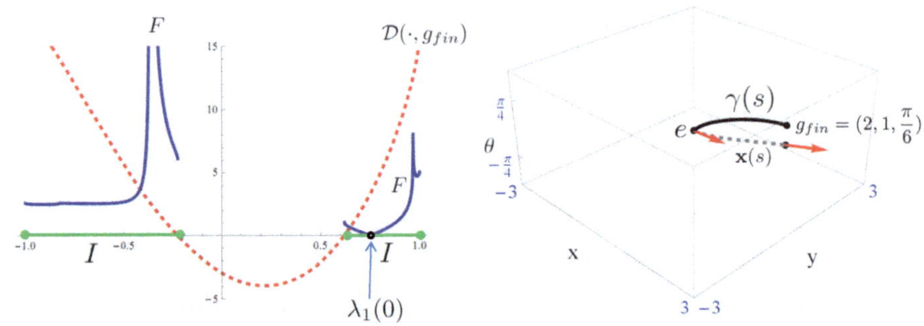

Fig. 5.8 The particular case where $g_{fin} = (2, 1, \pi/6)$, where $\text{sign}(g_{fin}) = -1$ and where unique root of $F(\cdot, g_{fin})$, whose domain I is indicated in green, is approximatively $\lambda_1(0) \approx 0.749551$ (and thereby $\lambda_2(0) = \sqrt{1 - |\lambda_1(0)|^2}$, $\lambda_3(0) \approx -0.809740$ and $L \approx 2.26253$).

5.3.5 Modeling Association Fields with Solutions of P_{curve}

Sub-Riemannian geometry plays a major role in the functional architecture of the primary visual cortex (V1) and more precisely its pinwheel structure, cf. [42]. In his paper [42], Petitot shows that the horizontal cortico-cortical connections of V1 implement the contact structure of a continuous fibration $\pi : R \times P \to P$ with base space the space of the retina and P the projective line of orientations in the plane. He applies his model to the Field's, Hayes' and Hess' physical concept of an association field, to several models of visual hallucinations [25] and to a variational model of curved modal illusory contours [33, 38, 54].

In their paper, Field, Hayes and Hess [26] present physiological speculations concerning the implementation of the association field via horizontal connections. They have been confirmed by Jean Lorenceau et al. [34] via the method of apparent speed of fast sequences, where the apparent velocity is overestimated when the successive elements are aligned in the direction of the motion path and underestimated when the motion is orthogonal to the orientation of the elements. They have also been confirmed by electrophysiological methods measuring the velocity of propagation of horizontal activation [29]. There exist several other low-level vision models and neuro-physiological measurements that have produced similar fields of association and perceptual grouping [31,39,58]. For an overview see [42, ch:5.5,5.6].

Subsequently, we discuss three models of the association fields: Legendrian geodesics, cuspless sub-Riemannian geodesics and horizontal exponential curves. W.r.t. the latter model, we recall that horizontal exponential curves [20,48] in the sub-Riemannian manifold $(SE(2), \Delta_2, G_\beta)$, Eq. (5.14), are given by circular spirals

$$\begin{aligned} r \mapsto g_0\, e^{r(c^1 A_1 + c^2 A_2)} &= \\ \left(x_0 + \tfrac{c^2}{c^1}(\sin(c^1 r + \theta_0) - \sin(\theta_0)), y_0 - \tfrac{c^2}{c^1}(\cos(c^1 r + \theta_0) - \cos(\theta_0)), \theta_0 + r c^1\right), \end{aligned} \tag{5.48}$$

for $c^1 \neq 0$, $g_0 = (x_0, y_0, \theta_0) \in SE(2)$ and all $r \geq 0$. If $c^1 = 0$ they are straight lines:

$$g_0 e^{rc^2 A_2} = (x_0 + rc^2 \cos\theta_0, y_0 + rc^2 \sin\theta_0, \theta_0).$$

Clearly, these horizontal exponential curves reflect the co-circularity model [36].

To model the association fields from neuropsychology and neurophysiology Petitot [42] computes "Legendrian geodesics", [42, ch.6.6.4,eq.49] minimizing Lagrangian $\sqrt{1+|y'(x)|^2+|\theta'(x)|^2}$ under the constraint $\theta(x) = y'(x)$. This is directly related[6] to the sub-Riemannian geodesics in

$$((SE(2))_0, \mathrm{Ker}(-\theta\, \mathrm{d}x + \mathrm{d}y), \mathrm{d}\theta \otimes \mathrm{d}\theta + \mathrm{d}x \otimes \mathrm{d}x), \tag{5.49}$$

where $(SE(2))_0$ is the well-known nilpotent Heisenberg approximation ([19, ch:5.4]) of $SE(2)$, which minimize Lagrangian $\sqrt{1+|\theta'(x)|^2}$ under constraint $\theta(x) = y'(x)$. The drawback of such curves is that they are coordinate dependent and not covariant[7] with rotations and translations. Similar problems arise with B-splines which minimize Lagrangian $1+|\theta'(x)|^2$ under constraint $\theta(x) = y'(x)$ which are commonly used in vector graphics.

To this end, Petitot [42] also proposed the "circle bundle model" which has the advantage that it is coordinate independent. Its energy integral

$$\int_0^{x_{fin}} \sqrt{1+|y'(x)|^2 + \frac{|y''(x)|^2}{(1+|y'(x)|^2)^2}}\, \mathrm{d}x \tag{5.50}$$

[6] The dual basis in $(SE(2))_0$ is equal to $(\mathrm{d}\theta, \mathrm{d}x, -\theta\, \mathrm{d}x + \mathrm{d}y)$ and thereby the sub-Riemannian metric on $(SE(2))_0$ does not include the $|y'(x)|^2$ term.

[7] The corresponding minimization problem (and induced sub-Riemannian distance) is left invariant in $(SE(2))_0$ and not left-invariant in $SE(2)$.

can be expressed as $\int_0^\ell \sqrt{1+\kappa^2}\mathrm{d}s$, where $s \in [0,\ell]$ denotes spatial arc-length parametrization. So in case one restricts problem $\mathbf{P_{curve}}$ to those admissible endpoints that allow a cuspless sub-Riemannian geodesic which can be well-parameterized by $(x,y(x),\theta(x))$ with $\theta(x) = \arctan y'(x)$, this restricted problem coincides with Petitot's circle bundle model of finding sufficiently smooth curves $x \mapsto (x,y(x))$ such that the functional in Eq. (5.50) is minimal.

In Figure 5.9, we have modeled the association field with sub-Riemannian geodesics ($\beta = 1$) and horizontal exponential curves (Eq. (5.48) as proposed in [5,48]). Horizontal exponential curves are circular spirals and thereby rely on "co-circularity", a well-known principle to include orientation context in image analysis, cf. [28,36].

On the one hand, a serious drawback arising in the co-circularity model for association fields is that only the spatial part (x_{fin}, y_{fin}) of the end-condition can be prescribed (the angular part is imposed by co-circularity), whereas with geodesics one can prescribe $(x_{fin}, y_{fin}, \theta_{fin})$ (as long as the ending condition is contained within $\mathfrak{R}$). This drawback is clearly visible in Figure 5.9, where the association field (see a) in Figure 5.9) typically ends in points with almost vertical tangent vectors.

On the other hand, the sub-Riemannian geodesic model describes less accurately the association field by Field and co-workers in the (much weaker) connections to the side (where the co-circularity model is reasonable). One could improve the modeling by varying β, but even then it is hard to approximate large circles: the ODE $\ddot{z} = \beta^2 z$ does not allow z to be constant and one can approximate large circles by resigning to large β.

In the more aligned connections in the association field the sub-Riemannian geodesics model the field lines remarkably well (in comparison to the exponential curves), as can be observed in part b) of Figure 5.9. Moreover, the field curves of the association field end with vertical tangent vectors, and these endpoints are very close to cusp points in the sub-Riemannian geodesics modeling these field lines. Following the general idea of Petitot's work [42] (e.g. the circle bundle model) and the results in this article on the existence set $\mathfrak{R}$, this puts the following conjecture:

Conjecture 5.1. The criterium in our visual system to connect two local orientations, say $g_0 = (x_0, y_0, \theta_0) = (0,0,0)$ and $g_{fin} = (x_{fin}, y_{fin}, \theta_{fin}) \in SE(2)$, could be modeled by checking whether g_{fin} is within the range $\mathfrak{R}$ of the exponential map.

This conjecture needs further investigation by neuro-physiological experiments. In any case, within the model $\mathbf{P_{curve}}$ (coinciding with Petitot's circle bundle model [42] and the sub-Riemannian model by Citti and Sarti [13,49]) a curve is globally optimal if and only if it is stationary, by the results in [8] (summarized in Theorem 5). Furthermore, the sub-Riemannian geodesics strongly deviate from horizontal exponential curves even if the end condition is chosen such that the co-circularity condition is satisfied (see c) in Figure 5.9).

Remark 5.11. Regarding association field models, we discussed 3 different models:

1. The cuspless sub-Riemannian geodesic model $\mathbf{P_{curve}}$, cf. [6,8,13,20,32], (extending Petitot's circle bundle model [42, ch:6.6.5])

2. The Legendrian geodesic model [42],
3. The horizontal exponential curve model in [48] given by Eq. (5.48).

These models relate as follows:

- The Legendrian geodesics follow from the cuspless sub-Riemannian geodesic model by contracting (e.g. [19]) the sub-Riemannian manifold on $(SE(2), \Delta_2, \mathscr{G}_\beta)$ towards its nilpotent approximation, cf. Eq. (5.49).
- The horizontal exponential curves keep the control variable in $\mathbf{P_{curve}}$ constant and they are rough local approximations of sub-Riemannian geodesics, see item c) in Figure 7. We also recall Theorem 5.2: The discrepancy between horizontal exponential curves and sub-Riemannian geodesics in $(SE(2), \Delta_2, G_\beta)$ is also intriguing from the differential geometrical viewpoint. Due to the presence of torsion in the Cartan connection auto-parallel curves (i.e. *the straight* curves in $(SE(2), \Delta_2, G_\beta)$ satisfying $\nabla_{\dot{\gamma}}\dot{\gamma} = 0$) do not coincide with the sub-Riemannian geodesics (i.e. *the shortest curves* in $(SE(2), \Delta_2, G_\beta)$ satisfying $\nabla_{\dot{\gamma}}\lambda = 0$).

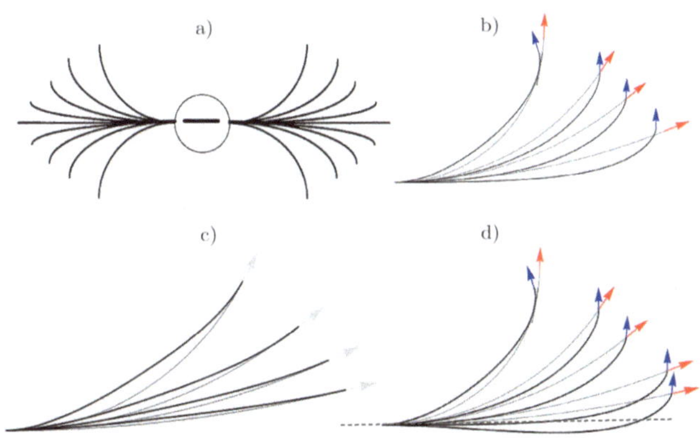

Fig. 5.9 Modeling the association field with sub-Riemannian geodesics and exponential curves, a) the association field [26, 42]. Compare the upper-right part of the association field to the following lines: in b) we impose the end condition (blue arrows) for the SR-geodesic model in black and the end condition (red arrows) for the horizontal exponential curve model in grey; c) comparison of sub-Riemannian geodesics with exponential curves with the same (co-circularity) ending conditions; d) as in b) including other ending conditions.

5.4 The Case $d = 3$: Sub-Riemannian Geodesics in $(\mathbb{R}^3 \rtimes S^2, \Delta_3, G_1)$

In order to obtain momentum along cuspless sub-Riemannian geodesics in $(\mathbb{R}^3 \rtimes S^2, \Delta_3, G_{\beta=1})$, we apply Theorem 5.1 to the case $d = 3$ with $\mathbf{a} = \mathbf{e}_z$. We find

$$
\begin{aligned}
\lambda_1(s) &= \lambda_1(0)\cosh s - \lambda_5(0)\sinh s, \\
\lambda_5(s) &= \lambda_5(0)\cosh s - \lambda_1(0)\sinh s, \\
\lambda_2(s) &= \lambda_2(0)\cosh s + \lambda_4(0)\sinh s, \\
\lambda_4(s) &= \lambda_4(0)\cosh s + \lambda_2(0)\sinh s, \\
\lambda_3(s) &= \sqrt{1-|\lambda_4(s)|^2-|\lambda_5(s)|^2},
\end{aligned}
\tag{5.51}
$$

and by using the horizontal part of the PMP, we now find explicit formulas for curvature and torsion of the spatial part of the sub-Riemannian geodesic:

$$
\begin{aligned}
\kappa(s) &= \frac{\sqrt{|\lambda_4(s)|^2+|\lambda_5(s)|^2}}{\lambda_3(s)} = \frac{\sqrt{1-|\lambda_3(s)|^2}}{\lambda_3(s)}, \\
\tau(s) &= -\frac{(\lambda_5(s)\lambda_2(s)+\lambda_4(s)\lambda_1(s))}{1-|\lambda_3(s)|^2} = \frac{W}{1-|\lambda_3(s)|^2}
\end{aligned}
\tag{5.52}
$$

where W denotes the Wronskian, recall Eq. (5.24), of $\lambda_5 = \frac{\kappa_1}{\sqrt{\kappa^2+1}}$ and $-\lambda_4 = \frac{\kappa_2}{\sqrt{\kappa^2+1}}$, where $\underline{\kappa} = (\kappa_1,\kappa_2) = \kappa_1\mathscr{A}_1 + \kappa_2\mathscr{A}_2$ denote the curvature components of the curvature vector-field $\ddot{\mathbf{x}} = \underline{\kappa}$ along the curve.

Consequently, we have

$$
\tau(s) = W\left(1+\kappa^{-2}(s)\right)
$$

whenever $\kappa(s) \neq 0$ for $s < s_{max}(\lambda(0))$ and W the constant Wronskian.

Furthermore, we see that if torsion is absent in any part of the spatial part of a sub-Riemannian geodesic it is absent everywhere, and we have

$$
\gamma(s) \text{ is planar } \Leftrightarrow \tau = 0 \Leftrightarrow W = 0 \Leftrightarrow \text{ the boundary conditions are co-planar}\,.
$$

The final equivalence, is non-trivial. For details on the proof, see [30, Cor.3.44].

Remark 5.12. The planar solutions $\mathbf{P}_{\mathbf{curve}}$ with $W=0$ coincide with the unique two dimensional sub-Riemannian geodesics connecting the corresponding points in $\mathbb{R}^2 \rtimes S^1$ (see [8, 15]) discussed in the previous section. As a result, the set

$$
\begin{aligned}
&\{\left(\mathbf{x}_{fin}, R_{\mathbf{n}_{fin}(\mathbf{x}_{fin},\theta_{fin})}\right) \mid (\mathbf{x}_{fin},\theta_{fin}) \in \mathfrak{R}_2\} \\
&\text{with } \mathbf{n}_{fin}(\mathbf{x}_{fin},\theta_{fin}) = (\sin\theta_{fin}\frac{x_{fin}}{\sqrt{x_{fin}^2+y_{fin}^2}}, \sin\theta_{fin}\frac{y_{fin}}{\sqrt{x_{fin}^2+y_{fin}^2}}, \cos\theta_{fin})^T,
\end{aligned}
$$

where $\mathfrak{R}_2$ denotes the set of admissible end conditions in $SE(2)$ allowing a connection via a globally minimal sub-Riemannian geodesic in $(SE(2),\Delta_2,G_1)$, is a subset of end conditions admitting a unique globally minimizing sub-Riemannian geodesic in $(SE(3),\Delta_3,G_1)$. See Figure 5.11.

Remark 5.13. A sub-Riemannian geodesic is co-planar if $W = -\lambda_5\lambda_2 - \lambda_4\lambda_1 = 0$, i.e. if its angular momentum $(\lambda_4,\lambda_5,0) \equiv \lambda_4\omega^4 + \lambda_5\omega^5$ is orthogonal to its spatial momentum $(\lambda_1,\lambda_2,\lambda_3) \equiv \lambda_1\omega^1 + \lambda_2\omega^2 + \lambda_3\omega^3$.

Now in order to compute the exponential map $\widetilde{\mathrm{Exp}}(\lambda(0),\ell) = \gamma(\ell) = g_{fin} \in SE(3)$, one can substitute Eq. (5.52) into Eq. (5.51) while setting $s=\ell$ and then integrate the well-known Frenet-Serret formulas for curves in $\mathbb{R}^3$. However, the last step in this procedure is somewhat cumbersome and here (again) Theorem 5.3 comes at hand.

5.4.1 Explicit Parameterizations of the Sub-Riemannian Geodesics

In order to integrate the Frenet-Serret formulas we apply Theorem 5.3 to the case $d=3$. This provides the following explicit formulas for the sub-Riemannian geodesics, where we use the short notation $\underline{\lambda}^{(1)} = (\lambda_1, \lambda_2)$ and $\underline{\lambda}^{(2)} = (\lambda_5, \lambda_4)$ from Section 5.2.

Theorem 5.9. *Let the momentum covector be given by Eq. (5.51). Then the spatial part of the cuspless sub-Riemannian geodesics*[8] *in $(SE(3), \Delta_3, G_1)$ is given by*

$$\mathbf{x}(s) = \tilde{R}(0)^T (\tilde{\mathbf{x}}(s) - \tilde{\mathbf{x}}(0)) \tag{5.53}$$

where, $\tilde{R}(0)$ and $\tilde{\mathbf{x}}(s) := (\tilde{x}(s), \tilde{y}(s), \tilde{z}(s))$ are given in terms of $\lambda^{(1)}(0)$ and $\lambda^{(2)}(0)$ depending on several cases. For all cases, we have

$$\tilde{x}(s) = \frac{1}{\mathfrak{c}} \int_0^s \lambda_3(\tau)\, \mathrm{d}\tau. \tag{5.54}$$

For the case $\underline{\lambda}^{(1)}(0) = \mathbf{0}$, we have

$$\tilde{R}(0) = \begin{pmatrix} 0 & 0 & 1 \\ 0 & 1 & 0 \\ -1 & 0 & 0 \end{pmatrix} \in SO(3), \tag{5.55}$$

$$\begin{pmatrix} \tilde{y}(s) \\ \tilde{z}(s) \end{pmatrix} = \frac{-1}{\mathfrak{c}} \begin{pmatrix} \lambda_4(s) \\ \lambda_5(s) \end{pmatrix}. \tag{5.56}$$

For the case $\underline{\lambda}^{(1)}(0) \neq \mathbf{0}$, we have

$$\tilde{R}(0) = \frac{1}{\mathfrak{c}} \begin{pmatrix} \lambda_1(0) & \lambda_2(0) & \lambda_3(0) \\ \mathfrak{c}\frac{-\lambda_2(0)}{\|\underline{\lambda}^{(1)}(0)\|} & \mathfrak{c}\frac{\lambda_1(0)}{\|\underline{\lambda}^{(1)}(0)\|} & 0 \\ \frac{-\lambda_1(0)\lambda_3(0)}{\|\underline{\lambda}^{(1)}(0)\|} & \frac{-\lambda_2(0)\lambda_3(0)}{\|\underline{\lambda}^{(1)}(0)\|} & \|\underline{\lambda}^{(1)}(0)\| \end{pmatrix} \in SO(3). \tag{5.57}$$

For the case $W=0$ along with $\underline{\lambda}^{(1)}(0) \neq \mathbf{0}$, we have

$$\begin{pmatrix} \tilde{y}(s) \\ \tilde{z}(s) \end{pmatrix} = \frac{\lambda_2(0)\lambda_4(s) - \lambda_1(0)\lambda_5(s)}{\mathfrak{c}\|\underline{\lambda}^{(1)}(0)\|} \begin{pmatrix} 0 \\ 1 \end{pmatrix}. \tag{5.58}$$

[8] which are the lifts of the stationary curves of $\mathbf{P_{curve}}$ for appropriate boundary conditions and which coincide with the solutions of $\mathbf{P_{mec}}$ for the same boundary conditions.

While for $W \neq 0$ *along with* $\underline{\lambda}^{(1)}(0) \neq 0$ *and*

$$A(s) = \frac{1}{\|\underline{\lambda}^{(2)}(0)\|^2 - \frac{W^2}{\mathfrak{c}^2}} \begin{pmatrix} \lambda_2(s)\lambda_4(s) - \lambda_1(s)\lambda_5(s) & -\frac{W}{\mathfrak{c}}\lambda_3(s) \\ \frac{W}{\mathfrak{c}}\lambda_3(s) & \lambda_2(s)\lambda_4(s) - \lambda_1(s)\lambda_5(s) \end{pmatrix}, \tag{5.59}$$

we have

$$\begin{pmatrix} \tilde{y}(s) \\ \tilde{z}(s) \end{pmatrix} = \frac{e^{\int_0^s A(s')\,ds'}}{\mathfrak{c}^2 \|\underline{\lambda}^{(1)}(0)\|} \begin{pmatrix} W\lambda_3(0) \\ \mathfrak{c}(\lambda_2(0)\lambda_4(0) - \lambda_5(0)\lambda_1(0)) \end{pmatrix}. \tag{5.60}$$

Proof. From Theorem 5.3, we have

$$d(\lambda m(\gamma)^{-1}) = 0 \tag{5.61}$$

with $\lambda = (\lambda_1, \ldots, \lambda_6)$, the Lagrange multipliers which are already known by Theorem 5.1, and with matrix representation m given by Eq. (5.30). Hence, as $\gamma(0) = \mathbf{e}$, we have that $\forall s \in [0, \ell]$ the geodesic must satisfy

$$\lambda(s) = \lambda(0) m(\gamma(s)). \tag{5.62}$$

To make calculations easier, we translate and rotate the curve and solve a slightly easier equation and transform it back to the original curve. With $m(\tilde{g}) = \begin{pmatrix} \tilde{R} & \sigma_{\tilde{\mathbf{x}}}\tilde{R} \\ 0 & \tilde{R} \end{pmatrix}$, we solve the system

$$\lambda(s) = (\mathfrak{c}, 0, 0, -\frac{W}{\mathfrak{c}}, 0, 0) m(\tilde{\gamma}(s)) \tag{5.63}$$

with $\tilde{\gamma}(s) = \tilde{\gamma}(0)\, m(\gamma(s))$ such that

$$\lambda(0) = (\mathfrak{c}, 0, 0, -\frac{W}{\mathfrak{c}}, 0, 0)\tilde{\gamma}(0). \tag{5.64}$$

Thus after having $\tilde{\gamma}(s)$, we retrieve the geodesic by using the relation

$$\gamma(s) = \tilde{\gamma}(0)^{-1}\tilde{\gamma}(s). \tag{5.65}$$

For the most general case, assuming non-vanishing denominators throughout, we see that choosing (5.57) and

$$\tilde{\mathbf{x}}(0) := \frac{1}{\mathfrak{c}^2\sqrt{|\lambda_1(0)|^2 + |\lambda_2(0)|^2}} \begin{pmatrix} 0 \\ W\lambda_3 \\ \mathfrak{c}(\lambda_2\lambda_4 - \lambda_5\lambda_1) \end{pmatrix},$$

(5.64) is satisfied. Then solving (5.63) for $\tilde{x}$, $\tilde{y}$ and $\tilde{z}$ we obtain $\tilde{x}$ and the following system:

$$\begin{pmatrix} \dot{\tilde{y}}(s) \\ \dot{\tilde{z}}(s) \end{pmatrix} = A(s) \begin{pmatrix} \tilde{y}(s) \\ \tilde{z}(s) \end{pmatrix}.$$

Noting that $A(s)$ and $A(t)$ commute for all pairs s and t, and hence using Wilcox formula [56], we get the desired results.

Clearly, the formulas are not valid as the denominators in some of the expressions become zero. Hence we do the whole procedure keeping in mind the special cases right from the start and get the required results.

The matrix $e^{\int_0^s A(s')\mathrm{d}s'}$ can be computed explicitly, for details see [30, Cor.4.11]. One has $e^{\int_0^s A(s')\mathrm{d}s'} = \sqrt{\frac{\|\underline{\lambda}^{(2)}(s)\|^2 - W^2\mathfrak{c}^{-2}}{\|\underline{\lambda}^{(2)}(0)\|^2 - W^2\mathfrak{c}^{-2}}} \begin{pmatrix} \cos\phi(s) & -\sin\phi(s) \\ \sin\phi(s) & \cos\phi(s) \end{pmatrix}$, with $\phi(s) = \int_0^s \frac{W\mathfrak{c}^{-1}\lambda_3(s')}{\|\underline{\lambda}^{(2)}(s')\|^2 - W^2\mathfrak{c}^{-2}}\mathrm{d}s'$.

5.4.2 *Explicit Definition of the Exponential Map of* $\mathbf{P}_{\mathbf{curve}}$

In this section, we provide the explicit definition of the exponential map which maps the pair $(\lambda(0), \ell)$ to the endpoint $g_{fin} = \widetilde{\text{Exp}}(\lambda(0), \ell)$ of the corresponding cusp-less sub-Riemannian geodesic in sub-Riemannian manifold $(SE(3), \Delta_3, G_1)$.

Definition 5.1. Using the arc-length parametrization and setting $t = s$ ($\Rightarrow \sigma = 1$), we consider the canonical ODE system for $\Gamma(s) = (g(s), \kappa(s), \lambda(s))$ given by

$$\begin{aligned} \dot{\Gamma}(s) &= F(\Gamma(s)) \qquad s \in [0, \ell] \\ \Gamma(0) &= (e, \kappa(0), \lambda(0)) \end{aligned}$$

with unity element $e = (\mathbf{0}, I) \in SE(3)$ and with $\kappa = (\kappa_1, \kappa_2)^T$ where $\kappa_1(0) = \frac{\lambda_5(0)}{\sqrt{1-(\lambda_4(0)^2+\lambda_5(0)^2)}}$ and $\kappa_2(0) = \frac{-\lambda_4(0)}{\sqrt{1-(\lambda_4(0)^2+\lambda_5(0)^2)}}$ where F denotes the corresponding flow field given as

$$F(\begin{pmatrix} R & \sigma_{\mathbf{x}}R \\ 0 & R \end{pmatrix}, \kappa(s), \lambda(s)) =$$

$$\left(\begin{pmatrix} RK & R\sigma_{e_z} + \sigma_x RK \\ 0 & RK \end{pmatrix}, \frac{\lambda_2\lambda_4 - \lambda_1\lambda_5}{\lambda_3^3} \begin{pmatrix} \lambda_5 \\ -\lambda_4 \end{pmatrix} - \frac{1}{\lambda_3} \begin{pmatrix} \lambda_1 \\ \lambda_2 \end{pmatrix}, \lambda \begin{pmatrix} K & \sigma_{e_z} \\ 0 & K \end{pmatrix} \right) \tag{5.66}$$

where $K = \frac{1}{\lambda_3} \begin{pmatrix} 0 & 0 & \lambda_5 \\ 0 & 0 & -\lambda_4 \\ -\lambda_5 & \lambda_4 & 0 \end{pmatrix}$. and $\sigma_{\mathbf{e}_z} \in \mathbb{R}^{3\times 3}$ such that $\sigma_{\mathbf{e}_z}\mathbf{x} = \mathbf{e}_z \times \mathbf{x}$.This ODE has a unique solution

$$\Gamma(s) = \Gamma(0)e^{sF} \qquad s \in [0, \ell].$$

Recall the definition of $\mathfrak{D}$ in Eq. (5.31) (for $d = 3$). On this set, we define $\widetilde{Exp}_e$: $\mathfrak{D} \to SE(3)$ by

$$\widetilde{Exp}_e(\lambda(0), l) = \pi \circ e^{lF}(e, \kappa(0), \lambda(0))$$

with π being the natural projection on $SE(3)$. Note that this exponential map is different from the Lie group valued exponential map defined on the Lie algebra.

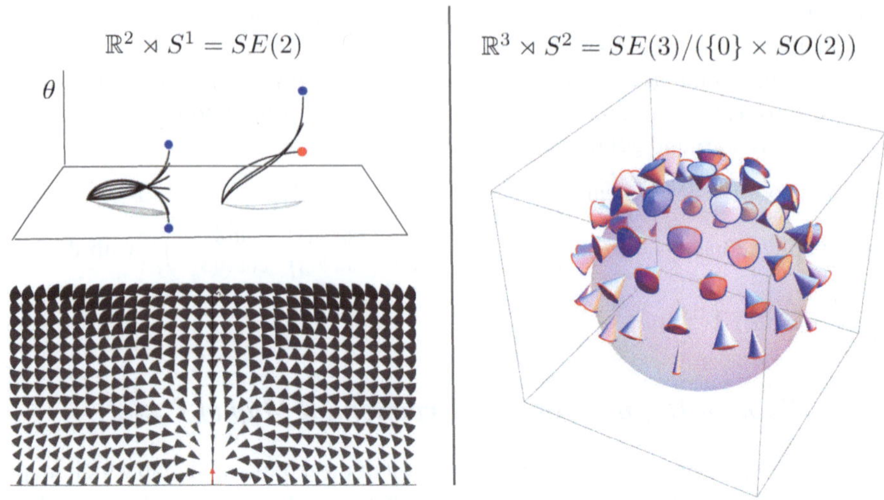

Fig. 5.10 A comparison of the possible end conditions of $\mathbf{P_{curve}}$ for the two dimensional and the three dimensional cases. Right: possible tangent directions are depicted of cuspless sub-Riemannian geodesics in $(SE(3), \Delta_3, G_1)$ with initial position at the origin and the initial direction along $\mathbf{e}_z$ and the final positions at unit distance from the origin. Left: cones of possible end conditions of cuspless sub-Riemannian geodesics in $(SE(2), \Delta_2, G_1)$. According to Theorem 5.7, these cones are obtained by considering the end conditions of sub-Riemannian geodesics that either begin with a cusp point (shown in red) or end at a cusp point (shown in blue). Figure 5.11 depicts the comparison in the special case when we set the end conditions on a unit circle containing the z-axis.

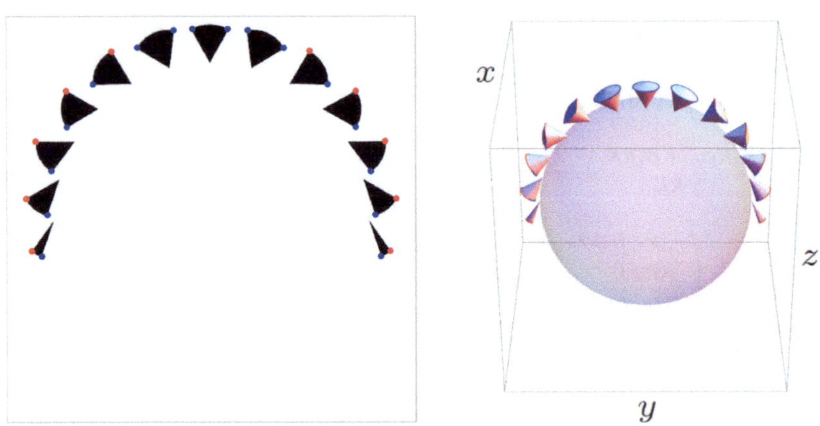

Fig. 5.11 A comparison of the cones of reachable angles by the cuspless sub-Riemannian geodesics in the two dimensional case as in [8, 15] (left) and those in the three dimensional case (right). It represents the special case in Figure 5.10, of the end conditions being on a unit circle containing the z-axis. The intersection of the cones in Figure 5.11 right with $x = 0$ coincides with the cones depicted in Figure 5.11 left.

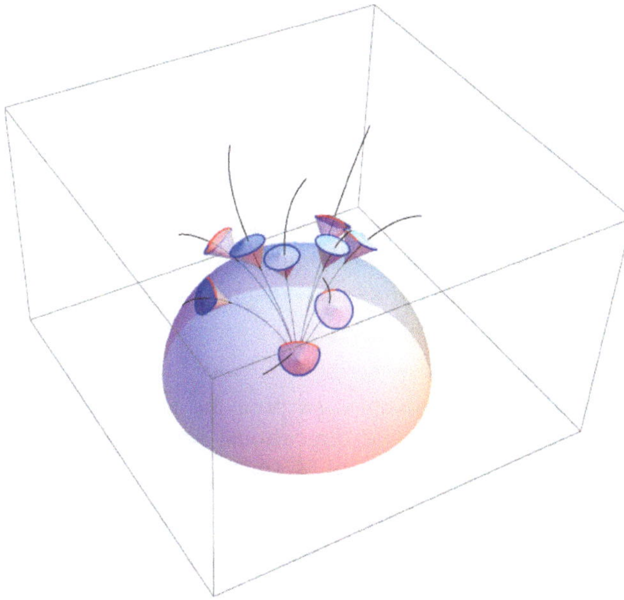

Fig. 5.12 An illustration of the spatial part of arbitrary cuspless sub-Riemannian geodesics in $(SE(3), \Delta_3, G_1)$ and the cones of reachable angles as depicted in Figures 5.10 and 5.11. The cuspless sub-Riemannian geodesics are always contained within the cones. We checked this for many more cases, which supports our Conjecture 5.2.

5.4.3 The Range of the $\widetilde{\text{Exp}}$ Map and Cones of Reachable Angles

There are various restrictions on the possible boundary conditions for which we can get a cuspless sub-Riemannian geodesic of problem $\mathbf{P_{curve}}$, see Fig. 5.10. We present some special cases which help us to get an idea about the range of the exponential map of $\mathbf{P_{curve}}$. Note that this set coincides with the set of end conditions for which $\mathbf{P_{curve}}$ is expected to be well defined, as we have shown for the 2D-case (recall Theorem 5.5). The next corollary gives us the possible final positions when the final direction is anti parallel to the initial direction.

Corollary 5.3. *Let $(\boldsymbol{x}_1, \boldsymbol{n}_1)$ be the end condition of $\mathbf{P_{curve}}$ with the initial condition being $(\boldsymbol{0}, \boldsymbol{e}_z)$. Then, given that $\boldsymbol{n}_1 = -\boldsymbol{e}_z$, a cuspless sub-Riemannian geodesic of problem $\mathbf{P_{curve}}$ exists only for $\boldsymbol{x}_1 \cdot \boldsymbol{e}_z = 0$. Moreover, this condition is only possible for curves departing from a cusp and ending in a cusp.*

Proof. Let $\mathbf{x}$ be a cuspless sub-Riemannian geodesic of problem $\mathbf{P_{curve}}$ with $\dot{\mathbf{x}}(0) = -\dot{\mathbf{x}}(\ell)$ for some $\ell \leq s_{max}$. This means that going to the tilde coordinates, we have $\dot{\tilde{\mathbf{x}}}(0) = -\dot{\tilde{\mathbf{x}}}(\ell)$, which implies $\dot{\tilde{x}}(0) = -\dot{\tilde{x}}(\ell)$. But this is possible only if $\dot{\tilde{x}}(0) = 0 = \dot{\tilde{x}}(\ell)$, which is possible only if $\|\underline{\lambda}^{(2)}(0)\| = 1$ and $\ell = s_{max}$, i.e., if the geodesic both starts and ends in cusp. $\square$

Now we recall from Subsection 5.3.2, that in the 2D-case cuspless sub-Riemannian geodesics in $(SE(2),\Delta_2,G_1)$ are contained in the half space $x \geq 0$ and $x = 0$ can only be reached with sub-Riemannian geodesics both departing from and ending in a cusp. In the 3D-case one expects a similar result, as it is confirmed by many numerical experiments, see e.g. Figure 5.12 and Figure 5.11. However, it turns out to be hard to prove for all cases. At least we have the following formal result.

Corollary 5.4. *If a cuspless sub-Riemannian geodesic departs from a cusp, then it can never have a negative component along the z-axis. Moreover, it can meet the* $z = 0$ *plane at non zero time only if* $s = s_{max}$ *and* $W = 0$.

Proof. See [30, Lemma 4.13].

Based on our numerical experiments, we pose the following conjecture which is analogous to a result in the two dimensional case of finding cuspless sub-Riemannian geodesics in $(SE(3),\Delta_3,G_1)$ [8,15].

Conjecture 5.2. Let the range of the exponential map defined in Definition 5.1 be denoted by $\mathfrak{R}$ and let $\mathfrak{D}$ be as defined in Definition 5.1.

- $\widetilde{Exp}_e : \mathfrak{D} \to \mathfrak{R}$ is a homeomorphism when $\mathscr{D}$ and $\mathscr{R}$ are equipped with the sub-space topology.
- $\widetilde{Exp}_e : \mathring{\mathfrak{D}} \to \mathring{\mathfrak{R}}$ is a diffeomorphism. Here $\mathring{S}$ denotes the interior of the set S.

The boundary of the range is given as

$$\partial\mathfrak{R} = \mathscr{S}_B \cup \mathscr{S}_R \cup \mathscr{S}_L \text{ with} \tag{5.67}$$
$$\mathscr{S}_B = \{\widetilde{Exp}_e(\lambda(0), s_{max}(\lambda(0)))|\lambda(0) \in \mathscr{C}\} \text{ and}$$
$$\mathscr{S}_R = \{\widetilde{Exp}_e(\lambda(0), s)|\lambda(0) \in \mathscr{C} \text{ and } \lambda_4(0)^2 + \lambda_5(0)^2 = 1 \text{ and } s > 0\}$$
$$\mathscr{S}_L = \{(\mathbf{0}, R_{\mathbf{n}}) \in SE(3)|\mathbf{n} \in S^2\}.$$

This conjecture would imply that no conjugate points, recall Remark 5.4, arise within $\mathfrak{R}$ and problem $\mathbf{P_{curve}}$ (5.3) is well posed for *all* end conditions in $\mathfrak{R}$.

The proof of this conjecture would be on similar lines as in Appendix F of [15]. If the conjecture is true, we have a reasonably limited set of possible directions per given final positions for which a cuspless sub-Riemannian geodesic of problem $\mathbf{P_{curve}}$ exists. Then likewise the $d = 2$ case, we have that every end condition in $\mathfrak{R}$ can be connected with a unique minimizer of a well-posed problem $\mathbf{P_{curve}}$. Moreover, the cones determined by $\mathscr{S}_B$ and $\mathscr{S}_R$ provide the boundaries of the field of reachable cones. Figure 5.11 shows the special case of the end conditions being on a unit circle containing the z-axis. The final tangents are always contained within the cones at each position. Numerical computations indeed seem to confirm that this is the case (see Figure 5.12). The blue points on the boundary of the cones correspond to $\mathscr{S}_B$ while the red points correspond to $\mathscr{S}_R$ given in Equality (5.67).

5.5 The Case $d = 4$: Sub-Riemannian Geodesics in $(SE(4), \Delta_4, G_1)$

Let us apply the results regarding sub-Riemannian geodesics in $(SE(d), \Delta_d, G_{\beta=1})$ to the special case $d = 4$. Here we will rely on the standard matrix group representation of $SE(4)$ given by $M(g) = \begin{pmatrix} R & \mathbf{x} \\ \mathbf{0} & 1 \end{pmatrix} \in \mathbb{R}^{5\times 5}$, for all $g = (\mathbf{x}, R) \in SE(4)$. In matrix-form, the Lie-algebra elements spanning $T_e(SE(4))$ are given by

$$A_k = \mathscr{A}_k\big|_e \equiv \begin{pmatrix} \mathbf{0} & \mathbf{e}_k \\ \mathbf{0} & 0 \end{pmatrix}$$

for $k = 1 \ldots 4$, with $(\mathbf{e}_k)^j = \delta_k^j$, and

$$A_5 = \begin{pmatrix} 0&0&0&0&0 \\ 0&0&0&0&0 \\ 0&0&0&-1&0 \\ 0&0&1&0&0 \\ 0&0&0&0&0 \end{pmatrix}, A_6 = \begin{pmatrix} 0&0&0&0&0 \\ 0&0&0&1&0 \\ 0&0&0&0&0 \\ 0&-1&0&0&0 \\ 0&0&0&0&0 \end{pmatrix}, A_7 = \begin{pmatrix} 0&0&0&1&0 \\ 0&0&0&0&0 \\ 0&0&0&0&0 \\ -1&0&0&0&0 \\ 0&0&0&0&0 \end{pmatrix},$$
$$A_8 = \begin{pmatrix} 0&0&0&0&0 \\ 0&0&-1&0&0 \\ 0&1&0&0&0 \\ 0&0&0&0&0 \\ 0&0&0&0&0 \end{pmatrix}, A_9 = \begin{pmatrix} 0&0&1&0&0 \\ 0&0&0&0&0 \\ -1&0&0&0&0 \\ 0&0&0&0&0 \\ 0&0&0&0&0 \end{pmatrix}, A_{10} = \begin{pmatrix} 0&1&0&0&0 \\ -1&0&0&0&0 \\ 0&0&0&0&0 \\ 0&0&0&0&0 \\ 0&0&0&0&0 \end{pmatrix}.$$

The commutator table is given by

$$[A_i, A_j]_{i,j=1\ldots 10} = \begin{pmatrix} 0&0&0&0&0&0&A_4&0&A_3&A_2 \\ 0&0&0&0&0&A_4&0&-A_3&0&-A_1 \\ 0&0&0&0&-A_4&0&0&A_2&-A_1&0 \\ 0&0&0&0&A_3&-A_2&-A_1&0&0&0 \\ 0&0&A_4&-A_3&0&A_8&-A_9&-A_6&A_7&0 \\ 0&-A_4&0&A_2&-A_8&0&A_{10}&A_5&0&-A_7 \\ -A_4&0&0&A_1&A_9&-A_{10}&0&0&-A_5&A_6 \\ 0&A_3&-A_2&0&A_6&-A_5&0&0&-A_{10}&A_9 \\ -A_3&0&A_1&0&-A_7&0&A_5&A_{10}&0&-A_8 \\ -A_2&A_1&0&0&0&A_7&-A_6&A_9&A_8&0 \end{pmatrix}$$

and the PMP produces the following ODE for the momentum components:

$$\dot{\lambda}_i(t) = -\sum_{j=4}^{7} \sum_{k=1}^{7} c_{ij}^k \lambda_j(t) \lambda_k(t),$$

or more explicitly, using the fact that $\lambda_8 = \lambda_9 = \lambda_{10} = 0$ yields

$$\begin{aligned}
&\dot{\lambda}_1(t) = -\lambda_4(t)\lambda_7(t), && \dot{\lambda}_6(t) = -\lambda_2(t)\lambda_4(t),\\
&\dot{\lambda}_2(t) = -\lambda_4(t)\lambda_6(t), && \dot{\lambda}_7(t) = -\lambda_1(t)\lambda_4(t),\\
&\dot{\lambda}_3(t) = \lambda_4(t)\lambda_5(t), && \dot{\lambda}_8(t) = 0,\\
&\dot{\lambda}_4(t) = -\lambda_3(t)\lambda_5(t) + \lambda_2(t)\lambda_6(t) + \lambda_1(t)\lambda_7(t), && \dot{\lambda}_9(t) = 0,\\
&\dot{\lambda}_5(t) = \lambda_3(t)\lambda_4(t), && \dot{\lambda}_{10}(t) = 0.
\end{aligned}$$

Along *cuspless* sub-Riemannian geodesics, this ODE-system simplifies to

$$\begin{aligned}
&\dot{\lambda}_1(s) = -\lambda_7(s), && \dot{\lambda}_6(s) = -\lambda_2(s),\\
&\dot{\lambda}_2(s) = -\lambda_6(s), && \dot{\lambda}_7(s) = -\lambda_1(s),\\
&\dot{\lambda}_3(s) = \lambda_5(s), && \dot{\lambda}_8(s) = 0,\\
&\dot{\lambda}_4(s) = -(\lambda_4(s))^{-1}(\lambda_3(s)\lambda_5(s) + \lambda_2(s)\lambda_6(s) + \lambda_1(s)\lambda_7(s)), && \dot{\lambda}_9(s) = 0,\\
&\dot{\lambda}_5(s) = \lambda_3(s), && \dot{\lambda}_{10}(s) = 0,
\end{aligned}$$

which is indeed a special case of Eq. (5.21) with $d=4$ and $\Lambda = \text{diag}\{-1,-1,1\}$, whose solutions are now given by Eq. (5.25) (again with $d=4$ and $\Lambda = \text{diag}\{-1,-1,1\}$).

Now the case $d=3$ with (cuspless) sub-Riemannian geodesics on $(SE(3), \Delta_3, G_{\beta=1})$ studied in detail in the previous section, by omitting the A_1, A_7, A_9 and A_{10} directions and relabeling the indices of the Lie-algebra elements and momentum components as follows: $(2,3,4,5,6,8) \to (1,2,3,4,5,6)$.

Acknowledgements. The authors gratefully acknowledge the Russian Foundation for Basic Research (project no. 12-01-00913-a), and the Ministry of Education and Science of Russia within the federal program "Scientific and Scientific-Pedagogical Personnel of Innovative Russia" (contract no. 8209), and the European Commission ITN-FIRST, agreement No. PITN-GA-2009-238702 and the European Research Council ("Lie Analysis", contract number Stg 553333) for financial support. The research to the results of this article has received funding from the European Research Council under the European Commityâs 7th Framework Program, ERC grant agreement no.335555.

Appendix A: Derivation of the Canonical Equations for Sub-Riemannian Geodesics in $(SE(d), \Delta_d, G_1)$ Using the Pontryagin Maximum Principle

Consider the sub-Riemannian manifold $M = (SE(d), \Delta_d, G_\beta)$ given by Eq. (5.14), with $\beta = 1$. Consider geometric control problem $\mathbf{P_{mec}}$ given by Eq. (5.15).

There exists a standard PMP for $\mathbb{L}_\infty([0,T])$ controls and there exists a recently generalized PMP for $\mathbb{L}_1([0,T])$ controls [55] which in this case produces (by a reparameterization argument) the same solutions, despite the fact that $\mathbb{L}_\infty([0,T]) \subset \mathbb{L}_1([0,T])$. Here we note that $\mathbf{P_{mec}}$ is equivalent to the solutions $\gamma\colon [0,T] \to SE(d)$, with prescribed boundary conditions $\gamma(0) = (0, I)$ and $\gamma(T) = (\mathbf{x}_1, R_{\mathbf{n}_1})$, of the control problem

$$\int_0^T L(u^d(t), \ldots, u^{2d-1}(t))\mathrm{d}t \to \quad \text{minimize (with fixed } T)$$

with

$$\dot{\gamma}(t) = \sum_{i=d}^{2d-1} u^i(t)\mathscr{A}_i|_{\gamma(t)} = \sum_{i=d}^{2d-1} \langle \omega^i|_{\gamma(t)}, \dot{\gamma}(t)\rangle \mathscr{A}_i|_{\gamma(t)}$$

where for $i = 1,2,3$, $u^i \in \mathbb{L}_1([0,T])$, and with Lagrangian

$$L(u^d(t), \ldots, u^{2d-1}(t)) = \sum_{i=d}^{2d-1} |u^i(t)|^2.$$

Applying the standard PMP to this problem with fixed time T, we have that there exists a Lipshitzian curve in the cotangent bundle given by $[0,T] \ni t \mapsto \mu(t) = (\gamma(t), \lambda(t)) \neq 0$ with $\lambda(t) \in T^*_{\gamma(t)}(M)$ such that

$$\begin{aligned}\dot{\mu} &= \mathbf{H}(\mu(t)) \\ H(\mu) &= \max_{u \in \mathbb{R}^d} \left(L(u^d, \ldots, u^{2d-1}) - \sum_{i=d}^{2d-1} \mu_i u^i \right)\end{aligned}$$

where $\mu = (\gamma, \lambda)$, and where the Hamiltonian is given by $H(\mu) \equiv H(\lambda) = \frac{1}{2}\sum_{i=d}^{2d-1} |\lambda_i|^2$. The Hamiltonian vector field $\mathbf{H}$ given by

$$\mathbf{H} = \sum_{i=1}^{2d-1} \alpha^i \frac{\partial}{\partial \lambda_i} + \beta^i \mathscr{A}_i \tag{5.68}$$

is such that it preserves the canonical symplectic structure

$$\sigma = \sum_{i=1}^{2d} \mathrm{d}\lambda_i \wedge \omega^i = \sum_{i=d}^{2d-1} \mathrm{d}\lambda_i \wedge \omega^i$$

and hence, we have

$$\sigma(\mathbf{H}, \cdot) = -\mathrm{d}H = -\sum_{i=d}^{2d-1} \mathscr{A}_i H \omega^i + \frac{\partial H}{\partial \lambda_i}\mathrm{d}\lambda_i. \tag{5.69}$$

From Equations (5.68), (5.69), we obtain for $i = d, \ldots, 2d-1$ that

$$\alpha^i = -\mathscr{A}_i H \text{ and } \beta^i = \frac{\partial H}{\partial \lambda_i} = \lambda_i.$$

Consequently, (noting that $\mathscr{A}_i H = 0$ for $i = 1, \ldots, d$ and $i = 2d, \ldots, d(d+1)/2$), we have the Hamiltonian vector field

$$\begin{aligned}\mathbf{H}(\mu) &= \sum_{i=d}^{2d-1} \beta^i \mathscr{A}_i - \sum_{i=d+1}^{2d-1} \sum_{k=1}^{2d-1} \sum_{j=d}^{2d-1} c_{i,j}^k \lambda_k \beta^j \omega^i \\ &= \sum_{i=d}^{2d-1} \lambda_i \mathscr{A}_i - \sum_{i=d+1}^{2d-1} \sum_{k=1}^{2d-1} \sum_{j=d}^{2d-1} c_{i,j}^k \lambda_k \lambda_j \omega^i\end{aligned}$$

So now let us consider the full canonical ODE in the PMP: $\dot{\mu} = \mathbf{H}(\mu)$.

Clearly, the horizontal part of PMP (where time derivatives are w.r.t. sub-Riemannian arclength t) is given as

$$\dot{\gamma} = \sum_{i=d}^{2d-1} \lambda_i \mathscr{A}_i|_\gamma \Rightarrow \lambda_i(t) = \langle \omega^i|_{\gamma(t)}, \dot{\gamma}(t)\rangle \text{ for } i = d, \ldots, 2d-1.$$

The vertical part of PMP gives

$$\frac{d}{dt} \sum_{i=1}^{2d-1} \lambda_i(t)\, \omega^i|_{\gamma(t)} = \sum_{i=1}^{2d-1} \dot{\lambda}_i(t)\, \omega^i|_{\gamma(t)} + \lambda_i(t) \frac{d}{dt}\, \omega^i|_{\gamma(t)} = -\sum_{i=d+1}^{2d-1} \sum_{k=1}^{2d-1} \sum_{j=d}^{2d-1} c_{i,j}^k \lambda_k \lambda_j\, \omega^i|_{\gamma(t)}$$

which is equivalent to

$$\sum_{i=1}^{2d-1} \dot{\lambda}_i\, \omega^i\big|_\gamma + \sum_{i=1}^{d} \lambda_i \left(\sum_{k=d}^{2d-1} \sum_{j=1}^{d} c_{j,k}^i \dot{\gamma}^k\, \omega^j\big|_\gamma \right) = -\sum_{i=d+1}^{2d-1} \sum_{k=1}^{2d-1} \sum_{j=d}^{2d-1} c_{i,j}^k \lambda_k \lambda_j\, \omega^i\big|_\gamma$$

and therefore using the horizontal part of PMP we obtain

$$\begin{aligned}\sum_{i=1}^{d} \left(\dot{\lambda}_i(t) + \sum_{k=d}^{2d-1} \sum_{j=1}^{d} c_{i,k}^j \lambda_k(t) \lambda_j(t) \right) \omega^i\big|_{\gamma(t)} &= 0, \\ \sum_{i=d+1}^{2d-1} \left(\dot{\lambda}_i(t) + \sum_{j=d}^{2d-1} \sum_{k=1}^{2d-1} c_{i,j}^k \lambda_k(t) \lambda_j(t) \right) \omega^i\big|_{\gamma(t)} &= 0.\end{aligned}$$

Now in the first equation above index j can as well run from 1 to $2d-1$, since if $i \leq d$ and $k > d$ then for all $j > d$ we have $c_{i,k}^j = 0$. As a result we obtain

$$\dot{\lambda}_i(t) = -\sum_{j=d}^{2d-1} \sum_{k=1}^{2d-1} c_{i,j}^k \lambda_k(t)\, \lambda_j(t), \text{ for all } i \in \{1, \ldots, 2d-1\}.$$

References

1. Agrachev, A.A., Sachkov, Y.L.: Control Theory from the Geometric Viewpoint. Springer (2004)
2. Agrachev, A., Boscain, U., Gauthier, J.P., Rossi, F.: The intrinsic hypoelliptic Laplacian and its heat kernel on unimodular Lie groups. J. Funct. Anal. 256, 2621–2655 (2009)
3. van Almsick, M.: Context Models of Lines and Contours, Eindhoven University of Technology. Department of Biomedical Engineering (2007), `http://alexandria.tue.nl/extra2/200711652.pdf`
4. Ardentov, A., Sachkov, Y.: Solution of Euler's elastic problem. Avtomatika i Telemekhanika (4), 78–88 (2009), English translation in Automation and remote control
5. Barbieri, D., Citti, G., Sanguinetti, G., Sarti, A.: An uncertainty principle underlying the functional architecture of V1. Journal of Physiology Paris 106(5-6), 183–193 (2012)
6. Ben-Yosef, G., Ben-Shahar, O.: A Tangent Bundle Theory for Visual Curve Completion. PAMI 34(7), 1263–1280 (2012)
7. Boscain, U., Charlot, G., Rossi, F.: Existence of planar curves minimizing length and curvature. Proceedings of the Steklov Institute of Mathematics 270(1), 43–56 (2010)
8. Boscain, U., Duits, R., Rossi, F., Sachkov, Y.: Curve cuspless reconstruction via sub-Riemannian geometry. SIAM (2013), `http://bmia.bmt.tue.nl/people/RDuits/1203.3089v2.pdf` (submitted)
9. Bruckstein, A., Holt, R., Netravali, A.: Discrete elastica. Appl. Anal. (78), 453–485 (2001)
10. Bryant, R., Griffiths, P.: Reduction for constrained variational problems and $(1/2)\int \kappa^2 ds$. American Journal of Mathematics 108(3), 525–570 (1986)
11. Bryant, R.L., Chern, S.S., Gardner, R.B., Goldschmidt, H.L., Griffiths, P.A.: Exterior differential systems. In: Mathematical Sciences Research Institute Publications. Springer (1991)
12. Chan, T.F., Kang, S.-H., Shen, J.: Euler's elastica and curvature based inpaintings. SIAM J. Appl. Math. (63), 564–594 (2002)
13. Citti, G., Sarti, A.: A cortical based model of perceptional completion in the roto-translation space. Journal of Mathematical Imaging and Vision 24(3), 307–326 (2006)
14. Dela Haije, T.C.J., Duits, R., Tax, C.: Sharpening fibers in diffusion weighted MRI via erosion. In: Visualization and Processing of Tensors and Higher Order Descriptors for Multi-Valued Data (2012), `www.bmia.bmt.tue.nl/people/RDuits/DDT.pdf`, Accepted for publication, bookchapter in Visualization and Processing of Tensors and Higher Order Descriptors for Multi-Valued Data. Dagstuhl Seminar 11501
15. Duits, R., Boscain, U., Rossi, F., Sachkov, Y.: Association felds via cuspless sub-Riemannian geodesics in $SE(2)$. Journal of Mathematical Imaging and Vision, 1–32 (2013), `http://bmia.bmt.tue.nl/people/RDuits/cusp.pdf` (submitted)
16. Duits, R., Dela Haije, T.C.J., Creusen, E.J., Ghosh, A.: Morphological and Linear scale Spaces for Fiber Enhancement in DW-MRI. Journal of Mathematical Imaging and Vision 46(3), 326–368 (2013)
17. Duits, R., Franken, E.M.: Line Enhancement and Completion via Left-invariant Scale Spaces on SE(2). In: Tai, X.-C., Morken, K., Lysaker, M., Lie, K.-A. (eds.) SSVM 2009. LNCS, vol. 5567, pp. 795–807. Springer, Heidelberg (2009)
18. Duits, R., Franken, E.M.: Left-invariant Diffusions on the Space of Positions and Orientations and their Application to Crossing-preserving Smoothing of HARDI images. IJCV 92(3), 231–264 (2011)

19. Duits, R., Franken, E.M.: Left-invariant parabolic evolution equations on SE(2) and contour enhancement via invertible orientation scores, part I: Linear left-invariant diffusion equations on SE(2). Quarterly on Applied Mathematics 68(2), 255–292 (2010)
20. Duits, R., Franken, E.M.: Left-invariant parabolic evolution equations on SE(2) and contour enhancement via invertible orientation scores, part II: Nonlinear left-invariant diffusion equations on invertible orientation scores. Quarterly on Applied Mathematics 68(2), 293–331 (2010)
21. Duits, R., van Almsick, M.: The explicit solutions of linear left-invariant second order stochastic evolution equations on the 2D Euclidean motion group. Quart. Appl. Math. 66, 27–67 (2008)
22. Duits, R., Franken, E.M.: Left-invariant Stochastic Evolution Equations on $SE(2)$ and its Applications to Contour Enhancement and Contour Completion via Invertible Orientation Scores. arXiv: 0711.0951v4, Also available as CASA report nr.35. Eindhoven University of Technology (2007),
`www.win.tue.nl/casa/research/casareports/2007.html`
23. Duits, R., Dela Haije, T.C.J., Creusen, E.J., Ghosh, A.: Morphological and Linear scale Spaces for Fiber Enhancement in DW-MRI. Journal of Mathematical Imaging and Vision 46(3), 326–368 (2013)
24. Euler, L.: Methodus inveniedi lineas curvas maximi minimive proprietate gaudentes, sive solutio problematis isoperimitrici latissimo sensu accepti, Lausanne, Geneva (1744)
25. Ermentrout, G.B., Cowan, J.D.: A mathematical theory of visual hallucinations. Kybernetic 34, 137–150 (1979)
26. Field, D.J., Hayes, A., Hess, R.F.: Contour integration by the human visual system:evidence for a local "association field". Vision Res. 33(2), 253–278 (1993)
27. Franken, E., Duits, R.: Crossing-Preserving Coherence-Enhancing Diffusion on Invertible Orienattion Scores. IJCV 85(3), 253–278 (2009)
28. Franken, E.M.: Enhancement of Crossing Elongated Structures in Images. Ph.D. thesis, Eindhoven University of Technology, Eindhoven (2008),
`http://www.bmi2.bmt.tue.nl/Image-Analysis/People/EFranken/PhDThesisErikFranken.pdf`
29. Frégnac, Y., Shulz, D.: Activity-dependent regulation of receptive field properties of cat area 17 by supervised Hebbain learning. J. Neurobiol. 41(1), 69–82 (1999)
30. Ghosh, A., Duits, R., Dela Haije, T.C.J.: On the Cuspless Sub-Riemannian Geodesics in $\mathbb{R}^3 \rtimes S^2$. submitted to JDCS (2013),
`www.bmia.bmt.tue.nl/people/RDuits/GhoshDuitsDelaHaije.pdf`
31. Grossberg, S., Mingolla, E.: Neural dynamics of form perception: boundary completion, illusory figures and neon color spreading. Psych. Rev. 92, 173–211 (1985)
32. Hladky, R.K., Pauls, S.D.: Minimal Surfaces in the Roto-Translation Group with Applications to a Neuro-Biological Image Completion Model. J. Math. Imaging Vis. 36, 1–27 (2010)
33. Horn, B.K.P.: The curves of least energy. ACM Trans. Math. Software 9 (4), 441–460 (1983)
34. Georges, S., Series, P., Fregnac, Y., Lorenceau, J.: Orientation-dependent modulation of apparent speed: psychophysical evidence. Vision Res. 42, 2557–2572 (2002)
35. Mashtakov, A., Ardentov, A., Sachkov, Y.: Parallel Algorithm and Software for Image Inpainting via Sub-Riemannian Minimizers on the Group of Rototranslations. Numerical Methods: Theory and Applications 6(1), 95–115 (2013)
36. Medioni, G., Lee, M.-S., Tang, C.-K.: A computational framework for segmentation and grouping. Elsevier Science (2000)

37. Moiseev, I., Sachkov, Y.L.: Maxwell strata in sub-Riemannian problem on the group of motions of a plane. ESAIM: COCV 16(2), 380–399 (2010)
38. Mumford, D.: Elastica and computer vision. In: Algebraic Geometry and Its Applications, pp. 491–506. Springer (1994)
39. Parent, P., Zucker, S.W.: Trace interference, curvature consistency, and curve detection. IEEE Trans. Pattern Anal. Math. Intell. II 8, 823–939 (1989)
40. Petitot, J.: Vers une Neuro-géomètrie. Fibrations corticales, structures de contact et contours subjectifs modaux. Math. Inform. Sci. Humaines (145), 5–101 (1999)
41. Petitot, J.: Neurogéomètrie de la vision - Modèles mathématiques et physiques des architechtures fonctionnelles. Les Éditions de l'École Polytechnique (2008)
42. Petitot, J.: The neurogeometry of pinwheels as a sub-Riemannian contact structure. Journal of Physiology - Paris 97, 265–309 (2003)
43. Pontryagin, L.S., Boltyansij, V.G., Gamkrelidze, R.V., Mishchenko, E.F.: The Mathematical Theory of Optimal Processes. Pergamon Press, Oxford (1964)
44. Sachkov, Y.: Maxwell strata in the Euler elastic problem. J. Dyn. Control Syst. 14(2), 169–234 (2008)
45. Sachkov, Y.: Conjugate and cut time in the sub-Riemannian problem on the group of motions of a plane. ESAIM: COCV 16(4), 1018–1039 (2010)
46. Sachkov, Y.: Cut locus and optimal synthesis in the sub-Riemannian problem on the group of motions of a plane. ESAIM: COCV 17(2), 293–321 (2011)
47. Sanguinetti, G., Citti, G., Sarti, A.: Image completion using a diffusion driven mean curvature flow in a sub-Riemannian space. In: Int. Conf. on Computer Vision Theory and Applications (VISAPP 2008), Funchal (2008)
48. Sanguinetti, G., Citti, G., Sarti, A.: A model of natural image edge co-occurance in the rototranslation group. Journal of Vision 10(14), 1–16 (2010)
49. Sarti, A., Citti, G.: A cortical based model of perceptual completion in the Roto-Translation space. In: Proceeding of the Workshop on Second Order Subelliptic Equations and Applications, Cortona (June 2003)
50. MomayyezSiahkal, P., Sidiqqi, K.: 3D Stochastic Completion Fields for Mapping Connectivity in Diffusion MRI. IEEE PAMI 35(4) (2013)
51. Tai, X.-C., Hahn, J., Chung, G.S.: A fast algorithm for Euler's elastica model using augmented Lagrangian method. SIAM J. Imaging Sciences 4(1), 313–344 (2011)
52. Hahn, J., Chung, G.J., Wang, Y., Tai, X.-C.: Fast algorithms for p-elastica energy with the application to image inpainting and curve reconstruction. In: Bruckstein, A.M., ter Haar Romeny, B.M., Bronstein, A.M., Bronstein, M.M. (eds.) SSVM 2011. LNCS, vol. 6667, pp. 169–182. Springer, Heidelberg (2012)
53. Thornber, K.K., Williams, L.R.: Characterizing the Distribution of Completion Shapes with Corners Using a Mixture of Random Processes. Pattern Recognition 33, 543–553 (2000)
54. Ullman, S.: Filling in the gaps: the shape of subjective contours and a model for their generation. Biological Cybernetics 25, 1–6 (1976)
55. Vinter, R.: Optimal Control. Birkhauser (2010)
56. Wilcox, R.M.: Exponential operators and parameter differentiation in quantum physiscs. Journal of Mathematical Physics (1966)
57. Wittich, O.: An explicit local uniform bound on Brownian bridges. Statistics and Probability Letters 73(1), 51–56 (2005)
58. Zucker, S.W., David, C., Dobbins, A., Iverson, L.: The organization of curve detection: coarse tangent fields and fine spline covering. In: Proc. of the 2nd Int. Conf. on Computer Vision, pp. 568–577. IEEE, New York (1988)

Chapter 6
Psychophysics, Gestalts and Games

José Lezama, Samy Blusseau, Jean-Michel Morel,
Gregory Randall, and Rafael Grompone von Gioi

Abstract. Many psychophysical studies are dedicated to the evaluation of the human gestalt detection on dot or Gabor patterns, and to model its dependence on the pattern and background parameters. Nevertheless, even for these constrained percepts, psychophysics have not yet reached the challenging prediction stage, where human detection would be quantitatively predicted by a (generic) model. On the other hand, Computer Vision has attempted at defining automatic detection thresholds. This chapter sketches a procedure to confront these two methodologies inspired in gestaltism.

Using a computational quantitative version of the non-accidentalness principle, we raise the possibility that the psychophysical and the (older) gestaltist setups, both applicable on dot or Gabor patterns, find a useful complement in a Turing test. In our perceptual Turing test, human performance is compared by the scientist to the detection result given by a computer. This confrontation permits to revive the abandoned method of gestaltic games. We sketch the elaboration of such a game, where the subjects of the experiment are confronted to an alignment detection algorithm, and are invited to draw examples that will fool it. We show that in that way a more precise definition of the alignment gestalt and of its computational formulation seems to emerge.

Detection algorithms might also be relevant to more classic psychophysical setups, where they can again play the role of a Turing test. To a visual experiment where subjects were invited to detect alignments in Gabor patterns, we associated

José Lezama · Jean-Michel Morel
École Normale Supérieure de Cachan, Cachan, France
e-mail: {lezama,morel}@cmla.ens-cachan.fr

Samy Blusseau
Centre de Mathématiques et de Leurs Applications - ENS de Cachan Paris, France
e-mail: sblusseau@gmail.com

Gregory Randall · Rafael Grompone von Gioi
Instituto de Ingegneria Eléctrica, Universidad de la República, Montevideo, Uruguay
e-mail: randall@fing.edu.uy, grompone@cmla.ens-cachan.fr

G. Citti and A. Sarti (eds.), *Neuromathematics of Vision*,
Lecture Notes in Morphogenesis,
DOI: 10.1007/978-3-642-34444-2_6,

a single function measuring the alignment detectability in the form of a number of false alarms (NFA). The first results indicate that the values of the NFA, as a function of all simulation parameters, are highly correlated to the human detection. This fact, that we intend to support by further experiments, might end up confirming that human alignment detection is the result of a single mechanism.

6.1 Introduction

Alan Turing advanced a controversial proposal in 1950 that is now known as the *Turing Test* [35]. Turing's aim was to discuss the problem of machine intelligence and, instead of giving a premature definition of thinking, he framed the problem in what he called the *Imitation Game*: A human interrogator interacts with another human and a machine, but only in typewritten form; the task of the interrogator is to ask questions in order to determine which of its two interlocutors is the human. Turing proposed that a machine that eventually could not be distinguished from humans by its answers should be considered intelligent. This influential suggestion sparked a fruitful debate that continues to this day [29].

Our concern here is however slightly different. We are studying perception and Turing precluded in his test any machine interaction with the environment other that the communication through the teletype; he concentrated on the pure problem of thinking and to that aim avoided fancy computer interactions, that anyway did not exist at his time. Yet, machine perception is still a hard problem for which current solutions are far from the capacities of humans or animals[1]. Our purpose is to discuss a variety of *perceptual imitation games* as a research methodology to develop machine vision algorithms on the one hand, and quantitative psychophysical protocols on the other.

Human perceptual behavior has been the subject of quantitative experimentation since the times of Fechner, the founder of Psychophysics. This relatively new science investigates the relationship between the stimulus intensity and the perceived sensation [33]. But this approach does not provide a perceptual theory in which machine vision and an imitation game could be based.

The Gestalt school, Wertheimer, Köhler, Koffka, Kanizsa among others [8, 18, 21, 25, 42], developed from the twenties to the eighties an original *modus operandi*, based on the invention and display to subjects of clever geometric figures [40, 41]. A considerable mass of experimental evidence was gathered, leading to the conclusion that the first steps of visual perception are based on a reduced set of geometrical grouping laws. Unfortunately these Gestalt laws, relevant though they were, remained mainly qualitative and led to no direct machine perception approach.

[1] It is a common practice in Internet services to use the so-called CAPTCHAs to ensure that the interaction is made with a human and not an automatic program. A CAPTCHA, which stands for *Completely Automated Public Turing test to tell Computers and Humans Apart*, usually consists in a perceptual task, simple to perform for humans but hard for known algorithms. This suggests that visual and auditive perception currently provides the most effective Turing test.

Since the emergence of the field of Computer Vision [24] about fifty years ago – initially as a branch of the Artificial Intelligence working with robots and its artificial senses – there have been many attempts at formalizing vision theories and especially Gestalt theory [31]. Among them one finds models of neural mechanism [15], theories based on logical inference [9], on information theory [22], invoking minimum description principles [45], or grammars of visual elements [16, 46]. Nevertheless, only a small fraction of these proposals has been accompanied by systematic efforts to compare machine and human vision. An important exception is the Bayesian theory of perception [27] that has attracted considerable attention in cognitive sciences, leading to several experimental evaluations [10,20]. A recent groundbreaking work by Fleuret et al. [13] compared human and machine performing visual categorization tasks. Humans are matched against learning algorithms in the task of distinguishing two classes of synthetic patterns. One class for example may contain four parallel identical shapes in arbitrary position, while the other class contains the same shapes but with arbitrary orientation and position. It was observed that humans learn the distinction of such classes with very few examples, while learning algorithms require considerably more examples, and nevertheless gain a much lower classification performance. The experimental design was more directed at pointing out a flaw of learning theory, though, than at contributing to psychophysics.

Such experiments stress the relevance of computer vision as a research program in vision, in addition to a purely technological pursuit. Its role should be complementary to explanatory sciences of natural vision by providing, not only descriptive laws, but actual implementations of mechanisms of operation. With that aim, perceptual versions of the imitation game should be the Leitmotiv in the field, guiding the conception, evaluation and success of theories.

Here we will present comparisons of human perception to algorithms based on the *non-accidentalness principle* introduced by Witkin, Tenenbaum and Lowe [23, 43, 44] as a general grouping law. This principle states that spatial relations are perceptually relevant only when their accidental occurrence is unlikely. We shall use the *a contrario* framework, a particular formalization of the principle due to Desolneux, Moisan and Morel [6, 7] as part of an attempt to provide a mathematical foundation to Gestalt Theory.

This chapter is intended to give an overview of our research program; for this reason we reduced the settings to the bare minimum, concentrating in one simple geometric structure, namely alignments. The methodology however is general. By using such a simple structure we will present two complementary aspects of the same program, each one with specific imitation games: a research procedure inspired in the methodology of the Gestalt school and the use of online games for psychophysical experimentation.

Gestaltism created clever figures in which humans fail to perceive the expected structures, generating illusions. In the *gestaltic game*, as we shall call our first proposed methodology, the experimenter tries to fool the algorithm by building a particular data set that produces unnatural results. This methodology is discussed in Sect. 6.2, along with a brief introduction to the *a contrario* methods.

The second part, in Sect. 6.3, is dedicated to a first attempt at a psychophysical evaluation of the same theory. There is a difference with classic psychophysical experiments in which detection thresholds are measured; here each stimulus will be shown to human subjects but also to an algorithm, and both will answer yes or no to the visibility of a given structure. In a second variation, both humans and machine will also have to point to the position of the observed structure. This last variation is proposed as an online game, used as a methodology to facilitate experimentation and the attraction of volunteers.

Being the result of a work in progress, no final conclusion will be drawn. Our overall goal is to advocate for new sorts of quantitative Gestalt and psychophysical games.

6.2 Detection Theory versus Gestaltism

Here and in most of the text we shall call "gestalt" any geometric structure emerging perceptually against the background in an image. We stick to this technical term because it is somewhat untranslatable, meaning something between "form" and "structure". According to Gestalt theory, the gestalts emerge by a grouping process in which the properties of similarity (by color, shape, texture, etc.), proximity, good continuation, convexity, parallelism, alignment can individually or collaboratively stir up the grouping of the building elements sharing one or more properties.

6.2.1 The Gestaltic Game

One of the procedures used by Gestalt psychology practitioners was to create clever geometric figures that would reveal a particular aspect of perception when used in controlled experiments with human subjects. They pointed out the grouping mechanisms, but also the striking fact that geometric structures objectively present in the figure are not necessarily part of the final gestalt interpretation. These figures are in fact counterexamples against simplistic perception mechanisms. Each one represents a challenge to a theory of vision that should be able to cope with all of them.

The methodology we propose in order to design and improve automatic geometric gestalt detectors is in a way similar to that of the gestaltist. One starts with a primitive method that works correctly in very simple examples. The task is then to produce data sets where *humans* clearly see a particular gestalt while the rudimentary method produces a different interpretation. Analyzing the errors of the first method gives hints to improve the procedure in order to create a second one that produces better results with the whole data set produced until that point. The same procedure is applied to the second method to produce a third one, and successive iterations refine the methods step by step. The methodology used by the Gestalt psychologist to study human perception is used here to push algorithms to be similar to their natural counterpart. Finding counterexamples is less and less trivial after some iterations and the counter-examples become, like gestaltic figures, more and more clever.

We decided to render this process interactive by drawing figures in a computer interface that delivers a detection result immediately. The exploration of counterexamples is in that way transformed into an active search where previous examples are gradually modified in an attempt to fool the detection algorithm. The figures are all collected to be later used at the analysis stage. The *gestaltic game* is at the same time a method to produce interesting data sets, a methodology to develop new detection algorithms and a collaborative tool for research in the computational gestalt community. Each detection game will only stop when it eventually passes the Turing test, the algorithm's detection capability becoming undistinguishable from that of a human.

6.2.2 Dot Alignments Detection

For its simplicity, dot patterns are often used in the study of visual perception. Several psychophysical studies led by Uttal have investigated the effect of direction, quantity and spacing in dot alignment perception [36,37]. The detection of collinear dots in noise was the target of a study attempting to assess quantitatively the masking effect of the background noise [34]. A recent work by Preiss analyzes various perceptual tasks on dot patterns from a psychophysical and computational perspective [30]. An interesting computational approach to detect gestalts in dot patterns is presented in [1], although the study is limited to very regularly sampled patterns. A practical application of alignment detection is presented in [38].

From a gestaltic point of view, a point alignment is a group of points sharing the property of being aligned in one direction. While it may seem a simple gestalt, Fig. 6.1 shows how complex the alignment event is. From a purely factual point of view, the same alignment is present in the three figures. However, it is only perceived as such by most viewers in the first one. The second and the third figures illustrate two occurrences of the *masking phenomenon* discovered by gestaltists [19]:

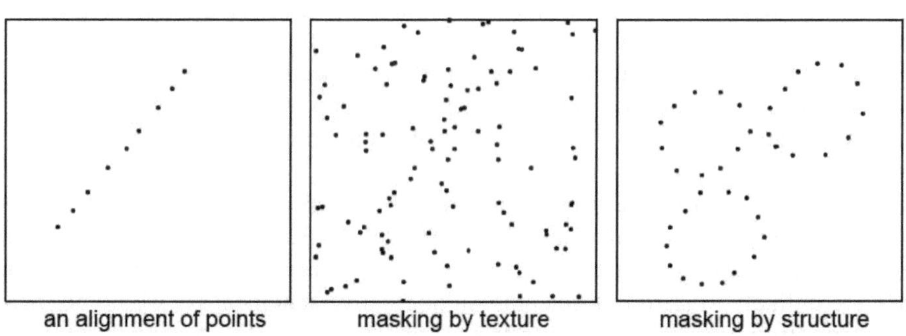

Fig. 6.1 Exactly the same set of aligned dots is present in the three images, but it is only perceived as such in the first one. The second one is a classic masking by texture case and the third a masking by structure one, often called "Gestalt conflict".

the *masking by texture*, which occurs when a gestalt is surrounded by a clutter of randomly distributed similar objects or *distractors*, and the *masking by structure*, which happens when the alignment is masked by other perceptually more relevant gestalts, a phenomenon also called *perceptual conflict* by gestaltists [18, 26, 45]. The magic disappearance of the alignment in the second and third figures can be accounted for in two very different ways. As for the first one, we shall see that a probabilistic *a contrario* model [7] is relevant and can lead to a quantitative prediction. As for the second disappearance, it requires the intervention of another more powerful grouping law, the *good continuation* [17].

These examples show that a mathematical definition of dot alignments is required before even starting to discuss how to detect them. A purely geometric-physical description is clearly not sufficient to account for the masking phenomenon. Indeed, an objective observer making use of a ruler would be able to state the existence of the very same alignment at the same precision on all three figures. But this statement would contradict our perception, as it would contradict any reasonable computational (definition and) theory of alignment detection.

This experiment also shows that the detection of an alignment is highly dependent on the context of the alignment. It is therefore a complex question, and must be decided by building mathematical definitions and detection algorithms, and confronting them to perception. As the patterns of Fig. 6.1 already suggest, simple computational definitions with increasing complexity will nevertheless find perceptual counterexamples. There is no better way to describe the ensuing "computational gestaltic game" than describing how the dialogue of more and more sophisticated alignment detection algorithms and counterexamples help build up a perception theory.

6.2.3 Basic Dot Alignment Detector

A very basic idea that could provide a quantitative context-dependent definition of dot alignments is to think of them as thin, rectangular shaped point clusters. In that case, the key measurements would be the relative dot densities inside and outside the rectangle. The algorithm described in this section follows the *a contrario* methodology [7, Sect. 3.2] according to which a group of elements is detectable as a gestalt if and only if it has a low enough probability of occurring just by chance in an *a contrario* background model.

We shall first introduce briefly the *a contrario* framework [4, 6, 7]. The approach is based on the *non-accidentalness principle* [5,32,39,44] (sometimes called *Helmholtz principle*) that states that structures are perceptually relevant only when they are unlikely to arise by accident. An alternative statement is "we do not perceive any structure in a uniform random image" [7, p.31]. The *a contrario* framework is a particular formalization of this principle adjusting the detection thresholds so that the *expected* number of accidental detections is provably bounded by a small constant ε. The key point is how to define *accidental* detections. This requires a

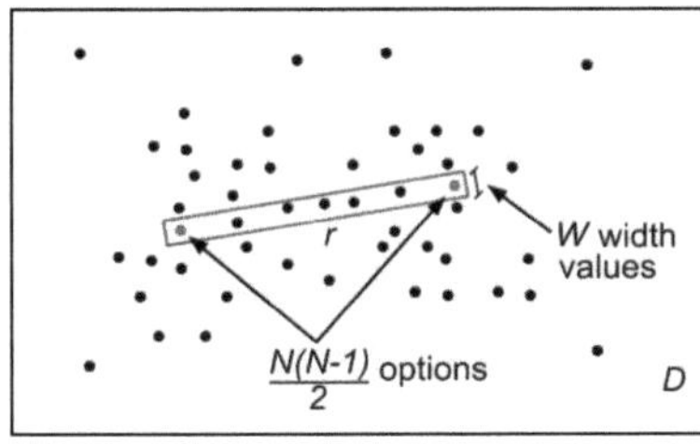

Fig. 6.2 A schematic representation of the evaluated rectangle. In an image with N points, there are $(N(N-1) \times W)/2$ possible rectangles defined by two dots. In the case shown in this figure, $N = 49$ and $k(r, \mathbf{x}) = 5$.

stochastic model, the so-called *a contrario* model, characterizing unstructured or random data in which the sought gestalt could only be observed by chance.

Consider a dot pattern defined on a domain D with total area S_D and containing N dots, see Fig. 6.2. We are interested in detecting groups of dots that are well aligned. A first reasonable *a contrario* hypothesis H_0 for this problem is to suppose that the N dots are the result of a random process where points are independent and uniformly distributed in the domain. The question is then to evaluate whether the presence of aligned points contradicts the *a contrario* model or not.

Given an observed set of N points $\mathbf{x} = \{x_i\}_{i=1\dots N}$ and a rectangle r (the candidate to contain an alignment), we will denote by $k(r, \mathbf{x})$ the number of those points observed inside r. We decide whether to keep this candidate or not based on two principles: a good candidate should be non-accidental, and any equivalent or better candidate should be kept as well. The degree of non-accidentalness of an observed rectangle r can be measured by how small the probability $\mathbb{P}\left[k(r,\mathbf{X}) \geq k(r,\mathbf{x})\right]$ is, where $\mathbf{X}$ denotes a random set of N dots following H_0. In the same vein, a rectangle r' will be considered at least as good as r given the observation $\mathbf{x}$, if $\mathbb{P}\left[k(r',\mathbf{X}) \geq k(r',\mathbf{x})\right] \leq \mathbb{P}\left[k(r,\mathbf{X}) \geq k(r,\mathbf{x})\right]$.

Recall that we want to bound the expected number of accidental detections. Given that N_{tests} candidates will be tested, the expected number of rectangles which are as good as r under H_0, is about [7]

$$N_{tests} \cdot \mathbb{P}\Big[k(r,\mathbf{X}) \geq k(r,\mathbf{x})\Big]. \tag{6.1}$$

The H_0 stochastic model fixes the probability law of the random number of points in the rectangle, $k(r, \mathbf{X})$, which only depends on the total number of dots N. The discrete nature of this law implies that (6.1) is not actually the expected value but an upper bound of it [7, 14]. Let us now analyze the two factors in (6.1).

Here the *a contrario* model H_0 assumes that the N points are i.i.d. with uniform density on the domain. Under the *a contrario* hypothesis H_0, the probability that one dot falls into the rectangle r is

$$p = \frac{S_r}{S_D}, \tag{6.2}$$

where S_r is the area of the rectangle and S_D the area of the domain. As a consequence of the independence of the random points, $k(r,\mathbf{X})$ follows a binomial distribution. Thus, the probability term $\mathbb{P}\left[k(r,\mathbf{X}) \geq k(r,\mathbf{x})\right]$ is given by

$$\mathbb{P}\left[k(r,\mathbf{X}) \geq k(r,\mathbf{x})\right] = \mathscr{B}(N, k(r,\mathbf{x}), p) \tag{6.3}$$

where $\mathscr{B}(n,k,p)$ is the tail of the binomial distribution

$$\mathscr{B}(n,k,p) = \sum_{j=k}^{n} \binom{n}{j} p^j (1-p)^{n-j}. \tag{6.4}$$

The *number of tests* N_{tests} corresponds to the total number of rectangles that could show an alignment, which in turn is related to the number of pairs of points defining such rectangles. With a set of N points this gives $\frac{N\times(N-1)}{2}$ different pairs of points.

The set of rectangle widths to be tested must be specified *a priori* as well. In the *a contrario* approach, a compromise must be found between the number of tests and the precision of the gestalts that are being sought for. The larger the number of tests, the lower the statistical relevance of detections. However, if the set of tests is chosen wisely, gestalts fitting accurately the tests will have a very low probability of occurrence under H_0 and will therefore be more significant.

At a digital image precision, the narrowest possible width for an alignment is 1 (taking the side of a pixel as length unit). The series of tested widths grows geometrically until it achieves a maximal possible width, which can be set *a priori* as a function of the alignment length. Since the number of tested widths depends on the length of the alignment, we cannot predict *a priori* (before the dots have been drawn) how many tests will be done. Fortunately the total number of widths can be estimated as the number of widths tested in an average rectangle times the number of evaluated rectangles. We call this quantity W. The impact of this approximation in the detector results is insignificant [7]. The total number of tested rectangles is then:

$$N_{tests} = \frac{N(N-1)\times W}{2}. \tag{6.5}$$

We will define now the fundamental quantity of the *a contrario* framework, the Number of False Alarms (NFA) associated with a rectangle r and a set of dots $\mathbf{x}$:

$$\mathrm{NFA}(r,\mathbf{x}) = N_{tests} \cdot \mathbb{P}\left[k(r,\mathbf{X}) \geq k(r,\mathbf{x})\right] = \frac{N(N-1)\times W}{2} \cdot \mathscr{B}\left(N, k(r,\mathbf{x}), p\right). \tag{6.6}$$

This quantity corresponds, as said before (Eq. 6.1), to the expected number of rectangles which have a sufficient number of *points* to be as rare as r under H_0. When the NFA associated with a rectangle is large, this means that such an event is to be expected under the *a contrario* model, and therefore is not relevant. On the other hand, when the NFA is small, the event is rare and probably meaningful. A *perceptual threshold* ε must nevertheless be fixed, and rectangles with $\mathrm{NFA}(r,\mathbf{x}) < \varepsilon$

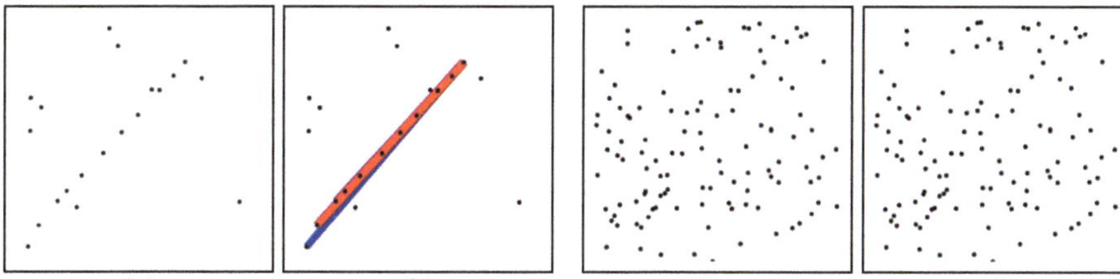

Fig. 6.3 Results from the basic dot alignment detector. **(a)** and **(c)** are the input data, and **(b)** and **(d)** are the corresponding results. Each detection is represented by a rectangle and its color indicates the NFA value. In **(b)** the algorithm correctly detects the obvious alignment. Notice that multiple and redundant rectangles were detected; this problem will be discussed in Sect. 6.2.5. The data set **(c)** contains the same set of points in **(a)** plus added noise dots, thus the aligned dots are still present. However, the algorithm handles correctly the masking by texture or noise and produces no detection.

will be called ε-*meaningful rectangles* [5], constituting the detection result of the algorithm.

Theorem 6.1 ([7])

$$\mathbb{E}\left[\sum_{R\in\mathscr{R}} \mathbb{1}_{\mathrm{NFA}(R,\mathbf{X})<\varepsilon}\right] \leq \varepsilon$$

where $\mathbb{E}$ *is the expectation operator,* $\mathbb{1}$ *is the indicator function,* $\mathscr{R}$ *is the set of rectangles considered, and* $\mathbf{X}$ *is a random set of points on* H_0.

The theorem states that the average number of ε-*meaningful rectangles* under the *a contrario* model H_0 is bounded by ε. Thus, the number of detections in noise is controlled by ε and it can be made as small as desired. In other words, this shows that our detector satisfies the non-accidentalness principle.

Following Desolneux, Moisan, and Morel [4, 7], we shall set $\varepsilon = 1$ once and for all. This corresponds to accepting on average one false detection per image in the *a contrario* model, which is generally reasonable. Also, the detection result is not sensitive to the value of ε, see [7].

Figure 6.3 shows the results of the basic algorithm in two simple cases. The results are as expected: the visible alignment in the first example is detected, while no detection is produced in the second. Actually, the dots in the first example are also present in the second one, but the addition of random dots masks the alignment, in accordance with human perception. Note that the first example produces many redundant detections; this problem will be handled in Sect. 6.2.5.

6.2.4 A Refined Dot Alignment Detector

Naturally, the simple model for dot alignment detection presented in the last section does not take into account many situations that can arise and significantly affect

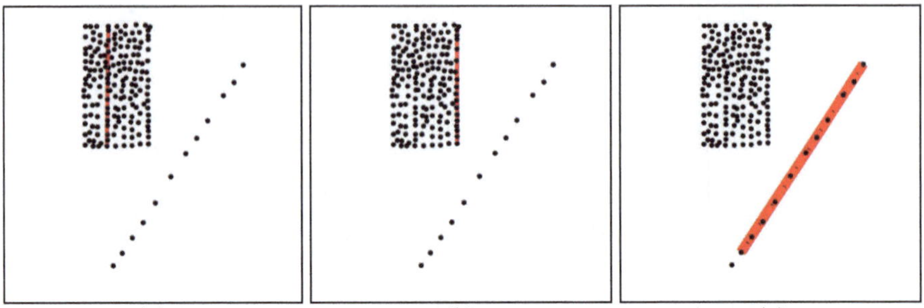

(a) (b) (c)

Fig. 6.4 Local vs. global density estimation. In each example, only the most meaningful detected alignment (the one with the lowest NFA) is shown for each algorithm. The algorithms (a), (b), (c) use a background model with growing complexity to avoid wrong detections. **(a)** global density estimation: the detected segment is not the most meaningful for our perception, but has nevertheless a high dot density compared to the average image density used as background model. **(b)** here a local density estimation gives the background model, but the local density is lower on the border of the big dot rectangle, hence the detection. **(c)** this last problem is avoided by computing a local density estimation taking the maximum density on both sides of the alignment.

the perception of alignments. For example: what happens if there are point clusters inside the alignment? What if the background image has a non uniform density? Should not the algorithm prefer alignments where the points are equally spaced? These questions, among others, arise when subjects play the gestaltic game and try to fool the algorithm with new drawings. There are two ways to fool the algorithm: One is by drawing a particular context that prevents the algorithms from detecting a conspicuous alignment. Inversely, the other sort of counterexample is a drawing inducing detections that remain invisible to the human eye. As more counterexamples are found, more sophisticated versions of the algorithm must be developed, and each new version will become harder to falsify than the previous one.

Using this methodology, we produced several refined versions of the basic algorithm. Here we will present the principal counterexamples that were found, and then describe the last version of the algorithm which takes all of them into account. This algorithm is therefore harder to fool. Ideally, the game should end when the *Turing test* [29] is satisfied, namely when a human observer will be unable to distinguish between the detections produced by a machine and by a human.

First, we noticed a deficiency in the detector when zones in the image have higher dots density. This problem arises naturally from the wrong *a contrario* assumption that the whole image has the same density of points. When it is not the case, the global density estimation can be misleading and produces poor detection results, as illustrated in Fig. 6.4 (a). The solution for this is to compute a *local density estimation* with respect to the evaluated rectangle. The algorithm uses a local window with size proportional to the width of the evaluated alignment.

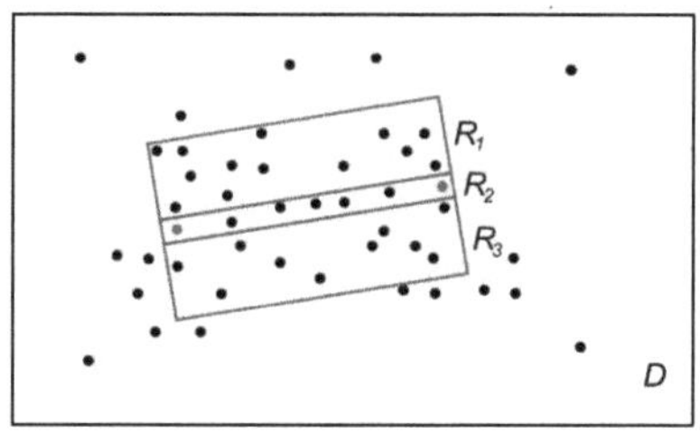

Fig. 6.5 In the refined version of the algorithm, the density of points is measured to each side of the evaluated rectangle. The maximum of the densities in R_1 and R_3 is taken and this value is used as an estimation of the dot density in both R_1 and R_3.

However, this local density estimation can introduce new problems such as a "border effect", as shown in Fig. 6.4 (b). Indeed, the density estimation is lower on the border of the dot rectangle than inside it, because outside the rectangle there are no dots. Thus, the algorithm detects on the border a non-accidental, meaningful excess with respect to the local density.

In order to avoid this effect, the version of the algorithm used in Fig. 6.4 (c) measures for the background model the *maximum of the densities* measured on both sides of the alignment. In short, to be detected, an alignment must show a higher dot density than in both regions immediately on its left and right. This local alignment detector is therefore similar to classic second order Gabor filters where an elongated excitatory region is surrounded by two inhibitory regions. The local points estimation is calculated in the following way, see Fig. 6.5. The local window is divided in three parts. R_1 is the rectangle formed by the area of the local window on the left of the alignment. R_3 is the area of the local window on the right of the alignment, and R_2 is the rectangle which forms the candidate alignment. Note that the length of the local window is the same as the alignment and that we can consider any arbitrary orientation for it. Next, the algorithm counts the numbers of dots M_1, M_2, and M_3 in R_1, R_2 and R_3 respectively. Finally the *a contrario* model assumes that the number of dots in the local window $R_1 \cup R_2 \cup R_3$ is

$$n(r,\mathbf{x}) = \max(M_1, M_3) \times 2 + M_2, \tag{6.7}$$

and that these dots are randomly distributed.

There is still an objection to this new algorithm, obtained in the gestaltic game by introducing small dot clusters, as shown in Fig. 6.6 (a). The detected alignment in Fig. 6.6 (b) seems clearly wrong. There is indeed a meaningful dot density excess inside the red rectangle, but this excess is caused by the clusters, not by what could be termed an alignment. While the algorithm counted every point, the human perception seems to group the small clusters into a single entity, and count them only once. Also, as suggested in other studies [30, 34, 37], the density is not the only property that makes an alignment perceptually meaningful; another characteristic to consider is the uniform spacing of the dots in it, which the gestaltists call the principle of *constant spacing*. These objections have led to a still more sophisticated version of the alignment detector. In order to take into account both issues (avoiding small clusters and favoring regular spacing) a more advanced version of the alignment detector was designed which divides each candidate rectangle into equal *boxes*. The algorithm counts the number of boxes that are occupied by at least one point, instead of counting the total number of points. In this way, the minimal NFA is attained when the dots are perfectly distributed along the alignment. In addition, a concentrated cluster in the alignment has no more influence on the alignment detection than a single dot in the same position.

The NFA calculation for this refined version of the algorithm is slightly different than for the basic one. The event for which we are estimating an expected number of occurrences in a background model is defined as follows. Given two points and a number of boxes c, the question is: What is the probability that the number of occupied boxes among the c is larger than the expected number under the *a contrario* model? Let us start by computing the probability of one dot falling in one of the boxes:

$$p_0 = \frac{S_B}{S_L}, \tag{6.8}$$

where S_B and S_L are the areas of the boxes and the local window respectively. Then, the probability of having one box occupied by at least one of the $n(r, \mathbf{x})$ dots (Eq. 6.7) is:

$$p_1 = \mathscr{B}\big(n(r, \mathbf{x}), 1, p_0\big). \tag{6.9}$$

We call *occupied* boxes the ones that have at least one dot inside, and we will denote by $b(r, c, \mathbf{x})$ the observed number of occupied boxes in the rectangle r divided into c boxes. Finally, the probability of having at least $b(r, c, \mathbf{x})$ of the c boxes occupied is

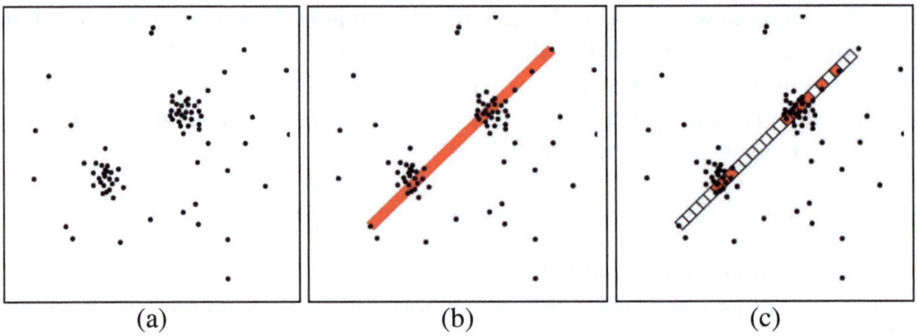

Fig. 6.6 Counting occupied boxes to avoid false detections from the presence of clusters. The dot pattern shown in image **(a)** presents two dot clusters but no alignment. However, the basic algorithm finds a thin rectangle with a high dot density, hence a false detection, as shown in **(b)**. Dividing the rectangle into boxes and counting the occupied ones, avoids this misleading cluster effect, as seen in **(c)**, where the occupied boxes are marked in red and no alignment is actually detected.

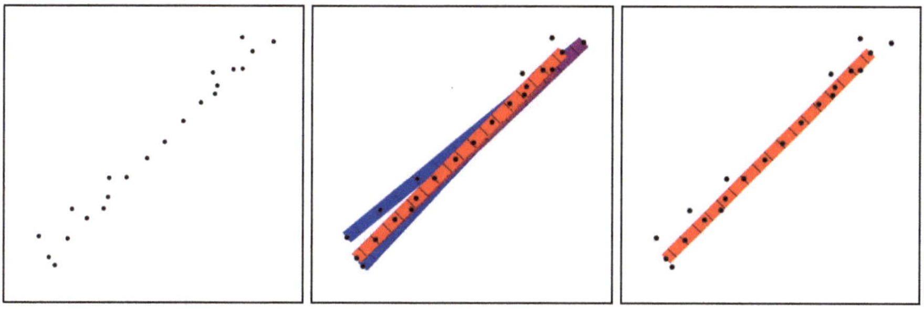

Fig. 6.7 Redundant detections. **Left:** dot pattern. **Center:** all significant alignments found by the refined dot alignment detector described in Sect. 6.2.4. The color represents the relative NFA value, where red is the most significant (smallest NFA value) and blue the least (highest NFA value). **Right:** Result of the masking process.

$$\mathscr{B}\big(c, b(r,c,\mathbf{x}), p_1\big). \tag{6.10}$$

A set $\mathscr{C}$ of different values are tried for the number of boxes c into which the rectangle is divided. Thus, the number of tests needs to be multiplied by its cardinal $|\mathscr{C}|$. In practice we set $|\mathscr{C}| = \sqrt{N}$ and that leads to

$$N_{tests} = \frac{N(N-1) \times W \times |\mathscr{C}|}{2} = \frac{N(N-1) \times W \times \sqrt{N}}{2}. \tag{6.11}$$

The NFA of the new event definition is then:

$$\mathrm{NFA}(r,\mathbf{x}) = \frac{N(N-1) \times W \times \sqrt{N}}{2} \cdot \min_{c \in \mathscr{C}} \mathscr{B}\big(c, b(r,c,\mathbf{x}), p_1\big). \tag{6.12}$$

Figs. 6.4 (c) and 6.6 (c) show two examples of the resulting algorithm, and we will show some more after discussing the masking problem.

6.2.5 *Masking*

As was observed in Fig. 6.3, all the described alignment detectors may produce redundant detections. The reason is that a relevant gestalt is generally formed by numerous elements and many subgroups also form relevant gestalts in the sense of the non-accidentalness principle. Every pair of dots defines a rectangle to be tested. Clearly, in a conspicuous alignment there will be many such rectangles that partially cover the main alignment and are therefore also meaningful. This redundancy phenomenon can involve dots that belong to the real alignment as well as background dots near the alignment, that can contribute to a rectangle containing a large number of dots, as illustrated in Fig. 6.7. However, in such cases humans perceive only one gestalt. Indeed, one could expect that there is only one causal reason leading to redundant detections and it makes sense to select the best rectangle to represent it.

A similar phenomenon is described in the Gestalt literature [19]. Most scenes contain other possible interpretations that are masked by the global interpretation. A simple example is shown in Fig. 6.8 where subsets of the grid of dots form a huge quantity of gestalts, but are invisible because they are masked by the rectangular matrix of dots. This fact is, after Vicario, called Kanizsa's paradox [7].

A simple model for this masking process was proposed by Desolneux et al. [7] under the name of "exclusion principle". The main idea is that each *basic element* (for example the dots) cannot contribute to more than one perceived group or gestalt. The process is as follows: The most meaningful observed gestalt (the one with smallest NFA) is kept as a valid detection. Then, all the basic elements (the dots in our case) that were part of that validated group are assigned to it and the remaining candidate gestalts cannot use them anymore. The NFA of the remaining candidates is re-computed without counting the excluded elements. In that way redundant gestalts lose most of their supporting elements and are no longer meaningful. On the other hand, a candidate that corresponds to a different gestalt keeps most or all of its supporting basic elements and remains meaningful. The most meaningful candidate among the remaining ones is then validated and the process is iterated until there are no more meaningful candidates.

This formulation of the masking process often leads to good results, removing redundant detections while keeping the good ones. However, the gestaltic game showed that it may also lead to unsatisfactory results as illustrated in Fig. 6.9. The problem arises when various gestalts have many elements in common. As one gestalt is evaluated after the other, it may happen that all of its elements have been removed, even if the gestalt is in fact not redundant with any of the other ones. In the example of Fig. 6.9, individual horizontal and vertical alignments are not redundant, but if all the vertical ones have been detected first, the remaining horizontal ones will be (incorrectly) masked. This example shows a fundamental flaw of the exclusion principle: it is not sound to impose that a basic element belongs to a single perceptually valid gestalt. There must be a global explanation of the organization of the basic elements in visible gestalts which is at the same time coherent with each individual gestalt (eliminating local redundancy) and with the general explanation of the scene

Fig. 6.8 A masking example by Kanizsa [19, p.155]: The "curve" in B is also present in the grid of dots A; nevertheless, it is not visible as it is masked by the global matrix configuration

in such a way that some basic elements can participate of several gestalts without contradiction. The solution seems to be in a sort of relaxation of the exclusion principle. The following definitions sketch a possible solution.

Definition 6.1 (Building Elements). We call *building element* any atomic image component that can be a constituent element of several gestalts. Valid examples of building elements are dots, segments, or even gestalts themselves, that can be recursively grouped in clusters or alignments. From that point of view any gestalt can be used as a building element for higher level gestalts.

Definition 6.2 (Masking Principle). A meaningful gestalt B will be said "masked by a gestalt A" if B is no longer meaningful when evaluated without counting its building elements belonging to A. In such a situation, the gestalt is not retained as detected.

In short, a meaningful gestalt will be detected *if* it is not masked by any other detected gestalt. The difference is that here a gestalt can only be masked by another *individual* gestalt and not by the union of several gestalts as is possible with the exclusion principle. Thus this masking principle is analogous to a Nash equilibrium, in the sense that every gestalt remains meaningful when separately subtracting from it the building blocks of any other gestalt. A procedural way to attain this result is to validate gestalts one by one, starting by the one with smallest NFA; before accepting a new gestalt, it is checked that it is not masked by any one of the previously detected gestalts. The masking principle applies easily to point alignments.

Fig. 6.10 shows some dot alignment detection results when combining the method of the previous section and the masking principle. The results obtained in these examples are as expected.

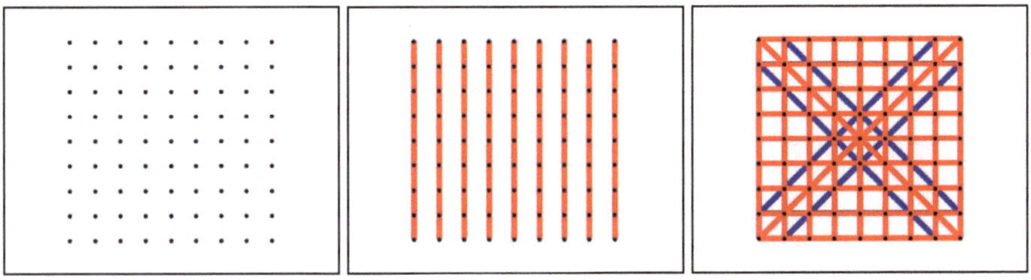

Fig. 6.9 Examples of two alternative formulations of the masking process. **Left:** Set of dots. **Center:** The Exclusion Principle as defined in [7], a validated gestalt prevents others from using its dots. The vertical alignments (that were evaluated first) mask almost all the horizontal ones. **Right:** The Masking Principle, described in the text, which solves the ambiguities without forbidding basic elements to participate of two different gestalts. In this example, no individual alignment can mask an individual one in another direction. Thus we get all oblique, horizontal and vertical meaningful alignments.

6.2.6 *Online Gestaltic Game*

The gestaltic game allowed us to discover examples of dot arrangements that the current algorithm is not able to handle correctly. The hardest ones we encountered to date belong to the "masking by structure" kind, as those presented in the right hand part of Fig. 6.1. Surely there are more cases than those discovered so far. To facilitate the search we created an online interface where everyone can easily play the gestaltic game inventing new counterexamples.[2]

Being interactive, the online gestaltic game is designed to be eventually published in the IPOL journal. It allows users to draw their own dot patterns and to see the output of the detection algorithm. Alternatively, the user can upload a set of dots, or modify an existing one by adding or removing individual dots or adding random dots. All the experiments are stored and accessible in the "archive" part of the site and may help improve the theory.

Current work is focused on the conflict between different gestalts with the objective of handling the masking by structure problem.

6.3 Detection Theory versus Psychophysics

In this second part we leave the question of a quantitative gestaltism and go back to more classic psychophysics. The question is whether a quantitative framework like the *a contrario* detection theory can also become a useful addition for human contour perception psychophysical experiments.

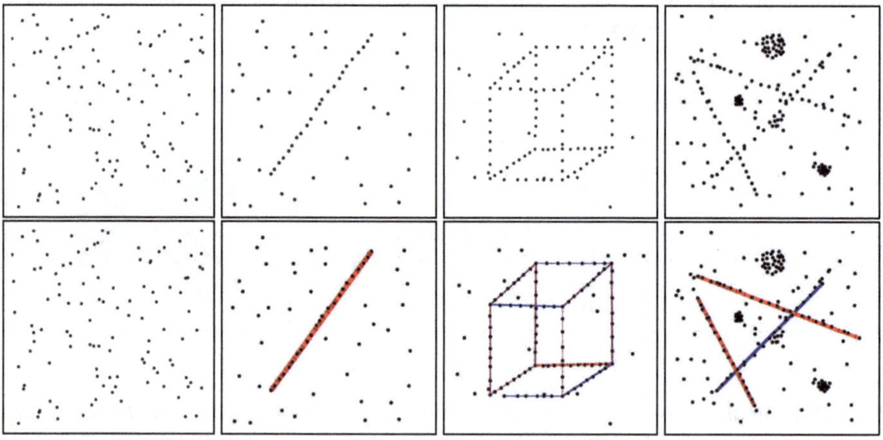

Fig. 6.10 Results of the final dot alignment detector, using the refined method described in Sect. 6.2.4 in conjunction with the Masking Principle (Def. 6.2). The top row is the input data; the bottom row shows the results.

[2] http://dev.ipol.im/~jlezama/dot_alignments

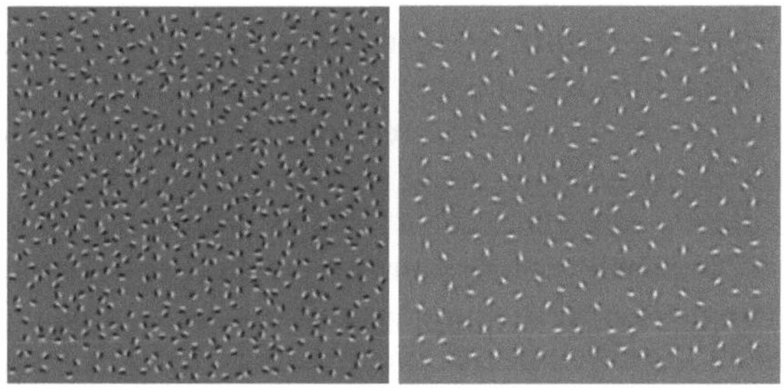

Fig. 6.11 **Left:** An image extracted from Nygård et al. [28]. **Right:** Example of an alignment detection experiment to be developed here.

Arrays of Gabor patches have become a classic tool for the study of the influence of good continuation in perceptual grouping [12,28]. Gabor functions ensure a control on the stimuli spectral complexity and on the spatial scale of the contours. They give a flexible and easy way for building a great variety of stimuli. It has been verified that the more aligned the Gabor patches are to the contour they lie on, the easier their perceptual grouping into a shape's outline [12,28]. Fig. 6.11 (left), shows an easy example where most subjects recognize a bottle. But the more freedom is left to the Gabor orientation, the harder it is to distinguish such contours from the background. For the influence of other perturbations of the contour such as its motion or its curvature on the object's identifiability, we refer to a recent study [28].

Can we hope for a quantitative interpretation to this experimental framework, namely a function of the stimuli parameters that would predict and explain the evolution of the detection performance? Probabilistic approaches (mainly Bayesian) exist for contour modeling from a perceptual point of view [2,11], and have sometimes been compared experimentally to human visual perception [10]; but none of these approaches proposed to compute *a priori* detection thresholds as functions of the stimuli parameters.

The influence of experimental factors such as the length of the alignment, the density of the patches, and the angular accuracy on human detection is a classic subject of psychophysical inquiry. But the question of whether human performance can be measured with only one adequate quantitative function of the parameters is still open. We shall explore here if the NFA furnished by the *a contrario* theory can play this role. Indeed, the NFA retains the remarkable property of being a scalar function of the three psychophysical parameters generally used in this kind of detection experiment. In classic experimental settings, these parameters are varied separately and independently, and no synthetic conclusion can be drawn; only separate conclusions on the influence of each parameter can be reached. If a function like the NFA could play the role of generic *detectability* parameter, the experimental parameters could for example be made to vary simultaneously in the very same experiment. In short, if the hypothesis of a single underlying detection parameter is validated, this would simplify the experimental setups and entail a new sort of quantitative analysis of the results, two stimuli being *a priori* considered as equivalent in difficulty if their NFA are similar.

The underlying hypothesis, that the reaction of the subjects to varying stimuli might be predicted as a single scalar function of the stimulus' parameters, is equivalent to the classic hypothesis of a "single mechanism" for contour detection. More precisely, we shall explore if this single mechanism might obey the non-accidentalness principle (the NFA being its probabilistic quantitative expression).

To keep the line of the previous section, this study will again focus on the same simple gestalt: straight contours, that is to say alignments of Gabor elements, as illustrated in Fig. 6.11 (right). The remainder of this section describes the patterns used, the *a contrario* method, the experiment performed on humans, and the result of the comparison.

6.3.1 The Patterns

Figure 6.12 shows three examples of the stimuli used in our experiments. All of them consist of symmetric Gabor elements with varying positions and orientations placed over a gray background. There are two kinds of stimuli: positive stimuli and negative stimuli. Negative stimuli contain elements with random orientations sampled in $[0,\pi)$, e.g. Fig. 6.12 (c). Positive stimuli, see Fig. 6.12 (a) and (b), contain a majority of random elements like in negative stimuli but also a small set of *foreground* elements. The latter lie on a straight line and are uniformly spaced; their orientations are randomly and uniformly sampled from an interval centered on the alignment direction. The size of this interval gives a measure of the *angular jitter* and will be noted by J. When the jitter is zero, the foreground elements have the exact same orientation as their supporting line. Inversely, a jitter of π leads to completely isotropic elements.

The experiment is designed to study how *angular* jitter affects visibility. Yet, a natural question arises about the contribution to the detection of the accuracy of

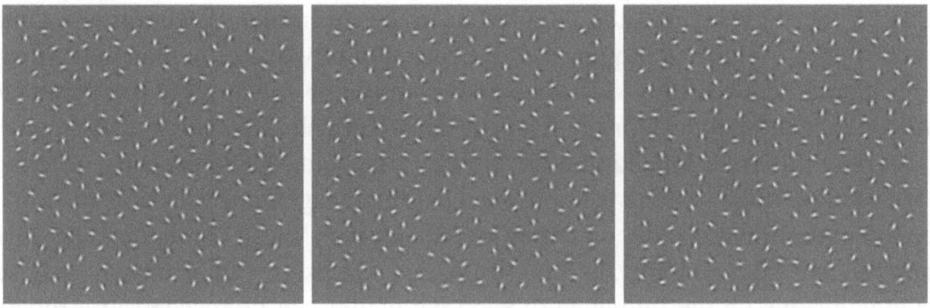

(a) (b) (c)

Fig. 6.12 Three examples of stimuli used in our experiments. **(a)** A jitter-free alignment with 10 elements. **(b)** A weakly jittered alignment with 10 elements. **(c)** A stimulus with no alignment, containing only elements with random orientations.

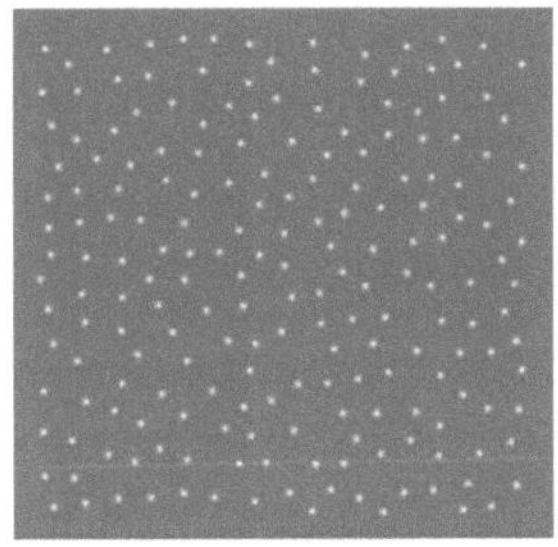

Fig. 6.13 Influence of stimuli position. Each dot represents the position where a Gabor element will be placed. The figure includes a perfectly regular and aligned set of dots, surrounded by random placed elements, all generated by the GERT package. It is very difficult to find the alignment, which shows that the position of the elements by itself conveys few cues about the presence of the alignment. (For a comparison, see the same stimulus with Gabor elements, Fig. 6.11 (right), where the alignment is easily spotted.)

the alignment and of the regular spacing of the aligned elements. All the stimuli presented in this section were generated with the software GERT (v1.1) that includes special algorithms for the generation of random placed and oriented Gabor elements that mask as much as possible the aligned Gabor elements structure [3]. Figure 6.13 shows an example displaying only the elements position; even if there is in fact a set of perfectly regularly aligned dots, it is very hard to spot them. This suggests that the position of the elements carries few useful cues about the alignment.

6.3.2 The Detection Algorithm

Let us now present the alignment detection algorithm that will be matched to human perception. The input to the algorithm is a set of Gabor elements $\mathbf{g} = \{(x_i, \theta_i)\}_{i=1...N}$, defined by the position and orientation of each element. We will further assume that the total number of elements is a fixed quantity N.

A candidate to alignment is defined as a rectangle r, see Fig. 6.14 (left), and the orientation of the Gabor elements inside it will determine whether the candidate is evaluated as a valid alignment or not. The orientation of each Gabor element is compared to the one of the rectangle and when the difference is smaller than a given tolerance threshold τ, the element is said to be τ-*aligned*, see Fig. 6.14 (right). Two quantities will be observed for each rectangle r: the total number of Gabor elements inside it, $n(r, \mathbf{g})$, and the number among them that are τ-aligned, $k_\tau(r, \mathbf{g})$. The *a contrario* validation is analogue to the one described in Sect. 6.2.3.

Due to the way the patterns are generated, the only relevant information to evaluate in an alignment is the orientation of the Gabor elements. Consequently, the *a contrario* model H_0 is defined with N random variables corresponding to the orientation of the elements and satisfying the following two conditions:

- the orientations Θ_i are independent from each other;
- each orientation Θ_i follows a uniform distribution in $[0, \pi)$.

Under these *a contrario* assumptions, the probability that a Gabor element be τ-aligned to a given rectangle is given by

$$p(\tau) = \frac{2\tau}{\pi}. \tag{6.13}$$

Notice that the symmetric Gabor elements are unaltered by a rotation of π rads. The independence hypothesis implies that the probability term $\mathbb{P}[k_\tau(r, \mathbf{G}) \geq k_\tau(r, \mathbf{g})]$, where $\mathbf{G}$ is a random set of Gabor elements following H_0, is given by

$$\mathbb{P}\left[k_\tau(r, \mathbf{G}) \geq k_\tau(r, \mathbf{g})\right] = \mathscr{B}\Big(n(r, \mathbf{g}), k_\tau(r, \mathbf{g}), p(\tau)\Big), \tag{6.14}$$

where as before $\mathscr{B}(n, k, p)$ is the tail of the binomial distribution.

We still need to specify the family of tests to be performed. Each pair of dots will define a rectangle of fixed width w, so the total number of rectangles is $\frac{N(N-1)}{2}$. Also, a finite number of angular precisions τ_i will be tested for each rectangle. Then,

$$N_{tests} = \frac{N(N-1)}{2} \cdot \#\mathscr{T}, \tag{6.15}$$

where $\#\mathscr{T}$ is the cardinality of the set $\mathscr{T}$ of precisions. The NFA of a candidate is defined by

$$\mathrm{NFA}(r, \mathbf{g}) = N_{tests} \cdot \min_{\tau \in \mathscr{T}} \mathscr{B}\Big(n(r, \mathbf{g}), k_\tau(r, \mathbf{g}), p(\tau)\Big). \tag{6.16}$$

A large NFA value corresponds to a likely (and therefore insignificant) configuration in the *a contrario* model; inversely, a small NFA value indicates a rare and interesting event. The proposed detection method validates a rectangle candidate

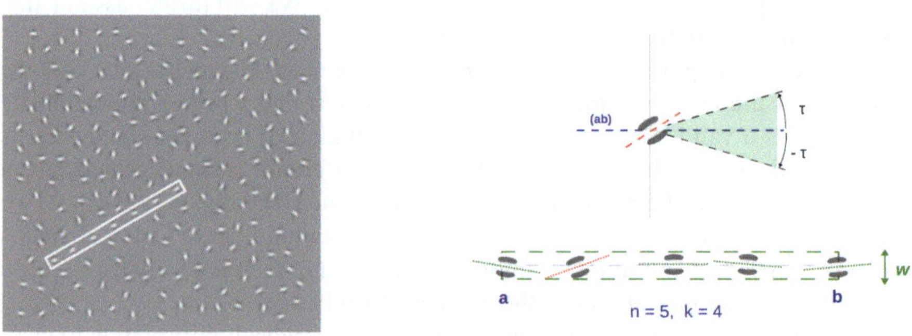

Fig. 6.14 Left: A candidate to alignment, defined by a rectangle R. **Right-Top:** A Gabor element whose angle with (ab) is larger than τ and thus it is not counted as an aligned point. **Right-Bottom:** A detailed example where we see a total of five Gabor elements inside the rectangle, $n(r, \mathbf{g}) = 5$, being τ-aligned with (ab), i.e. $k_\tau(r, \mathbf{g}) = 4$.

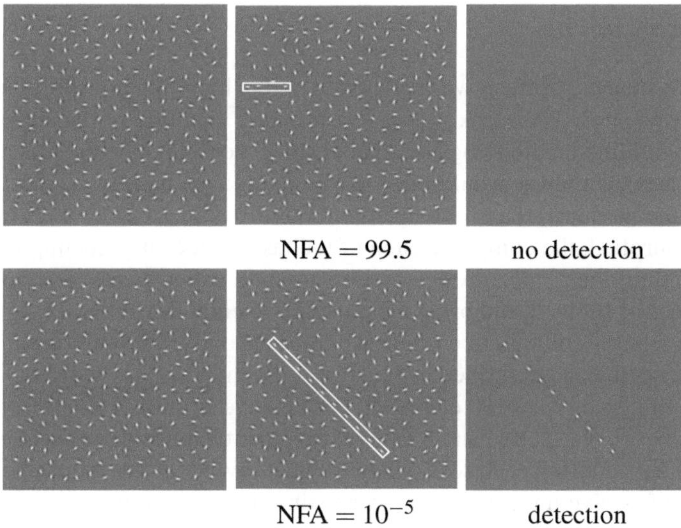

Fig. 6.15 Two examples of the proposed validation method for alignment of Gabor elements. The rectangle in the first row has three elements inside, all of them aligned; that number is nevertheless too small to produce a detection, as its NFA value is larger than one. In the second example, all ten Gabor elements are aligned, giving an NFA of 10^{-5} and producing a detection.

r whenever $\mathrm{NFA}(r, \mathbf{x}) < \varepsilon$. The following theorem shows that it satisfies the non-accidentalness principle.

Theorem 6.2

$$\mathbb{E}\left[\sum_{R \in \mathscr{R}} \mathbb{1}_{\mathrm{NFA}(R,\mathbf{G})<\varepsilon}\right] \leq \varepsilon.$$

where $\mathbb{E}$ *is the expectation operator,* $\mathbb{1}$ *is the indicator function,* $\mathscr{R}$ *is the set of rectangles considered, and* $\mathbf{G}$ *is a random set of Gabor elements on* H_0.

Once again we follow Desolneux et al. [4,7] and set $\varepsilon = 1$. In our experiments, we use the NFA as an indication of the visibility of the gestalt according to the proposed theory; a value considerably smaller than 1 is "non-accidental" and should imply a conspicuous gestalt. A value larger than 1 can occur just by chance and should therefore be associated to an irrelevant gestalt. Figure 6.15 shows two examples of detection by this method.

6.3.3 Experiment

A psychophysical experiment was performed online by voluntary subjects using an interactive web site[3]. Their task was to report on the visibility of the aligned Gabor patterns. The online methodology was necessarily more flexible and less controlled on various aspects than it would be in a laboratory: we had no reliable information about the subjects, their visualization conditions in front of their computers were not controlled, the comprehension of the task by the subjects might vary, etc. Notwithstanding their uncontrolled essence, online experiments give access to a larger number of subjects and bring a great experimental flexibility.

The data set used for this experiment is composed of over 14000 stimuli (negative and positive) as the one described in Sect. 6.3.1. Each image has a size of 496×496 pixels and containing $N = 200$ elements. For positive stimuli, 9 levels of jitter ($J \in \{0, \frac{\pi}{5}, \frac{\pi}{4}, \frac{\pi}{3}, \frac{\pi}{2}, \frac{2\pi}{3}, \frac{3\pi}{4}, \frac{4\pi}{5}, \pi\}$) and 8 different segment lengths were used, between 3 and 10 elements. During each session, the subject saw 35 of these images, one after another. The first five images were training stimuli and no results were recorded at this stage of the experiment. The following 30 images were randomly sampled over the data set, with constraints that ensured a balance between negative, positive, hard and easy stimuli. For each stimulus, the subject was asked to answer whether they saw or not a "straight line"; the answer and response time were recorded. There was no time limit to provide the answer but it was suggested to answer as soon as the subject made up their mind. At the end of the session, a feedback was given on false detections and on the consistency of the subject's answer through an "attention score". This score rewarded the fact that the subject answered better on easy stimuli than on hard ones and indicated if the task was well understood or not.

6.3.4 Results

In order to compare human and machine perception we precomputed the NFA for each rectangle on all the images of the data set. Each image was associated to its best (smallest) NFA. The hypothesis to be tested was that the NFA value should be related directly to the visibility by humans; if this is true, the average score given to an image by humans, namely the proportion of "Yes", should be related to the NFA of the most salient structure. In what follows we will analyze the data obtained from 7137 trials.

The NFA scale was divided into bins. To each bin were associated statistics on the trials whose NFAs belonged to this bin. Figure 6.16 shows the average answer rate and response times for nine $\log_{10}(\text{NFA})$ intervals. Note that $\text{NFA} < 1$ (or $\log_{10}(\text{NFA}) < 0$) means detection of the alignment by the algorithm.

The results significantly support the hypothesis that a single scalar function of all parameters predicts the detectability. Indeed, the answer rate follows a sigmoid shape roughly centered at $\log_{10}(\text{NFA}) = 0$. The second graph, plotting the response time versus the NFA, also agrees with the hypothesis: the less visible the stimuli

[3] `http://dev.ipol.im/~blusseau/aligned_gabors`

are, the more time is spent searching for valid gestalts. Statistical tests confirm this average tendency.

The experiment confirms the hypothesis that, at least in this restricted perceptual environment (formed of three parameters, the number of Gabor patches, the length of the alignment and its jitter on orientation), the value of NFA may account for the human "detectability" of an alignment. Surprisingly, the human detection (attentive) threshold is close to the best algorithm in this restricted environment. Indeed, alignments with NFA smaller than 1 were detected by a majority of subjects. Alignments with NFA larger than one, which are likely to occur just by chance, were detected by a minority of subjects. Furthermore, the detection curve is steepest when the NFA crosses 1. The curve is not as steep for the mean response time as a function of NFA. This can be simply explained by the fact that the patience of subjects undergoes a rapid temporal erosion; they are not ready to look long for a needle in a haystack.

6.3.5 *Consequence: An Online Game*

Online experimentation opens new possibilities that need to be explored farther, and in particular the use of computer games as an experimentation tool. A successful game may attract the attention of subjects and if the resulting mass of results is large enough, it could compensate for the lack of control on other aspects of the experimental setting.

The player of a computer game is usually directed toward an objective and faced with obstacles. To be attractive, a game cannot be too easy, but not too hard either; a

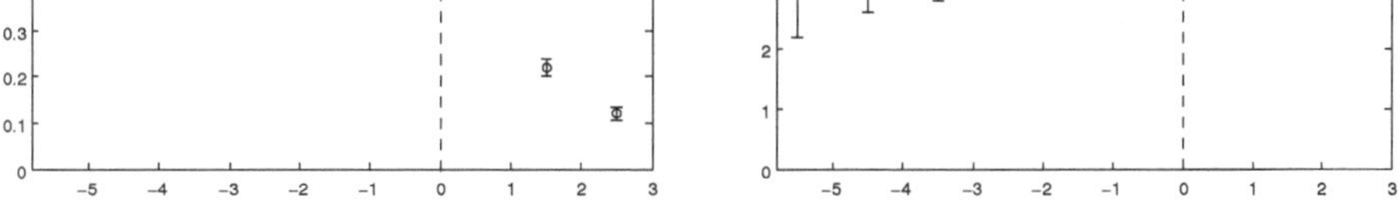

Fig. 6.16 Comparison of the subjects' responses to the NFA. **Left:** The average answer rate is plotted relative to $\log_{10}(\text{NFA})$. Each point indicates the proportion of positive answers to stimuli with best NFA in the corresponding bin. **Right:** The average response times in seconds per bin. In both cases, the abscissa represents the scale of $\log_{10}(\text{NFA})$ divided into 9 bins; the first bin is defined by $\log_{10}(\text{NFA}) < -5$, the last one by $\log_{10}(\text{NFA}) \geq 2$, and the other 7 bins by $k \leq \log_{10}(\text{NFA}) < k+1$ for $k = -5, \dots, 1$. The error bars give approximately 95 % confidence about the mean values (each interval is defined as $[\bar{x}-2\frac{s}{\sqrt{n}}, \bar{x}+2\frac{s}{\sqrt{n}}]$, where $\bar{x}, s$ and n are respectively the mean, standard deviation and number of trials of the bin).

good balance of this difficulty is the key to the popularity of the game. To use games for psychophysical purposes, the player should be directed to detect some pattern, the obstacles being the conditions that preclude this perception. Motivated players will do their best effort, revealing the limits of human perception.

To go in this direction we created a prototype version of an alignment game.[4] The player is presented with Gabor stimuli as described before. Only positive stimuli with variable difficulties are used. In this way one knows that there is an alignment gestalt; but its position is unknown and the assignment of the player is to spot it. The subject is asked to click in the image on any point of the straight line. The distance between the clicked point and the actual line segment is recorded and a score over 100 is computed as a function of this distance (the closer to the segment, the better the score). When the stimulus is quite visible, all subjects are able to point correctly to it; when it is not, the distance to the alignment becomes random. This rash transition should permit to pinpoint the human detection threshold.

The presentation of the stimuli is divided into several sequences of ten images. The first sequence is always supposed to be very easy (long segments with little jitter). Then the difficulty of the following sequences change according to the performance achieved on the previous one. The collected data will permit us to compare a detection method to human performance in the way described before. The game is still in a prototype phase but readers are invited to try it and provide feedback.

6.4 Conclusion

Needless to be said, the experimental devices and first results that we just described are not sufficient to make any rash conclusion on the existence of quantitative predictions of human perception. They will need to be extended to other gestalts commonly used in psychophysics, such as for example contours (good continuation), clusters, or symmetries. In the same way, the first described gestaltic game does not furnish an end algorithm modeling what we could call the human notion of alignment. Finally, we did not deliver a detection algorithm directly usable on any image, as required by the computer vision methodology. In short, this is work in progress, and our goal was to raise the attention of psychophysical researchers and computer scientists on the interest of introducing Turing tests in their methodology.

References

1. Ahuja, N., Tuceryan, M.: Extraction of early perceptual structure in dot patterns: integrating region, boundary, and component gestalt. Comput. Vision Graph. Image Process. 48(3), 304–356 (1989)
2. Attneave, F.: Some informational aspects of visual perception. Psychological Review 61(3), 183–193 (1954)
3. Demeyer, M., Machilsen, B.: The construction of perceptual grouping displays using GERT. Behavior Research Methods, 1–8 (2011), online first

[4] `http://dev.ipol.im/~blusseau/clickline`

4. Desolneux, A., Moisan, L., Morel, J.: Meaningful alignments. International Journal of Computer Vision 40(1), 7–23 (2000)
5. Desolneux, A., Moisan, L., Morel, J.: Computational gestalts and perception thresholds. Journal of Physiology – Paris 97, 311–324 (2003)
6. Desolneux, A., Moisan, L., Morel, J.: A grouping principle and four applications. IEEE Transactions on Pattern Analysis and Machine Intelligence (2003)
7. Desolneux, A., Moisan, L., Morel, J.: From Gestalt Theory to Image Analysis, a Probabilistic Approach. Interdisciplinary Applied Mathematics, vol. 34. Springer (2008)
8. Ellis, W. (ed.): A Source Book of Gestalt Psychology. Humanities Press (1967 (originally 1938))
9. Feldman, J.: Regularity-based perceptual grouping. Computational Intelligence 13(4), 582–623 (1997)
10. Feldman, J.: Bayesian contour integration. Attention, Perception, & Psychophysics 63, 1171–1182 (2001)
11. Feldman, J., Singh, M.: Information along contours and object boundaries. Psychological Review 112(1), 243–252 (2005)
12. Field, D.J., Hayes, A., Hess, R.F.: Contour integration by the human visual system: Evidence for a local association field. Vision Research 33(2), 173–193 (1993)
13. Fleuret, F., Li, T., Dubout, C., Wampler, E.K., Yantis, S., Geman, D.: Comparing machines and humans on a visual categorization test. Proceedings of the National Academy of Sciences 108(43), 17,621–17,625 (2011)
14. von Gioi, R.G., Jakubowicz, J.: On computational Gestalt detection thresholds. Journal of Physiology – Paris 103(1-2), 4–17 (2009)
15. Grossberg, S., Mingolla, E.: Neural dynamics of perceptual grouping: Textures, boundaries, and emergent segmentations. Attention, Perception, & Psychophysics 38(2), 141–171 (1985)
16. Han, F., Zhu, S.C.: Bottom-up/top-down image parsing with attribute grammar. IEEE Transactions on Pattern Analysis and Machine Intelligence 31(1), 59–73 (2009)
17. Kanizsa, G.: Organization in vision: Essays on Gestalt perception. Praeger, New York (1979)
18. Kanizsa, G.: Grammatica del vedere. Il Mulino (1980)
19. Kanizsa, G.: Vedere e pensare. Il Mulino (1991)
20. Kersten, D., Mamassian, P., Yuille, A.: Object perception as bayesian inference. Annual Review of Psychology 55(1), 271–304 (2004)
21. Köhler, W.: Gestalt Psychology. Liveright (1947)
22. Leclerc, Y.: Constructing simple stable descriptions for image partitioning. International Journal of Computer Vision 3(1), 73–102 (1989)
23. Lowe, D.: Perceptual Organization and Visual Recognition. Kluwer Academic Publishers (1985)
24. Marr, D.: Vision. Freeman and co. (1982)
25. Metzger, W.: Gesetze des Sehens, 3rd edn. Verlag Waldemar Kramer, Frankfurt am Main (1975)
26. Metzger, W.: Laws of Seeing. The MIT Press (2006 (originally 1936)), English translation of the first edition of [25]
27. Mumford, D.: Pattern theory: the mathematics of perception. In: Proceedings of the International Congress of Mathematicians, Beijing, vol. I, pp. 401–422 (2002)
28. Nygård, G., Van Looy, T.: Wagemans: The influence of orientation jitter and motion on contour saliency and object identification. Vision Research 49, 2475–2484 (2009)
29. Pinar Saygin, A., Cicekli, I., Akman, V.: Turing test: 50 years later. Minds and Machines 10(4), 463–518 (2000)

30. Preiss, K.: A theoretical and computational investigation into aspects of human visual perception: Proximity and transformations in pattern detection and discrimination. Ph.D. thesis, University of Adelaide (2006)
31. Sarkar, S., Boyer, K.L.: Perceptual organization in computer vision: A review and a proposal for a classificatory structure. IEEE Transactions on Systems, Man, and Cybernetics 23(2), 382–399 (1993)
32. Spelke, E.: Principles of object perception. Cognitive Science 14(1), 29–56 (1990)
33. Stevens, S.: Psychophysics. Transaction Publishers (1986)
34. Tripathy, S.P., Mussap, A.J., Barlow, H.B.: Detecting collinear dots in noise. Vision Research 39(25), 4161–4171 (1999)
35. Turing, A.: Computing machinery and intelligence. Mind 59, 433–460 (1950)
36. Uttal, W., Bunnell, L., Corwin, S.: On the detectability of straight lines in visual noise: An extension of frenchs paradigm into the millisecond domain. Perception and Psychophysics 8, 385–388 (1970)
37. Uttal, W.R.: The effect of deviations from linearity on the detection of dotted line patterns. Vision Res. 13(11), 2155–2163 (1973)
38. Vanegas, M.C., Bloch, I., Inglada, J.: Detection of aligned objects for high resolution image understanding. In: IGARSS, pp. 464–467 (2010)
39. Wagemans, J.: Perceptual use of nonaccidental properties. Canadian Journal of Psychology 46(2), 236–279 (1992)
40. Wagemans, J., Elder, J.H., Kubovy, M., Palmer, S.E., Peterson, M.A., Singh, M., von der Heydt, R.: A century of gestalt psychology in visual perception: I. perceptual grouping and figureground organization. Psychological Bulletin (2012)
41. Wagemans, J., Feldman, J., Gepshtein, S., Kimchi, R., Pomerantz, J.R., van der Helm, P.A., van Leeuwen, C.: A century of gestalt psychology in visual perception: II. conceptual and theoretical foundations. Psychological Bulletin (2012)
42. Wertheimer, M.: Untersuchungen zur Lehre von der Gestalt. II. Psychologische Forschung 4(1), 301–350 (1923), An abridged translation to English is included in [8]
43. Witkin, A.P., Tenenbaum, J.M.: On the role of structure in vision. In: Beck, J., Hope, B., Rosenfeld, A. (eds.) Human and Machine Vision, pp. 481–543. Academic Press (1983)
44. Witkin, A.P., Tenenbaum, J.M.: What is perceptual organization for? IJCAI-83 2, 1023–1026 (1983)
45. Zhu, S., Yuille, A.: Region competition: Unifying snakes, region growing, and bayes/mdl for multiband image segmentation. Pattern Analysis and Machine Intelligence 18(9), 884–900 (1996)
46. Zhu, S.C., Mumford, D.: A stochastic grammar of images. Foundations and Trends in Computer Graphics and Vision 2(4), 259–362 (2006)

Chapter 7
Remarks on Invariance in the Primary Visual Systems of Mammals

Daniel Bennequin

7.1 Introduction

Poincaré observed that the perception of space is based on active movements, and relies on the notions of invariance, covariation between sensors and environment, and active compensation ([179], [180], [181], [182]). The research of Piaget has proved the importance of various kinds of geometrical invariance in cognitive and behaviorial development ([173], [177], [176]). To him intelligence is a form of adaptation, the continuous process of using the environment for learning ([174]). Adaptation is a process that can happen at the scale of evolution, development or functioning. In ecology, or in population biology and genetics, it means the adjustment or change in behavior, physiology, and structure of an organism to become more suited to an environment, thus better fitted to survive and passing their genes on to the next generation (Darwin plus Mendel, [45]). In Neuroscience it often means the decline in the frequency of firing of a neuron in response to constantly applied environmental conditions, or more generally, any change in the relationship between stimulus and response that is induced by the level of stimulus (Laughlin, [121]). Adaptation is an ubiquitous essential property of sensory and motor processing, allowing the living systems to sense and anticipate what is changing in the world ([27]). As we will see, invariance can contribute to adaptation, and adaptation can create new invariance structures. Gibson gave a precise formulation of invariance and adaptation in psychology, with special emphasis on vision ([71], [72], [73], [74]). From the formal point point of view, the mathematical theory of groups, and its many extensions in algebra and analysis (cf. 2.3 below for definitions), offer a clear mathematical basis for discussing the notions of invariance ([84]). About Gibson and symmetry groups, see Shaw, McIntyre, and Mace [202]. For a general discussion of psycho-physics, groups and adaptation, see Shepard 1994 [203]. The link between symmetry, groups, perceptual invariance and *a priori* of the brain was discussed by J.Droulez and myself in [57].

Daniel Bennequin
Institut de Mathématiques de Jussieu, Université Paris VII, France

G. Citti and A. Sarti (eds.), *Neuromathematics of Vision,*
Lecture Notes in Morphogenesis,
DOI: 10.1007/978-3-642-34444-2_7,

In the following notes, we give examples of invariance, and of its multiform relation to adaptation, in the mammalian visual systems, in cases where the actions of groups can be made explicit. Most of these examples concern the primary cortical visual system of higher mammals: carnivores, scandentia (three shrew) and primates. The text is assembled from a collection of unpublished studies conducted during the years 2004 to 2012, mainly made in collaboration with students and other researchers in the laboratory LPPA (Collège-de-France, Paris). The emphasis is put on the mathematical equations.

The principle underlying these examples is that neurons and brain areas are better understood by describing their *structure of indifference*, i.e. what they neglect in the world for their function, and their *structure of ambiguity*, i.e. how they create new entities from their inputs, allowing internal transformations. We have proposed that group theory, in particular Galois theory, can be seen as an information theory, in the same manner that probability theory subtends several kind of information theories (Fisher, Shannon, Wiener, Kullback, Kolmogorov). The first mathematical appendix is added to explain this point of view.

In order to modelize invariance and adaptation in one neuron or in a set of neurons, we introduce a *virtual space* (that we propose to consider as a kind of *homology group*, in the mathematical sense, cf. [61]), made by combinations of equivalence classes of possible attractors of responses, modulated by relevant stimuli, for categorization. The symmetry groups and the ambiguity groups are acting on this virtual space (cf. 2.4. and Appendix 1 for the precise definitions). In general this space I corresponds to a dimensional reduction of the input stimuli, it is a skeleton of the rapid internal dynamics. In this model the possible responses (of the neuron or the area) correspond to the coupling of virtual classes with dynamical parameters of neurons or interactive sets of neurons. The space M of these parameters can be seen as an unfolding in the sense of Thom ([212]), which introduces structural stability in the process. It contains intrinsic parameters and preferences of the neurons, in general induced by connectivity or ambient activity. Invariance acts both on I and M. Then, in some cases, adaptation is generated by compensation of transformations in homology by transformations in the unfolding, including bifurcations. We suggest that, in many cases, adaptation is a dynamic on parameters that results from changing the attracting state. In turn this dynamic induces a change in rapid dynamics.
This model permits to use at the same time a notion of receptive field (cf. Marr [137]), with adaptable parameters, and a notion of affordance (Gibson), in the form of invariant categorization, thus we can combine the notions of hierarchical visual fields and the notions of sensori-motor loops for perception and action.

We propose that no meaning can emerge without some kind of invariance: at each step of the neuronal sensory process, the neuronal activity is an act of recognition, which, as every act of knowledge, either explicit or implicit, consists in neglecting most of the individual aspects of the input for selecting the output, combining *a priori* knowledge, novelties and inventions.

The exposition starts with a general introduction to invariance from the perspective of groups an categories. Then we summarize several examples of invariance in the visual system, mainly related to its adaptation to motions. The basic group comes from Galilean relativity, and is implemented in the visuo-vestibular system. However, the evolution of higher mammals, and specially carnivores and primates, used other fundamental groups for vision and visual perception: phase space symmetries (i.e. symplectic transformations), internal affine symmetries of colors, affine and projective symmetries of figures, diffeomorphisms of the visual plane, their extension by gauge transformations for colors, and so on. At the end of visual path, we find in particular the Euclidian geometry realized in abstract spaces, and topology.

In section 2, we insist on the role of eye movements for invariant perception. In section 3, we study affine invariance and adaptation in the color space. In 4.1 we describe models of the transformation from simple cells to complex cells and columns in $V1$, with application to the "energy model". Here we see an intervention of special functions theory. In 4.2 we prove that the principles of invariance and probability explain the geometry of the cortical map of orientations in $V1$. In 4.3 we put forward a probabilistic model of optimal gluing of neuronal activities, with applications to callosal connections and disparity. In 4.4 we study LGN and $V1$ in tree shrew, in particular the structure of callosal connections and orientation maps. In this part we give a review of the organization of the visual pathway from the retina to $V1$ through the LGN in mammals. In 5.1 we discuss the notion of general covariance in $V1$. In 5.2 we describe the projective invariance of the optic flow, for application to MT, MST. In this example we meet the classical invariant theory of linear group representations. In 5.3 we review geometric cells in the para-hippocampal region. In 5.4 we mention higher invariance in recognition. In 6 we briefly outline the way a topology of information quantities, as developed in collaboration with Pierre Baudot, could go deeper into the mechanisms of adaptation and formation of invariance structure.

The original developments are 1) the formal definition of homology, periods and adaptation, in 2.4; 2) the discussion in 3.2 of the affine structure on colors in LGN and $V1$ and the description in 3.4 of color constancy in higher visual areas (inspired from David Philipona and Kevin O'Regan); 3) the computations of non-linearities in complex cells and superficial micro-columns of $V1$, through Abel transforms, in 4.1 (work with Simon Capern and Jacques Droulez); in the same subsection, we introduce of a special class of non-linear pseudo-differential operators for transforming field activities from one brain area to another by taking frequencies in account; 4) the link between probability and invariance in the orientation map of $V1, V2$, in 4.2, implying that the map is a Brownian coefficient of a unitary irreducible representation of the Euclidian group (inspired by F.Wolf, T.Geisel, and D.Barbieri, Giovanna Citti, G.Sanguinetti, Alessandro Sarti); 5) the formulation of the gluing of the two eyes and the two hemispheres by callosal connections, in 4.3 (work with Luc Foubert, Jacques Droulez and Chantal Milleret); 6) the explanation of the impossibility for the tupaia to respect orientation preference in its callosal connection (and the possibility for the cat to do that), in 4.4; 7) the hypothesis of general covariance in the area $V1$, in 5.1; 8) a suggestion for the nature of objects and images in 5.4, modeled on color constancy and using co-cycles; 9) the perspective offered by information theory, in 6 (joint work with Pierre Baudot); 10) the link between the three roads of information geometry given by Shannon (probabilities), Galois (groups) and Thom (catastrophes) theories, in Appendix 1; 11) the general definition of invariant receptive fields, through coefficients of unitary representations of Lie groups, in Appendix 2.

An excellent reference in Neuroscience more than sufficient for all the applications we will present is the encyclopedia in two volumes, "The visual neuroscience", edited by L.M. Chalupa and J.S. Werner (MIT 2004); we will give in most subsection the related articles in this book, refereed as TVN or [44].

7.2 The Notion of Invariance

7.2.1 Mathematical Definition of Invariance

Given a set X (in applications the elements x of X can be things or events) and a set G of partial transformations acting on X (i.e. a set of maps $g : X' \to X$ where X' is a subset of X (in applications these transformations can be permutations but they also can be deformations). We note $x \mapsto g.x$ (or $x.g$) the effect of the map on the elements, and we say that an element x_0 of X is an *invariant point* of the action of G if, for each g acting on x_0 we have $g.x_0 = x_0$ (or $x_0.g = x_0$), i.e. x_0 is a fixed point of the action. When G is stable by composition of transformations, i.e. $g, h \in G \Rightarrow g \circ h \in G$, we assume that for the operation $g(h.x) = (gh).x$ (or $(x.g)h = x.(gh)$ respectively). For $g.x$ we say that we have a *left action* of G, and for $x.g$ we say that we have a *right action*.

We need also the more general notion of *co-variant*: given two sets X and Y and a set G acting by partial transformations of both X and Y, on the left for X and on the right for Y, we say that a function f_0 from X to Y is a *co-variant function* if we have $f_0(g.x) = f_0(x).g$, when both terms are defined. In particular a function $x \mapsto f(x)$ of X to a set Y where the action of G is trivial (i.e. for any $y \in Y$ one has $y.g = y$), is a co-variant if and only if, for any g and x such that $g.x$ is defined, we have $f(g.x) = f(x)$. In this last case we also say that f is an *invariant function*, or simply an *invariant*; in fact it is an invariant point of the natural (right) action of G on functions on X with values in Y: $f^g(x) = f(g.x)$.

Remark that covariance appears as a particular case of invariance when the transformations g of Y are everywhere defined and invertible: denote by g^{-1} the inverse operation of g, and consider the set F of all partially defined applications f from X to Y, with the action of G on F given by $f^g(x) = g^{-1}.f(g.x)$, then f_0 is a co-variant for the actions of G on X and Y if and only if f_0 is an invariant point of the action of G in F.

The main concept underlying invariance study is the concept of *group G*, that is a set equipped with a neutral element e and an internal law $(g,h) \mapsto gh$ satisfying associativity $g(hk) = (gh)k$, neutrality $ge = g = eg$ and invertibility (existence of g^{-1} such that $gg^{-1} = g^{-1}g = e$) (cf. section 2.3). For applications in Physics an important role is played by the theory of *harmonic analysis*, that is the study of unitary linear representations of groups in Hilbert spaces (cf. Appendix 2). Note that the definition we will suggest for *invariance of a receptive field* is based on coefficients of unitary linear representations of a group, thus belongs to harmonic analysis. However, for most applications in Biology we are forced to restrict the action to a certain subset of G, which relies in general on the notion of *groupuscule* (cf. [34]) or the notion of *category* with inverse of arrows, or *groupoids* ([133]).

The theory of groups was invented by Galois to formalize invariance, symmetry and ambiguity, that arize in the study of algebraic equations (cf. [70], [218]); Klein, Lie and Cartan have shown that all of geometry, with its geometrical transformations and curvature notions, can be understood in this framework. And the notion of groups has been extended by notions from categories, introduced by Eilenberg and Maclane (cf. [133]).

In another direction, the notions of signal analysis, in particular Fourier transform or wavelets transforms, have been extended to non-commutative groups, giving the modern harmonic analysis (cf. [222]).

Note that several authors already insisted that invariance in visual recognition is better understood through general Lie groups theory ([64], [143], cf. [168]).

The theory of groups and that of their representations are recognized as fundamental for the description of physical and chemical systems, from Classical Mechanics to Quantum Field Theory through Quantum Mechanics. In order to be useful in Biology these tools have to be adapted to the proper concepts of Biology, taking into account the discrete nature of neuronal networks, and their inherent variability. In particular, as was noticed by René Thom, the usual assumptions of analyticity are not generally valid in Biology. The study of Vision is an ideal occasion to test and adapt these ideas.

7.2.2 Invariance for Neurons or Brain Areas

Consider a set A of neurons a; the biological function of a neuron a is the transformation of an input into an output, thus the biological function of the set A is a vector of transformations. For instance the output of a *primary visual neuron* a is a function $F_a(I;t)$, where t denotes the time of response and I denotes the past images before t. By an image, here, we simply understand the local field of intensity of light captured in the retina. (For taking color in account, we must add a wavelength content, and I must be a vector.) By a response here, we understand the spiking activity -that is, a numerical function of time. However, in general, to understand the behavior of the neuron a we must also take in account internal parameters (of the neuron or the brain) and other inputs than the image (from other neurons or glial cells). Suppose now that a group G acts on both the collection of input activity and the collection of output activity, we say that the set A is *co-variant* for G if its biological function $in_a \mapsto out_a; a \in A$ is co-variant for the actions of G. Note that in most cases the elements g of G do not respect separately the activities of the units a, but they mix the components of the vector activities.

To be concrete let us give four examples of neuronal invariance in the visual information flow, from the retina to areas at the end of the visual path in the brain: 1) adaptation of light sensors to the intensity: here the transformation g is made by global shift and scaling of the luminosity $I \mapsto (I - I_0)/I_1$, it involves a cascade of chemical and electrical processes modifying the response of ganglion cells (cf. P. Sterling in [44]); 2) view-co-variant recognition: there is a group of permutations g of the cells in $V1$ such that, after any translation and rotation of a visual scene, there is a permutation of the cells in $V1$ that gives a response equivalent to the response for the initial scene (cf. G.C. DeAngelis, A. Anzai in [44]); 3) view-invariant recognition: an individual cell in the human fusiform gyrus can be invariant under the animated deformation of a face which is seen (cf. E.T. Rolls in [44]); 4) direction encoding: in the presubiculum of a rat, a head direction cell a is invariant by translation of the head of the rat to a parallel direction, but the collection A of all head direction cells in the presubiculum is co-variant under the group of rotations of the head in the horizontal plane. In fact the response after a rotation corresponds to a permutation in the population, moreover A is also co-variant for the group of rotations of distant visual cues; in this case all the cells a rotate their preference by the same angle (cf. [234]). In this last example, (that of the head direction cells), we see the interest of considering at least two different groups, one acting on the head, another one acting on the distant visual cues; this multiplicity of frames helps to precise the notions of ego-centric and allo-centric reference frames.

Each neuron, each area, obeys to a first form of invariance, which is defined by all the changes in the world or in the nervous system that have no influence on its function. This invites to consider neurons and their assemblies from the point of view of their *indifference* to the changes in the world. Based on this indifference, neurons and areas can share a co-variance with their inputs, defined by all changes in their function which can compensate changes in the world. This permits to consider neurons and their assemblies as operators detecting certain changes in the world.

However, we will also present a dual point of view, that invariance and covariance are first invented by the brain, from molecules to networks, by creating original *ambiguity* groups in internal spaces. This gives geometrical structures on inner spaces. These two kinds of operations, indifference and ambiguity, can be seen as convergence of invariance and divergence of invariance respectively, as there exist convergence and divergence in neuronal networks connectivity.

A basic example of divergence of invariance is given by *inhibition*: a given input can generate two kinds of answers, one excitatory, another one inhibitory, creating an involution in the brain exchanging excitation and inhibition. Remark that in reality, the symmetry is not exact, in general there is more excitation than inhibition (except some balanced cases), and in some brain systems inhibition largely dominates (as the output of Purkinje cells in the cerebellum), but this symmetry between excitation and inhibition gives a principle of organization, which explains part of the functioning of the brain. For instance see M.M. Slaughter in [44].

A basic example of convergence of invariance is given by the ON-OFF cells in the retina: opening or closing light, which are two different stimuli, generate the same response of the so-called ON-OFF cells. Cf. R. Nelson, H. Kolb in [44].

The groups underlying these two examples are isomorphic to the simplest non-trivial group, with two elements; it is noted $\mathbb{Z}/2\mathbb{Z}$ when considered additively, or $C_2 = \{+1, -1\}$ when considered multiplicatively.

In general the invariance structure is neither pure convergence nor pure divergence; the mixture must be described case by case. For instance global multiplication of intensity of light I by a positive non-zero number λ, has no effect on most ganglion cells, by re-scaling, that is pure convergence; but the same light wavelengths composition can produce different responses of the CC-cells in $V4$, for detecting reflectance of objects; this is a re-construction and constitutes a divergence. Both are called adaptation; changes in the entries are compensated by changes in the analysis to produce efficient signals.

We should note a great similarity between Quantum fields and neuronal functions $F_a(\varphi;t)$, where φ denotes a general field input, not necessarily an image, before the time response t and a denotes a cell in an area A. Each neuron at a given time is a function of an extended function, in the same manner that a state of quantum field is a function of all the classical fields, i.e. it is a *function of functions*. From one area to another, we get an operator transforming these functions of functions into other functions of functions, and symmetry groups act on these operators. In the simplest regime, or in first approximation, the operator appears to be linear. In the same manner interactions between fields are expressed in Quantum Field Theory by a linear operators transforming the states. The coefficients of this operator are the correlations (or amplitudes) between the states, input and output. More generally, considering a set of interacting areas $A_1, ..., A_n$ we can consider the correlation coefficients

$$A(a_1, t_1, a_2, t_2, ..., a_n, t_n) = \langle F_{a_1}(\varphi_1; t_1) F_{a_2}(\varphi_2; t_2) ... F_{a_n}(\varphi_n; t_n) \rangle, \qquad (7.1)$$

by taking the average over repetition of the events (cf. [96]).

It is known that the *renormalization group* is a necessary tool for giving a sense to the locality of Quantum objects; according to Wilson and Kadanoff ([235], [235]) this group describes the manner the injection of high momentum and high energy levels at small scales influences the theory at all larger spatial scales. We will see in the examples below that *adaptation* is something comparable: a change of the parameters in the receptive fields that are induced by changing the scale, the frame or the context. This subject is further developed in joint work with Pierre Baudot.

One of the first noticed visual adaptation was the tilt aftereffect of Gibson ([71]): when tilted lines are viewed, even during a short time, then a new set of viewed lines appear tilted on the other side. A long time elapsed till, Dragoi, Schummers, Sur et al. ([56], [201], [200]) showed that in V1, near the pinwheels, there exists a corresponding shift of orientation preference of cells.

Michael A. Webster ([227]) remarks that Gibson himself ceased to accord importance to this effect, because he got more and more interested in natural images. However, later on, researchers got a better understanding of the general role of adaptation, underlying vision and all other sensory modalities, even for natural data.

As other sensory systems, the visual system prepares future actions of the animal depending on the contexts. The paradigm of animal action is locomotion. The primitive function of the visual system is to transform the captured light into information on the situation of the animal in its environment and its future direction of displacement. Note that even protozoa can have eyespots (aggregation of light-sensitive pigments) to help them for swimming. So the visual system completes the vestibular system for sensing and controlling the translations and rotations of the head the world (cf. [24]). This information is combined with all other sensory modalities (in particular proprioceptive) for managing the invariance under the relativity group of the physical laws of movements in Euclidean space. This relativity group was described by Galilée (at the birth of Physics); it was formalized into the Galilean group (cf. J-M. Souriau [207]). The Galilean group has 10 dimensions; it is made by the $6D$ groups of rotations and uniform translations, extended by the $4D$ translations in space-time. Due to the principle of inertia, the last part, the change of origin of the frame, needs vision (or somatic sensation) to be detected. This global invariance is the first basic invariance to be controlled by vision from the point of view of evolution, because it allows the animal to internalize what comes from its action and what comes from the changes in the world. Cf. [11]. Galilean invariance was first implemented in subcortical vestibular and visual systems, for instance in the cerebellum, and it was further elaborated in the cortical system, for instance in the vestibular cortex or the parietal cortex. Note that the divergence in Galilean invariance can sustain what has been named *internal models* (cf. McIntyre, Berthoz and Lacquaniti [140]); a spectacular example is the separation of the gravity vector from the linear acceleration in the vestibular system, which appears in the vestibular cerebellum (cf. [245]).

The linear part of the Galilean group is isomorphic to the Euclidean group of $3D$ rigid motions. Poincaré suggested that the general notion of *object* is issued from the manipulation of rigid bodies, the rigidity property being defined by the possibility to compensate the time evolution of the object by a change of Euclidean frame. For Poincaré this compensation characterizes rigid motions, then the nature of ambient space. This approach was recently developed again by D. Philipona, K. O'Regan, J-P. Nadal et al. in the context of sensory-motor contingence with virtual robotic systems ([171]).

In the primary visual area $V1$, the *simple cells* form a system of co-variants under the $2D+1$ translation of an image (two coordinates for the visual plane and one coordinate for the time), and the *complex cells* have responses that are invariant under small translations (cf. section 4.1 below). This invariance was the reason invoked by Hubel and Wiesel ([88]) to explain the appearance of *complex cells* in $V1$; this produces an information on images that is stable under small displacement. Fourier analysis and its localization by wavelets permit to analyze the receptive fields in $V1$ from the point of view of $2D+1$ translation invariance. However a more complete invariance for all translations is achieved in the infero-temporal region (IT), cf. Rolls et al. [188], [191])

Hubel and Wiesel ([87], [88], [89]) also observed that the neurons in $V1$ of cats and monkeys, have a preferred orientation of stimuli, which is constant in vertical columns of the cortex and varies continuously along the cortical surface. Thus an action of $2D$ rotations in the visual plane is also implemented in $V1$. From the works of F.Wolf, T.Geisel ([238]), and D.Barbieri, Giovanna Citti, G.Sanguinetti, Alessandro Sarti ([16]), it appears that the organization of orientation maps results from the invariance by the group of $2D$ Euclidean displacements, rotations and translations together (see 4.2 below). In the same manner the *disparity* of binocular cells can be studied from the point of view $3D + 1$ invariance (cf. Ohzawa, Freeman [159]).

A further function of the visual system is to document the brain on the content of the environment, from the neighborhood of the body to the far landscape. However this information flow is not a passive registration of photons, it results from an active process. A variety of eye movements (saccades, smooth pursuit, VOR, nystagmus, ...) and internal adaptations (accommodation, efferent copy, ...) helps to stabilize space and objects perception during relative motions. In the following lines we only want to give an idea of the complexity of the brain system (in humans or monkeys) that helps to preserve space invariance during locomotion. We will see that invariance is distributed over many areas, not only visual. As said by Alain Berthoz in [27], this system intertwines motor and sensory information in such a manner that it is preferable to abandon the distinction between them.

The basic compensation movements of the eyes to stabilize the image in the retina when the head is moving, is the vestibulo-ocular reflex (VOR); it is composed of a rotational reflex, the angular VOR (aVOR) and a translational reflex (tVOR). To generate rapid (few *ms*) compensation, the aVOR is organized in the planes of the semi-circular canals, which register rotation acceleration in the inner ear. A wonderful example of adaptation for respecting geometry in space is given by the way flatfishes reorganize the connection from canals to eyes muscles during development ([77]). For aVOR and tVOR a path of three neurons exists in all vertebrates (first order afferent contacting hair cells in the labyrinth, second order neurons in vestibular nuclei, neurons in ocular motor nuclei activating extraocular muscle fibers), but many other brain centers in the brainstem and the cerebellum are used to control and adapt the VOR (cf. [237], [75]). Moreover in natural situation, this reflex must be counteracted by other eyes movements for visual exploration or smooth pursuit or motion anticipation, that involve for instance the saccadic system in the brainstem and the reticulate formation, the superior colliculus (*SC*), the basal ganglia (*BG*), the thalamus, the cerebellum and the neocortex (cf. [27]).

The cerebral network concerned with eye movement control involves a very large part of the brain, for instance in the cortex, it involves FEF (frontal eye field), SEF (supplementary eye field, known for its role in motor programs), $preSEF$ (for motor learning), CEF (for motivation), $MT+$, $V5$ (visual areas concerned by image movement), PEF (parietal eye field, in particular projecting to FEF), and intraparietal areas IPA (for visuo-spatial integration), but also the dorsolateral prefrontal cortex ($DLPFC$), involved in inhibition of saccades, visual prediction and short term spatial memory, plus several areas involved in visuo-spatial attention (as SMG, SPL). The three areas FEF, PEF and $DLPFC$ have strong projection to the SC, the projection of $DLPFC$ being inhibitory. These areas are interconnected with the parahippocampal region PHC and the hippocampal formation HF, known to play an important role in spatial memory, in the medium and long term. Cf. the review in [178], and [27], [28]. It is also known that vestibular information reaches PHC and HF ([223]) and the intraparietal areas (VIP in monkeys, IPA in humans) ([250]). Thus gaze behavior during locomotion, naturally influenced by visual and vestibular information, modulates the spatial perception and spatial memory, in particular through the interaction of the cortical network $PEF,FEF,DLPFC$ with PHC,HF. In addition there is a strong interaction of $PEF,FEF,DLPFC$ with the brainstem, SC and other sub-cortical nuclei ($NRTP$, OS,...). Moreover we cannot underestimate the important role of the cortico-cerebellar loops in producing the precise movements of the eyes during locomotion (cf. [12], [60], [51]).

An anticipation of the future visual scene is necessary for stable perception (cf. [65]). It relies on an unconscious visual remapping (made by a subset of the visual neurons) preceding the eye movement. In particular neurons in intra-parietal LIP (for monkeys) or its homolog for humans, have activities modulated by the future visual stimulus; in particular they are known to be informed by afference copy of voluntary eye movements ([244], [149]). Such remapping is also present in FEF,SC, and extra-striate visual cortex. This system is known to be modulated by attention. Studies in human primary visual areas have shown visual remapping in the ipsilateral field (cf. [142]), that is a probable feedback from the parietal cortex (normalized responses in visual areas are as follows: in $hV4$ 71%, $V3A$ 61%, $V3$ 35%, $V2$ 23%, $V1$ 17%). Neurons in $V3,V3A,V4$ react to saccades directed to their RFs (more than 20%, less than 24%), and even in $V1,V2$ (respectively 12% and 14%). The cited authors verified that the remapping effect cannot be explained by saccades effect, or by the apparition of the stimulus, or by the independent conjugation of both. (Note that in $V1$ of macaques, Nakamura and Colby ([151]) found less than one cell over 64 responding to remapping; this can be due to the small size of RFs in $V1$ of macaques, inducing error on the stimulus position, or to the difference between human and monkey for attentional effect.) In $V4$ (for monkeys and humans) there are influences of covert attention and oculomotor information, or stimulation of FEF.

On the interaction of motor and sensory systems in $V1$, we must also cite the experimental use of image movements that mimic the natural eye movements in cats (Baudot et al. [19]): these movements have profound effects on the statistics of individual activity of neurons, they modify the non-linearly in the RFs (mostly spiking) and augment the mutual information with the stimulus. In this case, we see not only compensation for stabilization but also adaptation of the RFs for constructing more visual information.

For all the above matter see the chapter XI in TVN, which contains ten articles on eye movements.

Object perception is the main subject in which ideas of invariance have been applied (cf. Rolls, and Riesenhuber, Poggio in TVN, [44]). We will come back to this vast subject in the last sections.

For a stable perception of space and objects, the visual system must compensate transformations of stimuli. For far space this involves $2D$-affine transformations in the visual plane. Note that this $2D$-affine invariance is used to explain the duration of hand movement and locomotion [25], [169]). For nearer space $2D$-projective deformation occurs. Certainly the way the sensory-motor system works with this extended affine and projective invariance plays a role in the learning of geometry by children, so useful to act in the world (Piaget). Computer vision has made a great use of affine and projective geometry, see Devernay, Faugeras [54] and Lowe et al. [130], [131].

Koenderink et al. have established that the projective geometry of the visual plane organizes the coherence of figure perception in space; cf. [114], [105], [163], [216]. For instance the Pappus theorem, and the Varignon theorem are implicitly used in perceptual judgments. More generally, pictures deformed by projective transformations are recognized (sometimes with difficulty) as the same picture seen from another place.

In the area MT (or $V5$) of primates neurons are sensitive to stimulus speed with respect to the retina frame. In MT and part of MST (or 19 and 37 Broadmann's areas of humans), the so called complex $MT+$, the geometry of the optic flow is exploited to give knowledge of the eyes (or body) position with respects to objects (cf. [119], [147], [11]). For instance, projective geometry describes the optic flow of one eye that appears when the subject moves in the Euclidean space E and looks at pictures drawn on a fixed plane H in it (cf. section 5.2 below). The vector field on H that describes the apparent movement of fixed points of H in the moving frame of the eye, belongs to the special eight dimensional family of projective vector fields, i.e. the infinitesimal transformations that generate the planar projective transformations. Analysis of the invariant characteristics of the optic flow (the characteristics that cannot be compensated by eye movements with respect to the head) relies on the theory of linear representations of the $3D$ rotation group ([115], [103]).

In the dorsal medial superior temporal area ($MSTd$) there are two types of visually responsive cells: 1) expansion/contraction cells, which selectively respond either to an expansion or to a contraction; and 2) rotation cells, which selectively respond either to a clockwise or to a counterclockwise rotation ([210]).

Neurons in the area MST have responses correlated to speeds with respect to a fixed frame in space, thus they compensate for smooth pursuit of the eye so as to extract higher invariant information from relative information in MT (cf. [92]).

It has been suggested that $V1, V2$ computations involve other Lie groups than the Euclidean group: for instance $4D$ and $6D$ symplectic transformations in the description of the energy flow (cf. [18]), or the group of isometries of the hyperbolic plane for analyzing edges and textures in the model proposed by Chossat and Faugeras ([46]). See Petitot's book [168] for a general discussion.

As we will see in 3.3 below the color space in LGN, and partly in $V1, V2$ and $V4$, offers a good example of invariance constructed by the brain in order to work with frequency of light and physical properties of objects; this invariance relies on affine geometry in nD spaces ($3D$ form most mammals because they are trichromatic).

A main subject in this text will be the *general co-variance* of $V1$: we suggest that in $V1$ local deformations of the images are represented, to sense contours and surfaces in motion. The invariance here relies on feed-forward connections from

LGN, on horizontal connections in $V1$ and on feedback from higher areas. The local deformations generate a group with an infinity of dimensions. However it is better in this case to consider limited distortions, thus the considered transformations are better represented by a category, not a group, which consists of open sets of the visual plane (as objects) and diffeomorphisms from an open set to another one (as morphisms), these diffeomorphisms being sufficiently close to the Identity, to insure that they do not deform too much the objects.

Note that considering categories in order to extend the implication of geometry in perception was suggested long time ago by Piaget ([175]).

7.2.3 Categories and Groups

A nice introduction to categories, functors and their use is [133].

A *category* $\mathscr{C}$ is specified by a set $\mathscr{C}_0$ of *objects* $a,b,c,...$, and for each pair of objects (a,b) a set $\mathscr{C}(a,b)$ of *arrows* $f,g,h,...$, equipped with an operation of composition

$$\mathscr{C}(a,b) \times \mathscr{C}(b,c) \to \mathscr{C}(a,c) \tag{7.2}$$

$$(f,g) \mapsto g \circ f \tag{7.3}$$

which satisfy the two following two axioms:

(i) for each a there is an element e_a in $\mathscr{C}(a,a)$ such that whenever it has a meaning $g \circ e_a = g$ and $e_a \circ f = f$;
(ii) whenever it has a meaning we have $(h \circ g) \circ f = h \circ (g \circ f)$.

An arrow $f \in \mathscr{C}(a,b)$ is noted $f : a \to b$ and is said to go from a to b, the object a being its *source* and the object b being its *target*. Another name for arrow is *morphism*. The axiom (ii) is expressed by saying that composition is an *associative* law. The morphisms e_a are called *identity elements*; the axiom (i) asserts they are *neutral elements* for composition to the right and to the left respectively.

For each category $\mathscr{C}$ we can form the dual category $\mathscr{C}^{op}$: it has the same objects but we decide that any arrow $f : a \to b$ of $\mathscr{C}$ becomes an arrow $f^{op} : b \to a$.

Consider two categories $\mathscr{C}, \mathscr{C}'$, by definition, a (covariant) *functor* F is the datum of map $a \mapsto F(a)$ from $\mathscr{C}_0$ to $\mathscr{C}'_0$, and for each pair (a,b) of objects of $\mathscr{C}$, of a map from $\mathscr{C}(a,b)$ to $\mathscr{C}(F(a),F(b))$, -usually denoted $f \mapsto F(f)$- such that whenever possible we have $F(g \circ f) = F(g) \circ F(f)$.
Note that, if F is a functor from $\mathscr{C}$ to $\mathscr{C}'$ and G is a functor from $\mathscr{C}'$ to $\mathscr{C}$", the composite maps form a functor from $\mathscr{C}$ to $\mathscr{C}$". A sub-category $\mathscr{D}$ of a category $\mathscr{C}$ is a category specified by a subset of $\mathscr{C}_0$ and subsets of morphisms, containing the identity elements and closed under composition. The embedding from $\mathscr{D}$ to $\mathscr{C}$ is a functor.

A functor F from $\mathscr{C}^{op}$ to $\mathscr{C}'$ is called a contra-variant functor from $\mathscr{C}$ to $\mathscr{C}'$.

A *natural transformation* T between two functors F,G from $\mathscr{C}$ to $\mathscr{C}'$ is a set of arrows $T(a) : F(a) \to G(a)$ defined for each object a of $\mathscr{C}$, satisfying the following commutativity relation for each morphism $f : a \to b$ in $\mathscr{C}$: $T(b) \circ F(f) = G(f) \circ T(a)$. It is a *natural equivalence* when, for every a, the arrow $T(a)$ is invertible. An *equivalence* between two categories $\mathscr{C}, \mathscr{C}'$ is a pair of functors, F from $\mathscr{C}$ to $\mathscr{C}'$ and G from $\mathscr{C}'$ to $\mathscr{C}$ such that there are natural equivalences T from $F \circ G$ to $Id_{\mathscr{C}}$ and T' from $G \circ F$ to $Id_{\mathscr{C}'}$.

Let $\mathscr{C}, \mathscr{E}$ be two categories, the set of contra-variant functors from $\mathscr{C}$ to $\mathscr{E}$ forms the set of objects of a category whose arrows are the natural transformations. This is named the *topos* of $\mathscr{E}$-valued pre-sheaves on $\mathscr{C}$; we denote it by $\mathscr{T}(\mathscr{C};\mathscr{E})$. The category $\mathscr{C}$ embeds naturally in this topos, if we associate to a the functor $b \mapsto \mathscr{C}(b,a)$. (In fact the complete notion of topos asks for a Grothendieck topology on a category, and considers *sheaves* (cf. Grothendieck and Verdier, SGA4, [79]); here we considered only the so called discrete topology.)

To any set X we can associate the category $\mathscr{M}(X)$ which has only one object, X itself, and whose morphisms are the maps from X to X; to any vector space V we can associate the sub-category $\mathscr{L}(V)$ with linear mappings as morphisms.

A *left action* of a category $\mathscr{C}$ on a set X is a functor F from $\mathscr{C}$ to the category $\mathscr{M}(X)$; a *linear representation* of $\mathscr{C}$ on a vector space V is a functor F from $\mathscr{C}$ to the category $\mathscr{L}(V)$. A right action is a left action of $\mathscr{C}^{op}$. When nothing is specified an action means a left action.

A *group* G is the set of arrows of a category $\mathscr{C}$ with one and only one object o, such that, for every arrow f there is an arrow g with $f \circ g = g \circ f = e_o$. In this case g is unique (because if g' satisfies the same equations, we have $g' = g' \circ (f \circ g) = (g' \circ f) \circ g = g$), it is called the *inverse* of f and it is written $g = f^{-1}$. It is traditional to identify G with $\mathscr{C}(o,o)$, to forget o and to write $e_o = e$. In general, the composition $g \circ f$ is written as a product gf.

By definition a morphism of a group G in a group G' is a functor from G to G'. It is a map φ from G to G' such that $\varphi(e_G) = e_{G'}$ and such that for any pair g,h of elements of G we have $\varphi(gg') = \varphi(g)\varphi(g')$.

Let H be a subgroup of G, two elements g, g' in G are said to be equivalent to the right modulo H if it exist an element h in H such that $g' = gh$. The set of equivalence classes is denoted by G/H. The group G acts on this set by left multiplication. According to Klein this constitutes the model of a geometry.

A Geometry on a set X is characterized by a group G of transformations of X which acts transitively on points (i.e. for any pair x,x' of elements of X there exists an element g of G such that $x' = g.x$), and a space for this geometry is characterized by a sub-group of this group (Klein, Lie, Cartan), thus we are following the invitation of Llinas and Pellionisz ([166] [167]) to visit the brain as a *geometric machine*. Let us stress that most of the involved geometries are not directly represented in the external world, they act inside the brain on internal spaces, which are in general dynamically distributed over many interacting brain areas.

Let us now set forth a scheme for such internal spaces.

7.2.4 Adaptation and Co-homology

Cf. the appendix 1 for mathematical sources of the definitions below.

Individual neurons, denoted a, have functions $F_a(\varphi;t)$ that transform incoming fields φ into numerical time functions, like electric potential or spiking time sequence or firing rate. In many cases it is possible to give precisions on the manner F_a depends on φ. Inspired by the theory of singularities (Whitney, Milnor, Thom, Arnold, cf. Appendix 1), we assume the existence of three sets M_a, I_a, and E_a, the first one has a structure of smooth manifold, it describes the parameters of dynamics of responses of the cell (and its preferences), the second one is discrete in nature and describes the categorization made by the cell (and circuits around it), in particular in relation with the external world, and the third one describes the internal context (that we consider as a sort of boundary condition). We assume the existence of a function σ which associates to any input field φ and context ε in E_a, an element $\sigma(\varphi,\varepsilon)$ of I_a, called a *vanishing cycle*. This can be seen as a virtual state of the neuron, a virtual attractor of the internal dynamics of the neuron, or a combination of them, affected by signs plus or minus, as could be given by reentries. We also assume the existence of a map $\widetilde{P}$ (named the *period map*) which associates to any μ in M_a a numerical function on I_a, that we name a *co-cycle* (and I_a^* denotes their space). Our first hypothesis on the (biological) functions of the neurons is that F_a factorizes through σ and $\widetilde{P}$, i.e. the equation for F_a as a function on $M_a \times E_a$ is given by

$$F_a(\varphi,\varepsilon,\mu;t) = \widetilde{P}(\mu_t)(\sigma(\varphi,\varepsilon_t)) \tag{7.4}$$

Commentary: each point in M_a represents a possible rapid dynamic of the cell; a point in I_a represents a combination of attractors of the dynamics, a skeleton of the dynamics which can represent memory or/and prediction and thus sustains ambiguity; a point in E_a plays the role of an initial condition for the dynamics. The map σ represents the manner the cell and the circuit around it, in particular feedback loops and reentries, integrate the signal φ with the boundary condition, or context, ε. We can imagine that rapid dynamics happens in a fixed set X_a and that I_a describe combinations of stable attractors in X_a. For μ fixed, the function $\widetilde{P}_\mu$ generalizes the notion of a probability law on the effective attractors. Cf. figure (7.1).

We consider M_a as the unfolding of the dynamics of the cell, in the sense of René Thom [212]. Examples of characters of neurons underlying all the processes in the brain were described by Rodolfo Llinas (cf. [125], [126]). Examples of unfolding for neuronal function were given by Izhikevitch [97]. The categorization process, underlying I_a in our model, is also a subject of study, for instance by Jean-Pierre Nadal et al. ([31]). At the level of evolution M,I,X,E are changed, for instance by coupling several systems. We can suppose that M, when fixed, is universal, in the sense of universal unfolding, which implies structural stability of the system, cf. [212].

For describing the role of invariance (or co-variance) in *cell adaptation*, we assume the existence of a group $\widetilde{G}_a$ which acts on M_a and I_a in such a way that, for any $\mu \in M_a$, $\sigma \in I_a$, $g \in \widetilde{G}_a$, we have

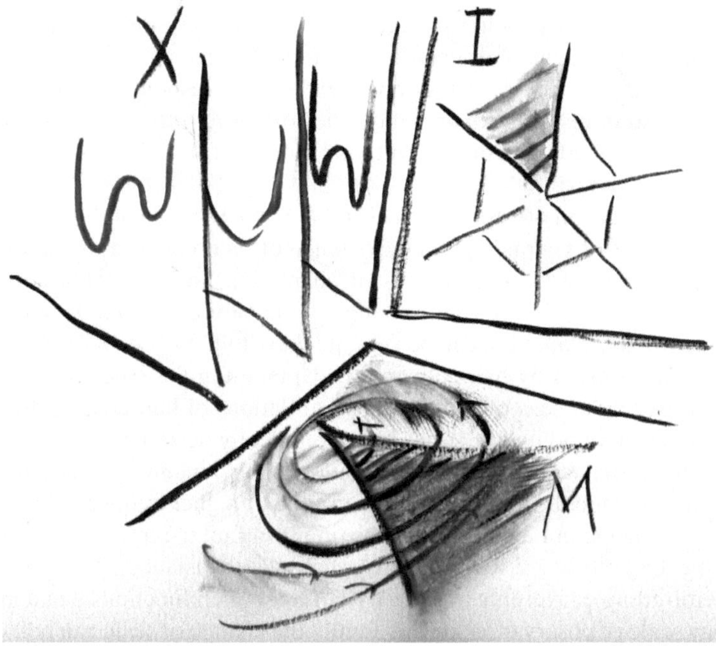

Fig. 7.1 The three dynamical spaces, the internal space X (with its variety of rapid dynamics), the virtual space I (ideal homology, with rigid structures) and the unfolding space M, equipped with a dynamics of adaptation.

$$\widetilde{P}(g.\mu)(\sigma) = \widetilde{P}(\mu)(g.\sigma). \tag{7.5}$$

This relation expresses the possibility to compensate a change of the state σ by a change of the parameter μ. It is more a co-variance than an invariance. In such a way a flow in M changing the parameters of the neurons can compensate a change in the input, the context in E or their interpretations by co-cycles in I.

It is the main suggestion of the present exposition, that it can exist on the unfolding M a dynamic induced by the dynamic in the internal space X of the cell, when expressed by the virtual state in I. This dynamic on M subtends adaptation.

In most interesting examples, the action of $\widetilde{G}_a$ on I_a corresponds to a set of transformations of the input φ and the context ε. But it can happen that some transformations exist only at the level of the cycles σ in I_a. This is the main trend of generation of internal symmetries: new symmetries are generated by ambiguities. We will see examples of both situations.

The action of $\widetilde{G}_a$ on M_a may describe a structural homeostasis, but in general it describes a structural variation, an evolution of the internal system. Moreover, in reality it can happen that a change in context ε or in stimulus φ is not compensated. In general the point μ does'nt change co-variantly, the equation (7.5) expresses a

virtuality, not a necessity. For instance, in $V1$ of cats only the cells at pinwheels follow the ambient change of orientation, not the other cells, cf. [200].
Importantly, a change in M_a can happen without corresponding changes in the context or the stimulus, this change plays the role of an *internal simulation*. This could be the basis of construction of "internal models" (cf. [27]).

Commentary: the origin of symmetries $\widetilde{G}_a$ on I_a is the inherent ambiguity of virtual attractors, they are of Galois type; the existence of its compensatory action on M_a expresses particular cases of dynamics of adaptation. In general these dynamics are slow dynamics that describes the change in the nature of rapid dynamics, rapid reactions. However, adaptation can be very quick, within milliseconds. A wonderful example of that is given by hair cells rapid adaptation in the inner ear ([59], [90], [91]). In the particular case of spontaneous oscillations of hair cells in the bullfrog sacculus the adaptation flow on M has been completely described ([122]).

A guide for understanding this compensation is the dynamics of renormalization in statistical mechanics: in this case the integration of higher frequency band is compensated by a change in Lagrangian for leaving invariant the truncated correlations, i.e. for respecting physics at low frequency scale (large distance). This process generates a semi-group converging to a family of Lagrangian functions that depends on a frequency scale of observation; on this family the change of scale can be expressed by a change in the strengths of interaction. Cf. [96], [235], [236].
There is a parallel between the invariance equation of a cell and the co-variance equation of movement production suggested in [25]. In this paper it was described how movements μ adapt their duration to the action of geometrical transformations g of the space on trajectories σ:

$$(g.\mu)(\sigma) = \mu(g.\sigma), \tag{7.6}$$

meaning that the movement in time $\mu(g.\sigma)$ on the transformed trajectory $g.\sigma$ is given by applying the transformation g to the movement μ on the initial trajectory σ. The groups considered in [25] were the group of affine transformations of the plane, the equi-affine group made by transformations preserving the area, and the Euclidean group. In [24] the Galilean group is considered. Thus invariance in motion production has the same form as invariance in perception; which is not surprising in the above mentioned view that perception is nothing else that overturned action ([27], [126]).

Let A be a brain area. Several cells $a \in A$ form a vector or responses, where groups can act linearly or not. On the union M of all M_a, the union I all I_a and the union E of all E_a, there is a structure of *bundle* over A. A section μ of the bundle $M \to A$ represents a collection of receptive fields in the area A, a section σ of $I \to A$ represent a collection of inputs and a section ε of $E \to A$ a state of contexts as seen by the a in A. However this point of view of bundles and sections is largely misleading, because it neglects the possible interactions in A, the horizontal connections that give structure to an area. What replaces M_a when we go to the area A is a much larger space M_A than the space of sections, in order to take into account the parameters of possible interactions of the dynamics between the cells a in A. For instance, consider two cells a and b; the dynamical systems on the Cartesian product $X_a \times X_b$ of their internal spaces cannot be reduced to separate dynamics on X_a and X_b, we must consider in addition a variety of couplings. We name the space M_A the parametric space of A. In the case of a finite number of neurons (sic) and finite dimensional spaces M_a the dimension of M_A approaches the product of the dimensions of the M_a, not the sum, as would be the case for the product of the spaces M_a.

However, instead of I_a we consider the product $\mathscr{I}_A$ of all I_a for $a \in A$ and the same for $\mathscr{E}_A$, the product of all E_a for $a \in A$; they are respectively the spaces of sections of the bundles $I \to A$ and $E \to A$. They will play the roles of the spaces of virtual attractors and contexts for the area A; we name them the vanishing homology and the boundary condition respectively.

Moreover we introduce here a vector space V_A to take in account the dimensions of the responses from A (they are the analog of *primitive forms* in Appendix 1). And we assume the existence of a *joint period mapping* which associates to any element μ of M_A and any vector v of V_A a numerical function on the product $\mathscr{I}_A$ of the I_a. Then, using the maps σ_a that associate to each input φ, each cell a and each context ε_a, an element $\sigma_a(\varphi, \varepsilon_a)$ in I_a, we can form the element $\sigma(\varphi, \varepsilon)$ of $\mathscr{I}_A$, and we get the *amplitude of responses*

$$F_v(\varphi, \varepsilon, \mu; t) = \widetilde{P}_v(\mu_t)(\sigma(\varphi, \varepsilon_t)) \tag{7.7}$$

If we assume linearity in σ and v, the applications $\widetilde{P}_v$ define a mapping $\widetilde{P}$ from M_A to $\mathscr{I}_A^* \otimes V_A^*$ (that is the space of co-cycles with multiplicity).

The numerical functions $a \mapsto \widetilde{P}_v((\mu)(a))(\sigma(a)); v \in V_A$, represent the collection of informative outputs of the area A.

If a stimulus φ and a context ε are given, we assume that a well defined cycle σ results, and that we have the response:

$$\rho_v(\mu, \varphi, \varepsilon) = \widetilde{P}_v(\mu)(\sigma(\varphi, \varepsilon)). \tag{7.8}$$

The invariance in the area A is expressed by a group $\widetilde{G}_A$ that acts on the spaces M_A and $\mathscr{I}_A$ and now V_A in such a way that, for any $\mu \in M_A$, $\sigma \in \mathscr{I}_A$, $g \in \widetilde{G}$, $v \in V_A$ we have

$$\widetilde{P}_{g.v}((g.\mu))(\sigma) = \widetilde{P}_v(\mu)(g.\sigma). \tag{7.9}$$

This relation expresses the compensation of a change of the global state σ by a change of the global parameter μ and the vector v, which can correspond to an *adaptation* (in wide sense) at the level of the area A. In particular a flow in M_A, changing the parameters of the joint activity of the neurons a in A, may compensate a change in the joint categorized input.

Take the example of orientation preferences of principal cells in $V1$: this preference is described by an angular coordinate in M_a for each a. After exposition to an environment where orientations are biased with a certain angle α, some cells change their preferred orientation by a turn of $-\alpha$, this modification is understandable from the action of $\widetilde{G}_A$. According to Dragoi, Schummers, Sur et al. ([56], [201], [200]) this adaptation concerns the neurons near the pinwheels singularities, thus we can consider that a separate subarea corresponds to pinwheel regions, with an invariance structure different from the regular complementary area. However horizontal projections inform the regular area of the changes in pinwheels integration.

If we rotate a given image by a rotation of angle α in the visual plane, there is a shift in activity of the cells, without changing the preferences of these cells, and this shift may produce no effect in certain higher areas; this can be interpreted as the invariance of object after rotation in those areas. This happens for instance in IT (cf. Tanaka et al. [209], [118]).

Remark that an enlargement of the above notions is certainly necessary to describe general covariance under diffeomorphisms; in this case we must replace groups by categories with invertible arrows, i.e. *groupoids*.

We do not pretend that the above model can describe all kinds of adaptation. For instance, at the level of evolution, adaptations invent new structures, or uses old structures for new functions, thus, in our terminology, it appears new spaces M_A, I_A, E_A and groups $\widetilde{G}_A$, and even new areas A (and new animal species). However, every adaptation seems to reflect an internalization of changing rules of interactions between the organism and the external world, motor and sensory, then it could be that an extended model of compensation can describe every adaptation: we should replace the spaces M_A, I_A, E_A by categories of spaces, parametric and virtual, equipped with a functor of period maps, and we should replace the groups $\widetilde{G}_A$ by sets of functors verifying equations of the type of (7.9), prescribing structural variations from reaction dynamics.

7.3 The Affine Space of Color

(The two first subsections below are taken from talks I gave in Collège-de-France, Paris, in 2005, as examples of co-homology operations in perception. The last one in section 4, was inspired by the thesis of David Philipona, January 14 in 2008.)

A full chapter in TVN is dedicated to color; for our discussion, cf. in particular the articles of Webster, Jacobs, Brainard, and De Valois.

7.3.1 The Subjective Manifolds of Color

Cones in the retina react more strongly for certain wavelengths of incoming light. In many mammalian species several kinds of cones exist, allowing the detection of several aspects of the distribution of wavelength. For each category of cones c_i, the detection is fairly well described by a scalar kernel $K_i(\lambda)$, where $\lambda \in \mathbb{R}_+^*$ denotes the wave-length: when a composed light of density $E(\lambda)d\lambda$ (called illuminance), illuminates a surface of reflectance $R(\lambda)$ the model of the response of a cone c_i is given by

$$R_i = \int E(\lambda)R(\lambda)K_i(\lambda)d\lambda. \tag{7.10}$$

This means that its activity is a function of time proportional to R_i.

Note that the response of a cone can only take in account the photons that it absorbs, thus everywhere in the following discussion we must have in mind that the analysis is local in the retina and the efferent areas. In the retina the spatial receptive field of cells that sense color is small (for instance in primate the cones are concentrated in the fovea), but in $V4$ it is large (several tens of degrees).

Remark, dichromatic models take in account the position of the eye with respect to the surface and the angle of incidence of the light beam on the surface, cf. Koenderink's book [113].

In mammals, the number of kinds of cones varies between one and six (sometimes more, but rarely), see Jacobs (in TVN and [98]). In invertebrates (in particular cephalopods) this number can be larger than ten. Nocturnal new-world monkeys for instance have only one. Rodents, ferrets, cats, dogs and tree threw have two types of cones. In general humans and old-world primates have three sorts of cones, they are denoted S, M, L in accord with the wave-length they prefer, short, medium or long respectively. However in humans a non-negligible proportion have only two, and some other (in particular among females) have four sort of cones ([100]). Consequently, in what follows we consider animals having N sorts of cones, not necessarily 3. For the interesting case of Cichlid fishes see [194].

Axiom 1: there exists a minimal number $n \leq N$ such that the set of responses in LGN and V1 can be explained by n functions of the N cone responses.

The number n is called the *chromatic dimension*. The axiom 1 gives a relation of equivalence between light compositions, each equivalence class is called a *color*. No special set of n linear functions is preferable, as soon as they are linearly independent. In particular, the color space is a smooth contractible manifold C of dimension n.

We speak of dichromatic animals when $n = 2$, this is the case for dogs and cats for instance; we speak of trichromatic animals when $n = 3$, this is the case for most old-world primates (or honeybees); and $n = 4$ is tetrachromacy. Tetrachromacy is known to exist for certain primates, in particular for a certain proportion of humans females ([100]).

Let us describe the principle of the experiment which justifies the axiom 1; this principle is attributed to the mathematician Grassmann (1853), which also invented most of the elements of linear algebra (see [116]): A set of m compositions of light $F_i(\lambda); i = 1, ..., m$ is chosen as reference, where m is *a priori* any natural number. Now, to the same subject is presented a light whose composition is a fixed $B(\lambda)$, and at the same time, another light whose composition $A(\lambda)$ is linear in the F_i, i.e.:

$$A(\lambda) = \sum_i a_i F_i(\lambda). \tag{7.11}$$

For instance one light composition is presented at the left eye and the other one at the right eye. Then the coefficients $a_i, i = 1, ..., m$ are varied, and the subject is required to choose the set which gives the better fit between the effects of A and B.

It is not true that positive numbers a_i can always be found for a good fit. First m can be too small, then we augment m until it works. But, in addition, to obtain a good fit, experiment shows that it is necessary to introduce negative combination, i.e. it is necessary to test similarities like

$$B(\lambda)+\sum_i b_i F_i(\lambda) \approx A(\lambda), \tag{7.12}$$

with b_i *a priori* larger than a_i for some index i. Now the first experimental fact is that a good fit always exists. And the second experimental fact is that, when m grows after a certain number n (dependent on the subject and *a priori* on the $F_i; i \in I$), always smaller than N, no gain is obtained and ambiguities appear.

A remarkable fact is that n can be strictly less than N. For instance Jordan et al. 2010 ([100]) have proved than for the women which are known to have at least four sorts of cones, almost all are trichromatic. But they also have proved that tetrachromatic women exist. Very few animal species in the world are proved to have $n \geq 4$: some shallow-water fishes, some birds (pigeons, budgerigars, ...), and insects (butterflies). However, even when $n < N$, the larger is N for a given subject, the better is its discrimination between different colors, at least in some region of C. This was established for monkeys ([192]). On the other side, it is known that subjects having less than the common N sorts of cones have deficits in color discrimination. For instance a non-negligible proportion of humans cannot distinguish red from green, due to the absence of separation between L and M cones. Note this red-green deficit is also a notable characteristic of dog's vision.

In colorimetry, specialists prefer to describe colors by positive coordinates, thus they have chosen (*CIE* 1931), a set of linear combinations E_X, E_Y, E_Z of cone responses (or fundamental colors, like B, G, R for blue, green and red), such that all detectable color have coordinates X, Y, Z which are positives. This is a non-trivial fact, but such a combination exists.

There is a beautiful curved and twisted curve H in C, made of the hues, which represents the set of equivalence classes of pure wave-length spectra $\delta(\lambda = \lambda_0)$, cf. figure (7.2). Another curve G, which looks like an axis, represents light composition without color, i.e. pure luminosity, made only with grays. In the case of dichromacy, the curve H is planar and meets G (cf. Jacobs, study of dog vision.) In the case of trichromacy, the curve H has a convex shape, surrounding G. The closure $\widehat{H}$ of H by a straight segment in C appears in $V1$, not before, and realizes the colors circle of Goethe. The segments issued from the particular white point of G and ending in $\widehat{H}$ are interpreted as saturation of hues, or washed colors. In tetrachromacy or more, the curves H and G (and probably $\widehat{H}$) continue to exist in the hyperspace C, but saturation is multi-dimensional.

The way the space C is defined by similarities (7.12) among all light combinations is typical of homological construction in Mathematics (cf. [61]).

The axiom 1 describes a subjective property, or the final effect in behavior of a neural process which starts in the retina; we will see now one of the basis of this process, at work in the thalamus.

Fig. 7.2 The spectral curve in the color space is the locus generated by light of pure wavelength at fixed energy (cf. [113]). It is curved and twisted and goes through 0. A central projection of this curve gives the CIE triangle of colors; the line of purple joins the two tangent lines that correspond to invisible infrared and ultraviolet respectively.

7.3.2 An Example of Covariance in the LGN

Ganglion cells send color information in the LGN by forming linear combinations through excitation and inhibition of the different cones responses.

In case of three kinds of cones, the standard combinations are *brightness* (or luminosity) $S+L+M$, the generator of red-green axis, $L-M$ and the generator of blue-yellow axis: $S-L-M$. We see the use of inhibition for introducing negative coefficients, a proper brain's algebra.

A more exact basis is given by a more complex $3\times n$ matrix of real numbers. Many different combinations exist for individual ganglion cells, and induce a variety of responses in LGN.

In case of more or less than three cones, it seems that ganglion cells also form n linear combinations of cones. We will consider generic choices of coordinates $(S_1,\dots,S_n)$ on C:

$$S_k(ER)=\sum_{i=1}^{n}S_k^iR_i(ER). \tag{7.13}$$

Here E and R are varying illuminance and reflectance respectively, and S^i_k are fixed real numbers.

Axiom 2: In LGN there exists a subset of relay cells $a \in A$, named *affine*, and there exists on C a canonical structure of affine space, such that for any choice of coordinates S_k and any affine cell $a \in A$, the response to incoming light ER is given by

$$A_a(ER) = B_a + \sum_{k=1}^{k=n} A^k_a S_k(ER). \tag{7.14}$$

Remark this axiom does'nt need axiom 1 to be formulated, thus it gives a stronger statement, probably underlying axiom 1, restricted to LGN. However apparently the affine structure has a strict sense only in LGN and $V1$ or $V2$, because it seems not fully valid for perception. Only the dimension remains for perception, with a more complex geometric structure on the space C, not flat. See [113].
For the affine structure on colors, and more, see [199].

Remark: the coefficients B_a, A^k_a can be modulated by the context and the corticothalamic feed-back. Note that other cells exist in LGN, in particular there exists cells which react non-linearly.

This axiom 2 is sustained by wonderful experimental results from the neurophysiology on monkeys (De Valois, Wiesel and Hubbel 1966, Derrington 1984, Lennie, Krauskopf, ...) (cf. in particular [53], [124]): inside the infinite dimensional vector space of local numerical continuous functions of (E,R), the set of possible response functions of the element a in LGN forms, in first approximation, a linear subspace of dimension $n+1$.

Definitions: An incident light E generate a marginal distribution $\rho(c)dc$ of colors, having a *mean* $m(\rho)$, which is a point in C, and a *covariance form* $Q(\rho)$, which is a quadratic form on the vector space associated to C (isomorphic to C but centered in $m(\rho)$). In particular the decomposition of *sun light* (or other natural light) introduces a mean point, named 0, corresponding to the *white*, and an Euclidean metric Q_0, characterizing the decomposition of this light at the order two, which we name the *spectrum* of sun light. Choosing this point 0 and this metric Q_0 as references, the space C becomes an *Euclidean vector space* that we denote by E_0. Using this space, which depends on a specific light, we can associate to every point $m(\rho)$ a vector Om and to every quadratic form Q a co-variance operator Σ which is a symmetric endomorphism of E_0. These data became a numerical vector and a symmetric matrix respectively, when we choose in addition special linear coordinates S_k on C.

Note that in general experimenters prefer to introduce coordinates on C, giving to C a linear structure and a basic quadratic metric. This is also a manner to define a structure of Euclidean vector space on C. For example Von Kries suggested to work in the coordinates given by the cone themselves, which gives three preferred axis, and distinguishes the co-variance matrices that are diagonal. However, the only intrinsic structure of C is its affine structure, not the basis. The structure of the adaptation of affine color cells in *LGN* is precisely based on this fact:

Axiom 3: When, around the receptive field of the cell a, and during a certain interval of time, the ambient color distribution $\rho(\lambda)$ departs from the one of the sun light, we get a mean vector m and a covariance Σ for ρ, then the responses of the cells a are changed by an affine transformation τ of C, i.e. R_a becomes $R_a \circ \tau^{-1}$, to compensate for the change of sun light in ρ.

More precisely, there exists a linear transformation R of E_0 such that Σ is equal to $R^*.R$, and the cell a which had the response $R_a(c)$ (an affine function on C) is transformed into the response $\tau_*(R_a)(c) = R_a(R^{-1}(c-m))$ (another affine function on C). This has the effect to reproduce the world as if it were illuminated by the sun light, at least at the second order of the moments. Then the compensation equation is:

$$R_a(c) = \tau_* R_a(m + Rc). \tag{7.15}$$

Experimental justification of this axiom is given by Stiles 1959 [208], see Webster in *TVN*. Cf. figure (7.3).

For τ negative coefficients are necessary. A remarkable fact, that we suggest to be a common denominator of all co-homological constructions in the brain (in the sense of 2.4), is that *excitation and inhibition* are used carefully to produce the structure (here affine) and its transformations (here affinities) together.

Observe that a transformation of the moments of ρ is not sufficient to define a unique element τ. Many other elements do the job, they differ by the rotations preserving Q_0 if we compose to the left or by the rotations preserving $\underline{Q}$ if we compose to the right. A natural choice consists in taking for R the root $\sqrt{\Sigma}$, i.e. the unique positive symmetric operator R such that $R_\Sigma^2 = \Sigma$, but we can also compose this R by any rotation S i.e. take $S\sqrt{\Sigma}$. Thus the ambiguity after adaptation is an orthogonal group. There is a possibility to use higher order statistical invariants of ρ to better determine the adaptation, but I know of no experimental basis for that.

We see that color in *LGN* gives a perfect example of the general structure we have introduced for describing invariance: In this case the parametric space M describing the kind of response of affine color cells is the space of affine forms for each cell a, and the affine group G_C acts on M by affine change of variables. We have a special point in M, which is 0, corresponding to the mean information on the world under sun light. The vanishing co-cycles are the colors themselves, they form the space $C = I^*$, and the Galois group G_C acts in the obvious way.

Experiments show the effect of a continuous path of transformation, opposed to sudden changes of color context. Then $\widetilde{P}$ appears.

Remark that most affine cells behave under natural illumination as linear cells, and undergo a translation after adaptation, plus a linear deformation. However several cells (more found in $V1$ but also found in *LGN*) are better explained by absolute values of affine functions on C ([53], [124]), which indicates the pertinence of the n dimensional projective space associated to M.

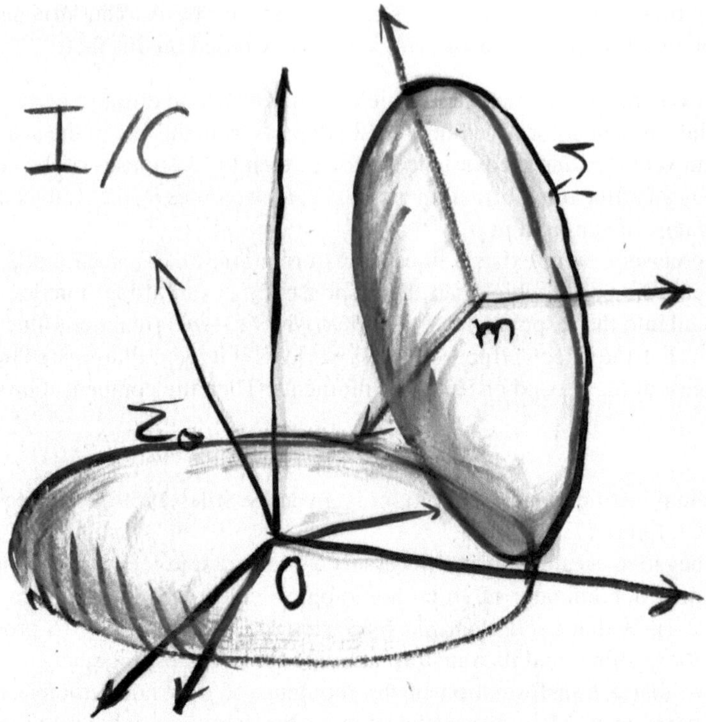

Fig. 7.3 Two ellipsoids for mean and variance of color distribution in the color space I or C; adaptation can deform one into the other, cf. Webster in TVN.

7.3.3 Color in $V1$ and $V2$

The color representation in $V1$ is based on the same principles as in LGN, but in $V1$ there is less concentration of cells along special axis (Lennie et al, De Valois et al., cf. [44]). Remarkably the *purple* appears in $V1$, where it is even over-represented for macaques. Cf. [225].

7.3.4 Color Constancy

More advanced areas introduce curvature in the color space, taking in account perception, for instance personal and cultural dimensions. The Riemannian structure on the color space is studied in [199]. In $V4$ context has large influence, cf. Zeki in TVN. The responses become dependent of all other colors. In $V8$ a larger spaces for color is introduced, which represents the distributions of illuminance and reflectance separately.

David Philipona and Kevin O'Regan ([172], [99], [170]) reported a fundamental experimental result for the coding of the relation between illuminance and surfaces reflectance:

Axiom 4: For most homogeneous surface S, there exists a linear operator A^S of C_0 (C pointed in 0), such that for most illuminant $E(\lambda)$, the following relation holds with very good accuracy

$$[SE] = A^S[E], \tag{7.16}$$

where $[E]$ and $[SE]$ denote the equivalence classes in C of the spectrum of a light $E(\lambda)$, and of the same spectrum, once reflected by the surface of reflectance $S(\lambda)$, i.e. $S(\lambda)E(\lambda)$.

The variance accounted by this linear law is larger than 90% for both Munsell chips and natural surfaces ([172]).

Philipona and O'Regan gave a clear justification for this law, following the observations of Judd et al. 1964 [101], and Dixon 1978 [55], that natural illuminants are well explained linearly by three dimensions. Thus, if U denotes the operator sending a basis E_1, E_2, E_3 of this illuminant space to its representation in C_0, and V^S denotes the matrix sending SE_1, S_2, SE_3 in C_0, the operator can be take as

$$A^S = V^S U^{-1}. \tag{7.17}$$

In fact Judd et al. [101] found a basis of spectra made by three or four vectors. Let us observe that even if the illuminants where not well explained by three dimensions, it would be possible to obtain a formula for A^S by taking a pseudo-inverse, as soon as we could assume a linear explanation for E. That is, we consider a generator system $E_1, ..., E_M; M \geq 3$, the operator U_M sending it to C_0 and the operator V_M^S sending $SE_1, ..., SE_M; M \geq 3$ to C_0, and we choose a pseudo-inverse U_M' of U_M (i.e. an operator from C_0 to $\mathbb{R}^M$, such that $U_M' U_M U_M' = U_M'$ and $U_M U_M' U_M = U_M$) and we put

$$A^S = V_M^S U_M'. \tag{7.18}$$

Once normalized, the operator U_M' can be interpreted as an *a priori* probability for each illuminant when the color is given.

Remark that with respect to reflectance, objects appear as transformations in the space C. This invites us to suggest that more generally, *objects*, in the world for a brain, are operators in (co)homology. This suggestion connects a circle of ideas of David Philipona that we discussed during the preparation of his thesis.

Now we will use both the adaptation operator τ and the illumination transforms A^S to propose a formula for constancy.

A surface S is given. Let $E^0(\lambda)$ be the composition of an ambient light, this gives a reflected light $S(\lambda)E^0(\lambda)$ defining a color $[SE^0]$. A new illuminant $E^1(\lambda)$ produces in an analog manner a color $[SE^1]$. We also introduce the pure homology classes $[E^0]$ and $[E^1]$. As we saw before in subsection 3.2, the adaptation from *LGN* to higher areas is realized by a linear operator R_1^0 from C_0 to C_0 such that

$$[E^0] = R_1^0[E^1]. \tag{7.19}$$

Consequently

$$[SE^0] = A^S[E^0] = A^S R_1^0 [E^1] = A^S R_1^0 (A^S)^{-1} [SE^1]. \tag{7.20}$$

This gives the following result:

Theorem 3.4.1: The operator $A^S R_1^0 (A^S)^{-1}$ obtained by conjugating the adaptation by the A transform of reflectance compensates the change of color induced by a change of incident light.

A discussion of different matrix models underlying color constancy, along different lines of thought, was made by Maloney-Wandell [135].

Our model is compatible with the result of Yang and Shevell [246], that the presence of two spatially separated illuminants in the scene has the effect of reducing the color constancy. This is due to the fact that in the case of several illuminants, the adaptation R_1^0 cannot be formed.

Considering color distribution in complex natural images, the above formulas can explain only little of the activity of cells and even less of the perception; for that purpose non-linearities and higher correlations in the image have to be involved. These aspects were the main subject of the thesis of P.Baudot (2006) [19], which was the starting point of the research on information geometry and topology he has developed with me, see section 6 below. However, the above linear computations cannot be rejected without care, their existence in simplified situations is not here by hazard. The same is true for the properties of primary visual receptive fields that we will expose now; even if they explain a few percent of the responses of most visual neurons, the fact they exist in simplified contexts is a main pillar for any future theory; see also the section 5.4 on higher invariance.

From the point of view of mathematical education, it is admitted that linear algebra is a good way for beginning. A second step should be differential calculus, as we will encounter in the following section.

7.4 Covariance in $V1$

7.4.1 Simple Cells and Complex Cells

(This chapter is issued from works that grew around the thesis of Simon Capern, Paris VI, September 29 in 2008, in particular it reflects discussions with Simon Capern and Jacques Droulez during ACI NIM (2005-2009).)

Consider a neuron a; its receptive field is better understood as an operator, which transforms a stimulus in a response, respecting causality. The incoming stimulus is represented by a function $\varphi(s,\omega,x)$ where $s \in]-\infty,t]$ represents the time before t, $\omega \in \Omega$ represents chance and $x \in X$ represents variables in the external world; the signal sent to other cells by the neuron a is a function $\psi_a(t,\omega,y)$ of time, chance and other parameters $y \in Y$ describing internal parameters of the brain. This gives a transformation of functions into functions; in general this transformation is non-linear. The usual way to analyze such a transformation is a Wiener-Volterra expansion:

$$\psi_a(t,y,\omega;\varphi) = f_a(\omega, \sum_{n=0}^{\infty} \psi_{a,n}(t,\omega,y;\varphi)), \tag{7.21}$$

where f_a is a sigmoid function in the second variable and where, for each n,

$$\psi_{a,n}(t,\omega,y;\varphi) = \int_{X^n} dx_1 ... dx_n \int_{\mathbb{R}^n} ds_1 ... ds_n K_{a,n}(\omega,y,x_1,...,x_n,s_1,...,s_n) \varphi(t-s_1,\omega,x_1)...\varphi(t-s_n,\omega,x_n) \tag{7.22}$$

for a certain function $K_{a,n}$, named the kernel of order n of the transformation. Cf. [1], DeAngelis and Anzai in [44].

In the simplest model of simple cells in the primary cortical visual system, the function $\varphi_a(s,x)$ represents the normalized intensity of the image at time s and x denotes a point in the visual plane, then the function $\psi_a(t,y;\varphi)$ is the probability of spiking at time t, and the variable y represents the internal context modulating the response, for instance the eyes vergence. In the kernels appear the preferred parameters of the cell, for example a preferred center x_a, a certain disparity vector δ_a depending on the eyes vergence, plus a $2D$ spatial frequency vector ξ_a and a time frequency ω_a. The kernel of order 1 is given by a Gabor wavelet

$$K_1(\omega,y,x,s) = G_a(x-x_a,s)e^{-i(x\xi_a+s\omega_a+\varphi_a)}, \tag{7.23}$$

where G_a localizes $x-x_a$ around 0 and depends on time, giving in particular a delay τ_a where it is maximum and another delay τ'_a where inhibition occurs. In general this distribution is elongated in the direction of frequency vector.

The fact that the Fourier transform of a Gaussian is a Gaussian with inverse variance implies that a strong localization of activity on x_a or t is incompatible with a strong precision on ξ_a or ω_a; thus some cells give good information on the local intensity of the image, while other cells give more information on its frequency structure. Layers *IV* and *V* in $V1$ seem to belong to the first type, layers *III* and *II* to the second type.

The direction selectivity, i.e. the fact that most cells in $V1$ for cats or monkeys a.s.o. are selective for the direction of the stimulus motion, is well explained by the spatio-temporal structure of the RF, named *inseparability*, i.e. the fact that the vector (ξ,η,ω) and the phase φ form interferences which are not parallel to the time axis, joined to the fact that the time axis is not a principal axis of the RF. Prediction of this model were given for instance by Adelson and Bergen [3], and confirmation is exposed for instance by DeAngelis and Anzai in *TVN* ([44]). It was also confirmed that the model of linearity followed by non-linearity (i.e. "static non-linearity") explain a non-negligible part of data, see Albrecht, Geisler and Crane in [44].

This receptive field expresses the co-variance for translation in space and time: For simple cells of $V1$, we can start with the group $V_{2,1} = \mathbb{R}^2 \times \mathbb{R}$ of $2D+1$ translations; it is an abelian group and the frequency co-vectors $(\xi_0, \eta_0, \omega_0)$ parametrize the irreducible *unitary* representations of this group. Cf. for instance [222]. Unitarity comes from the reality of the frequency vector. These representations are all of dimension 1, they are given by complex numbers of modulus 1:

$$\chi_0(x,y,t) = e^{i(x.\xi_0+y.\eta_0+t.\omega_0)} \tag{7.24}$$

so, they are pure changes of phases.

Let us choose a smooth function K_0 on $V_{2,1}$. By definition an *image* is a distribution I on the space $V_{2,1}$, and a linear simple response of a cell, associated to the kernel K_0, is given by smoothing the real part of an imaginary virtual response represented by

$$E_a(t) = \int e^{-i(x.\xi_0+y.\eta_0+s.\omega_0)} K_0(x-x_0, y-y_0, s-t_0) I(x,y,t-s)\,dxdyds,$$

or equivalently, by using the standard notations,

$$E_a(t) = \langle \overline{\chi_0} \tau_{(x_0,y_0,t_0)} K_0, I_t \rangle = \mathscr{F}(\tau_{(x_0,y_0,t_0)} K_0.I_t)(\xi_0, \eta_0, \omega_0). \tag{7.25}$$

It represents the element of response at the order one before non-linear rectification. This is the starting point for all what follows: $E_a(t)$ is given by the Fourier transform of $I.K_0$ at the point $(\xi_0, \eta_0, \omega_0)$.

Moreover K_0 localizes images I around the center (x_0, y_0, t_0).

Thus the group of $2D+1$ translations acts on the simple receptive fields by changing the phase, and the visual center.

An important property for Infomax is the *orthogonality of characters* (cf. [222]):

By definition, a character χ on a group G is a morphism form G to the circle group $U(1)$ of complex units. In the case of a finite commutative group, the characters form an orthogonal basis of functions on the group; this is the precise statement of orthogonality. In the present case, because $V_{2,1}$ is an infinite group, then the characters are not square integrable, however something remains of this property of the characters, that is, when we compute the mean over space-time round domain of volume N of scalar products of different characters, we have

$$\frac{1}{N}\int \overline{\chi(x)}\chi'(x)dx \to 0,$$

as N tends to ∞.

Three principles underline the form of the receptive field: 1) Observation of space-time is local and causal; 2) Locality requires commutativity; 3) Elementarity requires irreducibility.

Observe that the same analysis can be applied to all terms of the Wiener-Volterra series, because $I \otimes I, I \otimes I \otimes I, \ldots$ are distributions on the commutative groups obtained by Cartesian products $V_{2,1}^2, V_{2,1}^3, \ldots$ respectively, and their irreducible representations are also given by one dimensional characters; they are parametrized by multi-covectors $(\xi_{\alpha_1}, \eta_{\alpha_1}, \omega_{\alpha_1}, ..., \xi_{\alpha_n}, \eta_{\alpha_n}, \omega_{\alpha_n})$.

Remark: The interpretation of the response formulas by harmonic analysis is not new, it has its origin in Gabor, see Petitot ([168]). Much of the work in this direction were inspired by image analysis, or signal analysis. Several groups have been used, the Heisenberg group ([102]), the group of plane displacements $E(2)$, through its regular representation (Sarti, Petitot, Bressloff, ... cf. [16], and [168]), or the hyperbolic group ([46]) for taking textures into account. The wavelet aspect is fundamental in these studies. It is known that wavelets have much to do with information theory, their efficiency for optimal coding, or maximal compression is well established (or well known); see for instance Atick with Redlich [14], or Dan, Reid, Li Zhaoping, ..., cf. [251]) who explained the mexican hat of Ganglion cells. We should also note that Olshausen and Field [161], [162], Nadal, Parga, Brunel [150] have studied the Infomax principle at natural frequencies. Cf. [1].

In fact a more efficient model of RFs, when compared to experimental data, is given by a finite sum over index $j \in J$ of static-non-linear function u_j (Naka-Rushton, with sign $+$ for excitation and $-$ for inhibition) of kernels centered on points x_j, τ_j, with spatio-temporal extension R_j (non-separable co-variance), evaluated on frequency ξ_j, ω_j with phase φ_j. Cf. [217].

$$R_a(t) = \sum_j \lambda_j u_j(Re[e^{-i\varphi_j} \mathscr{F}(G_j I_t)(\xi_0, \eta_0, \omega_0)]) \tag{7.26}$$

$$u_m(x) = C_0 \frac{x_+^m}{1 + C_\infty x_+^m}; \tag{7.27}$$

However things are not so simple, because experiment also shows that all the parameters in (7.26) (except perhaps x_j, φ_j) depend in a non-linear way of the whole image, through semi-local norms of I and its contrast C. Cf. [42], and Albrecht et al. in [44]:

Lowering the contrast expands spatially the RF and the latency, augments the saturation, and shifts the preferred time frequency to lower values. (Apparently there is no effect on the preferred spatial frequency, and no effect also on the structure of the cortical maps.) There is an advance in spatial phase when contrast augments. But orientation tuning or spatial frequency selectivity vary little with contrast, cf. [204], [8].

Remark that the measured spatial frequency seems higher than the one predicted by Gaussian models, and the same holds for direction selectivity , orientation selectivity and acuity; see Albrecht et al.in [44]. Remark that all these effects can be attributed to interactions, and could theoretically be included in the higher order kernels.

Remark: the shift to the right of time frequency with contrast augmentation can be explained in part by a dilatation of the time difference $t - \tau$: if contrast is multiplied by λ, the transformation of $t - \tau$ into $\lambda^\alpha (t - \tau)$ has the same effect on the response as the transformation of spatial frequency ω_i into $\lambda^{\alpha}\omega_i$ joined with a change of φ_i into $(\lambda^{\alpha})\varphi_i$. By applying this time dilatation to the windows G_i we also obtain the effect of scaling τ_i and the effect of scaling u_i by a negative power $\lambda^{-\gamma}$ according to Naka-Rushton formula.

The non-linearity u_j, the frequency ξ_j, ω_j and the phase φ_j could also depend non-linearly of time course. For instance there is a general tendency from the first ms to 200ms to shift the cells preference in the direction of higher spatial frequency. There is also a reported shift to higher selectivity with time course which can be explained by the greater expansion of spatial integration as time flows.

A dominant non-linearity comes from the curvature of the *RF*: starting from the linear (or static-non-linear) *RF* we can apply a diffeomorphism of space time (see Dali's watches, and experimental figures of DeAngelis et al., cf. [44]); this introduces shifts in frequency and phase and a change in Naka-Rushton exponents. Moreover this diffeomorphism depends on the contrast in the image over larger and larger domains when time flows. This can be described by the following formula at the order one:

$$R^r_a(t) = \sum_j \lambda_j u_j(Re[e^{-i\varphi_j}\mathscr{F}((G_j I_t)\circ \Phi_j)(\xi_0,\eta_0,\omega_0)]);$$

where Φ_j is a change of space-time variables.

All the above non-linearities describe a particular of adaptation. Thus, we can say that in this case: invariance (given here by translation, and symplectic geometry of light rays acting on the set of irreducible representations of the translations) imposes the form of RF's, and adaptation acts on this set of RF's.

Complex cells can be seen as composition of several simple cells, with comparable frequency preferences but a variety of phases. The same principle can explain in part the response of columns that are seen in optical imaging.

To describe the effect of phase distribution, we consider first the case of *uniform* distribution.

We introduce the hypothesis that among cells in the neighbor of a, we have cells a', a", ... whose subunits possess the same characteristics (G_j, u_j) of a, with the same $(\xi_0, \eta_0, \omega_0)$, but where all angles φ between 0 and 2π are represented with the same probability. References: Pollen and Ronner [183], DeAngelis et al. [49].

The cells form assemblies A with response

$$R_A(t) = \int_0^{2\pi} (u_+(\rho_+(t)\cos(\varphi-\varphi_+(t))) - u_-(\rho_-(t)\cos(\varphi-\varphi_-(t)))d\varphi \quad (7.28)$$

This integral is the difference between two integrals that can be simplified by changes of variables $\varphi \mapsto \varphi - \varphi_+(t)$ and $\varphi \mapsto \varphi - \varphi_-(t)$ respectively, giving

$$R_A(t) = \int_0^{2\pi} (u_+(\rho_+(t)\cos(\varphi))d\varphi - \int_0^{2\pi} u_-(\rho_-(t)\cos(\varphi))d\varphi \quad (7.29)$$

An essential role is played by the transformation $u \mapsto A(u)$, which send the bounded increasing continuous function $u : \mathbb{R} \to [0,\infty[$ to the continuous function $A(u) : [0,\infty[\to [0,\infty[$ defined by the formula:

$$A(u)(\rho) = \int_0^{2\pi} u(\rho \cos\varphi) d\varphi \tag{7.30}$$

This transformation A is named the *Abel transform*.

The integral transform which was originally considered by Abel in 1823 (cf. [2]) is:

$$A_0(u)(\rho) = \int_0^{\rho} \frac{u(r)dr}{\sqrt{\rho^2 - r^2}} \tag{7.31}$$

But A can be easily deduced from A_0 through the change of variable $x = \cos\varphi$, followed by the change of variable $r = \rho x$:

$$\begin{aligned} A(u)(\rho) &= 2\int_{-1}^{1} \frac{u(\rho x)}{\sqrt{1-x^2}} dx \\ &= 2\int_{-\rho}^{\rho} \frac{u(r)}{\sqrt{\rho^2 - r^2}} dr \\ &= 2\int_0^{\rho} \frac{u(r)}{\sqrt{\rho^2 - r^2}} dr + 2\int_0^{\rho} \frac{u(-r)}{\sqrt{\rho^2 - r^2}} dr \end{aligned}$$

then, denoting $\check{u}(r) = u(-r)$, we find $A(u) = 2A_0(u) + 2A_0(\check{u})$. Moreover in the cases we examine, $u(x)$ is identically zero for $x \leq 0$, and we simply find $A(u) = 2A_0(u)$.

This shows that the usual rectification Re_+ made by simple cells is changed by summation over phases into the modulus non-linearity. Then the most important conclusion for us is:

Assertion: *around the center of its receptive field, the response of a complex cell is invariant by small translations.*

This invariance of complex cells under small translation corresponds to what Hubel and Wiesel had suggested in 1962 ([88]). It is certainly useful in image inspection during micro-saccades of the eyes.

The same result holds for the average of activity in a cortical column, where the phase is uniformly represented.

Remark, in reality the distribution of phases integrated by complex cells is not fully uniform; this gives them a residual phase.

If $u(x) = x_+^{\gamma}$, which annihilates for $x \leq 0$ and coincides with x^{γ} when $x \geq 0$, we have:

$$A(u)(\rho) = [2I_{\gamma}]\rho_+^{\gamma}, \tag{7.32}$$

where I_{γ} is given by the Wallis formula:

$$I_\gamma = \int_0^{\frac{\pi}{2}} (\cos\varphi)^\gamma d\varphi = \frac{\sqrt{\pi}}{2}\frac{\Gamma(\frac{1}{2}+\frac{\gamma}{2})}{\Gamma(1+\frac{\gamma}{2})}.$$

Here Γ denotes the Euler function:

$$\Gamma(z) = \int_0^\infty e^{-t}t^{z-1}dt$$

This also permits to obtain $A(u)$ for the most-often-used non-linearities in neurophysiology, introduced by Naka and Rushton (1966):

$$u_m(x) = C_0\frac{x_+^m}{1+C_\infty x_+^m}; \tag{7.33}$$

where m is a strictly positive real number. A value which is justified by experiments in V1 is $m = 1,6$ for simple cells, and $m = 2,5$ for complex cells. Thus m is not far from half-integer value, $3/2$ and $5/2$ respectively.

We could also consider the following functions:

$$\tilde{u}_m(x) = C_0\frac{x_+^m}{(1+C_1x_+^2)^{\frac{m}{2}}}; \tag{7.34}$$

their Abel transforms can be computed in terms of usual special functions. Note that for $m > 2$, the functions u_m and $\tilde{u}_m$ differ notably (cf. below).

The function u_m is analytic on $]0,\infty[$ given by a uniformly convergent series on $[0, C_\infty^{-1/m}[$:

$$u_m(x) = C_0x^m\sum_{k=0}^\infty(-1)^kC_\infty^kx^{m(k+1)};$$

thus, using (7.32) with $\gamma = m(k+1)$, we deduce for $\rho \in [0, C_\infty^{-1/m}[$:

$$A(u_m)(\rho) = C_0\frac{\sqrt{\pi}}{2}\sum_{k=0}^\infty(-1)^kC_\infty^k\frac{\Gamma(\frac{1}{2}+\frac{m(k+1)}{2})}{\Gamma(1+\frac{m(k+1)}{2})}\rho^{m(k+1)} \tag{7.35}$$

This expression furnishes a good approximation for $A(u_m)(\rho)$ when ρ is small.

(This function $A(u_m)(\rho)$ is a parent of the generalized hypergeometric series studied by E.M. Wright ([240], [241], [242]) when ρ goes to infinity).

Recall that Γ is log-convex, and satisfies the functional equation $\Gamma(1+x) = x\Gamma(x)$; it follows that the coefficients $\Gamma(\frac{1}{2}+\frac{m(k+1)}{2})/\Gamma(1+\frac{m(k+1)}{2})$ decrease when k increases, they satisfy:

$$(\frac{1}{2}+\frac{m(k+1)}{2})^{-1/2} \leq \frac{\Gamma(\frac{1}{2}+\frac{m(k+1)}{2})}{\Gamma(1+\frac{m(k+1)}{2})} \leq (\frac{m(k+1)}{2})^{-1/2}.$$

From (7.31), we get another formula for $A(u_m)$:

$$A(u_m)(\rho) = 2\rho^m \int_0^1 \frac{t^m}{\sqrt{1-t^2}} \frac{dt}{1+\rho^m t^m} \tag{7.36}$$

We can also start with (7.30), and make the change of variable $t = \tan(\varphi/2)$, this gives:

$$A(u_m)(\rho) = 4C_0\rho^m \int_0^1 \frac{dt}{(1+t^2)} \frac{(1-t^2)^m}{(1+t^2)^m + C_\infty \rho^m (1-t^2)^m} \tag{7.37}$$

This can be integrated with elementary functions when m is integer and with elliptic functions (associated to the lemniscate) when m is half-integer.

For $m = 1$ and $C_\infty \rho < 1$, we pose

$$\kappa(\rho) = \sqrt{\frac{1 - C_\infty \rho}{1 + C_\infty \rho}},$$

then the transform $f_1 = A_0(u_1)$ is given by:

$$f_1(\rho) = 2\frac{C_0}{C_\infty}[\frac{\pi}{2} - \frac{2\kappa}{1 - C_\infty \rho} \arctan(\kappa)]. \tag{7.38}$$

Proof: we consider the function in two variables given by

$$g_1(\rho, \varphi) = \varphi - \frac{2\kappa}{1 - C_\infty \rho} \arctan(\kappa \tan(\varphi/2));$$

Its derivative with respect to φ is

$$\begin{aligned}\frac{\partial g_1}{\partial \varphi}(\rho, \varphi) &= 1 - \frac{2\kappa}{1 - C_\infty \rho} \frac{\frac{1}{2}(1 + (\tan(\varphi/2))^2)}{1 + \kappa^2 \tan(\varphi/2)^2} \\ &= \frac{1 - C_\infty \rho + (1 - C_\infty \rho)\kappa^2 \tan(\varphi/2)^2 - \kappa^2(1 + (\tan(\varphi/2))^2)}{(1 - C_\infty \rho)(1 + \kappa^2 \tan(\varphi/2)^2)}.\end{aligned}$$

By developing the numerator and denominator in function of $C_\infty \rho$ and $\cos(\varphi) = (1 - \tan(\varphi/2)^2)/(1 + \tan(\varphi/2)^2)$ we get

$$\frac{\partial g_1}{\partial \varphi}(\rho, \varphi) = \frac{C_\infty \rho \cos(\varphi)}{1 + C_\infty \rho \cos(\varphi)},$$

therefore

$$g_1(\rho, \varphi) = \int_0^\varphi \frac{C_\infty \rho \cos(\phi)}{1 + C_\infty \rho \cos(\phi)} d\phi,$$

and, by making $\varphi = \frac{\pi}{2}$ in this formula we obtain the desired result.

If ρ tends to zero $f_1(\rho)$ also tends to zero. If $C_\infty\rho$ tends to 1, $f_1(\rho)$ tends to $2\frac{C_0}{C_\infty}(\pi/2-\sqrt{2})$.

When $C_\infty\rho > 1$ we define κ as the square root of $(C_\infty\rho-1)/(C_\infty\rho+1)$, a similar computation shows that $f_1 = A_0(u_1)$ is given by

$$f_1(\rho) = 2\frac{C_0}{C_\infty}[\frac{\pi}{2} + \frac{2\kappa}{C_\infty\rho - 1}\operatorname{arg\,tanh}(\kappa)]. \tag{7.39}$$

Now consider the functions $\tilde{u}_m$. For every positive value of m, we have

$$A(\tilde{u}_m)(\rho) = 2C_0\int_0^1 \frac{\rho^m x^m}{(\sqrt{1-x^2})(1+C_1\rho^2x^2)^{\frac{m}{2}}}dx. \tag{7.40}$$

By putting $t = x^2$ we get

$$A(\tilde{u}_m)(\rho) = C_0\rho^m\int_0^1 t^{\frac{m}{2}-\frac{1}{2}}(1-t)^{-\frac{1}{2}}(1+C_1t\rho^2)^{-\frac{m}{2}}dt \tag{7.41}$$

The hypergeometric function was introduced by Euler (in 1769) by the following integral:

$$F(a,b,c;z) = \frac{\Gamma(c)}{\Gamma(b)\Gamma(c-b)}\int_0^1 t^{b-1}(1-t)^{c-b-1}(1-tz)^{-a}dt. \tag{7.42}$$

This formula has a meaning for any a if we have $Re(c) > 0, Re(b) > 0, z \in]-\infty,+1[$.

If we put $z = -C_1\rho^2, a = m/2, b = m/2+1/2, c = m/2+1$, we have

$$A(\tilde{u}_m)(\rho) = C_0\rho^m\sqrt{\pi}\frac{\Gamma(\frac{m}{2}+\frac{1}{2})}{\Gamma(\frac{m}{2}+1)}F(\frac{m}{2},\frac{m}{2}+\frac{1}{2},\frac{m}{2}+1;-C_1\rho^2). \tag{7.43}$$

The hypergeometric function was studied by Gauss (in 1823) by using series expansions. he found in particular that $A(\tilde{u}_m)$ is given by:

$$A(\tilde{u}_m)(\rho) = C_0\rho^m\frac{\sqrt{\pi}}{\Gamma(\frac{m}{2})}\sum_{k=0}^{\infty}(-1)^k\frac{\Gamma(\frac{m}{2}+k)\Gamma(\frac{m}{2}+\frac{1}{2}+k)}{\Gamma(\frac{m}{2}+1+k)k!}C_1^k\rho^{2k}; \tag{7.44}$$

which converges uniformly for $\rho \in [0, C_1^{-1/2}]$.

By using another formula of Euler we obtain the following formula of $A(\tilde{u}_m)$:

$$A(\tilde{u}_m)(\rho) = C_0C_1^{-\frac{m}{2}}\sqrt{\pi}\frac{\Gamma(\frac{m}{2}+\frac{1}{2})}{\Gamma(\frac{m}{2})}\int_0^{C_1\rho^2} t^{\frac{m}{2}-1}(1+t)^{-\frac{m+1}{2}}dt. \tag{7.45}$$

When m takes integer values, explicit formulas for $A(\tilde{u}_m)$ can be obtained easily from this last formula. For instance if $m = 1$, from (7.45) we get:

$$\frac{d}{d\rho}A(\tilde{u}_1)(\rho) = 2C_0(1+C_1\rho^2)^{-1}$$

Which integrates to

$$A(\tilde{u}_1)(\rho) = \frac{2C_0}{\sqrt{C_1}}\arctan\sqrt{C_1}\rho. \tag{7.46}$$

When ρ tends to zero we verify that $A(\tilde{u}_1)(\rho)$ is equivalent to $2C_0\rho$, and when ρ tends to infinity we see that $A(\tilde{u}_1)(\rho)$ tends to $\pi C_0/\sqrt{C_1}$, which is coherent with the fact that the limit of $\tilde{u}_1(x)$ when x tends to infinity is equal to $C_0/\sqrt{C_1}$.

When $m=2$, by putting $C_1 = C_\infty$ we obtain $\tilde{u}_2 = u_2$, and in this case the formula (7.45) gives

$$A(u_2)(\rho) = \pi\frac{C_0}{C_1}\int_0^{C_1\rho^2}(1+t)^{-\frac{3}{2}}dt.$$

Which can be computed as

$$A(u_2)(\rho) = \pi\frac{C_0}{C_1}(1-(1+C_1\rho^2)^{-1/2}). \tag{7.47}$$

Let us assume that there is a constant $C>0$, such that u/C is the repartition function of a real positive random variable X (i.e. u is zero out of $[0,\infty[$, positive, increasing and bounded, and C is the limit of $u(x)$ when x tends to $+\infty$). Then the same is true for $A(u)/C\pi$, it is the repartition function of a real positive random variable R. However, it may happen that X has a finite expectation and R has not, or that X has finite variance and R has not. This is the case in the examples we just saw:

1)If X is uniform in $[0,L]$, the function R has a density equivalent to $C/(L\rho^2)$, when ρ tends to ∞, thus it has no mean value.
2) When $u=u_m$, with $C_0=C_\infty=1$ to simplify, we have $u'(x)=mx^{m-1}/(1+x^m)^2$, thus X has a finite mean when $m>1$ and a finite variance when $m>2$. However the derivative $A(u)'(\rho)$ is larger than a multiple of ρ^{-m}. For $m=1$ the derivative is equivalent to $C'\rho^{-2}\ln\rho$, and for $m=2$ it is equivalent to $C''\rho^{-2}$, (cf. (7.39) and (7.47)). Then for $1\le m\le 2$ the expectation of R est infinite, and for $2<m\le 3$, even if the mean of R exists, its variance is infinite.
3) When $u=\tilde{u}_m$, with $C_0=C_1=1$ to simplify, the differences between X and R are even more pronounced: the derivative of u in x is $mx^{m-1}(1+x^2)^{-1-m/2}$, then X possesses a finite mean and an infinite variance. (This makes a deep difference between $\tilde{u}_m$ and u_m pour $m>2$.) But, from (7.45) we see that the derivative of $A(\tilde{u}_m)$ is of order ρ^{-2} at $+\infty$. Thus R never possesses a finite expectation.
All this can have non-negligible consequences on the kind of integration made by the columns of V1, and even on the integration made by the complex cells when the phase disappears.

Remarks: Abel had defined transformations of functions A_α generalizing $A=A_{1/2}$ for real numbers $\alpha\in]0,1[$:

$$A_\alpha(u)(\rho) = \frac{1}{\Gamma(\alpha)}\int_0^\rho (\rho - t)^{\alpha-1}u(t)dt, \tag{7.48}$$

These operators satisfy:

$$A_\alpha \circ A_{\alpha'} = A_{\alpha+\alpha'}. \tag{7.49}$$

Hardy et Littlewood have proved that, for any $p \in]1, 1/\alpha[$, if we write $q_\alpha = p/(1-\alpha p)$, there exists a constant $C_{\alpha,p}$ such that, for any function u defined on the interval $[0,N]$:

$$\|A_\alpha(u)\|_{q_\alpha} \leq C_{\alpha,p}\|u\|_p \tag{7.50}$$

Under the same hypotheses A_α defines a compact operator from $L^p(0,N)$ to $L^{q_\alpha}(0,N)$. In particular the original Abel transform $A = A_{1/2}$ defines a compact operator from $L^{3/2}(0,N)$ to $L^6(0,N)$.

When $p > 1/\alpha$, the transform A_α becomes compact from $L^p(0,N)$ to the Hölder space $C^{\alpha-1/p}(0,N)$.

In particular $A = A_{1/2}$ defines a compact operator from $L^3(0,N)$ into the space of continuous function $C^0(0,N)$ equipped with the uniform norm.

More generally, the result of push-froward integration by columns in $V1$ is described by summing the activities $R_A(t)$ of (7.29) on sets of cells with several phase distributions. This corresponds to peculiar distributions of the parameters $G_\pm, u_\pm, \xi_0, \eta_0, \omega_0$, given by density functions $D_+(\sigma,m,\xi,\eta,\omega)$ and $D_-(\sigma,m,\xi,\eta,\omega)$. Here we suppose that the various non-linearities are of the Naka-Rushton type u_m (or $\tilde{u_m}$) and the kernels $G_\pm$ mostly vary with the variance σ^2, with comparable center (x_0,y_0); this gives the following formula for the responses in optical imaging:

$$R_C(I)(t) = \int D_+(\sigma,m,\xi,\eta,\omega)d\sigma dm d\xi d\eta d\omega A(u_m)(|\widehat{I_t G_+^\sigma}(\xi,\eta,\omega)|)$$
$$- \int D_-(\sigma,m,\xi,\eta,\omega)d\sigma dm d\xi d\eta d\omega A(u_m)(|\widehat{I_t G_-^\sigma}(\xi,\eta,\omega)|). \tag{7.51}$$

This gives a non-linear operator from tempered distributions I over plane-time to continuous functions on this plane-time, which is a difference between two operators of the following type:

$$P(I)(x,y,t) = \int D(\sigma,m,\xi,\eta,\omega)d\sigma dm d\xi d\eta d\omega$$
$$A(u_m)(|\int G_\sigma(X-x,Y-y,s)I(X,Y,t-s)e^{-i(X\xi+Y\eta+s\omega)}dY\,dY\,ds|) \tag{7.52}$$

The preference of the column for an orientation and a spatial frequency corresponds to a particular concentration of the density D in the vector (ξ,η); there also exists concentration for the temporal frequency ω.

It is tempting to suggest that operators from a cortical area Ξ to another one Υ are of the same type, i.e. given by a non-linear pseudo-differential operator.

We could introduce, the places x and y in the area Ξ and Υ respectively, the frequency vectors of spatial activation λ and μ and importantly the time frequencies ν and ω; we would have kernels $D(y,x,\sigma,m,\nu)$ non-necessarily positive, where σ denotes the diffusion factor and m a parameter for non-linearity of functions f_m : $\mathbb{C} \to \mathbb{R}$, in such a way that for any activity $\varphi(x,s)$ in the area Ξ, the impact of φ on the activity in Υ is represented by the formula

$$P(\varphi)(y,t) = \int D(\sigma,m,\lambda,\omega)d\sigma dm d\lambda d\omega$$

$$f_m(\int G_\sigma(y,x,s)\varphi(x,t-s)e^{-i(x\lambda+s\omega)}dxds) \quad (7.53)$$

This operator has to be composed with an operator of the same type describing the horizontal dynamics in Υ, and a feedback from Υ to Ξ, and so on, so we are face with a graph of propagation. But even this graph is an oversimplification, because quickly a set of areas should be treated as being in interaction, and we come back to an operator from activity in the set of areas to activities in the same set, modulated by external input.

However, we should remark that the dynamics cannot be fully reduced to such a semi-local non-linear pseudo-differential operator, for a more fundamental reason: contextual effects propagate to modify the parameters of the kernels in a non-linear way. The simplest effect is the non-linearity due to contrast: it integrates the contrast by semi-norms in the image, which modify m, σ, saturation and time delays. Thus, at the level of complex operators, adaptation can be represented by a transformation of the density $D(\sigma,m,\lambda,\omega)$.

To conclude this subsection, we come back to the original motivation of the above work on Abel transform, which was the thesis of Simon Capern. We tried to understand the remarkable phenomenon of invariance discovered by Basole, White and Fitzpatrick 2003 ([18]):

Basole et al. observed that several stimuli of different geometric designs, different velocities (in speed and direction), different contrasts produced the same activity of micro-columns in $V1$, as seen by optical imaging. They invoked a motion energy model and suggested that the cortical maps do not represent distinct parameters (position, frequency, orientation, direction) but mixtures of them. However, in his thesis Capern showed numerically that all the results of Basole et al. can be reproduced by the model of columnar responses exposed in this subsection, with m=1.

Note that Mante and Carandini [136] established a similar result without giving explicit formulas. In addition, let us note that these formulas are showing that the results of Basole et al. 2003 are compatible with the existence of geometrically independent maps, but this does'nt prove their independency.

In micro-columns of $V1$ and $V2$ the preferred orientation of neurons do not change, and the same is true of the preferred spatial frequency (cf. [95], [184]), but it seems that the preferred temporal frequency is not organized in the same way (cf. [109]), consequently the invariance of columns is not for $2D+1$ translations but for $2D$ translations only.

7.4.2 Orientation Maps, Probability, Co-variance and the Number Pi

References: Wolf and Geisel ([238], [239]), Kaschube et al. [106], Barbieri, Citti, Sanguinetti, Sarti, 2011 ([16]); Bednar ([20]).

In this third example, we will use Euclidean geometry, probability (bivariate Gaussian fields) and harmonic analysis (which in some sense unifies geometry and probability).

I thank Antoine Grappin, Gaëtan Hadjeres, Adrien Laversanne-Finot, Etienne Levilion, Viet-anh Nguyen and Alexandre Afgoustidis for their participation to this study, and Alessandro Sarti for his explanations.

In $V1$ of all mammals the neurons have preferred orientation (and spatial and temporal frequencies) of stimuli in the visual plane ([87], [88], [89]). In carnivores and primates these preferences are arranged in regular maps along the cortical surface ([30]). Note this is not the case for all species of mammals, for instance rodents, even squirrels which have a fine vision (and a large V1), do not have orientation maps. See [248], [155], [156].

As frequently reminded by Alain Berthoz (cf. [27], [28]), Merleau-Ponty suggested that "To see is to palpate with gaze", thus noticing the parallel between touch and vision, and insisting on the active process that is used in vision as in touch. In the case of the skin of fingers, we see lines in relief (named dermatoglyphes), forming a foliation, which allows for instance to use prints for identification of humans. A biological function of these lines is to help discriminate between touched textures. In exactly the same manner we use small eye displacements (like microsaccades) to vary the cortical input of images and we sense the image with a foliation of inscribed lines in $V1$ (or $V2$), whose tangent vectors are the preferred orientation of the micro-columns. In both cases, fingertips and visual cortex, the singularities are typical singular points of non-oriented fields of straight lines, i.e. either tripods or caps. Cf. figure (7.4).

The singularities of the orientation map are named *pinwheels* (cf. [30], [168]). Around these points there exist polar coordinates (ρ,θ) on the cortical surface such that the preferred orientation is constant when θ is constant.

Following the suggestion of Wolf and Geisel ([238], [239]), we consider that the pinwheels in $V1$ are given by the zeros of a smooth complex function ψ on an Euclidean plane E that represents the visual plane in the cortex.

Remark that orientations (or non-oriented directions) differ from directions of oriented lines in the oriented Euclidean plane, one orientation corresponding to two directions. If we choose in E a given oriented direction δ, the orientations are measured by a real number modulo π and the directions are measured modulo 2π. Thus, given the complex map ψ, we decide that the preferred orientation at the place (x,y) of E is the argument of the complex number $\psi_\omega(x,y)$ divided by two.

Fig. 7.4 Palping/Looking. The curves can represent dermatoglyphics of fingers tips or a map of preferred orientations in $V1$.

The modulus of the function ψ can represent the selectivity level (see for instance [153]).

Taking into account the variability of animals, the variability of their developments and the variability of the map under translation on the cortical surface, we can assume *a priori* that ψ is a random Gaussian field (cf. [4]), of class C^1, on $\mathbb{R}^2$ with values in $\mathbb{C}$. Then the real part $\xi(x,y)$ and the imaginary part $\eta(x,y)$ of $\psi(x,y)$ as well as their derivatives with respect to x and y are continuous Gaussian fields. In accordance with the data and analysis presented by Niebur and Wörgötter, [154], Wolf and Geisel, as Barbieri et al. have assumed that ψ is stationary and isotropic, and has a simple spectral structure.

By the spectral theorem (cf. [4]) we can write the complex field as a sum of random waves:

$$\psi(x,y) = \int d\mu(\overrightarrow{K}) R_\omega(\overrightarrow{K}) e^{i\overrightarrow{K}.\overrightarrow{r} + i\Phi_\omega(\overrightarrow{K})}; \tag{7.54}$$

where $\overrightarrow{r}$ denotes the vector (x,y), where $\overrightarrow{K}$ denotes a dual vector, and μ a positive bounded measure(the spectral measure of ψ), where $\Phi_\omega(\overrightarrow{K})$ denotes a family of random variables with values in $[0,2\pi]$, two by two independent with uniform laws, and with random moduli $R_\omega(\overrightarrow{K})$, of expectation 1.

It is remarkable that this hypothesis alone should be sufficient to estimate the density of pinwheels d_2. Berry and Dennis (2000, [26]), using a Kac-Rice formula (cf. [4]), found that $d_2 = 4\pi/K_2$ where $K_2/2$ is the variance of any one of the derivatives $\xi_x, \xi_y, \eta_x, \eta_y$.

Now, in the particular case of monochromatic waves, i.e. when μ is a uniform probability on the circle Γ_K of fixed radius $K_0 = 2\pi/\Lambda$ centered on 0 in the dual space V^*, Berry and Dennis proved

$$d_2 = \frac{\pi}{\Lambda^2}, \tag{7.55}$$

This is a consequence of the relation $K_2 = K_0^2$.

Note that a similar result was written as soon as 1998 in the supplementary document of [238], but a proof appeared in [239] after [26].

Remarkably this formula corresponds to the spectacular experimental results of Kaschube et al. ([106]) (cf. also the discussion in Science 2012).

The fact that ψ is monochromatic is essential to get π. It is also probably essential for defining a precise scale in the map.

We should also remark that the monochromaticity could also come from constraints on the spontaneous activity during development, having fixed wave length in each brain region dependent on the wiring. Note that the scale is different for $V1$ and $V2$, and depends on the animal studied ([106]).

Consequently we get the following statement:

Proposition 4.2.1: The density π of the pinwheels with respect to the scale Λ of approximate translation invariance, that is experimentally observed by Kaschube et al., is a corollary of the three hypotheses: 1) the fact that the map comes from the argument of a Gaussian field, 2) the invariance of the probabilistic law of the field by Euclidean displacements, 3) the monochromaticity.

Note that Niebur and Worgotter (1994, [154]) gave arguments for the concentration of the spectral measure on an annulus, that were based on an hypothesis of *local correlation*. When the field is not purely monochromatic the density becomes lower than π ([238]).

Barbieri, Citti, Sanguinetti and Sarti ([16]) have suggested that the formula for ψ, in particular the monochromaticity, should be justified by a principle of minimization of uncertainty for two operators analog to conjugate observable quantities in Quantum Mechanics. This principle expresses the fact that the orientation map is optimal for discriminating between two rigid motions; thus it is closely related to Euclidean invariance.
We can say that in this case adaptation is the adaptation of the cortical map for maximal invariance and information, and that it is realized at the scale of evolution, probably using development scale for complete installation (cf. [20]).

Let us give now a presentation of this work of Barbieri, Citti, Sanguinetti and Sarti:

We start with the structure of invariance; thus we consider the natural action of $E(2)$ on the Hilbert space of L^2 complex functions on the plane $\mathbb{R}^2$. Then we choose an irreducible factor H of this representation ([222]). Note that this irreducibility will appear later as the origin of the monochromaticity. The elements ψ of H give do not give probability densities on the plane; in this case they give probability laws on the momentum space, after Fourier transform. This is analogous to what happens with the representation of translations where pure waves are localized in frequency, not in space. However the unitarity of the representation of $E(2)$ in H allows for correlation computations: If Y is an Hermitian endomorphism of H, i.e. a real observable, the *mean* of Y in the state ψ is well defined as $\mu = \langle\psi|Y\psi\rangle$, and the variance of Y in the state ψ is well defined by

$$\sigma^2 = \|(Y-\mu)\psi\|^2. \tag{7.56}$$

The uncertainty principle in this framework is a theorem (not very difficult to prove, see [48], [195]) that states that, if X,Y are two observables, and if $[X,Y]$ denotes as usual their commutator $X \circ Y - Y \circ X$, we have

$$\sigma(X)\sigma(Y) \geq \frac{1}{2}\sigma^2([X,Y]). \tag{7.57}$$

Moreover this inequality becomes an equality if and only if there exists a real number λ such that

$$(Y-\mu(Y))\psi = i\lambda(X-\mu(X))\psi. \tag{7.58}$$

Let Y_1 the infinitesimal operator associated to the translation in the y direction, and Y_2 the infinitesimal operator associated to the rotation around $(0,0)$; we denote by $\widehat{\psi}$ the Fourier transform of ψ and by $\widehat{Y_1}, \widehat{Y_2}$ the operators conjugated by the Fourier transformation. Then the uncertainty principle implies that ψ gives the minimum of the product of variances of Y_1 and Y_2 in state ψ, if and only if there exists a real constant λ and a complex constant ζ, such that $\widehat{\psi}$ satisfies:

$$(\widehat{Y_2} - i\lambda\widehat{Y_1})\widehat{\psi} = \zeta\widehat{\psi} \tag{7.59}$$

which in our case, is equivalent to the differential equation:

$$\frac{\partial\widehat{\psi}}{\partial\theta} - i\lambda(iK\sin\theta)\widehat{\psi} = \zeta\widehat{\psi}, \tag{7.60}$$

where the vector $\overrightarrow{K}$ in the dual of $\mathbb{R}^2$ is described in polar coordinates $K\exp(i\theta)$. We consider this equation on the universal covering of the dual plane minus the origin in the usual plane of Cartesian coordinates K,θ. From the Cauchy theorem in distributions spaces we get the general solution depending on an arbitrary distribution $C(K)$:

$$\widehat{u}_{(\lambda,\zeta)}(R,\theta) = C(K)e^{\lambda K\cos\theta + \zeta\theta}. \tag{7.61}$$

But u must be periodic in θ so ζ should be written $i.n$ with $n \in \mathbb{Z}$. Thus the solution is

$$\widehat{u}_n(R,\theta) = C(K)e^{\lambda K\cos\theta + in\theta}. \tag{7.62}$$

From harmonic analysis, $C(K)$ must be a bounded measure, otherwise ψ cannot belong to a factor of the standard representation on the plane.

But the irreducibility of H implies that $C(K) = \delta(K = K_0)$, cf. [222]. This concentration of $C(K)$ is equivalent to the monochromaticity of u_n.

If we replace the operators Y_1, Y_2 by another infinitesimal translation and another rotation respectively we obtain in place of u_n the function

$$u_n(x,y;\alpha,a,b) = \int_0^{2\pi} d\theta e^{\lambda K\cos(\theta-\alpha)-in(\theta-\alpha)} e^{iAK\cos(\theta-\phi)} e^{irK\cos(\theta-\varphi)}. \quad (7.63)$$

adapted to the direction $\exp(i\alpha)$ and the rotation around $\overrightarrow{A} = (a,b) = A\exp(i\phi)$, i.e. the change of variables transforming z into $a+ib+z\exp(i\alpha)$.

Now we should realize that there is no deterministic function ψ which is invariant by displacement and which is at the same time optimal for discrimination between translation and rotation. However, as we will see, if we look for a random field with these properties there do exist solutions. In particular there are solutions obtained by summing functions u_n over an infinity of centers and phases:

Let us take a sum of an infinite number of independent random functions $u_n(x,y;\alpha,a,b)$ associated to random vectors $\overrightarrow{A}_k(\omega)$. After integration over α, this gives:

$$\Psi_n(x,y;\omega) = \sum_k F_n(K,k) \int d\theta e^{irK\cos(\theta-\varphi)-iA_k(\omega)K\cos(\theta-\phi_k(\omega))}. \quad (7.64)$$

If we choose the $\overrightarrow{A}_k(\omega)$ to be independent uniformly distributed in the plane and the series $F_n(K,k)$ to be in l^2, the central limit theorem implies the existence of a constant C_n such that the renormalized sum, obtained by dividing F_n by C_n, converges to a Gaussian random field of class C^1, which is stationary and isotropic, then we can apply the results of Berry and Dennis on d_2.

Remark: we could as well take a random sum

$$\Psi_n(x,y;\omega) = \sum_k C_n F_n(K,k) \int d\theta e^{\lambda_k K\cos(\theta-\alpha_k(\omega)} e^{irK\cos(\theta-\varphi)-iA_k(\omega)K\cos(\theta-\phi_k(\omega))}, \quad (7.65)$$

where the α_k are uniformly distributed with values in $[0,2\pi]$.

Now consider the irreducible unitary representation associated to the number K; it is given by:

$$T_K(\alpha,a,b)(\widehat{u})(\theta) = e^{iAK\cos(\theta-\phi)}\widehat{u}(\theta-\alpha); \quad (7.66)$$

where as before $a+ib = A\exp(i\phi)$ (cf. [222], see Appendix 2).

If v_1, v_2 are two square integrable functions on the unit circle, the associated coefficient is

$$\langle v_1 | T_K(\alpha,a,b) v_2 \rangle = \int_0^{2\pi} d\theta v_1(\theta) e^{iAK\cos(\theta-\phi)} v_2(\theta-\alpha). \quad (7.67)$$

Thus $u_n(x,y;\alpha,a,b)$ is the coefficient of T_K associated to $v_1(\theta) = \exp(irK\cos(\theta - \varphi))$ and $v_2(\theta) = \exp(\lambda K\cos\theta - in\theta)$. The function v_2 obtained when $n = 0$ is the Von-Mises density $v(\theta) = \exp(\lambda K\cos\theta)$.
The final function of the Barbieri et al. model is also a coefficient of T_K:

$$\Psi_n(x,y;\omega) = \langle C_n(K)T_K(0;x,y)V(\omega)\rangle; \tag{7.68}$$

The Gaussian field itself is given by the stochastic integral:

$$\Phi(re^{i\varphi};\omega) = \frac{1}{2\pi}\int_0^{2\pi} C(K)e^{irK\cos(\theta-\varphi)}dZ_\theta(\omega). \tag{7.69}$$

It is also a coefficient of T_K:

Proposition 4.2.2

$$\Phi(x,y;\omega) = \langle C_n(K)T_K(0;x,y)\frac{dZ}{d\theta}(\omega)\rangle, \tag{7.70}$$

Remark, by using the Fourier expansion of a Brownian bridge Z, this gives

$$\Phi(x,y;\omega) = \frac{iC(K)}{2\pi}\int_0^{2\pi} d\theta e^{irK\cos(\theta-\varphi)} \sum_{n\in\mathbb{Z}^\times} \frac{\zeta_n(\omega)}{\sqrt{2\pi}} e^{in\theta}, \tag{7.71}$$

and by commuting the sum and the integral, which is justified in $L^2(\Omega)$ for each $(x,y) = r\exp(i\varphi)$, we get

$$\Phi(x,y;\omega) = C'(K)\sum_{n\in\mathbb{Z}^\times}\zeta_n(\omega)\int_0^{2\pi} d\theta e^{irK\cos(\theta-\varphi)+in\theta}. \tag{7.72}$$

By taking the new integration variable $\theta - \varphi$, and using the definition of Bessel functions of integer indexes we find

$$\Phi(x,y;\omega) = C\text{''}(K)\sum_{n\in\mathbb{Z}^\times}\zeta_n(\omega)e^{in\varphi}J_n(rK). \tag{7.73}$$

Note that $J_{-n}(z) = \overline{J_n(-z)}$.

We must remark that monochromaticity appeared independently from the uncertainty principle. However irreducibility of the representation of isometries is a natural hypothesis from the invariance point of view: it gives more coherence to the visual area (as do characters for simple cells), and it gives a precisely defined *scale* of translational invariance ([154]). Also, it is remarkable that uncertainty minimization is compatible with irreducibility; that is simply because the principle of uncertainty can be enunciated abstractly, in every unitary representation of the Euclidean group.

7.4.3 Gluing, Two Eyes and Two Hemispheres

(From unpublished works that grew around the thesis of Luc Foubert, September 27 in 2007; see [68]. These works reflect discussions with Jacques Droulez, Luc Foubert, Nathalie Rochefort and Chantal Milleret; thanks to ACI NIC 2003-2007.)

The Mathematics we shall use rely on differential topology, probability and stochastic processes. Here, the invariance that comes into play is topological.

A necessary condition for successful action of an animal in the world is the coherence of the information flow in its brain that controls end effectors. We claim that geometrical co-variance facilitates this coherence (cf. Pellionisz and Llinas ([166], [167]). In fact the information flow follows disconnected routes from the sensory end organs to the internal loops and the end effectors, and co-variance helps to accord these routes. The example below concerns coherence between the left and right hemispheres, for treating visual information coming from the four hemi-retina. Cf. Bullier in *TVN* [44], [160], Olavarria in [165], [69].

In general visual systems divide the image-time in several pieces, and reconstruct a global invariance from there; wonderful examples are offered by the facet eye of the common fly, or by the eight eyes of its enemy the spider (cf. [120]). In mammals the two eyes see different but overlapping parts of the world, and in most of them each retina is divided vertically in two overlapping parts sending the left and right world to the right and left hemisphere respectively. Then the visual areas have to combine these four pieces of information for a unified perception. The bundle of neurons that join the right and left primary visual areas is the *corpus callosum*. Its activity must be conjugated with ocular dominance to exploit the four pieces of information in an optimal way.

Most primary sensory areas respect the right-left symmetry, i.e. the two corresponding hemispheres are covariant for right-left symmetry. Note this is not necessarily true for higher visual areas, for instance in birds or humans, at the end of the visual system, the para-hippocampal region is known to have different functions in the left and right hemispheres.

The invariance structures in visual areas depend on the gluing of information between the left and right eyes and the left and right hemispheres. For instance translations of the image from right to left (or the converse) cannot be analyzed in a co-variant way without correct prolongation across the central vertical meridian. Remark that the neurons of each subarea analyze all components in the image-time: position, duration, time and space frequency, energy flow, colors; therefore all these elements are *a priori* concerned by the re-combination. However the gluing can fail sometimes, as we will see on the example of the tupaia (cf. 4.4), because it can appear conflicts between different kind of invariance; these conflicts depend on the manner invariance is implemented. The example we will develop in subsection 4.4 is ON-OFF symmetry and orientation geometry, as in 4.2, in the case of tupaia. In the following discussion we will consider the role of horizontal connections for the construction of invariance, and the adaptation subtending invariance will depend on development, and on rapid adult adaptation.

We now introduce a general abstract model that we name *random gluing*, which aims at explaining how several brain areas can combine divergent information to preserve invariance structures. In this model we take as a guide the definition of manifolds by charts and gluing, the charts being identifications of pieces with certain models, and gluing being smooth maps between these models (cf. for instance [211] for the concept of charts and manifolds), but in our case the charts represent neuronal functions, that are input-output transformations, and the gluing maps are also stochastic operators supported by neurons. Everything here is described with random variables. The resulting inner model E of the external world will be described by parameters that modulate the neuronal activities.

Precisely, we assume the existence of sets $E_i; i \in I$ and random operators Φ_i which transforms the elements φ_i of E_i in neuronal activities ψ_i of brain areas $X_i; i \in I$. (For random processes and operators kernels see [52].) We also assume the existence of random internal operators Φ_{ji}, supported by sets of neurons X_{ji}, transforming the activity ψ_i in X_i into an activity $\Psi_{ji}(\psi_i)$ in X_j. The neurons in X_{ij} can be inter-area or intra-area, and they can be feed-forward or feed-back or horizontal; examples are commissural axons going from one hemisphere to the other, or back projection from $V1$ to LGN, or inside LGN or pulvinar, or internal to $V1$, from layer IV to III and II to V. Here we mainly consider callosal neurons between hemispheres and binocular neurons in each hemisphere.

The input ϕ_{ji} in X_j (coming from X_i) produces an output ψ_{ji}, which can again be transformed by X_{kj} for $k \in I$ into another input φ_{kji} of the set X_k, and so on.

The *compatibility axioms* for gluing are the following:

1) for each pair (i, j) there exist subsets E_{ij} and E_{ji} in E_i and E_j respectively such that for each input $\varphi_i \in E_{ij}$ there exists a unique input $\varphi_j \in E_{ji}$ denoted by $\varphi_j = \Phi_{ji}(\varphi_i)$ such that $\Phi_i(\varphi_i)$ and $\Psi_{ij}(\Phi_j(\varphi_j))$ have the same probabilistic laws;
2) for each triple (i, j, k) we have $\Phi_{ij} \circ \Phi_{jk} = \Phi_{ik}$ on the subset E_{ikj} in E_{ik} which corresponds to both E_{ij} and E_{jk}.

In particular, for each pair (i, j), we have $\Phi_{ij} \circ \Phi_{ji} = Id$.

These axioms are saying, that *in law* the activities represent a coherent world E made by gluing E_{ij} with E_{ji} for all the pairs E_i, E_j. Thus they are in the same spirit than the hypothesis of invariance in law of the orientation map.

Importantly, the satisfaction of these compatibility axioms require in general an active behavior, for example the convergence of the two eyes in the case of binocularity. In fact the preceding axioms can be understood as a particular case of adaptation, requiring that internal connections Ψ_{ji} are able to reproduce some activations created by direct stimuli. This is a form of simulation (cf. [27]).

From that we can formulate the two principles of random gluing:

(i) the compatibility axioms insure that the random variables corresponding to different roads of information are compatible in law when conditioned by the stimuli;
(ii) given (i), the optimality of the gluing process is obtained by maximizing the joint information.

When the axioms are satisfied each area X_i benefits from several compatible entries φ_i, φ_{ij}, and so on, but the information on a parameter e_i transmitted by the waves ψ_i and ψ_{ij} and so on is *a priori* larger than the information transmitted by each process alone, so that the gluing can augment the information on the common output.

The fundamental property of information that underlies this fact is the possible negativity of the 3 by 3 information of three random variables. A,B,C (for instance $A = \psi_i, B = \psi_{ij}, C = e_i$):

$$I(A;B;C) = I(A;C) + I(B;C) - I((A,B);C). \tag{7.74}$$

(Cf. [47], and section 6, Appendix 1 below.)

In fact $I(A;B;C) \leq 0$ is interpreted (in Physics and in Information theory) as a positive *synergy* between the two observables A,B over C (cf. [139]). For instance, when A and B are two independent variables, it is easy to see that $I(A;B;C) \leq 0$. This is a refinement of the well known fact that the variance of a mean of independent variables is half of the sum of the variances of the two variables.

This relative independence implies the apparently paradoxical fact, that something presented to the central vertical meridian in the visual sphere is more accurately detected with the two open eyes than with only one. In fact it can be twice better, and not only $\sqrt{2}$ times, i.e. the variance on the position can be divided by four, and not only 2, because, for this central region, we have in the brain (at least) four independent observations of the event: by two eyes and two hemispheres. For optimality of observation, the four pathways have to coincide in law and to be statistically independent when conditioned by the stimulus.

With respect to the general model, the set I of indexes i has four elements $(R,R),(R,L),(L,R),(L,L)$, the first letter for the eye, right or left, the second for the hemisphere, right or left, cf. figure (7.5). They can be considered as the vertexes of a tetrahedron. The internal processes Φ_{ji} between two eyes come from vertical connections inside the cortex, and between the two hemispheres they come from the callosal connections, joining a zone of one hemisphere to a corresponding zone of the contralateral hemisphere. Note that the central vertical meridian is not represented in the cortex exactly at the boundary of $V1$ and $V2$, but there is a *transition zone*, named TZ, extending the representation over the meridian in the ipsilateral visual field. Let us name $V1$-proper the complement of TZ in $V1$, then there exist neurons sending their axons through the corpus callosum from $V1$-proper to TZ (contralateral to ipsilateral) and other sending their axons from TZ to $V1$-proper (ipsilateral to contralateral). See [160], [165].

The uni-ocular hypothesis of Olavarria says that the guiding callosal projections connects regions that mainly prefer the same eye. If this is true we see four dominant gluing processes: two interhemispheric between (R,R) and (R,L), or between (L,L) and (L,R), and two intra-hemispheric between (R,R) and (L,R), or between (L,L) and (L,R). Moreover in majority, callosal projecting neurons are not direct projections of callosal neurons. This arrangement permit to satisfy the first compatibility axiom for the edges $((R,R),(R,L))$, $((L,L),(L,R))$, $((R,L),(L,L))$ and $((R,R),(L,R))$ in the tetrahedron. (They form a so called Petrie polygon in the

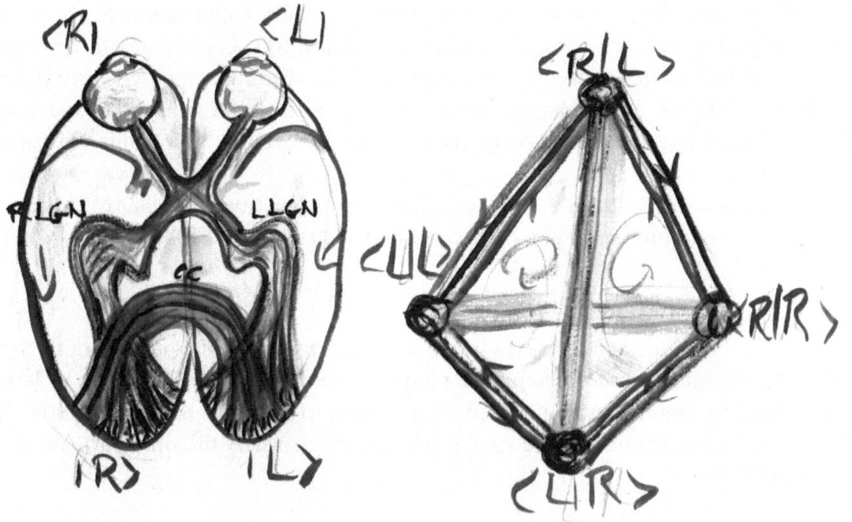

Fig. 7.5 Gluing. Left: the visual pathway form retinae to V1 through LGN, with interhemispheric connections through the corpus callosum. Remark the division of the optical nerve. Right: the symbolic tetrahedron for gluing two eyes and two hemispheres.

tetrahedron.) As those edges form a closed loop, there is one higher relation with four terms to satisfy:

Proposition 4.3.1: A necessary and sufficient condition for the coherent gluing along all edges, faces and volume of the tetrahedron made by two eyes and two hemispheres is

$$((R,L),(R,R),(L,R)) \sim ((R,L),(L,L),(L,R)) \qquad (7.75)$$

In fact, as soon as this relation holds, all the others follow: between (R,R) and (L,L) or between (R,L) and (L,R), we conform to one of the two adjacent faces. This way of satisfying all the constraints by satisfying only one is optimal.

Remark that (7.75) relies on an agreement between the two ipsilateral pathways (R,R) and (L,L) that develop later than contralateral pathways (R,L) and (L,R), so a natural conjecture on the development of ipsilateral connections is that it is constrained by this relation and callosal activity. It is a kind of learning, probably supervised. This hypothesis is in accord with the role of the corpus callosum in the installation of binocularity (cf. [62]).

Another application of the gluing process is the relation between *stereoscopy* and *disparity*. Cf. the four articles written by Freeman, Parker, Schor and Blake in *TVN*.

Disparity measures the difference between the field kernels of the left eye and the right eye. (The callosal connections contribute to disparity too, but we will neglect this effect now.) Note that the binocular fusion of information cannot be a static absolute gluing: The eye muscles adapt perception to a certain depth, by *convergence*; this makes the phase and orientation of the stimulus change (more than the modulus of frequency). Consequently it is not completely meaningful to speak of an absolute disparity: depending on the position of the eyes in their orbits the parameters change.

The rotation between the two eyes gives two projections of a $3D$ object on two planes, which *theoretically* can be used to re-construct the $3D$ object. This will give a $3D$ kernel:

$$G(x-x_a, y-y_a, z-z_a, t)e^{i(x\xi+y\eta+z\zeta+t\omega+\varphi)} \tag{7.76}$$

As a product of Gaussian densities is a Gaussian density and the product of exponential is the exponential of the sum, the simplest attempt for an interaction of kernels is provided by multiplication before rectification of separate receptive fields. Note this multiplication concerns the spatial part, not necessarily the time part. We define the projections

$$X_R - X_R^a = A_R(x-x_a)+B_R(y-y_a)+C_R(z-z_a), \tag{7.77}$$
$$Y_R - Y_R^a = D_R(x-x_a)+E_R(y-y_a)+F_R(z-z_a), \tag{7.78}$$
$$t_R = t - t_a^R, \tag{7.79}$$
$$X_L - X_L^a = A_L(x-x_a)+B_L(y-y_a)+C_L(z-z_a)), \tag{7.80}$$
$$Y_L - Y_L^a = D_L(x-x_a)+E_L(y-y_a)+F_L(z-z_a), \tag{7.81}$$
$$t_L = t - t_a^L. \tag{7.82}$$

Then for the binocular cell we define

$$G(x-x_a, y-y_a, z-z_a, t) = \\ = G(X_R - X_R^a, Y_R - Y_R^a, t_R)G(X_L - X_L^a, Y_L - Y_L^a, t_L). \tag{7.83}$$

And we get the following assertion:

Proposition 4.3.2: If the receptive field of a binocular cell is obtained by multiplication of two independent uni-ocular linear receptive fields, then the preferred $3D$ frequency vector is given as a function of the Euler angles between the two eyes by the following linear formulas:

$$\xi = A_R\xi_R + A_L\xi_L + D_R\eta_R + D_L\eta_L, \tag{7.84}$$
$$\eta = B_R\xi_R + B_L\xi_L + E_R\eta_R + E_L\eta_L, \tag{7.85}$$
$$\zeta = C_R\xi_R + C_L\xi_L + F_R\eta_R + F_L\eta_L. \tag{7.86}$$

In these formulas, we have four fixed frequencies, expressing disparity when the eyes are parallel, and twelve parameters coming from six independent parameters of rotations of the two eyes. Remark that multiplication of the complex waves added the phases of the uni-ocular RFs; a more reasonable model would be provided by multiplication of the real parts, giving two components of frequency, one for addition of the phases and the other one for substraction. Also, observe that this simple multiplication of linear kernels would give $\omega = \omega_R + \omega_L$, which is difficult to justify.

More elaborate models, based on the same idea of non-linear composition of two linear fields were studied by Ohzawa, DeAngelis and Freeman; see [157], [158].

Experiments have put in evidence cortical maps for disparity (ex. in area 18), it would be nice if, in cats or macaques, they subtend a grouping by similar $3D$ orientation, or at least an invariance under $3D$ displacements.

Remark: many examples of subcortical gluing exist for the visual system. For instance ([58]) in frogs each retina sends axons to the contralateral tectum only, and there is no direct connection between the two parts of the tectum, but the nucleus isthmi (IN) connects both tectal lobes (T). It receives from ipsilateral tectal lobe and projects bilaterally. Remark: this is not the case in most species, where projection from IN to T is ipsilateral only, for example most fishes (exception for electric fish) and reptiles (exception turtle) and birds. In birds binocular information is already present in the thalamus. There is an homolog in mammals: a bilateral projection to the superior colliculus SC from the parabigeminal nucleus ([78] for cats, [15] for monkeys).

7.4.4 The Strange Gluing of the Tupaia

(This model was discussed during the ACI NIM (2005-2009) and profited in particular of discussions with Chantal Milleret and Nathalie Rochefort.)

The question was: why do the callosal connections of the Tupaia disregard orientation preference?

Tupaia, or three shrew, is a strange animal, which was considered a long time the direct ancestor of primate (cf. [63]). Although it has laterally situated eyes, by turning the retina inside the orbits it has developed a fine frontal vision, with a large binocular domain, apparently useful to detect and select its smaller prays. Thus we have here a strong example of adaptation, enforced by anatomy. The tupaia has precise callosal connections respecting retinotopy. Its area $V1$, including TZ (transition zones) has superb orientation maps, which are respected by numerous intra-hemispheric horizontal connections.

However amazingly Bosking et al. [32] have shown that the callosal connections of the tupaia do not set up any correspondence between the orientation maps in the two hemispheres.

To the contrary the cat has no problem setting up this correspondence (Rochefort et al. [186]).

We will show now that this apparent neglect of invariance results from the insistence of the tupaia to maintain a separation between the ON and OFF pathways in the cortex. This can be seen as the preservation of another structure of invariance, the symmetry between ON and OFF, which is a discrete structure, and which was abandoned in the cortex by other animals, like cats and monkeys. We will show that this ON/OFF structure blocks the callosal gluing of the action of translations and rotations on the representation of orientation, which is a continuous invariance structure on internal parameters.

Because the discussion is somewhat involved we give a summary of the argumentation: (1) In the tupaia the ON/OFF division implies a contralateral ocular dominance in the layers II/III; (2) the orientation preference and its selectivity rely on parallel arrangements of axons of ON and OFF cells and horizontal connections; from (1) and (2) it results that (3) at the time callosal connections are selected the orientation preference is already stabilized in $V1$ proper and in TZ, thus (4) it is too difficult for a callosal connection to respect both position and orientation.

Remarks: (1) means that the discrete symmetry between ON/OFF enters in conflict with the binocular symmetry in adult animals; (2) relies on the connectivity underlying the emergence of the internal representation of orientations; (3) underlines the role of time in development; (4) underlines possible competitions between two continuous representations, here retinotopy and orientation. We saw in section (on pi), that orientation maps has an invariance in law under planar displacements, here we see the difficulty to extend this invariance for both hemispheres and both ocular dominations. This shows that evolution had to fight for invariance.

(1) The organization of the visual systems of mammalian species lies on common principles, but it also shows a number of variations. Cf. Casagrande and Xu in TVN. One of the common principles is the division ON/OFF of the visual information flow. Certain bipolar cells in the retina specialize to be center-ON (excited by light in the center of the RF), others specialize to be center-OFF (excited when light is absent in the center of the RF) (cf. Nelson and Kolb in TVN), then two information pathways take birth: the ganglion cells, sending axons out of the retina, are principally of three types: ON-cells, excited when the intensity of light augments in the center (of RF), OFF-cells excited when the intensity of light diminishes in the center, and ON-OFF-cells excited when light augments or diminishes in the center. The first two sorts emit spikes as long as the intensity does'nt change, but the third one emits spikes only when intensity varies in time. Another division superposes to the ON/OFF division: ganglion cells can be tonic (permanent spiking) or phasic (transient spiking); in carnivores (as cat, dog, ferret or mink) this corresponds to X and Y cells respectively. There exists a third class, named W of smaller and slower neurons. Note the homology with primates is difficult to precise, because X and Y are large and quick as are the magnocellular M cells, and W cells correspond to parvocellular P cells, however P and M neurons of primates have physiological properties which correspond to the characters of X and Y respectively, and the K cells (koniocellular) of primates are similar to W cells.

On the road between retina and cortex, in the dorsal lateral geniculate nucleus of the thalamus (dLGN) of most species of mammals, there exist layers (less precise in small rodents but already visible in rats; they are six in the carnivores and two in the rats), which continue to segregate ON, OFF or ON-OFF, and phasic or tonic. However these layers are disposed differently according to the species. For instance the LGN of cat has six layers A, A1, CM, C1, C2, C3 (plus a MIN and a Wing), cf. [165] ch.I, and Sherman and Guillery in TVN; the three first ones are dorsally situated containing X and Y cells, the three others (collectively named CP) are more ventrally situated and containing W cells. The layer A receives both X and Y from the contralateral eye, it has a gradient from X outside-dorsal to Y inside-ventral, A1 is homolog to A for the ipsilateral eye, CM is contralateral but receives only Y, the CPs are all of type W: C1 ipsilateral, C2 contralateral and C3 binocular. From the ON/OFF point of view (cf. [35]), there is in A and A1 a gradient in the Y-region from Y-ON outside-dorsal to Y-OFF inside-ventral, and CM is dominated by OFF. However most specialists consider that the division ON, OFF in the LGN of cats is subtle. In ferret or mink the division is more pronounced, there exist two sub-layers in A and in A1, a dorsal dedicated to ON and a ventral dedicated to OFF. In macaques (as in known old world monkeys and probably in humans, but not in new world monkeys) from dorsal to ventral the layers are denoted from 6 to 1; observations in macaques show that the layer 6 is contralateral and ON, the 5 is ipsilateral and ON, the 4 is contralateral and OFF, the 3 is ipsilateral and OFF, the 2 is is ipsilateral and mixed ON-OFF, the 1 is is contralateral and mixed ON-OFF ([144]). The layers 1 and 2 receive M and K ganglion cells and project to superficial regions of layer IV in V1.

In cats and monkeys, when the information flow arrives in the cortex, it is difficult to observe a separation between ON and OFF; in ferrets it is possible to discern patches which are OFF (cf. [249]). But the tree shrew decided to maintain a strict division between ON and OFF in LGN and to maintain this division in the cortex. In this animal species the layers are organized as in primates from 6 to 1, but only the 6 (which is contralateral) presents a mixture ON-OFF, the 5 is ipsilateral and OFF, the 4 is contralateral and OFF, the 3 is ipsilateral and ON, the 2 is is contralateral and ON, the 1 is is ipsilateral and ON. Thus 4 behaves as in primates but 3, 2, 1 behave at the opposite. The layer 5 is OFF in tree shrew but ON in monkey; the mixing in 6 replaces the mixing in 2 and 1.

The primary visual cortex of the tupai is divided in many layers and sub-layers (cf. Fitzpatrick [66]); the most external sub-layer in the layer IV, which is named IVa, receives LGN axons from 1 and 2 which are exclusively center-ON (remark in the macaque the homologous sub-layer receives also from 1 and 2 but these layers are mixed ON-OFF), the sub-layer IVb receives from 4 and 5 which are exclusively OFF (in the macaque the homologous sub-layer receives also from 4 and 5 but 4 is OFF and 5 is ON thus the ON/OFF information is mixed). Consequently the tupai is ON in IVa and OFF in IVb. Moreover the up of IVa and the bottom of IVb are binocular, but the middle (bottom of IVa and up of IVb) is uni-ocular, exclusively contralateral; this is due to the fact that 2 and 4 are contralateral (which is not the case in macaque, where 4 is contra and 2 is ipsi). And it is only at this frontier between IVa and IVb that ON and OFF are mixing in tupai's V1. Cf. figure (7.6).

(2) The origin of columns of orientation preference, even if not fully understood, takes advantage on alignments of center-ON projections and center-OFF projections, as it was suggested initially by Hubel and Wiesel. Mooser, Bosking and Fitzpatrick have proved in 2004 ([146]) that orientation preferences in $V1$ of the tupai are established inside the cortex, not at the arrival from LGN. This result is compatible with the separation of ON and OFF until there. In tupai, the alignments of ON and OFF are formed by horizontal connections in $V1$; they can exist only higher than IVa, in sub-layers IIIb and higher, where the neurons of IVa (which are ON) and IVb (which are OFF) project. Note that certain neurons in IV of the tupai are selective for orientation, but the layer IV doesn't contain long horizontal connection. Moreover orientation selectivity is stronger in layers II/III than in IV (cf. [66]). The long distance connections joining similar orientation columns are mostly present in II/IIIa,b (Bosking, Fitzpatrick et al. [33]et al. and before them Rockland and Lund, [187]). Note these results confirmed the important role of horizontal connections for orientation selectivity. (Remark in the tupai the patches of axonal terminations are preferentially aligned with the preferred orientation ([76]), which is not sure in the cat or the macaque ([134]).)

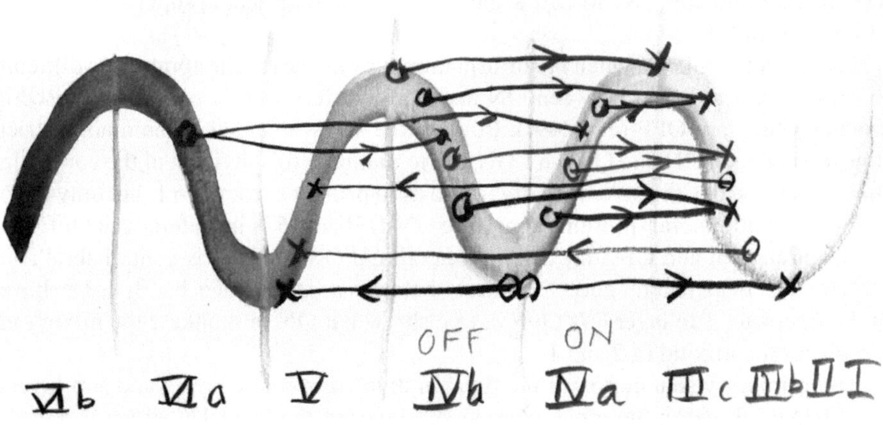

Fig. 7.6 Folding. The six layers of V1 of tupaia, with its vertical projection, according to Fitzpatrick [66]. Remark the functional folds, invariant by the connections.

To summarize what we wrote until now, the choices made by the evolution for the tree shrew have implied two consequences: (1) in the layers II/III the cortex is strongly dominated (if not exclusively) by the contralateral eye; (2) the orientation selectivity is mostly present in these layers II/III (or in the symmetric layers V/VI), which are also the layers where intra-hemispheric connections exist.

(3) The corpus callosum has its own developmental rules. Kretz and Rager ([117]) have studied this development in the tupai; they found that, from P7 to P9 few callosal connections appear in layer III but more appear in layer VI (where there is minor orientation selectivity at that time), from P10 to P13 supra-granular connections grow, their extension attains a maximum around P13, then between P13 and P15, 95% of these connections disappear! After this date the configuration is almost the same as the one in adult animals. Most of the connections now are belonging to III, they have disappeared from VI after P15, but the installation was driven by the connections in layer VI. Note that in the infra-granular layer VI the information on the place in visual plane and the orientation selectivity are installed between P7 and P17. At the end most callosal connections rely $V1$ (or 17) with the transition zone TZ (or 17/18), but there is also in tupai, contrarily to what happens in the cat and monkey, many heterotopic connections from TZ to 17 or 18, covering all the binocular field. Thus retinotopical gluing is also a little perturbed in tupaia.

In cats the construction period of callosal connections is much longer (cf. [145], [5]), but the principle is the same: elimination of exuberant connections which appear during the first days after birth (see [93]). In the first days after birth, callosal connections (or better varicosities) appear in all layers I,...,VI ([62]); then during two months these connections regress and subsist only between TZ and $V1$ or reciprocally. The same happens for the rat. In the case of ferret, inside the binocular region, which is exceptionally large (20 degrees in the contralateral field), most neurons are binocular but the ocular dominance has a patchy structure. The callosal connections are also patchy ([187]). There are more infra-granular callosal connections in ferrets than in cats ([36]). In old world monkeys (as us) callosal connections are formed before birth ([108]). Note this is an elegant way to solve the gluing problem: make connections before you choose preferences for position and orientation, and use these connections for coherence of preferences. Monkeys do that by establishing callosal connections before birth, and cats by forming the ipsilateral map by copying the contralateral one through callosal connections with TZ.

(4) Thus, in $V1$ of the tupaia the inter-hemispheric connections connect cells already dedicated to a well defined place, respecting visuotopy (cf. Olavarria [160]), however all these cells at that time have already chosen their preferred orientation, thus the callosal connections have few chance to respect orientation if it respect position.

The fact that orientation preference was prematurely determined is due to the dominance of the area by the contralateral eye, which results from the logic of the ON/OFF repartition.

So the impossibility of adaptation of connection for respecting orientation implies a defect of invariance.

(5) Supplementary discussion: we examine now the link between what precedes and the hypothesis that callosal connections are mainly driven by information coming

from one eye. This driving is the uni-ocular hypothesis of Olavarria that we discussed in section 4.3.

At first sight this hypothesis seems to contradict the well established fact that horizontal connections respecting the orientation preference exist between neurons which different ocular dominance as frequently as between neurons with the same ocular dominance (cf. [165]). However, in all species we are considering, the orientation maps have began to form before the eyes are opening almost without visual experience (cf. [233]), thus orientation maps begin to form in a cortex which is dominated by the contralateral eye. Then the orientation map for the columns that are dominated by the ipsilateral eye have to follow the contralateral map, extending it by continuity, with a relative freedom for variability. Note this second map can become dominant after occlusion of the contralateral eye.

A convincing explanation for the reinforcement of a synapse is the idea of Hebb, that the synapse becomes stronger when the postsynaptic neuron is excited after the presynaptic one was excited. According to Löwel and Singer, [132], cats which undergone a unilateral resection of the medial rectus muscle between age P15 and P21, are developing long range connections between domains which are controlled by the same eye, and according to Schmidt et al. [198], in cats with strabismus, the selectivity to orientation is not modified but the long range connections connect domains with similar orientation preference.

The biochemical and genetical processes, involving cadherins, NMDA-NRs, CREB, PKA,..., which are responsible of the stabilization of intra-cortical synapses, are changing their expression with developmental time; the fact that horizontal connections between opposite eyes arrive later than the ones between the same eye could be a sufficient reason for a different chemical expression, not relying on coincidence.

Before eye opening there exists a cortical activity which is modulated by visual stimuli and by natural motions ([6]). Spontaneous activity happens in the retina, in LGN and in the cortex (cf. [165]). For instance, in the LGN, it exist inter-hemispheric correlations of activities (cf. [231]), however most of these correlations should imply uni-ocular correlations and the pathway with same preference ON or OFF. Importantly there exists a difference of activity that distinguishes between contralateral and ipsilateral domination: there is more correlation for neurons informed by the same eye, and even more in the case of contralateral eye, and they can group ON or OFF separately or ON and OFF together; this activity can contribute to the formation of orientation columns. After eye opening exchanges of information exist between neurons of different ocular dominance: an ipsilateral map extends the contralateral map, and hetero-ocular connections happen.

According to Weliky and Katz [230], studying the development of the ferret, the horizontal connections begin to form at the moment where eyes are opening, at a time where the contralateral eye dominates. After eyes opening the orientation maps development becomes dependent on visual experience (NMDA regulation and CREB expression).

In tupaias, the layers II/III in V1 are monocular, thus intra-hemispheric horizontal connections are necessarily contralateral-contralateral, and inter-hemispheric connections are necessarily contralateral-ipsilateral, consequently they cannot be homo-ocular and going from II/II to II/III. Casagrande and Condo ([43]) have shown that, after unilateral enucleation at birth, the LGN layers of the tupaia controlled by the resting eye are preserved, but in the cat they are not; this indicates that the visual system of the tupaia is more uni-ocularly driven than the visual system of the cat. In cats and monkeys interaction between eye's domination appears important for normal development.

7.5 Higher Levels of Invariance

7.5.1 General Covariance in $V1$

(Prepared during ACI NIM 2005-2009, profited of many discussions with Jean Petitot and Pierre Baudot, and in other contexts with Jean-Pierre Nadal and Alessandro Sarti.)

Our main hypothesis about $V1$ is the presence of a *continuous invariance*, compatible with topological invariance. This invariance, at the opposite of a global invariance (as it happens for example in the fusiform gyrus for faces), has a local character: it postulates that stimuli which can be deduced one from each other by smooth deformations generate equivalent activities of different assemblies. For instance, it is known that a static line of dots L with convenient orientations generate a rebound activity $s \mapsto A(s,L)$ of the neurons in $V1$; cf. [111]. And there is no evidence that in $V1$ straight lines give more activation than curved lines, with not too large curvature. Then our hypothesis is that, if φ is a sufficiently small and sufficiently smooth homeomorphism of the visual plane, in particular not disturbing too much the curvature, there exists a bijective remapping $T(\varphi)$ of the $V1$ subunits such that the activity $s \mapsto A(s,\varphi(L))$ in $V1$ which is generated by the presentation of the line $\varphi(L)$, is (almost) equal to the activity $s \mapsto A(T(\varphi)^{-1}(s),L)$. The term "almost" here can be understood in a probabilistic sense, in accord with models by random processes.

The movement of a dots line L_t (where t denotes the time coordinate) generates an activity in time A_t, which cannot in general be deduced by just adding the above functions $A(L_t)$ (cf. [107]); our hypothesis in this case is the existence for any family of small deformations φ_t in the visual plane, of a family of bijections T_t such that for almost all s we get $A_t(T_t(s),\varphi_t(L_t)) = A_t(s,L_t)$. Another example is given by a sequence in time, i.e. the dots are not presented all together but one after each other in a certain order (cf. [129]). Here it is more convenient to speak of co-variance than invariance.

In $V1$ we find an approximate symplectic co-variance; it correspond to the symplectic group in dimension $2(2D+1)$ acting on the co-tangent space of the product of the visual plane by the time axis, and it is extended to $2(3D+1)$ if we take in account disparity. The parameters that are concerned are position and frequencies, spatial and temporal; they form a vector in a space of dimension 8. These parameters are almost uniformly distributed; in fact there are two populations, one for

for high spatial frequency and low temporal frequency, another one for the opposite. But this symplectic repartition of receptive fields is not sufficient to explain the diffeomorphism invariance; we need for that an organization of ensembles, for instance a convenient map of preferences, and horizontal connections, lateral exchanges, non-linear excitation-inhibition, loops with the Thalamus and feed-back return from the cortex. See Bullier and Van Essen in TVN. For instance $V2$ activity commands illusory lines that can appear in $V1$. Cf. [123]. Remark that an orientation maps seems to be unnecessary for high discrimination, as shows the example of the squirrel (cf. [219]).

Schematically, one can say that $V1$ (and probably $V2$) has a continuous horizon for its functional invariance, and is aimed to construct a sort of differential topology and differential geometry by itself, taking in account information of other areas. This doesn't exclude elaborate individual responses of neurons in $V1$, whom receptive field take in account complex characteristics of natural images. Cf. P.Baudot results ([19]). To the contrary, a high dimensional covariance allows the computation of higher correlations (or higher mutual information) in images, for recognition of complex forms. From its differential topology and differential geometry the area $V1$ contributes to the recognition of discrete invariants.

The areas $V1, V2$ can be seen as vectors of operators acting on image-movements (and transforming it into spikes fields), thus the more adapted pseudo-group for them is not the pseudo-group of diffeomorphisms but its extension, the pseudo-group of symplectomorphisms , acting on the cotangent space over space-time, by respecting the natural symplectic structure (cf. [207], [228]). This is also the natural symmetry of light rays bundles. As we suggested in 3.1 pseudo-differential operators represent function of assemblies in $V1, V2$, and the natural (pseudo)-group of symmetry for them is made by symplectomorphisms ([228]). In certain problems a more natural choice could be the Contact transformations, i.e. homogeneous symplectomorphisms; in this case the role of space is played by contact elements, i.e. pairs of a point and a co-direction.

This general covariance in $V1$ could also be extended to color information, a good candidate for that being a pseudo-group of Gauge transformations over the symplectomorphisms pseudo-group. Note this pseudo-group can be seen as a pseudo-group of transformations that respect a Poisson structure. Cf. [229].

7.5.2 Projective Invariance in $MT+$

The invariance in areas MT, MST, which are interested by movements in the image, is probably of high dimension, as in $V1, V2$; however in these areas it appears also a finite dimensional invariance given by projective deformations that we want to describe. It would be interesting to look at the possibility of a localization of this projective geometry in MT or MST.

A simple model of linear summation of non-linear activities of $V1$, followed by a static non-linearity is sufficient to explain how receptive fields in MT of primates detect the speed of objects with respect to the retina (see [193]).

Neurons in the MT complex (MT, MST) integrate the *optic flow*.

Our presentation of the optic flow is inspired by Koenderink and VanDoorn [112], [115] and Ken-Ichi Kanatani (IEEE 1988) [103].
We consider the Euclidean space $E = \mathbb{R}^3$, I_3 denotes the canonical frame, and G the group of displacements (direct isometries).

The movement of an orthonormal frame is described by a smooth map $R : I \to G$, where I is an open interval in $\mathbb{R}$. We note O_t the center of the moving frame, and $\rho(t)$ in SO_3 the linear part of $R(t)$. It is easy to show that the derivative of R in $t_0 \in I$ can be written

$$DR(t_0).s = sV(t_0) + sA(t_0)\rho(t_0), \tag{7.87}$$

where $V(t_0)$ is a $3D$ vector, named the linear derivative of R en t_0 and where $A(t_0)$ is a 3×3 antisymmetric matrix, named the angular speed of R in t_0.

We introduce the $3D$ vector Ω whose components in the fixed frame I_3 are $\Omega_1 = -a_{23}, \Omega_2 = +a_{13}, \Omega_3 = -a_{12}$; then for every $X \in \mathbb{R}^3$ we have:

$$A.X = \Omega \times X, \tag{7.88}$$

where the crux denotes the usual vector product.

This vector Ω is named the angular speed vector (viewed in the fixed frame).

An orthonormal base change P that transforms A into $A' = PA^tP = PAP^{-1}$ changes Ω into $\Omega' = P\Omega$.

Let Q be a fixed point in E; denote by $q(t)$ the trajectory of Q in the moving frame R; we get

$$\frac{dq}{dt}(t) = -v(t) - \omega(t) \times q(t); \tag{7.89}$$

where $R(t)v(t) = V(t)$ and $\rho(t)\omega(t) = \Omega(t)$.

Proof: we have $q(t) = R(t)^{-1}Q$, thus

$$\frac{dq}{dt}(t_0) = -R(t)^{-1}DR(t)q(t) \tag{7.90}$$

$$= -R^{-1}(V + A\rho q) \tag{7.91}$$

$$= -R^{-1}V - \rho^{-1}A\rho. \tag{7.92}$$

We finish by using the correspondence between A and Ω.

When Q is moving in I_3, its trajectory $Q(t)$ has a speed $\dot{Q}(t)$; we obtain

$$\frac{\partial Q}{dt}(t) = -V(t) + \dot{Q}(t) - \Omega(t) \times Q(t). \tag{7.93}$$

This gives in the moving frame

$$\frac{dq}{dt}(t) = -v(t) + R^{-1}(t)\dot{Q}(t) - \omega(t) \times q(t). \tag{7.94}$$

We will write $v_q = v - R^{-1}\dot{Q}$.

We assume $Q \neq O_t = R(t)(O)$, then we define the *proximity* $\mu = 1/\|q\|$, and the *visual point* $\sigma = \mu q$. The norm of this vector is equal to 1, we consider it as a moving point in the unit sphere Σ of R_t, it represents how appears Q for a cyclopean observer in O_t with eye moving like $\rho(t)$ with respect to the fixed reference frame I_3.

Proposition 5.2.1: in the moving frame we have

$$\frac{d\sigma}{dt}(t) = \mu(t)(v_q(t) \times \sigma(t)) \times \sigma(t) - \omega(t) \times \sigma(t). \tag{7.95}$$

Proof: the derivative of $\|q\|$ is $\mu q.\dot{q}$; then the derivative of μ is $-\mu^3 q.\dot{q}$, consequently

$$\frac{d\sigma}{dt} = -\mu^3(q.\dot{q})q + \mu\dot{q} \tag{7.96}$$

$$= \mu((-\sigma.\dot{q})\sigma + (\sigma.\sigma)\dot{q}). \tag{7.97}$$

However, for any triple of vectors (a,b,c) we have

$$(a \times b) \times c = (a.c)b - (b.c)a. \tag{7.98}$$

Thus

$$\frac{d\sigma}{dt} = \mu(\sigma \times \dot{q}) \times \sigma \tag{7.99}$$

$$= \mu((v_q + \omega \times q) \times \sigma) \times \sigma \tag{7.100}$$

$$= \mu(v_q \times \sigma) \times \sigma + ((\omega \times \sigma) \times \sigma) \times \sigma \tag{7.101}$$

$$= \mu(v_q \times \sigma) \times \sigma - \omega \times \sigma. \tag{7.102}$$

Observe that $v_q \times \sigma) \times \sigma$ is the projection of v_q on the tangent plane of Σ.

By considering a point Q in each direction σ of Σ the formula of prop.1 gives a vector field Φ, depending on time, tangent to the sphere Σ:

$$\Phi(\sigma) = \mu(\sigma)(v(\sigma) \times \sigma) \times \sigma) - \omega \times \sigma. \tag{7.103}$$

It is this vector field which is called the *optic flow*.

In general Φ has discontinuities along the apparent contour of visible surfaces, then it cannot be integrated in transformations of Σ. But we will consider a natural case where where Φ is of class C^1.

Remark that to compute with Φ it is better to project Σ on a plane attached to the moving frame. This corresponds to the usual rules of optics: the gaze is directed along e_3 and the image is interpreted as a projection from O_t on a plane perpendicular to e_3.

Let ρ be the distance from O_t to this plane, and (x,y) the Cartesian coordinates of the image of σ in a frame e_1, e_2 (fixed with respect to $R(t)I_3$). Let us note v_1, v_2, v_3 and $\omega_1, \omega_2, \omega_3$ the coordinates of v and ω respectively. If the coordinates of σ in $R(t)I_3$) are (x_1,x_2,x_3), we have

$$x = \rho \frac{x_1}{x_3}, \quad y = \rho \frac{x_2}{x_3}. \tag{7.104}$$

Then

$$\frac{dx}{dt} = \frac{\rho}{x_3}\frac{dx_1}{dt} - \frac{x}{x_3}\frac{dx_3}{dt}, \quad \frac{dy}{dt} = \frac{\rho}{x_3}\frac{dx_2}{dt} - \frac{y}{x_3}\frac{dx_3}{dt}. \tag{7.105}$$

On another side Φ can be written in the coordinates x_k as follows:

$$\frac{dx_1}{dt} = \mu[(v.\sigma)x_1 - v_1] - \omega_2 x_3 + \omega_3 x_2, \tag{7.106}$$

$$\frac{dx_2}{dt} = \mu[(v.\sigma)x_2 - v_2] - \omega_3 x_1 + \omega_1 x_3, \tag{7.107}$$

$$\frac{dx_3}{dt} = \mu[(v.\sigma)x_3 - v_3] - \omega_1 x_2 + \omega_2 x_1. \tag{7.108}$$

Thus

$$\frac{dx}{dt} = \mu[(v.\sigma)x - \rho\frac{v_1}{x_3}] - \rho\omega_2 + \omega_3 y \tag{7.109}$$

$$-x\mu[(v.\sigma) - \frac{v_3}{x_3}] + \frac{1}{\rho}\omega_1 xy - \frac{1}{\rho}\omega_2 x^2 \tag{7.110}$$

$$\frac{dy}{dt} = \mu[(v.\sigma)y - \rho\frac{v_2}{x_3}] + \rho\omega_1 + \omega_3 x \tag{7.111}$$

$$-y\mu[(v.\sigma) - \frac{v_3}{x_3}] + \frac{1}{\rho}\omega_1 y^2 - \frac{1}{\rho}\omega_2 xy. \tag{7.112}$$

or in other terms

$$\frac{dx}{dt} = \mu\frac{xv_3 - \rho v_1}{x_3} - \rho\omega_2 + \omega_3 y + \frac{x}{\rho}(\omega_1 y - \omega_2 x), \tag{7.113}$$

$$\frac{dy}{dt} = \mu\frac{yv_3 - \rho v_2}{x_3} + \rho\omega_1 + \omega_3 x + \frac{y}{\rho}(\omega_1 y - \omega_2 x). \tag{7.114}$$

Here x_3 is given by the formula

$$\frac{1}{x_3} = \sqrt{1 + x^2 + y^2}, \tag{7.115}$$

which says that σ has norm 1.

A first use of the registering of an optic is for estimating the distance of the subject (assimilated to O_t) to objects in the space E. We simplify by considering only N points Q_i.

Theorem 5.2.1: Assume $N \geq 6$, all σ_i distinct, and O_t does'nt displace in the direction of at least 6 of the Q_i; then the optic flow equations allows to determine v, ω and all the μ_i such that O_t does'nt displace in the direction of Q_i.

This an application of the implicit function theorem.

Now we consider the projective invariance hidden in the optic flow. For that purpose we look at points Q that belong to a fixed plane Π of the space E. Then, for each time, fro each point σ in Σ, the proximity $\mu(\sigma)$ is equal to the quotient $\delta(\sigma)/\delta$ of the distance $\delta(\sigma)$ of O_t to the plane $\Pi(\sigma)$ that is parallel to Π containing σ, by the distance δ from O_t to Π. (Just apply the Thalès theorem.)

But $\delta(\sigma)$ is a linear combination of the coordinates x_k:

$$\delta(\sigma) = \alpha x_1 + \beta x_2 + \gamma x_3. \tag{7.116}$$

By combining with the above equation of Φ in the coordinates x, y, this gives

$$\frac{dx}{dt} = \frac{1}{\delta}(xv_3 - \rho v_1)(\alpha x + \beta x + \gamma) - \rho\omega_2 + \omega_3 y + \frac{x}{\rho}(\omega_1 y - \omega_2 x), \tag{7.117}$$

$$\frac{dy}{dt} = \frac{1}{\delta}(yv_3 - \rho v_2)(\alpha x + \beta x + \gamma) + \rho\omega_1 + \omega_3 x + \frac{y}{\rho}(\omega_1 y - \omega_2 x). \tag{7.118}$$

We obtain an element of the eight dimensional family of vector fields that can be written

$$\frac{dx}{dt} = u + Ax + By + x(Ex + Fy) \tag{7.119}$$

$$\frac{dy}{dt} = v + Cx + Dy + y(Ex + Fy). \tag{7.120}$$

This family has a simple geometrical interpretation: the projective group $PGL(E)$ (that is the quotient of the linear group $GL(E)$ by the scalar multiplications) acts on the set $\mathbb{P}(E)$ of straight lines through O_t in E. This set is a (non-orientable) manifold of dimension 2. If we consider the coordinates x_1, x_2, x_3 on E, an open set U_3 of $\mathbb{P}(E)$ is formed by the lines that meet the plane $x_3 = \rho$, where x, y are coordinates. Every matrix in $SL_3(\mathbb{R})$ (of determinant equal to 1) sufficiently near the identity can be written $\exp(X)$ where X has trace zero, i.e. $X \in sl_3(\mathbb{R})$. Let us define

$$X = \begin{pmatrix} a & b & c \\ d & e & f \\ g & h & k \end{pmatrix} \tag{7.121}$$

The fact that $X \in sl_3(\mathbb{R})$ is expressed by $k = -a - e$.

Proposition 5.2.2: The derivative $d(\exp(\varepsilon X))/dt$ in $\varepsilon = 0$ induces in U_3 a vector field of the preceding form.

Proof:

$$\frac{dx}{d\varepsilon}|(\varepsilon = 0) = \frac{\rho}{x_3}\frac{dx_1}{d\varepsilon}(0) - \frac{\rho x_1}{x_3^2}\frac{dx_3}{d\varepsilon}(0) \tag{7.122}$$

$$= \rho(ax + by + c) - \frac{1}{\rho}x(gx + hy + k). \tag{7.123}$$

$$\frac{dy}{d\varepsilon}|(\varepsilon=0) = \frac{\rho}{x_3}\frac{dx_2}{d\varepsilon}(0) - \frac{\rho x_2}{x_3^2}\frac{dx_3}{d\varepsilon}(0) \tag{7.124}$$

$$= \rho(dx+ey+f) - \frac{1}{\rho}y(gx+hy+k). \tag{7.125}$$

Remark: by taking in account the relation $k = -a-e$, and introducing $\nu = \rho+\rho^{-1}$, we get

$$\frac{dx}{d\varepsilon} = \rho c + (\nu a + \rho e)x + \rho by - \frac{x}{\rho}(gx+hy) \tag{7.126}$$

$$\frac{dy}{d\varepsilon} = \rho f + \rho dx + (\rho a + \nu e)y - \frac{y}{\rho}(gx+hy) \tag{7.127}$$

The above proposition tells that the optic flow for a fixed plane is the generator of a projective transformation.

An orthonormal change of frame P transforms $X \in sl_3(\mathbb{R})$ into $PXP^{-1} = PX^tP$; this defines a linear action of SO_3 on sl_3, that is a vector space of dimension 8. This linear representation is the direct sum of two irreducible representations, the first one, of dimension 3, on the antisymmetric matrices, identified to the rotation speed vector ω, the second one, of dimension 5, on the vector space of symmetric matrices with trace zero. We will note them respectively E_3 and E_5, and we will denote by a and B respectively their generic elements.

We also introduce the five following functions: $f(a) = \|a\|^2$, $F(B) = \|B\|^2 = Tr(B^2)$, $G(B) = Tr(B^3)$, $g(a,B) = a.Ba$, $h(a,B) = a.B^2a$. These functions are smooth and invariant under the action of SO_3. We will note $\varphi = (f,F,G,g,h)$ the map from sl_3 to $\mathbb{R}^5$.

Note there is a natural prolongation of the action of SO_3 on $E_3 \oplus E_5$ to the full orthogonal group O_3; this prolongation is the natural action of O_3 on E_3 and the action by conjugation on E_5; only this second one comes from the action by conjugation of O_3 on sl_3. It is easy to show that O_3 acts transitively on each fiber of the map φ from sl_3 to $\mathbb{R}^5$. We deduce that all the fibers of φ are smooth submanifolds of sl_3, because they identify to homogeneous spaces O_3/H. In general this manifold has dimension 3. However there are points in sl_3 where φ is singular (i.e. its derivative is not surjective), for instance when $a=0$ and B has a symmetry of revolution, in this case the orbit is a 2D sphere, or when $a=B=0$ where the orbit reduces to a point.

Let W be the set of elements a,B of sl_3 such that a,Ba,B^2a are linearly independent in $\mathbb{R}^3$, and that the principal values $\lambda_1,\lambda_2,\lambda_3$ of B are two by two distinct and non-zero; then W is an open and dense subset of sl_3, invariant under the action of SO_3, and in this open set all points are regular for $\varphi = (f,F,G,g,h)$, thus the application φ from W into $\mathbb{R}^5$ defines a fibration.

Remark that in W there is a discrete invariant for the action of SO_3 in addition of the five functions f,g,h,F,G; this is the orientation of E defined by the frame a,Ba,B^2a. This defines the two connected components of W. The existence of this discrete invariant implies that the fibers of φ are not orbits under SO_3, and that there exists smooth functions that are invariant under SO_3 that are not functions of f,g,h,F,G. However it is easy to prove from restriction to W the following result:

Theorem 5.2.2: the smooth functions that are invariant under the action of O_3 are the C^∞ function of the five fundamental invariant f,g,h,F,G.

All that connects nicely invariance in vision with classical invariant theory (cf. [232]).

7.5.3 Examples of Geometric Cells

The perception of position in space is organized by the group G of Euclidean displacements. The geometric cells for this geometry are cells that gives an information in a space G/H for a subgroup H of G (see Appendix 1). This sub-group describes the changes in rigid spatial frames that let the cell indifferent.

The examples below are issued from studies in the para-hippocampal system of rats, and rats have developed in their para-hippocampal region a system that is adapted to horizontal displacements thus we will restrict ourselves to the group formed by displacements in the horizontal plane, and we denote by G this group. However primates, in particular humans, possess in the same region cells adapted to displacements in the visual plane (Rolls, [189], and Eckstrom, Fried, cf. [148]), and that bats have a complete $3D$ system ([247]).

The para-hippocampal region stays at the end of the temporal visual pathway. It also belongs to the limbic system, containing the entorhinal cortex (EC), the perirhinal cortex (PRC) and nuclei of Amygdala and Septum. It is a set of areas implicated in sentiment expression.

A very nice reference for this section is [10].

The first invariance that was proved experimentally is *place invariance*: the *place cells* (PCs) were discovered by O'Keeefe and Dostrowski in 1971, in the Hippocampus proper, CA1, CA3 et DG (dentate gyrus). Such cells exist also in the Subiculum, but apparently not in the entorhinal cortex (EC). Here the group H of a cell depends on a place P in the horizontal plane E, it is the subgroup of rotations around P. The quotient $G/H(P)$ identifies canonically with E, but this plane is marked with a point. All the groups $H(P)$ are conjugated in G. The translation from P to P' give a canonical isomorphism from $H(P)$ to $H(P')$.

The second system is the *direction invariance*; here the indifference is the subgroup T of translations. Cells with this invariance are called head direction cells (HDCs), they were found first by Jim Ranck in 1985. Then Taube, Muller and Ranck (1990) conducted a systematic study of HDCs. Such cells exist in the anterior dorsal nucleus of the Thalamus (ADN), the lateral mamillary nucleus (LMN), the medial dorsal entorhinal cortex (dMEC), in the parasbiculum (PaS) and the postsubiculum (PoS), which is part of the presubiculum (PS). A node for the HDC formation is the tegmental dorsal nucleus (DTN). Remarkably this node is at the intersection or three rods of information: a) on proper movement (Habenula), b) on vestibular flow (medial nucleus), c) on visual information (via PoS). All the HDCs share the same indifference group. The information quotient G/T is the circle of oriented directions of the head in the horizontal plane; it is a group isomorphic to $\mathbb{R}/2\pi\mathbb{Z}$. Then this case is Galoisian: the information itself forms a group. We suggest that this is the reason why direction remapping is coherent, contrarily to the case of place cells: when *HDCs* change their preferences, in general they change by the same angle (there are exception in ambiguous cases where part of the population changes and part does'nt), but when PCs change their preferences, in general there is no relation between the new places of different cells.

Grid invariance is stranger; it was discovered in 2005 in dMEC by Torkel Hafting et al. [80] in the laboratory of M-B.Moser and E.Moser in Trondheim. They are also present in PaS and PoS. In this case the indifference groups of the grid cells (GCs) are discrete subgroups of G, they are the groups H_Γ of direct isometries that fix an equilateral lattice Γ. The group H_Γ is generated by a discrete subgroup of translations T_Γ and by the rotations whose angle are integer multiples of $2\pi/3$ around the vertexes of Γ. The Galois information G/G_Γ is a smooth oriented closed $3D$ manifold that fibers over a two dimensional torus of phases T/T_Γ. Recent data indicate that the directions in T_Γ are the same for all GCs in the same hemisphere; these directions undergo a coherent remapping after change in the environment, for instance rotations of visual cues, but the phases change incoherently like places for PCs.

Frame invariance also exists in the dMEC, PaS and PoS; in this case indifference is reduced to the identity Id of G. The corresponding cells have a preferred place and a preferred direction. There are also cells which prefer a grid and a direction; we could call them *Toric cells* because the information content of the population coincide with a torus T/T_Γ; for this population the life plane is periodic. (See [10].)

Topological invariance is represented by cells coding for coins, boundaries or open sets; they were discovered by Trygve Solstad et al. [206], also in Trondheim with M-B.Moser and E.Moser, in dMEC, PaS, PS. In this case we cannot continue to work with the group of displacements as Galois group, we must start with the group G of diffeomorphisms of the plane, and the indifference groups H are subgroups that fix topological subsets, like coins or boundaries. Note the additional symmetry $ON - OFF$ for cells active on a boundary and cells active in the complementary set of this boundary.

7.5.4 View-Invariant Recognition, Manifolds of Sections

In TVN ([44]), many papers are concerned with this topic, for instance R.F. Hess (p. 1043), C.E. Connor (p. 1080), E.T. Rolls (p. 11665), N. Kanwisher (p. 1179), L. Spillman and W.H. Ehrenstein (p.1573), M. Riesenhuber and T. Poggio (p. 1640), W. Singer (p. 1665), C. Koch and F. Crick (p. 1682).

A fundamental property of visual processing and visual perception is the existence of invariant recognition (cf. Bienenstock and Von der Malsburg [29], Wallis, Rolls and Foldiak [226], Lowe [130], Riesenhuber and Poggio [185], Masquelier and Thorpe [138], ...): different views of an object (or a person or an event), give rise to the perception of this object as a unit (or this person or this event respectively), its identification as a specific entity. A related property is the recognition of similarities between forms or objects. In particular in humans the performances in recognition and detection of similarities are exceptionally good. A lot of researches focused on modeling these invariance properties; the proposed mechanisms were frequently based on temporal association (Foldiak), synchrony (Singer), or polychrony (Tallon-Baudry), ..., cf. the collection edited by Von der Malsburg, Phillips and Singer [224]. A multiplicity of factors are contributing here: compensation, illusions, learning and so on. Most studies insisted on hierarchical processing (a gradual emergence of view-invariant representation, by successive network's computations). Other studies insisted on the structure that prepares invariance in the visual areas, and the variety of forms of invariance beyond (cf. in particular Rolls and Stringer [191], Cadieu, Poggio et al. [41] and Lowe [130]).

In the following paragraphs we try to join our schema of (co-homological) adaptation (2.4) with the remarks we have made on general co-variance in $V1$, for proposing a model of what could be a "stable image" in the brain, and how it could refer to "stable objects". Note we assume that "images" are nothing else than collections of brain cells activities in visual (or multisensory) areas, and that the relations of "images" to "objects" are co-activations with other brain areas, in the sense of temporal association sequences, or at least coherent modulation.

The primary visual areas $V1,V2$ are the reservoir of all images. It is established that even for visual illusions, the primary areas become active. For poor stimuli, the cells in $V1,V2$ behave according to a small number of parameters, but these parameters are non-linearly modulated by horizontal and feedback connections, taking in account larger part of the image. For complex stimuli, as natural images in motion, the cells in $V1,V2$ behave differently, detecting more complex characteristics in the image (cf. [19]). Thus, collectively, the primary activity is adapted to a high dimensional invariance (practically infinite) that we have named general covariance. However, the inner representation of a "stable image of the world around" seems to lack in $V1,V2$. In fact this representation can correspond to the joint activity of higher visual areas, even without units cells for a total view. Note that every image must be the "image of something". We suggest that the "things" in the image correspond to "vanishing cycles", in higher visual areas, in the sense of the internal spaces of 2.4, equivalence classes of attractors combinations. The following paragraphs try to further formalize this suggestion.

1) Let M denotes the manifold of functional parameters of all the cells in $V1,V2$. Remind the parameters of each individual cell form an homogeneous space G/H, for instance the product of the color space with the co-tangent space of a $3D$ model vector space; the color space is homogeneous for the affine group (cf. 3), and the co-tangent spaces made by the frequencies, temporal or spatial and position (retinotopy), is homogeneous for the symplectic group of light rays; then the group G can be taken as the product of these two groups. By retinotopy and disparity (managing stereoscopic depth for binocular cells), the manifold M is fibered over

the $3D$ ambient space E. This allows a mirror in the brain of localization of the images and the objects in the world (cf. [19]). The trace of an image in $V1,V2$ is a mapping from $V1,V2$ to M, i.e. an element of the space of sections $\mathscr{M}$ (cf. 2.4). By general co-variance the spaces of categories $\mathscr{I}$ and the space of contexts $\mathscr{E}$ for $V1,V2$ are infinite dimensional in $V1,V2$. An example of element in $\mathscr{I}$ is a curved contour.

2) Let A be a higher area, for instance $V3,V4,V5,V6$ or $V8$ or IT in the inferotemporal cortex. The cells in A have in general large receptive fields allowing them to integrate correlations. Our hypothesis is that these cells can adapt their activity to a specific global invariance structure described by a finite dimensional Lie groups. For example this group can be the Euclidean group in area OT, the conformal group in $V4,V8,IT$ or the projective group in $V5,MT,MST$. To express this hypothesis, we introduce the sets M_A that represent parameters of the activity of A (cf. 2.4). The vector space V_A describes the degrees of freedom of the response of the area. We also introduce the sets I_a of characteristic cycles of dynamics and their product $\mathscr{I}_A$. *A priori* the activities are functions on the product of $M_A \times \mathscr{I}_A \times V_A$, which is infinite dimensional. However our hypothesis will ask that these activities depend only on the projection to a finite dimensional quotient $M'_A \times \mathscr{I}'_A \times V'_A$. Moreover we will require the existence of a Lie group G'_A acting on M'_A, $\mathscr{I}'_A$ and V'_A, and of a period mapping $\widetilde{P}'$ satisfying the equation of adaptation (7.9) with respect to this group. In particular the hypothesis implies a dimensional reduction in each higher visual area. Remark that, to get sub-sets $M'_A, \mathscr{I}'_A, V'_A$ and residual symmetries G'_A, a natural procedure would consist to introduce systems of differential equations on M_A and $\mathscr{I}_A$, such that M'_A, $\mathscr{I}'_A$ and V'_A are their solutions. These systems should express "association rules" in A.

Note that reduction of invariance to a Lie group G'_A can be accompanied by a large indifference; an example is in the anterior temporal lobe, where certain cells detect only topological aspects of figures (cf. [252].)

The "image in A" itself corresponds to a vector valued function s_A on M'_A, representing the response associated to the element in $\mathscr{I}'_A$ corresponding to the visual input coming through the areas $V1,V2$, and a contextual modulation in $\mathscr{E}_A$. Such a function on $\mathscr{I}'_A$ can be seen as a co-homology class for the space $\mathscr{I}$ that is associated to the primary areas.

We must also consider families of operators that relate different higher areas A and B, to express how different areas interact. These operators are induced by non-linear pseudo-differential operators (of the type introduced in section 2.1), and compatible with the invariance structures. They represent the internal work on the image in the brain. In particular there contain feedback information from A to $V1,V2$; it is known that, even at the first instants of analysis, the primary activations are modulated by contextual effects, and higher recognition through feed-forward precision, horizontal connections, inhibition and feed-back connections. (Cf. [164] for the representation of curvature in $V4$.)

Example 1: in $V4$ or $V8$ a cell can represent a color in a region that takes in account all the colors in $\mathscr{I}$; this resulting color is closer to sensation than the colors in $\mathscr{I}$.

Example 2: in IT a cell can represent a set of parallel lines localized in a strip of the horizontal plane.

Interestingly, the structure of variability in $V1$ and $V4$ are similar ([205]).

3) The next step for modeling recognition consists in describing how the above operators can define more abstract and stable equivalences classes, forgetting about the details in the various representations s_A and giving unity to the "image". We suggest that this step needs a further co-homological construction. Note that the result of this construction is not necessarily new information in other brain areas, it could be that, but it could be more simply a sort of synergy between sensory areas corresponding to a specific distribution of activities in these areas, as described by Hebb (cf. [82], [224]).

An example from Mathematics for this second level of identification is given by the concept of isomorphism classes of geometric objects. For instance for smooth manifolds: at the first level, the notion of atlas on a set X is introduced, an atlas is a covering by compatible charts, then a manifold is defined as an equivalence class of atlases on X; at the second level appear the isomorphism classes of manifolds. But these classes can also be defined as elements in the first co-homology of a sheaf over the category of open sets in numerical spaces with value in the pseudo-group of local smooth isomorphisms; with this road, at first level are co-cycles, at second level are co-homology classes.

A paradigm for the ultimate equivalence relation is the De Rham operator d on differential forms over a smooth manifold (cf. [50]): in this case co-homology classes are forms α verifying $d\alpha = 0$ modulo forms which are $d\beta$. The analog of the condition $d\alpha = 0$ should say that form is a *good form*, i.e. well formed in the sense of Gestalt, the analog of quotient by the elements boundaries $\alpha = d\beta$ should say that perception consists in forgetting something. Note that in the usual case of De Rham the general covariance and the naturality for all manifolds select the operator d among all operators on differential forms. We shall hope that in the case of images, also covariance and naturality would select the wanted "operator d".

Thus we suggest to consider the stabilization of an image as a class for a convenient co-homology, as we have done for internal categorization spaces in section 2.4 (cf. Appendix 1). Consequently the stabilization of an image would appear as co-homology of co-homology because the sets $\mathscr{I}'_A$ were already co-homology spaces. The dimensional reductions to the $\mathscr{I}'_A$ in the higher areas and the final rules of selection by $d = 0$ give together the expression in our model of the Gestalt rules of good formation: *symmetry*, *parallelism*, *smooth continuation*, *proximity*, *similarity* and *closure*. See Spillman and Ehrenstein in TVN, [44].

4) Finally for going to object recognition, we suggest to copy the relation of constancy of colors in the Theorem 3.4.1:

$$[SI^1] = A_S R A_S^{-1} [SI^0] \tag{7.128}$$

But now $[SI]$ denotes the equivalence class of the image of the object S, A_S is the deformation operator telling how the object changes the image at the level of classes, and is R the adaptation operator between classes of images, describing the invariance.

However, for objects, there is a necessity to use all sensory modalities, not only vision, and also many other modalities, like motor control. But the above model can be extended to any representation-of-objects, not only visual images, for instance sounds or odors, or motor plans:

Suggestion 5.4.1: 1) a *representation-of-object* is an equivalence class S in a co-homology space of virtual dynamical attractors for a system of brain areas, it is subject to adaptation R in some invariance structure G; 2) an *object-in-itself* (perhaps coming from a physical object in the external world) is an operator A of deformation of co-homology classes; 3) an *object*, as internalized entity, is an orbit for the conjugate group $ARA^{-1}; R \in G$.

7.6 Perspective Today: Information Topology

To understand better the nature of the internal spaces at works in the brain and the invariance structures on them for adaptation, there is a need of principles for a pure dynamic of information.

Several attempts were made in this direction (Barlow 1961, Von der Malsburg 1981, Rolls and Treves, Bialek 1992, Amari, Shompolinski, Nadal, Sejnowski, Friston et al. 2007, 2009, ... see Olshausen in TVN). These works started with the research of Schrödinger, "What is life", on negentropy.

An attractive principle is Infomax, postulating that mutual information between the brain and the external world $I(S;E)$ is maximized. However (see Baudot thesis) this principle has to be complemented with an over-complete basis representation in order to account for the sparseness of the code usually observed in the cortex.

One hope is to find evolution rules for the joint probability of neurons that depend on the set of all multiple information quantities (or higher order mutual information, see [47]).

A first step in this direction is to understand the structures that emerge from the distributions of multi-information on a family of probabilities and a compatible family of random variables.

This is what we explored with Pierre Baudot in a series of papers "to be submitted", but exposed in several seminars. Without pretending we solved the problem of writing dynamical equations for adaptation, I indicate in what sense our work could be relevant. Note the cited thesis of Pierre Baudot contained most of the questions and ingredients for this joint work. This thesis linked the theoretical approach with impressive experimental results of Baudot et al. showing the adaptation of probability of responses in $V1$ to higher information in images, for instance the decrease of spike variability for natural images and natural eye motions (also unpublished result in other form than the thesis, but frequently exposed).

Note that many recent works contain similar results, for instance [40], [39], [81], [85].

We found a way to define a kind of geometry and a kind of topology for a set of probability laws and a compatible set of variables. The first idea was to put on an equal footing random variables and probability laws forming the pairs (X,P). The second ingredient was a definition of multi-information quantities; this was achieved by a co-homology theory inspired by the co-homology of groups. Examples of co-cycles are Shannon higher information quantities. Examples of non-zero co-homology classes are furnished by entropy and Kullback-Leibler distance. The third ingredient is the study of convexity-concavity properties of binary multi-information functions.

Most of our computations dealt with probabilities on finite sets, where the probability space is a standard affine simplex, and the set of variables corresponds to a set of partitions of the simplex.

Remark in special cases it is also possible to decompose the Kolmogorov entropy of a conservative dynamical system into higher information quantities. This extends the point of view of higher correlations, in relation to Lyapounov exponents.

In our opinion, the most notable theorem on the distribution of information quantities (close to our approach), was established by Hu Kuo Ting ([215]). It allows to associate already known topological forms to specific information distribution; for instance standard Borromean links are associated to the (negative) minimum of the third order mutual information (Matsuda [139]). Standard Markov chains do not give rise to non-trivial link ([215]).

This constitutes our point of departure: by considering only the information quantities a geometry appears on the set of probability laws. This geometry is organized locally by special subsets of variables (like generalized Borromean links), representing local attractors of the information distribution in the simplex.

Let M denotes a variety of probabilities. The main phenomenon we describe is the appearance of "topological forms" on the collection of random variables, that are stable configurations of part of the variables having singular distributions of information, reminiscent of the Gestalt rules. These forms depend on the probability laws, thus we obtain a kind of geometry on M, given by the change of forms of observation.

We propose that the multiple information quantities of a probability law provide its Gestaltic-shape description. Then special configurations of information distribution, like Borromean links, correspond to the expression of Gestalt rules (as it must be for co-cycles).

Let us come back to the nervous system, or a subsystem, for instance the visual system of mammals.

We must distinguish (as before in the text) between the family of stochastic processes more or less deterministic that constantly exchange energy and information between the system and the world and inside the system, and the family of laws of these processes $\mu \in M$. Thus we propose to identify M with a set of probabilities on the attractors of the neuronal dynamics. This is not so original Cf. Olshausen, [161], [162], Sejnowski [21], [197]. The fact that parameters of activity of neurons identifies to a space of probability laws is essential in the approach of Amari (cf. [243]); accordingly this gives a differential geometry of information that is related to the geometry of forms of mutual information when statistical methods for large samples are considered, cf. [38]. And we consider the dynamics of adaptation on M. We assume that knowledge on μ comes only from certain variables $X, Y, \ldots$. These variables have a specific structure (a category of commutative algebras) which describes the possibility of observation. Then the discussion of "forms of information" applies. The ideal space I represents special events, or past events that could be doubtful, or future events with ambiguity; they correspond to observable functions, i.e. random variables. The virtual attractors that are represented in I coincide with the special configurations of the observable variables. It is important to remark that they correspond to physical flows in the brain.

To link this approach with Galois information, or with the compensation equation between M and I that we exposed in this text, we need a natural group G (or a category) acting on the co-cycles. We do not exclude that the probabilistic information structure is sufficient to generate this Galois actions, but we do not know. It could also happen that Galois theory and probability theory belong to a common wider theory. (Note that Quantum probability introduces naturally Lie groups of symmetries as subgroups of the unitary group, and that ordinary finite probability plays with finite subgroups of the symmetric group.) However, in the case of neuronal systems, the one-dimensional *scaling* semi-group must act on I; it contracts space and dilates frequency. We shall assume that this semi-group expends the local entropies of the joint probability of signals expression, and that it is compensated by adaptation on M.

7.7 Appendix 1: Invariance, Catastrophes and Information

7.7.1 Groups and Information

Group theory can be seen as a genuine theory of information, with the same rights as the Shannon theory, for describing structures of ambiguity. This fact would deserve a complete exposition (a text with Martin Devautour is in preparation), however a few words can help to understand the general co-variance appearing in $V1$ and some higher areas. Our aim is to explain how the point of view of groups and geometry (Galois, Riemann, Grothendieck) relates to statistical information theory (Shannon, Fisher, Wiener) and dynamical systems (Thom, Milnor, Smale).

Evariste Galois considered group theory as part of a general theory of ambiguity, which continues to inspire mathematicians today. The main idea of Galois is simple: let a problem to solve be given, then it is often more useful to understand the form of the ambiguity between the possible solutions, than to try to solve by force the equation. Galois considered the case of algebraic equations E in one unknown and he defined a group G of permutations of the solutions, attached to a set of known numbers S (representing the *a priori* knowledge). The definition of G relies on a basic lemma: the functions $F(x_1, x_2, \ldots, x_n)$ of the solutions of E which can be determined from the data in S are exactly the functions which are invariant under the action of a certain group G. Thus G measures the ambiguity on the solutions of E, given S. Moreover, Galois described how the group behaves when S is changed: the addition of new data, for instance the knowledge of a solution of an auxiliary equation E', transforms G into a subgroup H of G, reducing the ambiguity. The choices of other solutions of E' change H into conjugate groups gHg^{-1}. Thus we can say that the quotient set G/H describes the *information*, in the sense of Galois, produced by the auxiliary problem E'. The Galois correspondence associates to each subgroup H of G the set of numbers K_H, generated from the solutions of E, which are fixed by H; it establishes a bijection between the subgroups of G and the subfields of K containing the field of numbers k generated by S. It can happen that G/H itself possesses a structure of group, this occurs when H is a *normal* (or invariant, or distinguished) subgroup, which means that for every $g \in G$ we have $gHg^{-1} = H$; in this case we have $gH.g'H = gg'H$. In this case, named Galoisian, we can consider G/H as the ambiguity structure of a new problem.

Poincaré (La science et l'hypothèse, and other books) explained the sensorimotor origin of the group G of the Euclidean displacements: G can be seen as the Galois group of a joint system made by a moving subject S and its environment E.

By definition, in ordinary Galois theory the Galois group G" of a joint KK' of two number fields K,K' of respective Galois groups G,G', over a basis field k, can be identified with the group of pairs (g,g') where $g \in G$ and $g' \in G'$ act on $K \cap K'$ in the same manner (cf. Bourbaki, Algèbre VII). Thus, in our case it is natural to pretend that the Galois group corresponds to pairs of one transformation g of the sensations in S that are induced by motor action and one transformation g' in E, which can compensate reciprocally. As Poincaré said, from the experiment we see that G" is the group of displacements G. The Galois *mutual information* of S and E is defined by the quotient $G_S \times G_E/G$. Note that, in the same way as two random variables X and Y are independent in the probabilistic sense if and only if $I(X;Y) = 0$, the two fields are linearly independent (i.e. KK' is isomorphic to the tensor product $K \otimes K'$) if and only if $G" = G \times G'$.

This approach of Poincaré was implemented on a virtual robot by Philipona et al.. To deduce the Euclidean *space* E from its isometry group Poincaré invoked specific sensors, like finger tips, whose ambiguity is a rotation group. For the visual space we can do the same, using vergence and accommodation, the properties of visual cells sensing a light spot in a place P; their ambiguity inside the group G is the subgroup H_P of rotations around P. Thus the Galois information of the cell is the quotient G/H_P, which is an exemplar E_P of E. The existence of translations inside G guarantees the existence of canonical isomorphisms between two exemplars E_P and E_Q. This is the origin of an objective space around us.

As we see, continuous groups, not only discrete groups, can appear as Galois groups; this is not exceptional in the infinite dimensional setting, for instance in physical Field Theory, seen as generalization of a theory of singularity (as in the study of phase transition). We can expect that every symmetry from a certain point of view becomes Galoisian.

Information theory and ambiguity theory are synonymous: the information measures the reduction of ambiguity.

Aside from Galois theory there exist other theories of information, in particular the well known theory of Shannon, which is based on probability theory. For Shannon the measure of the ambiguity contained in a probability law $\mathbb{P}$ on a set Ω is given by the *entropy* $H(\mathbb{P})$, which, when Ω is discrete, is the expectation of the function $-\log\mathbb{P}(\omega)$. Then the knowledge that can be expected from a random variable X is given by the entropy $H(X)$ of the image probability X_*P. If Y is another random variable, Shannon defined the mutual information of X and Y by $I(X,Y) = H(Y) - H(Y|X)$ where $H(X|Y)$ denotes the average over y of the entropy of the conditional laws $X_*P|(Y=y)$. We also have $I(X;Y) = H(X) + H(Y) - H(X,Y)$. For Gaussian laws the entropy is a constant multiple of $\log Det\sigma$, where σ is the co-variance matrix and Det the determinant (more intrinsically, $Det\sigma = DiscQ$, where Q is the co-variance quadratic form). Therefore the mutual information is given by the norm of the cross-correlation matrix.

The theorems of Shannon (cf. Khinchin 1957) show that the structures of possible codings of a message rely on entropy computations.

The theories of Galois and Shannon are directly connected when a group G acts on the considered set of probabilities on Ω, for instance if it acts on Ω itself. In general, the higher is the symmetry of the law $\mathbb{P}$ (the subgroup fixing $\mathbb{P}$), the larger the entropy.

There exist other theories of information based on probability, for instance the theory of Fisher (1925, the first in date), studying statistical metrics, the theory of Wiener (1948) studying prediction. All these theories are related to the entropy or its generalization, the Kullback-Leibler distance between two probabilities $\mathbb{P},\mathbb{Q}$ on the same set:

$$H(\mathbb{P},\mathbb{Q}) = -\mathbb{E}_{\mathbb{P}}(\log\frac{\mathbb{Q}}{\mathbb{P}}). \tag{7.129}$$

This is not a true metric because it is not symmetric in $\mathbb{P}$ and $\mathbb{Q}$. But when the considered probability laws depend smoothly on the points in a manifold M, and when $\mathbb{P},\mathbb{Q}$ become close; this formula can be normalized to give a (possibly degenerate) Riemannian metric on M, the "Fisher information metric". In fact this metric was defined in general by Rao (1945): At any point of a statistical manifold of probability laws $\mathbb{P}_\xi$ it is the tensor

$$\mathbb{F}_\xi = \mathbb{E}_\xi(d\ln(\mathbb{P}_\xi)\otimes d\ln(\mathbb{P}_\xi)); \tag{7.130}$$

As we see it is independent of coordinate choices in M, and this general invariance is its main virtue for statisticians. In a coordinate system $\theta_i; i = 1,...,m$ on M the coefficients of the Fisher-Rao metric $\mathbb{F}$ are given by

$$g_{ij} = \mathbb{E}_\theta(\partial_i\ln(p_\theta)\partial_j\ln(p_\theta)). \tag{7.131}$$

The inverse matrix defines a canonical metric on the co-tangent bundle $T^*(M)$, which appears as a lower bound in the Cramer-Rao inequality for any test $\Phi : H \to \mathbb{R}$ of a given function $\varphi : M \to \mathbb{R}$, at any point $\xi \in M$:

$$\mathbb{E}_{\xi}(|\Phi-\varphi(\xi)|^2)\geq(1+|d_{\xi}(\mathbb{E}_{\xi}(\Phi)-\varphi(\xi))|^2)\mathbb{F}_{\xi}^{-1}(d_{\xi}\varphi,d_{\xi}\varphi). \quad (7.132)$$

Remark that when the test is unbiased, i.e. when for each point ξ we have $\mathbb{E}_{\xi}(\Phi)=\varphi(\xi)$, the inequality is a universal lower bound on the variance of Φ:

$$\mathbb{E}_{\xi}(|\Phi-\mathbb{E}_{\xi}(\Phi)|^2)\geq\mathbb{F}_{\xi}^{-1}(d_{\xi}\varphi,d_{\xi}\varphi). \quad (7.133)$$

Thus $\mathbb{F}$ appears as a limit of possible precision for a measurement. It has been widely used in Neuroscience (Amari, Nadal). In problems with large sets of data, for instance many independent but cooperative neurons, it appears that maximizing mutual information is equivalent to maximizing the Fisher metric (Nadal, ...).

A remarkable fact is that traditional families of probabilistic laws possess large invariance groups. In fact all classical Lie groups appear as symmetry groups for Fisher-Rao metrics (unpublished folklore, but cf. Amari book).

Example: With the natural coordinates $x=\mu, y=\sigma^2$ of a normal law in one dimension the matrix $\mathbb{F}$ is diagonal and its diagonal coefficients are y^{-1} and $(1/2)y^{-2})$ respectively. This is a model of the hyperbolic plane of Gauss, Bolyai and Lobatchevski; its isometry group is the unimodular group $PSL_2(\mathbb{R})$. For Gaussian families in higher dimensions we find the family $PSL_n(\mathbb{R})$.

It is not astonishing that the whole theory of statistics is traversed by the concept of invariance (cf. Lehmann).

Hartley (1928), and Kolmogorov (1933), have proposed that the whole theory of probabilities can be seen as a theory of ambiguity. As Bayes and Boole anticipated, probability furnishes the basis of a general theory of knowledge. We can say the same thing of the group theory, cf. Souriau "The grammar of the world".

However a third mathematical view on ambiguity is given by the study of the dynamics of competition. René Thom held that fundamentally, all information is of topological nature: to him every information is a *form* and "the meaning of a message is a relation of topological nature between the form of the message and the characteristic forms of the receptor (i.e. the forms able to generate an excitation of the receptor) ..." ([212] p. 156)". We shall show that this view completes the two other views, for applications in Neuroscience and especially for the visual system.

Catastrophe theory puts forward a science of morphologies. Its principle is simple: every form expresses a conflict. More precisely, every phenomenon in the world takes place in a manifold W (perhaps of high dimension, including time), and for each point p in W there is a dynamical system $D(p)$ (for rapid evolution) in an auxiliary space $X(p)$ (varying smoothly with p as a fiber bundle), which decides what happens over p. For us, who are interested by neuronal systems, $D(p)$ will represent the rapid neural activity, contributing to the emergence of a form in the brain for the phenomenon at p. (We admit the existence of several time scales of neuronal activity: a rapid one generating spikes for instance and a slower one for modulating the rapid one; note that at the molecular level also, such a dichotomy holds frequently between the rapid dynamics of pathways and the slower dynamics of gene expression.) The *visible forms* in W are generated by the *catastrophe locus* K, which is the set of points p where the dynamics $D(p)$ bifurcates. According to

Thom, only forms that correspond to generic bifurcations having some sort of stability can be identified. For certain dynamics $D(p_0)$ there exists a canonical family, the *unfolding* of $D(p_0)$, parameterized by a finite dimensional manifold U, in which all accidents deforming $D(p_0)$ can be found. We shall consider only this kind of dynamics. Examples are given by some gradient flows on X, their unfolding describes the possible forms of competition between a finite number of attractor points. For instance, if $X = \mathbb{R}$, the potential $\varphi(x) = x^3 + ux$ represents the birth or the death of an attractor, depending on the sign of u, U has dimension 1 and the set K is the point $u = 0$. The potential $\varphi(x) = x^4 + ux^2 + vx$ represents the competition between two attractors, and in this case the set K is a cuspidal curve in the plane U. Cf. figure ....

The competition between attractors is a model of ambiguous situation. A simple link with probability theory can be obtained, by choosing a positive measure of the sizes of the basins of the attractors; once normalized by the total sum, this gives a probability law $\mathbb{P}(p)$ on the finite set $\Omega(p)$ of attractors of $D(p)$, for each $p \in U$; this quantifies the ambiguity.

Let us consider the richer variant where the value of the potential φ is taken into account; for instance, if $X = \mathbb{R}$, we consider $\varphi(x) = x^3 + ux + c$, or $\varphi(x) = x^4 + ux^2 + vx + c$; then we replace the space U by the extended space $\Lambda = U \times \mathbb{R}$. For $\lambda \in \Lambda$ we define X_λ as the subset of X where $\varphi_\lambda(x) = 0$. The discriminant Σ is the subset of Λ where X_λ is singular. Now suppose we consider all that over complex numbers, i.e. replace everywhere real coefficients by complex numbers, and work over the complexification $\Lambda_\mathbb{C}$ of Λ (cf. [13]). Then each non-singular level X_λ is retractable on a bouquet of μ spheres that are named the *vanishing cycles*, where μ is an integer called the *multiplicity* or Milnor number. Thus the homology of X_λ is a lattice $I(\lambda)$ in a vector space $I_\mathbb{R}(\lambda)$ (resp. $I_\mathbb{C}(\lambda)$) over the real (resp. complex) numbers. Arnold et al. in [13] discuss the case where μ is constant. Now, choose an analytic family of volume forms ω_λ on X_λ, the integrals of ω_λ on the tubes generated by the vanishing cycles can be defined along paths in $\Lambda_\mathbb{C}$; this gives a *period* map $\widetilde{P}$ from the universal covering $M_\mathbb{C}^\times$ of the complement $\Lambda_\mathbb{C}^\times$ of the complex discriminant $\Sigma_\mathbb{C}$ in $\Lambda_\mathbb{C}$ to the space $I_\mathbb{C}^*$ of linear maps from I to $\mathbb{C}$ (cf. [127], [221]). If we now take into account the real structure, depending on the connected component of the point λ in $\Lambda^\times$ (the complement of the real discriminant Σ in the real basis Λ), certain vanishing cycles are real other while others are complex conjugates. For convenient forms ω, the image of the real part $\Lambda^\times$ by $\widetilde{P}$ describes a real vector subspace $I_\mathbb{R}$ in $I_\mathbb{C}$ ([128]). In particular, every generic minimum of φ defines a real vanishing cycle, which is a regular level near the minimum. For some λ the period map for the minima of φ, once normalized, coincides with a probability law $\mathbb{P}_\lambda$.

The following picture is due to Looijenga [127] for simple singularities (i.e. when the unfolding contains only a finite numbers of inequivalent functions) and to Saito [196] and Varchenko [220] in general. First suppose X of odd dimension and the singularity simple; then the ambiguity of $\widetilde{P}$, i.e. the existence of different values for different paths joining the same points, is given by a finite group G (Looijenga, cf. [13], [22]). Moreover, the quotient of $\widetilde{P}$ by G gives an analytic morphism from $M_\mathbb{C}^\times$ to the quotient of $I_\mathbb{C}$ by G: this morphism corresponds to an isomorphism from $\Lambda_\mathbb{C}$ to $I_\mathbb{C}/G$, sending the discriminant $\Sigma_\mathbb{C}$ onto the quotient by G of the set $H_\mathbb{C}$ made by the cycles which have non-trivial isotropy under G (i.e. which are invariant by some $g \neq Id$). Thus the discriminant is isomorphic to the quotient a finite union of hyperplanes. Finally in the simple case, Looijenga proved that each component of the complement of the *real* discriminant Σ in the real extended unfolding Λ corresponds to a finite union of real copies of $I_{\mathbb{R}}$ disposed inside $I_{\mathbb{C}}$; these real copies correspond to linear combinations of minima equipped with special phases.

In the general case of the universal deformation of an holomorphic function with isolated singularity, Saito (cf. [196]) defined the notion of *primitive form* ω_λ, and for such forms established that the period mapping extends to a surjective morphism also generically one to one from $M^{\times}_{\mathbb{C}}$ to $I^{*}_{\mathbb{C}}$. Note that the space of primitive forms for a joint sum of simple singularities has finite dimension; in general the dimension of the space V of primitive forms behaves additively under joins. Thus the best thing in the general case is to consider not only one period mapping but a vector of period mappings, i.e. $\widetilde{P}; v \in V$.

Remark: the origin of the above theory was in theorems of Brieskorn, Grothendieck, on the equations of maximal unipotent orbits of simple Lie groups, showing that continuous groups are also related to the picture.

7.7.2 Homological Forms

Based on these results, showing the compatibility of the points of views of Galois, Shannon and Thom on what is called ambiguity, and information, we suggest a generalized picture for neuronal systems:

Let A be a brain area; a real smooth manifold M_A (playing the role of the covering $M^{\times}_{\mathbb{C}}$ or a subset that corresponds to reality conditions) is used for parameterizing the stable joint receptive fields in the area A, i.e. the way the activity in the area is prepared to change under the influence of the external world or the afferent areas, taking in account the intrinsic properties of neurons in A and their interactions (for instance through interneurons or glial cells). Thus a point μ in M_A represents a certain dynamical system ξ_μ in an internal space X_A (giving a model of the possible responses of A, from molecular events to spikes). Boundary conditions are described by an auxiliary space E_A.

The complete details of the dynamics of the assembly of neurons are not essential to give a definite form in the brain for the external phenomena; for each μ only a stable skeleton I_μ issued from ξ_μ does matter. We assume that all the topological information in ξ_μ is coded by the set I_μ, which we shall call the *vanishing homology* (or simply the homology), whose elements we name the *vanishing cycles*. The homology I_μ represents the way A categorizes the responses when it is in the state μ. This skeleton is defined up to deformation; it has a discrete nature, thus we assume that over M_A it is possible to follow I_μ and identify it with a fixed set I_A.

The set I_μ is issued from a subset Ω_μ it contains, which collects possible attractor regimes of ξ_μ and from memory or context or anticipation, providing ambiguity. We can think to an element of I_μ as a virtual attractor with a phase, like inhibited regimes; only a subset of the phases gives real attractors, belonging to Ω_μ.

The output signal going from A depends on a $\sigma \in I_A$ and a $\mu \in M_A$. A family of probability laws $\mathbb{P}_\mu$ is assumed to be given on Ω_μ, for describing the law of spiking activity in A. Let us denote by I^* the space of numerical functions on I, and name it the *vanishing co-homology* (or simply the co-homology); when we want to specify the coefficients of functions we note $I^*_{\mathbb{R}}$ or $I^*_{\mathbb{C}}$. And let us denote by V^* the space which parameterizes the output dimensions (it plays the role of the dual vector space of the the space V of primitive forms of Saito). We assume the existence of a map $\widetilde{P}$ from M to $I^*_{\mathbb{C}} \otimes V^*$, such that, for any μ and $\omega \in \Omega$ we have $P(\mu)(\omega) = \mathbb{P}_\mu(\omega)$. We name $\widetilde{P}$ the *period map*.

Now a group $\widetilde{G}$ acts on M, and also acts on I, in such a manner that, for any $\mu \in M$, $\sigma \in I$, $g \in \widetilde{G}$, we have

$$\widetilde{P}(g.\mu)(\sigma) = \widetilde{P}(\mu)(g.\sigma). \tag{7.134}$$

This relation expresses the compensation of a change of the parameter μ by a change in the internal state σ. This compensation corresponds to an *adaptation*: a flow in M changing the parameters of the neurons to compensate a change in context.

Note that the brain area A can be composed of many neurons a, each of them has dynamics $\xi_{a,\mu}$ parameterized by a manifold M_a and having homological skeleton $I_{a,\mu}$; the manifold M_A associated to A is a sub-manifold of the joint of the M_a over all the cells a, the space I_A is the product of the I_a, and the vanishing co-homology I^* is embedded in the tensor product of the $I^*_{a,\mu}$ over a. In general the group G acts on I^* but not on the individuals $I^*_{a,\mu}$. Let us nor forget that M is much larger than the set of union of the M_a over a; it parameterizes the interactions between dynamics and not only marginal dynamics.

The response of the area A to a stimulus φ is assumed to be given by the value of $\widetilde{P}$ at a cycle σ that depends on φ and the context ε:

$$\rho(\mu, \varphi, \varepsilon) = \widetilde{P}(\mu)(\sigma(\varphi, \varepsilon)). \tag{7.135}$$

It depends on time t through φ and ε and μ. Note that ρ is a vector in V^*_A; to get scalars we must consider $v \in V_A$ and take the bracket $\langle \rho, v \rangle$. This is an important point because the output of a brain area is a field of responses of neurons, in general not reducible to a scalar function.

The group $\widetilde{G}$ describes the ambiguity of the period map on the quotient $M/\widetilde{G}$. Its action on I^* defines the Galois group G of the system. We see that, for each v in V_A, the map $\widetilde{P}$ induces a map from the quotient $M/\widetilde{G}$ to the quotient I^*/G. To conform with catastrophe theory we can require the inverse of this correspondence to extend to a bijection from I^*/G to a completion Λ of $M/\widetilde{G}$.

With respect to the inputs, the space I^* creates an internal space with a new geometry, described by G. This space represents the external world, but at the same time, it invents the observable properties of this world.

The intelligible forms according to Thom come from the prolongation to the set Σ of unstable dynamics; this corresponds in I^* to the subset Δ where the isotropy, describing the structure of indetermination, is non-trivial. The full dynamics of X, Λ, I^*

permits to give precision an intriguing observation of René Thom about the human brain ([212] XIII.3). "... in the human brain was realized a dispositive simulator of the self-reproducing singularities of epigenesis, which permits, in presence of a catastrophe of internal space Y and unfolding U, to send the unfolding U in the internal space Y, realizing the confusion between internal and external variables." (See also [214]) What we have suggested in the present text, is that, first this confusion is not reserved to humans, but it appeared in all living entities, and second, this confusion is ambiguous, in the sense that the correspondence between neural processes and thoughts contents is multi-valued; the ambiguity structure being the place where symmetries are invented.
The homological skeleton adds ambiguity to the output with respect to the input.

The prototype of $\widetilde{G}$ is the group of rigid motions, it first appears as a Galois group, describing the ambiguity between what moves in the world and what results from body's movements, then it is imposed as a geometry on the world, for isotropy and homogeneity, that do not exist in reality, but help perception. An example of a system where this group appears explicitly in the entry is the parahippocampal region cf. section 5.3.
To summarize the invariance expressed in a given area A, we start with the group collecting the transformations of the environment E that have no effect on the responses in A; this is the *indifference group* written $G(E)$. Next we have a group $G(I)$ acting on the quotient $I = E/G(E)$ representing categories in A, and a group $G(M)$ acting on the set M of parameters describing the possible dynamics M; a subgroup $\widetilde{G}$ of the product $G(I) \times G(M)$ is made with the transformations that have no effect on the response; from the above discussion we could name it the *ambiguity group*, or the *adaptation group*. The full group $G(I) \times G(M)$ acts on the response space R, thus the notion of co-variance of the response applies to this group.

7.8 Appendix 2: Unitary Representations

Harmonic analysis permits to describe how discrete units can work with continuous symmetries.

References: [222], [110], [83].

Suppose G is a Lie group and choose a left invariant measure on G (named Haar measure).

A unitary representation is a morphism $g \mapsto U^{\lambda}(g)$ from G to the group U_{λ} of bijective isometries of a complex Hilbert space H_{λ}; the representation is said to be irreducible when any invariant subspace of H_{λ} is $\{0\}$ or H_{λ} itself. See [110].

Definition: For any integrable function $\phi : G \to \mathbb{C}$ we define the Fourier-Plancherel transform by the equation

$$\widehat{\phi}(\lambda) = \int_G U^{\lambda}(g)\phi(g)dg \tag{7.136}$$

where λ describes the set $\widehat{G}$ of unitary representations of G. Then $\widehat{\phi}(\lambda)$ is an operator, an element of U_λ.

In what follows we assume that the Haar measure dg is also invariant by right translations in G. When such a measure exists the group is said to be unimodular. Any compact group, any discrete group, any commutative group is unimodular. The affine group (see 2.1), is non-unimodular. The Lie groups of interest for us, the Euclid group (displacements) and the Galilean group are non-compact but they are unimodular. Other non-compact examples are semi-simple Lie groups (cf. [83]), or nilpotent Lie groups (cf. [110]).

The Plancherel Theorem, valid for any unimodular Lie group, asserts the existence of a measure $d\pi(\lambda)$ on $\widehat{G}$, such that for any square integrable function ϕ on G, which is also integrable, we have

$$\int_G |\phi(g)|^2 = \int_{\widehat{G}} Tr[\widehat{\phi}(\lambda)^* \widehat{\phi}(\lambda)] d\pi(\lambda), \tag{7.137}$$

and the inversion formula holds true:

$$\phi(g) = \int_{\widehat{G}} Tr[U_\lambda^*(g) \widehat{\phi}(\lambda)] d\pi(\lambda) \tag{7.138}$$

The proof of this theorem for finite groups is due to I.Schur, for compact Lie groups to H.Weyl, and the general case of unimodular separable locally compact groups(in particular Lie groups) is due to Mautner and Segal (cf. [37]). The coefficients of the irreducible unitary representations have strong orthogonality properties (cf. [83]), but they rarely form a generating system of square integrable functions on G; for compact groups they do.

We suggest that the coefficients of unitary representations of a group G are good candidates for describing the receptive fields of neurons that are modulated by trajectories in G. This requires a definition of "windows in the groups", generalizing domains in the visual plane that can generate spikes.

Given a distribution μ on G depending on time t, and a smooth function ϕ on G, we define the translation operation on distributions by

$$\langle \tau_t(\mu), \phi \rangle (s) = \langle \mu, \phi \rangle (s-t),$$

and the inversion operation on distributions by

$$\langle \mu^\vee, \phi \rangle = \langle \mu, \phi \circ \varsigma \rangle,$$

where $\varsigma(\phi)(g) = \phi(g^{-1})$.We also write $\varsigma(\phi) = \phi^\vee$.

The first ingredient of a *simple window* c is a pair (σ_c, K_c) of one sigmoidal function from $\mathbb{R}$ to $\mathbb{R}$, and one C^∞ function on G, such that activity of c at time t induced by a stimulus μ is (at least in part) explained by the formula

$$R_t(\mu) = \sigma_c[K_c * (\tau_t \mu)^\vee]. \tag{7.139}$$

In general we will suppose that K_c respects causality, implying that R_t depends on the restriction of μ to values of s before t. When μ is a function we obtain

$$K_c * (\tau_t \mu)^\vee = \int_G dh K_c^0(h g_0^{-1}, s)\mu(h^{-1})(t-s).$$

In the case of a unimodular group, this also gives

$$K_c * (\tau_t \mu)^\vee = \int_G dh K_c^0(g_0 h^{-1}, s)\mu(h)(t-s).$$

In the hope of maximizing invariance and information together, we take a function $K_0(s)$, like a Gaussian density, centered on the identity element of the group G, and we choose an irreducible unitary representation $\lambda(s)$, and two vectors $\psi_1(s), \psi_2(s)$ in the Hilbert space $H_{\lambda(s)}$. The function K_0 gives the *size* to the window, the representation λ generalizes the preferred *frequency*, and the vectors ψ_1, ψ_2 generalize the phase. Then we define

$$K_c^0(g,s) = K_0(g,s)\langle \psi_1(s)|U^{\lambda(s)}(g)|\psi_2(s)\rangle,$$

so we obtain the following generic formula for the elementary linear operators underlying the receptive fields:

$$K_c * (\tau_t \mu)^\vee = \int_{\mathbb{R}} ds \int_G dh K_0(g_0 h^{-1}, s)\langle \psi_1(s)|U^{\lambda(s)}(g_0 h^{-1})|\psi_2(s)\rangle \mu(h, t-s)$$

The corresponding receptive fields could be named *harmonic receptive fields.*

In [24], we suggested that the elementary windows associated to the Galilean group are well adapted to describe non-linear receptive fields of the vestibular system. A subgroup G_0 of the Galilean group is isomorphic to the Euclid group, $E(3)$ of displacements in 3*D* Euclidean space. The unitary representations of G_0 were determined by the induction method of Mackey (1958). See also Thoma (1958), Ito 1952,1953, Mackey 1952,1953, 1958.

The sequel of this work, aimed at studying the grouping of the harmonic receptive fields, is the current thesis of Alexandre Afgoustidis.

Example: In two-dimension, for the Euclid group $E(2)$, the representations come in two families. First, the *special representations* are described by a relative integer number N, they are of dimension 1 and see only the rotation part in g.
Second, the *generic representations* are classified by a positive real number ρ, the Hilbert space is the space of square integrable functions on the circle; the transformation $g(\mathbf{a}, \alpha)$, whose parameters are the rotation of angle α around 0 and the gliding of velocity $\mathbf{a}$ of modulus r and argument φ, is represented by

$$U^\rho(g)(f)(\theta) = e^{i\rho r \cos(\theta - \varphi)} f(\theta - \alpha) \tag{7.140}$$

The coefficients in this case are associated to the canonical Fourier basis $\exp in\theta$; they give the following functions on G

$$t(g) = \langle \psi_1 | U(g) | \psi_1 \rangle \tag{7.141}$$

That are

$$t^{\rho}_{mn}(g) = i^{m-n} e^{-i(n\alpha + (m-n)\varphi)} J_{m-n}(\rho r) \tag{7.142}$$

See [222].

In this model we obtain the following formula for describing the response to a periodic stimulus ω in a direction θ:

$$R(t) = \int k_0(s) ds \cos(m(\theta - \theta_0)) J_m(\rho_0 \cos(\omega(t-s))); \tag{7.143}$$

where J_m denotes the Bessel function with integer index m.

Definition: in general the Bessel function of index t in $\mathbb{R}$ is given by

$$J_t(z) = \sum_0^{\infty} \frac{(-1)^n (z/2)^{2n+t}}{n! \Gamma(n+t+1)}. \tag{7.144}$$

It is a solution of the Bessel equation

$$\frac{d^2 y}{dz^2} + (1 + \frac{1-4t^2}{z^2}) y = 0 \tag{7.145}$$

At the first order approximation in ρ_0 this gives:

$$R(t) = \rho_0 \int k_0(s) ds \cos(m(\theta - \theta_0)) \cos(\omega(t-s)). \tag{7.146}$$

When k_0 is of Dirac type Dirac, $\delta(t - t_0)$, we find the classical model:

$$R(t) = \rho_0 \cos(m(\theta - \theta_0)) \cos(\omega(t - t_0)). \tag{7.147}$$

However, when k_0 is a real Riemann-Liouville derivative of order D_0, we find a "dynamical response":

$$R(t) = \rho_0 \omega^{D_0} \cos(m(\theta - \theta_0)) \cos(\omega(t - t_0')). \tag{7.148}$$

We recover the known characteristics of a central vestibular cell, which are the gain ρ_0, the phase delay t_0' and the dynamical index D_0.

In the $3D$ space, for the generic series, the Hilbert space is a convenient subspace H_λ of the space H of square integrable functions on the orthogonal group $O(3)$. We fix a number ρ, and a vector $\overrightarrow{e_0}$ of norm 1; then, if g is the composition of the rotation R centered in 0 withe Galilean gliding of constant velocity $\overrightarrow{v}$, we define

$$U^{\rho}(g)(f)(S) = e^{i\rho \overrightarrow{v}.S(\overrightarrow{e_0})} f(R^{-1}S) \tag{7.149}$$

Cf. [222].

The modification of the response due to space effects consists in replacing the circular function $\cos(m(\theta - \theta_0))$ by a spherical harmonic and the Bessel function of integer index by a Bessel function with semi-integer index.

Acknowledgments. I thank Alessandro Sarti and Giovanna Citti for their invitation to contribute to this book, and for their infinite patience. They also contributed to my understanding of the subject. The works that are presented could not have been made without a part of my collaborators in the Collège-de-France (LPPA), Jacques Droulez, Chantal Milleret, Luc Foubert, Nathalie Rochefort, Simon Capern; I thank them for the very good times we had together and for their great help. I also thank the students in Mathematics who were or are involved in these researches, in particular Raphael Lachièze-Rey, Martin Devautour, Alexandre Afgoustidis. A special thank is due to Alexandre for his careful reading of the text. In marge of these works I have been very influenced by friends (that are also great scientists); I thank them, Alain Berthoz, Rodolfo Llinas, Sisir Roy, Tamar Flash, Jean-Pierre Nadal, Sid Wiener, Kevin O'Regan, David Philipona, Jean Petitot and Pierre Baudot. Given the subjects that I have exposed, of course I thank specially Piero, for all he his and does.

References

1. Abbott, L.H., Dayan, P.: Theoretical Neuroscience. In: Computational and Mathematical Modeling of Neural Systems. Computational Neuroscience series, vol. 15, MIT Press, Cambridge (2001)
2. Abel, N.H.: Oeuvres complètes, vol. 1. C. Gröndahl (1839)
3. Adelson, E.H., Bergen, J.R.: Spatiotemporal energy models for the perception of motion. J. Opt. Soc. Am. A 2(2), 284–299 (1985)
4. Adler, R.J., Taylor, J.E.: Random fields and geometry, vol. 115. Springer (2007)
5. Aggoun-Zouaoui, D., Kiper, D.C., Innocenti, G.M.: Growth of callosal terminal arbors in primary visual areas of the cat. European Journal of Neuroscience 8(6), 1132–1148 (1996)
6. Akerman, C.J., Smyth, D., Thompson, I.D.: Visual experience before eye-opening and the development of the retinogeniculate pathway. Neuron 36(5), 869–879 (2002)
7. Albrecht, D.G., Geisler, W.S., et al.: Motion selectivity and the contrast-response function of simple cells in the visual cortex. Visual Neuroscience 7(6), 531–546 (1991)
8. Alitto, H.J., Usrey, W.M.: Influence of contrast on orientation and temporal frequency tuning in ferret primary visual cortex. Journal of Neurophysiology 91(6), 2797–2808 (2004)
9. Amari, S.-I.: Differential geometric methods in statistics. Lecture Notes in Statistics, vol. 28 (1985)
10. Andersen, P., Morris, R., Amaral, D., Bliss, T., O'Keefe, J.: The hippocampus book. Oxford University Press (2006)
11. Angelaki, D.E., Gu, Y., DeAngelis, G.C.: Multisensory integration: psychophysics, neurophysiology, and computation. Current Opinion in Neurobiology 19(4), 452–458 (2009)
12. Armstrong, D.M., Marple-Horvat, D.E.: Role of the cerebellum and motor cortex in the regulation of visually controlled locomotion. Canadian Journal of Physiology and Pharmacology 74(4), 443–455 (1996)
13. Arnold, V.I., Gusein-Zade, S.M., Varchenko, A.N.: Singularities of Differentiable Maps: Monodromy and asymptotics of integrals, vol. 83. Springer (2012)

14. Atick, J.J., Redlich, A.N.: What does the retina know about natural scenes? Neural Computation 4(2), 196–210 (1992)
15. Baizer, J.S., Whitney, J., Bender, D.: Bilateral projections from the parabigeminal nucleus to the superior colliculus in monkey. Experimental Brain Research 86(3), 467–470 (1991)
16. Barbieri, D., Citti, G., Sanguinetti, G., Sarti, A.: An uncertainty principle underlying the functional architecture of V1. Journal of Physiology-Paris 106(5), 183–193 (2012)
17. Barlow, H.B.: Possible principles underlying the transformation of sensory messages. Sensory Communication, pp. 217–234 (1961)
18. Basole, A., White, L.E., Fitzpatrick, D.: Mapping multiple features in the population response of visual cortex. Nature 423(6943), 986–990 (2003)
19. Baudot, P.: Natural computation, much ado about nothing? Doctorat sciences cognitives. University Paris VI, Pierre et Marie Curie (November 2006)
20. Bednar, J.A.: Building a mechanistic model of the development and function of the primary visual cortex. Journal of Physiology-Paris 106(5), 194–211 (2012)
21. Bell, A.J., Sejnowski, T.J.: An information-maximization approach to blind separation and blind deconvolution. Neural Computation 7(6), 1129–1159 (1995)
22. Bennequin, D.: Caustique mystique. Séminaire Bourbaki 27, 19–56 (1984)
23. Bennequin, D.: Questions de physique galoisienne. In: Porte, M. (ed.) Passion des Formes, à René Thom (ENS Editions, Fontenay-Saint Cloud), pp. 311–410 (1994)
24. Bennequin, D., Berthoz, A.: Non-linear galilean vestibular receptive fields. In: 2011 Annual International Conference of the IEEE Engineering in Medicine and Biology Society, EMBC, pp. 2273–2276. IEEE (2011)
25. Bennequin, D., Fuchs, R., Berthoz, A., Flash, T.: Movement timing and invariance arise from several geometries. PLoS Computational Biology 5(7), e1000426 (2009)
26. Berry, M.V., Dennis, M.R.: Phase singularities in isotropic random waves. Proceedings of the Royal Society of London. Series A: Mathematical, Physical and Engineering Sciences 456(2001), 2059–2079 (2001)
27. Berthoz, A.: Le sens du mouvement. Odile Jacob (1997)
28. Berthoz, A.: La décision. Odile Jacob (2003)
29. Bienenstock, E., von der Malsburg, C.: A neural network for invariant pattern recognition. EPL (Europhysics Letters) 4(1), 121 (1987)
30. Bonhoeffer, T., Grinvald, A.: Iso-orientation domains in cat visual cortex are arranged in pinwheel-like patterns. Nature 353(6343), 429–431 (1991)
31. Bonnasse-Gahot, L., Nadal, J.-P.: Perception of categories: from coding efficiency to reaction times. Brain Research 1434, 47–61 (2012)
32. Bosking, W.H., Kretz, R., Pucak, M.L., Fitzpatrick, D.: Functional specificity of callosal connections in tree shrew striate cortex. The Journal of Neuroscience 20(6), 2346–2359 (2000)
33. Bosking, W.H., Zhang, Y., Schofield, B., Fitzpatrick, D.: Orientation selectivity and the arrangement of horizontal connections in tree shrew striate cortex. The Journal of Neuroscience 17(6), 2112–2127 (1997)
34. Bourbaki, N.: Lie groups and lie algebras, vol. 7, ch. 7-9. Springer (2008)
35. Bowling, D.B., Michael, C.R.: Terminal patterns of single, physiologically characterized optic tract fibers in the cat's lateral geniculate nucleus. The Journal of Neuroscience 4(1), 198–216 (1984)
36. Bressoud, R., Innocenti, G.M.: Typology, early differentiation, and exuberant growth of a set of cortical axons. Journal of Comparative Neurology 406(1), 87–108 (1999)
37. Bruhat, F., Ramanan, S.: Lectures on Lie groups and representations of locally compact groups, vol. 14. Tata Institute of Fundamental Research (1958)

38. Brunel, N., Nadal, J.-P.: Mutual information, fisher information, and population coding. Neural Computation 10(7), 1731–1757 (1998)
39. Butts, D.A., Weng, C., Jin, J., Alonso, J.-M., Paninski, L.: Temporal precision in the visual pathway through the interplay of excitation and stimulus-driven suppression. The Journal of Neuroscience 31(31), 11313–11327 (2011)
40. Butts, D.A., Weng, C., Jin, J., Yeh, C.-I., Lesica, N.A., Alonso, J.-M., Stanley, G.B.: Temporal precision in the neural code and the timescales of natural vision. Nature 449(7158), 92–95 (2007)
41. Cadieu, C., Kouh, M., Pasupathy, A., Connor, C.E., Riesenhuber, M., Poggio, T.: A model of v4 shape selectivity and invariance. Journal of Neurophysiology 98(3), 1733–1750 (2007)
42. Carandini, M., Demb, J.B., Mante, V., Tolhurst, D.J., Dan, Y., Olshausen, B.A., Gallant, J.L., Rust, N.C.: Do we know what the early visual system does? The Journal of Neuroscience 25(46), 10577–10597 (2005)
43. Casagrande, V.A., Condo, G.J.: The effect of altered neuronal activity on the development of layers in the lateral geniculate nucleus. The Journal of Neuroscience 8(2), 395–416 (1988)
44. Chalupa, L.M., Werner, J.S., Barnstable, C.J.: The visual neurosciences, vol. 1. MIT Press, Cambridge (2004)
45. Charlesworth, B., Charlesworth, D.: Darwin and genetics. Genetics 183(3), 757–766 (2009)
46. Chossat, P., Faugeras, O.: Hyperbolic planforms in relation to visual edges and textures perception. PLoS Computational Biology 5(12), e1000625 (2009)
47. Cover, T.M., Thomas, J.A.: Elements of information theory. John Wiley & Sons (2012)
48. Dahlke, S., Maass, P.: The affine uncertainty principle in one and two dimensions. Computers & Mathematics with Applications 30(3), 293–305 (1995)
49. DeAngelis, G.C., Ghose, G.M., Ohzawa, I., Freeman, R.D.: Functional micro-organization of primary visual cortex: receptive field analysis of nearby neurons. The Journal of Neuroscience 19(9), 4046–4064 (1999)
50. de Rham, G.: Variétés différentiables. Hermann, Paris (1955)
51. De Zeeuw, C.I., Hansel, C., Bian, F., Koekkoek, S.K., van Alphen, A.M., Linden, D.J., Oberdick, J.: Expression of a protein kinase C inhibitor in purkinje cells blocks cerebellar ltd and adaptation of the vestibulo-ocular reflex. Neuron 20(3), 495–508 (1998)
52. Dellacherie, C., Meyer, P.A.: Probability and potential. Hermann, Paris (1978)
53. Derrington, A.M., Krauskopf, J., Lennie, P.: Chromatic mechanisms in lateral geniculate nucleus of macaque. The Journal of Physiology 357(1), 241–265 (1984)
54. Devernay, F., Faugeras, O.: From projective to euclidean reconstruction. In: Proceedings of the 1996 IEEE Computer Society Conference on Computer Vision and Pattern Recognition, CVPR 1996, pp. 264–269. IEEE (1996)
55. Dixon, E.R.: Spectral distribution of australian daylight. JOSA 68(4), 437–450 (1978)
56. Dragoi, V., Rivadulla, C., Sur, M.: Foci of orientation plasticity in visual cortex. Nature 411(6833), 80–86 (2001)
57. Droulez, J., Bennequin, D.: Perception des symétries et invariances perceptives. In: Siksou, M. (ed.) Symétries, Symétries et Asymétries du Vivant, pp. 155–171. Lavoisier (2005)
58. Dudkin, E.A., Sheffield, J.B., Gruberg, E.R.: Combining visual information from the two eyes: the relationship between isthmotectal cells that project to ipsilateral and to contralateral optic tectum using fluorescent retrograde labels in the frog, rana pipiens. Journal of Comparative Neurology 502(1), 38–54 (2007)

59. Eatock, R.A., Corey, D.P., Hudspeth, A.J.: Adaptation of mechanoelectrical transduction in hair cells of the bullfrog's sacculus. The Journal of Neuroscience 7(9), 2821–2836 (1987)
60. Eccles, J.C., Llinas, R., Sasaki, K.: The excitatory synaptic action of climbing fibres on the purkinje cells of the cerebellum. The Journal of Physiology 182(2), 268–296 (1966)
61. Eilenberg, S., Steenrod, N.: Foundations of algebraic topology, Princeton NJ, vol. 952 (1952)
62. Elberger, A.J.: The functional role of the corpus callosum in the developing visual system: a review. Progress in Neurobiology 18(1), 15–79 (1982)
63. Emmons, L.H.: Tupai: a field study of Bornean treeshrews, vol. 2. University of California Pr (2000)
64. Ferraro, M., Caelli, T.M.: Lie transformation groups, integral transforms, and invariant pattern recognition. Spatial Vision 8(1), 33–44 (1994)
65. Findlay, J.M., Gilchrist, I.D.: Active vision: The psychology of looking and seeing. Oxford University Press (2003)
66. Fitzpatrick, D.: The functional organization of local circuits in visual cortex: insights from the study of tree shrew striate cortex. Cerebral Cortex 6(3), 329–341 (1996)
67. Földiák, P.: Learning invariance from transformation sequences. Neural Computation 3(2), 194–200 (1991)
68. Foubert, L.: Spatio-temporal characteristics of the visual interhemispheric integration via the corpus callosum: computational modeling & optical imaging approaches. PhD thesis, Université Pierre et Marie Curie-Paris VI (2007)
69. Foubert, L., Bennequin, D., Thomas, M.-A., Droulez, J., Milleret, C.: Interhemispheric synchrony in visual cortex and abnormal postnatal visual experience. Frontiers in Bioscience: A Journal and Virtual Library 15, 681 (2010)
70. Galois, É.: Œuvres mathématiques. Gauthier-Villars et fils (1897)
71. Gibson, J.J.: Adaptation, after-effect and contrast in the perception of curved lines. Journal of Experimental Psychology 16(1) (1933)
72. Gibson, J.J.: Perception of the visual world. Houghton-Mifflin, Boston (1950)
73. Gibson, J.J.: Optical motions and transformations as stimuli for visual perception. Journal of Experimental Psychology 64(5), 288–295 (1957)
74. Gibson, J.J., Gibson, E.J.: Continuous perspective transformations and the perception of rigid motion. Journal of Experimental Psychology 54(2), 129–138 (1957)
75. Goldberg, J.M., Wilson, V.J., Cullen, K.E.: The Vestibular System: A Sixth Sense. Oxford University Press (2012)
76. González-Burgos, G., Barrionuevo, G., Lewis, D.A.: Horizontal synaptic connections in monkey prefrontal cortex: an in vitro electrophysiological study. Cerebral Cortex 10(1), 82–92 (2000)
77. Graf, W., Baker, R.: Adaptive changes of the vestibulo-ocular reflex in flatfish are achieved by reorganization of central nervous pathways. Science 221(4612), 777–779 (1983)
78. Graybiel, A.M.: Organization of the nigrotectal connection: an experimental tracer study in the cat. Brain Research 143(2), 339–348 (1978)
79. Grothendieck, A., Verdier, J.-L.: Théorie des topos (sga 4, exposés i-vi). Springer Lecture Notes in Math. 269, 270 (1972)
80. Hafting, T., Fyhn, M., Molden, S., Moser, M.-B., Moser, E.I.: Microstructure of a spatial map in the entorhinal cortex. Nature 436(7052), 801–806 (2005)
81. Haider, B., Krause, M.R., Duque, A., Yu, Y., Touryan, J., Mazer, J.A., McCormick, D.A.: Synaptic and network mechanisms of sparse and reliable visual cortical activity during nonclassical receptive field stimulation. Neuron 65(1), 107–121 (2010)

82. Hebb, D.O.: The organization of Behavior (1949)
83. Helgason, S.: Groups & Geometric Analysis: Radon Transforms. Invariant Differential Operators and Spherical Functions. Academic Press (1984)
84. von Helmholtz, H.: The Origin and Meaning of Geometrical Axioms. Mind (3), 301–321 (1876)
85. Herikstad, R., Baker, J., Lachaux, J.-P., Gray, C.M., Yen, S.-C.: Natural movies evoke spike trains with low spike time variability in cat primary visual cortex. The Journal of Neuroscience 31(44), 15844–15860 (2011)
86. Houzel, J.-C., Milleret, C., Innocenti, G.: Morphology of callosal axons interconnecting areas 17 and 18 of the cat. European Journal of Neuroscience 6(6), 898–917 (1994)
87. Hubel, D.H., Wiesel, T.N.: Receptive fields of single neurones in the cat's striate cortex. The Journal of Physiology 148(3), 574–591 (1959)
88. Hubel, D.H., Wiesel, T.N.: Receptive fields, binocular interaction and functional architecture in the cat's visual cortex. The Journal of Physiology 160(1), 106–154 (1962)
89. Hubel, D.H., Wiesel, T.N.: Receptive fields and functional architecture of monkey striate cortex. The Journal of Physiology 195(1), 215–243 (1968)
90. James Hudspeth, A.: Mechanical amplification of stimuli by hair cells. Current Opinion in Neurobiology 7(4), 480–486 (1997)
91. Hudspeth, A.J., Choe, Y., Mehta, A.D., Martin, P.: Putting ion channels to work: mechanoelectrical transduction, adaptation, and amplification by hair cells. Proceedings of the National Academy of Sciences 97(22), 11765–11772 (2000)
92. Inaba, N., Shinomoto, S., Yamane, S., Takemura, A., Kawano, K.: MST neurons code for visual motion in space independent of pursuit eye movements. Journal of Neurophysiology 97(5), 3473–3483 (2007)
93. Innocenti, G.M., Price, D.J.: Exuberance in the development of cortical networks. Nature Reviews Neuroscience 6(12), 955–965 (2005)
94. Isik, L., Leibo, J.Z., Poggio, T.: Learning and disrupting invariance in visual recognition with a temporal association rule. Frontiers in Computational Neuroscience 6 (2012)
95. Issa, N.P., Trepel, C., Stryker, M.P.: Spatial frequency maps in cat visual cortex. The Journal of Neuroscience 20(22), 8504–8514 (2000)
96. Itzykson, C., Zuber, J.B.: Quantum field theory (1980)
97. Izhikevitch, E.M.: Dynamical systems in neuroscience (2007)
98. Jacobs, G.H.: The distribution and nature of colour vision among the mammals. Biological Reviews 68(3), 413–471 (1993)
99. Johnson, K., Wright, W.: Reply to philipona and O' regan. Visual Neuroscience 25(02), 221–224 (2008)
100. Jordan, G., Deeb, S.S., Bosten, J.M., Mollon, J.D.: The dimensionality of color vision in carriers of anomalous trichromacy. Journal of Vision 10(8) (2010)
101. Judd, D.B., MacAdam, D.L., Wyszecki, G., Budde, H.W., Condit, H.R., Henderson, S.T., Simonds, J.L.: Spectral distribution of typical daylight as a function of correlated color temperature. JOSA 54(8), 1031–1040 (1964)
102. Kalisa, C., Torrésani, B.: N-dimensional affine Weyl-Heisenberg wavelets. Ann. Inst. H. Poincaré, Physique Théorique 59, 201–236 (1993)
103. Kanatani, K.-I.: Transformation of optical flow by camera rotation. IEEE Transactions on Pattern Analysis and Machine Intelligence 10(2), 131–143 (1988)
104. Kanizsa, G.: Organization in vision: Essays on Gestalt perception. Praeger, New York (1979)
105. Kappers, A.M.L., Todd, J.T., Oomes, S., Koenderink, J.J.: The intrinsic geometry of perceptual space: its metrical, affine and projective properties. In: Proceedings of the Fifteenth Annual Meeting of the International Society for Psychophysics, pp. 169–174. Arizona State University, Department of Psychology & Industrial Engineering (1999)

106. Kaschube, M., Schnabel, M., Löwel, S., Coppola, D.M., White, L.E., Wolf, F.: Universality in the evolution of orientation columns in the visual cortex. Science 330(6007), 1113–1116 (2010)
107. Kastner, S., Nothdurft, H.-C., Pigarev, I.N.: Neuronal correlates of pop-out in cat striate cortex. Vision Research 37(4), 371–376 (1997)
108. Kennedy, H., Dehay, C., Bullier, J.: Organization of the callosal connections of visual areas V1 and V2 in the macaque monkey. Journal of Comparative Neurology 247(3), 398–415 (1986)
109. Khaytin, I., Chen, X., Royal, D.W., Ruiz, O., Jermakowicz, W.J., Siegel, R.M., Casagrande, V.A.: Functional organization of temporal frequency selectivity in primate visual cortex. Cerebral Cortex 18(8), 1828–1842 (2008)
110. Kirillov, A.A.: Introduction to the theory of representations and noncommutative harmonic analysis. In: Representation Theory and Noncommutative Harmonic Analysis I, pp. 1–156. Springer (1994)
111. Knierim, J.J., Van Essen, D.C.: Neuronal responses to static texture patterns in area v1 of the alert macaque monkey. Journal of Neurophysiology 67(4), 961–980 (1992)
112. Koenderink, J.J.: Optic flow. Vision Research 26(1), 161–179 (1986)
113. Koenderink, J.J.: Color for the Sciences. MIT Press (2010)
114. Koenderink, J.J., van Doorn, A.J.: Pictorial space, pp. 239–299 (2003)
115. Koenderink, J.J., van Doorn, A.J.: Facts on optic flow. Biological Cybernetics 56(4), 247–254 (1987)
116. Krantz, D.H.: Color measurement and color theory: I. representation theorem for grassmann structures. Journal of Mathematical Psychology 12(3), 283–303 (1975)
117. Kretz, R., Rager, G.: Postnatal development of area 17 callosal connections in tupaia. Journal of Comparative Neurology 326(2), 217–228 (1992)
118. Kriegeskorte, N., Mur, M., Ruff, D.A., Kiani, R., Bodurka, J., Esteky, H., Tanaka, K., Bandettini, P.A.: Matching categorical object representations in inferior temporal cortex of man and monkey. Neuron 60(6), 1126–1141 (2008)
119. Lagae, L., Maes, H., Raiguel, S., Xiao, D.K., Orban, G.A.: Responses of macaque STS neurons to optic flow components: a comparison of areas MT and MST. Journal of Neurophysiology 71(5), 1597–1626 (1994)
120. Land, M.F., Nilsson, D.-E.: Animal eyes. Oxford University Press (2012)
121. Laughlin, S.B.: The role of sensory adaptation in the retina. Journal of Experimental Biology 146(1), 39–62 (1989)
122. Goff, L.L., Bozovic, D., Hudspeth, A.J.: Adaptive shift in the domain of negative stiffness during spontaneous oscillation by hair bundles from the internal ear. Proceedings of the National Academy of Sciences of the United States of America 102(47), 16996–17001 (2005)
123. Lee, T.S.: Dynamics of subjective contour formation in the early visual cortex. Proceedings of the National Academy of Sciences 98(4), 1907–1911 (2001)
124. Lennie, P., Krauskopf, J., Sclar, G.: Chromatic mechanisms in striate cortex of macaque. The Journal of Neuroscience 10(2), 649–669 (1990)
125. Llinás, R.R.: The intrinsic electrophysiological properties of mammalian neurons: insights into central nervous system function. Science 242(4886), 1654–1664 (1988)
126. Llinás, R.R.: I of the vortex: From neurons to self. MIT press (2002)
127. Looijenga, E.: A period mapping for certain semi-universal deformations. Compositio Mathematica 30(3), 299–316 (1975)
128. Looijenga, E.: The discriminant of a real simple singularity. Compositio Mathematica 37(1), 51–62 (1978)

129. Lorenceau, J., Baudot, P., Series, P., Georges, S., Pananceau, M., Frégnac, Y.: Modulation of apparent motion speed by horizontal intracortical dynamics. Journal of Vision 1(3), 400–400 (2001)
130. Lowe, D.G.: Object recognition from local scale-invariant features. In: Computer Vision, vol. 2, pp. 1150–1157. IEEE (1999)
131. Lowe, D.G.: Distinctive image features from scale-invariant keypoints. International Journal of Computer Vision 60(2), 91–110 (2004)
132. Lowel, S., Singer, W.: Selection of intrinsic horizontal connections in the visual cortex by correlated neuronal activity. Science 255(5041), 209–212 (1992)
133. Lane, S.M.: Categories for the working mathematician, vol. 5. Springer (1998)
134. Malach, R., Amir, Y., Harel, M., Grinvald, A.: Relationship between intrinsic connections and functional architecture revealed by optical imaging and in vivo targeted biocytin injections in primate striate cortex. Proceedings of the National Academy of Sciences 90(22), 10469–10473 (1993)
135. Maloney, L.T., Wandell, B.A.: Color constancy: a method for recovering surface spectral reflectance. JOSA A 3(1), 29–33 (1986)
136. Mante, V., Carandini, M.: Mapping of stimulus energy in primary visual cortex. Journal of Neurophysiology 94(1), 788–798 (2005)
137. Marr, D.: Vision: A computational investigation into the human representation and processing of visual information. Henry holt and co. Inc., New York (1982)
138. Masquelier, T., Thorpe, S.J.: Unsupervised learning of visual features through spike timing dependent plasticity. PLoS Computational Biology 3(2), e31 (2007)
139. Matsuda, H.: Physical nature of higher-order mutual information: Intrinsic correlations and frustration. Physical Review E 62(3), 3096 (2000)
140. McIntyre, J., Berthoz, A., Lacquaniti, F.: Reference frames and internal models for visuo-manual coordination: what can we learn from microgravity experiments? Brain Research Reviews 28(1), 143–154 (1998)
141. Merleau-Ponty, M.M.: L'oeil et l'esprit. Gallimard (1964)
142. Merriam, E.P., Genovese, C.R., Colby, C.L.: Remapping in human visual cortex. Journal of Neurophysiology 97(2), 1738–1755 (2007)
143. Miao, X., Rao, R.P.N.: Learning the Lie groups of visual invariance. Neural Computation 19(10), 2665–2693 (2007)
144. Michael, C.R.: Retinal afferent arborization patterns, dendritic field orientations, and the segregation of function in the lateral geniculate nucleus of the monkey. Proceedings of the National Academy of Sciences 85(13), 4914–4918 (1988)
145. Milleret, C., Houzel, J.-C., Buser, P.: Pattern of development of the callosal transfer of visual information to cortical areas 17 and 18 in the cat. European Journal of Neuroscience 6(2), 193–202 (1994)
146. Mooser, F., Bosking, W.H., Fitzpatrick, D.: A morphological basis for orientation tuning in primary visual cortex. Nature Neuroscience 7(8), 872–879 (2004)
147. Morrone, M.C., Tosetti, M., Montanaro, D., Fiorentini, A., Cioni, G., Burr, D.C.: A cortical area that responds specifically to optic flow, revealed by fMRI. Nature Neuroscience 3(12), 1322–1328 (2000)
148. Mukamel, R., Ekstrom, A.D., Kaplan, J., Iacoboni, M., Fried, I.: Single-neuron responses in humans during execution and observation of actions. Current Biology 20(8), 750–756 (2010)
149. Munuera, J., Morel, P., Duhamel, J.-R., Deneve, S.: Optimal sensorimotor control in eye movement sequences. The Journal of Neuroscience 29(10), 3026–3035 (2009)
150. Nadal, J.-P., Brunel, N., Parga, N.: Nonlinear feedforward networks with stochastic outputs: infomax implies redundancy reduction. Network: Computation in Neural Systems 9(2), 207–217 (1998)

151. Nakamura, K., Colby, C.L.: Updating of the visual representation in monkey striate and extrastriate cortex during saccades. Proceedings of the National Academy of Sciences 99(6), 4026–4031 (2002)
152. Nandy, A.S., Sharpee, T.O., Reynolds, J.H., Mitchell, J.F.: The fine structure of shape tuning in area v4. Neuron 78(6), 1102–1115 (2013)
153. Nauhaus, I., Busse, L., Carandini, M., Ringach, D.L.: Stimulus contrast modulates functional connectivity in visual cortex. Nature Neuroscience 12(1), 70–76 (2008)
154. Niebur, E., Wörgötter, F.: Design principles of columnar organization in visual cortex. Neural Computation 6(4), 602–614 (1994)
155. Ohki, K., Chung, S., Kara, P., Hübener, M., Bonhoeffer, T., Reid, C.: Highly ordered arrangement of single neurons in orientation pinwheels. Nature 442(7105), 925–928 (2006)
156. Ohki, K., Reid, R.C.: Specificity and randomness in the visual cortex. Current Opinion in Neurobiology 17(4), 401–407 (2007)
157. Ohzawa, I., DeAngelis, G.C., Freeman, R.D.: Stereoscopic depth discrimination in the visual cortex: neurons ideally suited as disparity detectors. Science 249(4972), 1037–1041 (1990)
158. Ohzawa, I., DeAngelis, G.C., Freeman, R.D.: Encoding of binocular disparity by complex cells in the cat's visual cortex. Journal of Neurophysiology 77(6), 2879–2909 (1997)
159. Ohzawa, I., Freeman, R.D.: The binocular organization of complex cells in the cat's visual cortex. Journal of Neurophysiology 56(1), 243–259 (1986)
160. Olavarria, J.F.: Non-mirror-symmetric patterns of callosal linkages in areas 17 and 18 in cat visual cortex. Journal of Comparative Neurology 366(4), 643–655 (1996)
161. Olshausen, B.A., et al.: Emergence of simple-cell receptive field properties by learning a sparse code for natural images. Nature 381(6583), 607–609 (1996)
162. Olshausen, B.A., Field, D.J.: Sparse coding with an overcomplete basis set: A strategy employed by v1? Vision Research 37(23), 3311–3325 (1997)
163. Oomes, A.H.J., Todd, J.T., Koenderink, J.J., Kappers, A.M.L.: Varignon in visual space. In: Perception, vol. 29, p. 51. Pion LTD 207 Brondesbury Park, London NW2 5JN (2000)
164. Pasupathy, A.: Neural basis of shape representation in the primate brain. Progress in Brain Research 154, 293–313 (2006)
165. Payne, B., Peters, A.: The cat primary visual cortex. Academic Press (2001)
166. Pellionisz, A., Llinás, R.: Tensorial approach to the geometry of brain function: cerebellar coordination via a metric tensor. Neuroscience 5(7), 1125–1136 (1980)
167. Pellionisz, A., Llinás, R.: Tensor network theory of the metaorganization of functional geometries in the central nervous system. Neuroscience 16(2), 245–273 (1985)
168. Petitot, J.: Neurogéométrie de la vision: modèles mathématiques et physiques des architectures fonctionnelles. Editions Ecole Polytechnique (2008)
169. Pham, Q.-C., Bennequin, D.: Affine invariance of human hand movements: a direct test. arXiv preprint arXiv:1209.1467 (2012)
170. Philipona, D., O'Regan, K.: Reply to johnson and wright. Visual Neuroscience 25(2), 225–226 (2008)
171. Philipona, D., O'Regan, K., Nadal, J.-P.: Is there something out there? inferring space from sensorimotor dependencies. Neural Computation 15(9), 2029–2049 (2003)
172. Philipona, D.L., O' Regan, K.: Color naming, unique hues, and hue cancellation predicted from singularities in reflection properties. Visual Neuroscience 23(3/4), 331 (2006)

173. Piaget, J.: Les notions de mouvement et de vitesse chez l'enfant. Presses Univ. de France (1946)
174. Piaget, J.: La naissance de l'intelligence chez l'enfant, 2nd edn. Delachaux et Niestlé (1948)
175. Piaget, J., Henriques, G., Ascher, E., Coll: Morphismes et catégories: comparer et transformer. Delachaux et Niestlé, 2nd edn (1990)
176. Piaget, J., Inhelder, B.: La représentation de l'espace chez l'enfant. Presses Universitaires de France (1948)
177. Piaget, J., Inhelder, B., Szeminska, A.: La géométrie spontanée de l'enfant. Presses Universitaires de France (1948)
178. Pierrot-Deseilligny, C., Milea, D., Müri, R.M.: Eye movement control by the cerebral cortex. Current Opinion in Neurology 17(1), 17–25 (2004)
179. Poincaré, H.: La Science et l'hypothèse. Flammarion (1902)
180. Poincaré, H.: La Valeur de la science. Bibliothèque de philosophie scientifique. Flammarion (1905)
181. Poincaré, H.: Science et méthode. Bibliothèque de philosophie scientifique. Flammarion (1908)
182. Poincaré, H.: Dernières Pensées. Bibliothèque de philosophie scientifique. Flammarion (1913)
183. Pollen, D.A., Ronner, S.F.: Phase relationships between adjacent simple cells in the visual cortex. Science 212(4501), 1409–1411 (1981)
184. Ribot, J., Aushana, Y., Bui-Quoc, E., Milleret, C.: Organization and origin of spatial frequency maps in cat visual cortex. The Journal of Neuroscience 33(33), 13326–13343 (2013)
185. Riesenhuber, M., Poggio, T.: Hierarchical models of object recognition in cortex. Nature Neuroscience 2(11), 1019–1025 (1999)
186. Rochefort, N., Buzás, P., Quenech'Du, N., Koza, A., Eysel, U.T., Milleret, C., Kisvarday, Z.F.: Functional selectivity of interhemispheric connections in cat visual cortex. Cerebral Cortex 19(10), 2451–2465 (2009)
187. Rockland, K.S., Lund, J.S.: Widespread periodic intrinsic connections in the tree shrew visual cortex. Science (1982)
188. Rolls, E.T.: Brain mechanisms for invariant visual recognition and learning. Behavioural Processes 33(1), 113–138 (1994)
189. Rolls, E.T.: Spatial view cells and the representation of place in the primate hippocampus. Hippocampus 9(4), 467–480 (1999)
190. Rolls, E.T., Deco, G.: Computational neuroscience of vision. Oxford University Press (2002)
191. Rolls, E.T., Stringer, S.M.: Invariant visual object recognition: a model, with lighting invariance. Journal of Physiology-Paris 100(1), 43–62 (2006)
192. Rowe, M.P., Jacobs, G.H.: Naturalistic color discriminations in polymorphic platyrrhine monkeys: Effects of stimulus luminance and duration examined with functional substitution. Visual Neuroscience 24(1), 17 (2007)
193. Rust, N.C., Mante, V., Simoncelli, E.P., Movshon, A.: How MT cells analyze the motion of visual patterns. Nature Neuroscience 9(11), 1421–1431 (2006)
194. Sabbah, S., Laria, R., Gray, S., Hawryshyn, C.: Functional diversity in the color vision of cichlid fishes. BMC Biology 8(1), 133 (2010)
195. Sagiv, C., Sochen, N.A., Zeevi, Y.Y.: The uncertainty principle: group theoretic approach, possible minimizers and scale-space properties. Journal of Mathematical Imaging and Vision 26(1-2), 149–166 (2006)

196. Saito, K.: Period mapping associated to a primitive form. Publications of the Research Institute for Mathematical Sciences 19(3), 1231–1264 (1983)
197. Salinas, E., Sejnowski, T.J.: Correlated neuronal activity and the flow of neural information. Nature Reviews Neuroscience 2(8), 539–550 (2001)
198. Schmidt, K.E., Kim, D.-S., Singer, W., Bonhoeffer, T., Löwel, S.: Functional specificity of long-range intrinsic and interhemispheric connections in the visual cortex of strabismic cats. The Journal of Neuroscience 17(14), 5480–5492 (1997)
199. Schrödinger, E.R.J.A.: Outline of a theory of color measurement for daylight vision. Annalen der Physik 63(397-447), 481–520 (1920)
200. Schummers, J., Cronin, B., Wimmer, K., Stimberg, M., Martin, R., Obermayer, K., Koerding, K., Sur, M.: Dynamics of orientation tuning in cat v1 neurons depend on location within layers and orientation maps. Frontiers in Neuroscience 1(1), 145 (2007)
201. Schummers, J., Mariño, J., Sur, M.: Synaptic integration by V1 neurons depends on location within the orientation map. Neuron 36(5), 969–978 (2002)
202. Shaw, R.E., McIntyre, M., Mace, W.M.: The Role of Symmetry in Event Perception. In: Perception: Essays in Honor of J.J. Gibson. Cornell University Press, Ithaca (1974)
203. Shepard, R.N.: Perceptual-cognitive universals as reflections of the world. Psychonomic Bulletin & Review 1(1), 2–28 (1994)
204. Skottun, B.C., Bradley, A., Sclar, G., Ohzawa, I., Freeman, R.D.: The effects of contrast on visual orientation and spatial frequency discrimination: a comparison of single cells and behavior. Journal of Neurophysiology 57(3), 773–786 (1987)
205. Smith, M.A., Sommer, M.A.: Spatial and temporal scales of neuronal correlation in visual area v4. The Journal of Neuroscience 33(12), 5422–5432 (2013)
206. Solstad, T., Boccara, C.N., Kropff, E., Moser, M.-B., Moser, E.I.: Representation of geometric borders in the entorhinal cortex. Science 322(5909), 1865–1868 (2008)
207. Souriau, J.-M.: Structure des systèmes dynamiques. Dunod (1970)
208. Stiles, W.S.: Color vision: the approach through increment-threshold sensitivity. Proceedings of the National Academy of Sciences of the United States of America 45(1), 100 (1959)
209. Tanaka, K.: Inferotemporal cortex and object vision. Annual Review of Neuroscience 19(1), 109–139 (1996)
210. Tanaka, K., Fukada, Y., Saito, H.A.: Underlying mechanisms of the response specificity of expansion/contraction and rotation cells in the dorsal part of the medial superior temporal area of the macaque monkey. Journal of Neurophysiology 62(3), 642–656 (1989)
211. Taubes, C.H.: Differential geometry: bundles, connections, metrics and curvature, vol. 23. Oxford University Press (2011)
212. Thom, R.: Stabilité structurelle et morphogenèse: essai d'une théorie générale des modèles. Interédition (1977)
213. Thom, R.: Esquisse d'une sémiophysique. physique aristotéliciene et théorie des catastrophes (1988)
214. Thom, R., Lejeune, C., Duport, J.P.: Morphogenèse et imaginaire, vol. 8. Lettres Modernes Minard (1978)
215. Ting, H.K.: On the amount of information. Theory of Probability & Its Applications 7(4), 439–447 (1962)
216. Todd, J.T., Oomes, A.H.J., Koenderink, J.J., Kappers, A.M.L.: On the affine structure of perceptual space. Psychological Science 12(3), 191–196 (2001)
217. Touryan, J., Felsen, G., Dan, Y.: Spatial structure of complex cell receptive fields measured with natural images. Neuron 45(5), 781–791 (2005)
218. van der Waerden, B.L., Artin, E., Noether, E.: Algebra, vol. 1. Springer (1966)

219. Van Hooser, S.D., Heimel, A., Nelson, S.B.: Functional cell classes and functional architecture in the early visual system of a highly visual rodent. Progress in Brain Research 149, 127–145 (2005)
220. Varchenko, A.N.: Asymptotic Hodge structure in the vanishing cohomology. Izvestiya: Mathematics 18(3), 469–512 (1982)
221. Varchenko, A.N., Givental, A.B.: Mapping of periods and intersection form. Functional Analysis and Its Applications 16(2), 83–93 (1982)
222. Vilenkin, N.Y.: Special Functions and the Theory of Group Representations, vol. 22. AMS Bookstore (1968)
223. Vitte, E., Derosier, C., Caritu, Y., Berthoz, A., Hasboun, D., Soulie, D.: Activation of the hippocampal formation by vestibular stimulation: a functional magnetic resonance imaging study. Experimental Brain Research 112(3), 523–526 (1996)
224. Von der Malsburg, C., Phillips, W.A., Singer, W.: Dynamic coordination in the brain: from neurons to mind. MIT Press (2010)
225. Wachtler, T., Sejnowski, T.J., Albright, T.D.: Representation of color stimuli in awake macaque primary visual cortex. Neuron 37(4), 681–691 (2003)
226. Wallis, G., Rolls, E., Foldiak, P.: Learning invariant responses to the natural transformations of objects. In: Proceedings of 1993 International Joint Conference on Neural Networks, IJCNN 1993, vol. 2, pp. 1087–1090. IEEE (1993)
227. Webster, M.A.: Pattern selective adaptation in color and form perception. The Visual Neurosciences 2, 936–947 (2003)
228. Weinstein, A.: Lectures on symplectic manifolds, vol. (29). American Mathematical Soc. (1977)
229. Weinstein, A.: Poisson geometry. Differential Geometry and its Applications 9(1), 213–238 (1998)
230. Weliky, M., Katz, L.C.: Functional mapping of horizontal connections in developing ferret visual cortex: experiments and modeling. The Journal of Neuroscience 14(12), 7291–7305 (1994)
231. Weliky, M., Katz, L.C.: Correlational structure of spontaneous neuronal activity in the developing lateral geniculate nucleus in vivo. Science 285(5427), 599–604 (1999)
232. Weyl, H.: The Classical Groups, Their Invariants and Representations. Princeton University Press (1946)
233. White, L.E., Fitzpatrick, D.: Vision and cortical map development. Neuron 56(2), 327–338 (2007)
234. Wiener, S.I., Taube, J.S.: Head Direction Cells and the Neural Mechanisms of Spatial Orientation (Bradford Books). The MIT Press (2005)
235. Wilson, K.G.: Renormalization group and critical phenomena. I. renormalization group and the Kadanoff scaling picture. Physical Review B 4(9), 3174 (1971)
236. Wilson, K.G.: Renormalization group and critical phenomena. II. phase-space cell analysis of critical behavior. Physical Review B 4(9), 3184 (1971)
237. Wilson, V.J., Jones, G.M.: Mammalian vestibular physiology. Plenum Press, New York (1979)
238. Wolf, F., Geisel, T.: Spontaneous pinwheel annihilation during visual development. Nature 395(6697), 73–78 (1998)
239. Wolf, F., Geisel, T.: Universality in visual cortical pattern formation. Journal of Physiology-Paris 97(2), 253–264 (2003)
240. Wright, E.M.: The asymptotic expansion of the generalized hypergeometric function. Journal of the London Mathematical Society 1(4), 286–293 (1935)
241. Wright, E.M.: The asymptotic expansion of the generalized hypergeometric function. Proceedings of the London Mathematical Society 2(1), 389–408 (1940)

242. Wright, E.M.: A recursion formula for the coefficients in an asymptotic expansion. In: Proc. Glasgow Math. Assoc., vol. 4, pp. 38–41. Cambridge Univ Press (1958)
243. Wu, S., Amari, S.-I., Nakahara, H.: Population coding and decoding in a neural field: a computational study. Neural Computation 14(5), 999–1026 (2002)
244. Wurtzand, R.H., Goldberg, M.E.: Activity of superior colliculus in behaving monkey. iii. cells discharging before eye movements. J. Neurophysiol. 35(4), 575–586 (1972)
245. Yakusheva, T.A., Shaikh, A.G., Green, A.M., Blazquez, P.M., Dickman, J.D., Angelaki, D.E.: Purkinje cells in posterior cerebellar vermis encode motion in an inertial reference frame. Neuron 54(6), 973–985 (2007)
246. Yang, J.N., Shevell, S.K.: Surface color perception under two illuminants: The second illuminant reduces color constancy. Journal of Vision 3(5) (2003)
247. Yartsev, M.M., Ulanovsky, N.: Representation of three-dimensional space in the hippocampus of flying bats. Science 340(6130), 367–372 (2013)
248. Yu, H., Farley, B.J., Jin, D.Z., Sur, M.: The coordinated mapping of visual space and response features in visual cortex. Neuron 47(2), 267–280 (2005)
249. Zahs, K.R., Stryker, M.P.: Segregation of on and off afferents to ferret visual cortex. Journal of Neurophysiology 59(5), 1410–1429 (1988)
250. Zhang, T., Britten, K.H.: Parietal area vip causally influences heading perception during pursuit eye movements. The Journal of Neuroscience 31(7), 2569–2575 (2011)
251. Zhaoping, L.: Theoretical understanding of the early visual processes by data compression and data selection. Network: Computation in Neural Systems 17(4), 301–334 (2006)
252. Zhuo, Y., Zhou, T.G., Rao, H.Y., Wang, J.J., Meng, M., Chen, M., Zhou, C., Chen, L.: Contributions of the visual ventral pathway to long-range apparent motion. Science 299(5605), 417–420 (2003)

Chapter 8
Hebbian Learning of the Statistical and Geometrical Structure of Visual Input

James A. Bednar

Abstract. Experiments on the visual system of carnivorous mammals have revealed complex relationships between the geometry and statistical properties of the visual world, and the geometry and statistical properties of the primary visual cortex. This review surveys an extensive body of modelling work that shows how a relatively simple set of general-purpose neural mechanisms can account for a large fraction of this observed relationship. The models consist of networks of simple artificial neurons with initially unspecific connections that are modified by Hebbian learning and homeostatic plasticity. Given examples of internally generated or visually evoked neural activity, this generic starting point develops into a realistic match to observations from the primary visual cortex, without requiring any vision-specific circuitry or neural properties. We show that the resulting network reflects both the geometrical and statistical structure of the input, and develops under constraints provided by the geometrical structure of the cortical and subcortical regions in the model. Specifically, the model neurons develop adult-like receptive fields and topographic maps selective for all of the major local visual features, and realistic topographically organized lateral connectivity that leads to systematic surround modulation effects depending on the geometry of both the visual input and the cortical representations. Together these results suggest that sensory cortices self-organize to capture the statistical properties of their inputs, revealing the underlying geometry using relatively simple local rules that allow them to build useful representations of the external environment.

James A. Bednar
Institute for Adaptive and Neural Computation,
The University of Edinburgh,
10 Crichton St., EH8 9AB, Edinburgh, UK
e-mail: jbednar@inf.ed.ac.uk

G. Citti and A. Sarti (eds.), *Neuromathematics of Vision*,
Lecture Notes in Morphogenesis,
DOI: 10.1007/978-3-642-34444-2_8,

8.1 Introduction

Over the past half-century, experiments on the visual system of carnivorous mammals have revealed complex relationships between the geometry and statistical properties of the visual world, and the geometry and statistical properties of the visual cortex. For instance, an oriented line projected onto the back of the eye will evoke responses in a topographically mapped region of the primary visual cortex (V1), but in a discontinuous fashion grouped locally by orientation preference rather than retinotopic location. Figure 8.1 illustrates this mapping for V1 in a tree shrew, a primate-like species where the geometrical relationships are clearer because it lacks the fovea/periphery distinctions common to humans and other primates. Figure 8.1 shows that tree shrew V1 is organized much like the retina, with location on the retina mapping to corresponding locations in V1. But overlaid on this retinotopic map is an orientation map [Blasdel(1992b)], with different patches of V1 neurons responding within the retinotopically mapped area, depending on the orientation of the input.

This patchy pattern of activity and orientation preference has been understood as the result of a dimension-reduction process [Durbin and Mitchison(1990), Carreira-Perpiñán et al.(2005)Carreira-Perpiñán, Lister, and Goodhill, Ritter et al.(1992)Ritter, Martinetz, and Schulten], wherein the many dimensions in which a small patch of visual input could vary are mapped continuously onto the two-dimensional surface of the cortex. If the cortex had as many geometrical dimensions as the ways in which the input varied, this mapping could be straightforward. E.g. if the input varied only in retinal location (X,Y), a simple retinotopic map onto the cortical surface would suffice. Instead, the various combinations of retinotopic position and other features like orientation are flattened onto the cortical surface in a way that achieves good coverage of the inputs while maintaining local continuity [Swindale et al.(2000)Swindale, Shoham, Grinvald, Bonhoeffer, and Hubener, Miikkulainen et al.(2005)Miikkulainen, Bednar, Choe, and Sirosh, Ritter et al.(1992)Ritter, Martinetz, and Schulten]. Figure 8.2 illustrates this folding and flattening process for the case of ocular dominance (OD), with a cortical ocular dominance pattern interpreted as a two-dimensional view of preferences that cover a three-dimensional (X,Y,OD) space.

Cortical OR and OD maps illustrate geometric relationships between input and output spaces, but the relationships also take statistics into account. Specifically, the area of the cortical maps devoted to each feature value reflects the frequency of occurrence of that feature [Sengpiel et al.(1997)Sengpiel, Sen, and Blakemore, Tanaka et al.(2006)Tanaka, Ribot, Imamura, and Tani]. Figure 8.3 shows examples of this phenomenon in cat visual cortex, for kittens reared with special goggles that blur non-vertical patterns. Similar effects occur for the OD map, when input from one eye is disrupted [Wiesel(1982)]. These results raise the possibility that the observed geometrical relationships could at least in part be the result of an underlying process of adaptation to the statistics of the input, where the input geometry constrains the possible input samples and the cortex organizes around the patterns seen on its

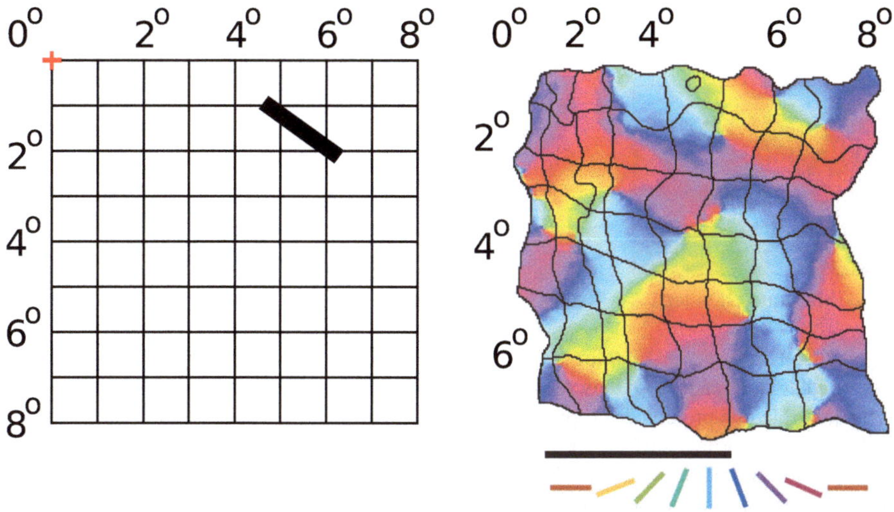

Fig. 8.1 Retinotopic and orientation map in V1. Given a particular fixation point (marked with a red + symbol above), the visual field seen by an animal can be divided into a regular grid, with each square representing a $1^\circ \times 1^\circ$ area of visual space. In cortical area V1 of mammals, neurons are arranged into a retinotopic map, with nearby neurons responding to nearby areas of the retina. As an example, the image on the right shows the retinotopic map on the surface of V1 of a tree shrew for an $8^\circ \times 7^\circ$ area of visual space (adapted from [Bosking et al.(2002)Bosking, Crowley, and Fitzpatrick]; scale bar is 1mm). A stimulus presented in a particular location in visual space (such as the thick black bar shown) evokes a response centered around the corresponding grid square in V1 (6°,2°). Which specific neurons respond within that general area, however, depends on the orientation of the stimulus. The V1 map is color coded with the preferred orientation of neurons in each location; e.g. the black bar shown at left will primarily activate neurons colored in purple in the corresponding V1 grid squares. Similar maps could be plotted for this same area showing preference for other visual features, such as motion direction, spatial frequency, color, disparity, and eye preference (depending on species). Other cortical areas are arranged into topographic maps for other sensory modalities, such as touch and audition, and for motor outputs.

inputs [von der Malsburg(1973), Miikkulainen et al.(2005)Miikkulainen, Bednar, Choe, and Sirosh, Ritter et al.(1992)Ritter, Martinetz, and Schulten].

Although the basic dimension-reduction and folding idea is now widespread, the links between dimension reduction, input statistics, and the actual machinery and circuits in the visual cortex remains obscure. This review surveys results from a large family of closely related *mechanistic* models of V1 development, which show how the observed map patterns can arise from plausible approximations to the mechanisms present in the subcortical visual pathways and in V1. Unlike other models of the map patterns, the resulting systems can then process actual visual images, and

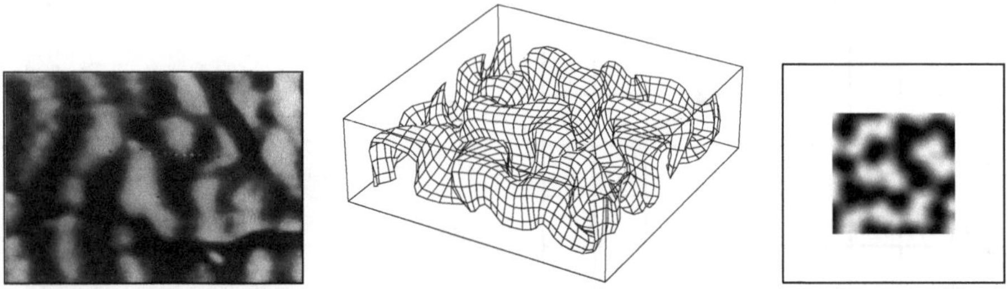

(*a*) Macaque OD map (*b*) Model preferences in (X, Y, OD) (*c*) Model OD

Fig. 8.2 Retinotopic and ocular dominance maps. (*a*) Just as for orientation, eye preference (ocular dominance; OD) is represented within the overall retinotopic map, with both of the possible eye preferences represented near any particular cortical location (here visualized across the cortical surface with black for one eye and white for the other eye); data for macaque V1 from [Blasdel(1992a)]. This pattern can be understood as a 2D projection onto the cortical surface of an underlying set of preferences in 3D: for X, Y, and ocular dominance. (*b*) shows the results of a self-organizing map (SOM) model of this organization, visualizing the 3D preference of each neuron (model from Ritter et al. [Ritter et al.(1991)Ritter, Obermayer, Schulten, and Rubner], [Ritter et al.(1992)Ritter, Martinetz, and Schulten]; figure and data from [Miikkulainen et al.(2005)Miikkulainen, Bednar, Choe, and Sirosh]). The 2D sheet of neurons has covered the 3D input space (delineated by the box outline) by folding in the third dimension, such that every value of (X, Y, OD) is well approximated by some neuron. The resulting pattern is a type of Peano (space-filling) curve. (*c*) When the OD preference is plotted in grayscale for each neuron in their cortical locations, projecting this 3D pattern space down to the 2D cortical space, the resulting pattern is similar to animal OD maps, suggesting that animals do a similar process of representing input spaces by folding in the non-retinotopic dimensions to fill a multidimensional input space, and that cortical feature maps are the result.

can thus be used to relate the map patterns, connectivity within the underlying networks, and observed visual and physiological phenomena. The models suggest that a wide and diverse range of observations about the visual cortex can be explained by a small set of general-purpose mechanisms. These mechanisms are not specific to vision, and should be applicable to most cortical regions.

Section 8.2 outlines the basic principles of these mechanistic models. Section 8.3) presents an implementation of a simple GCAL (Gain Control, Adaptive, Laterally connected) model [Stevens et al.(2013)Stevens, Law, Antolik, and Bednar], and describes both how it relates to earlier models on which it is based, and how it relates to a more realistic but more complex variant that covers all the phenomena reported here. Section 8.4 surveys results from GCAL and related models. Section 8.5 explores implications of the model, and areas for further work.

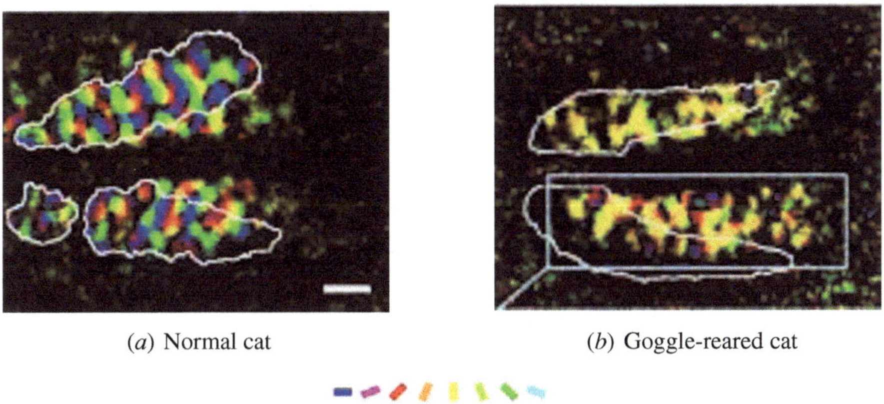

(*a*) Normal cat (*b*) Goggle-reared cat

Fig. 8.3 Maps reflect input statistics. Comparison between orientation maps for (*a*) cats raised in a normal environment and (*b*) cats reared wearing goggles that blur non-vertical orientations shows that the distribution of orientation preferences reflects the input statistics. Thus the relationship between input and output is not merely geometrical, but is based on the statistical structure of the inputs.

8.2 GCAL Model Overview

The models considered in this chapter are each based on the following biologically grounded principles and mechanisms (description follows [Bednar(2012),Bednar(2013)]):

1. Single-compartment (point neuron) firing-rate (i.e., non-spiking) retinal ganglion cell (RGC), lateral geniculate nucleus (LGN), and V1 model neurons (see figure 8.4),
2. Hardwired subcortical pathways to V1, including the main LGN or RGC cell types that have been identified,
3. Initially roughly retinotopic topographic projections from the eye to the LGN and from the LGN to V1, connecting corresponding areas of each region,
4. Initially roughly isotropic (i.e., radially uniform) local connectivity to and between neurons in layers in V1, connecting neurons non-specifically to their local and more distant neighbors,
5. Natural images and spontaneous subcortical input activity patterns that lead to V1 responses,
6. Hebbian (unsupervised activity-dependent) learning with normalization for synapses on V1 neurons,
7. Homeostatic plasticity (whole-cell adaptation of excitability to keep the average activity of each V1 neuron constant), and
8. Various modeller-determined parameters associated with each of these mechanisms, eventually intended to be set through self-regulating mechanisms.

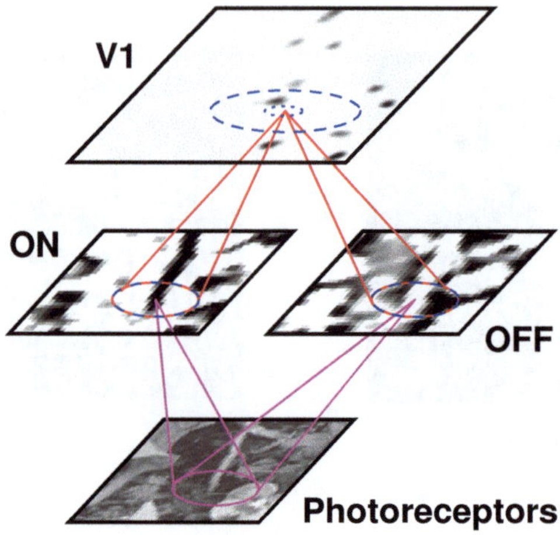

Fig. 8.4 Basic GCAL model architecture. In the simplest case, GCAL consists of a grayscale matrix representing the photoreceptor input, a pair of neural sheets representing the ON-center and OFF-center pathways from the photoreceptors to V1, and a single sheet representing V1. Each sheet is drawn here with a sample activity pattern resulting from one natural image patch. Each projection between sheets is illustrated with an oval showing the extent of the connection field in that projection, with lines converging on the target of the projection. Lateral projections, connecting neurons within each sheet, are marked with dashed ovals. Projections from the photoreceptors to the ON and OFF sheets, and within those sheets, are hardwired to mimic a specific class of response types found in the retina and LGN, in this case monochromatic center-surround neurons with a fixed spatial extent. Connections to and between V1 neurons adapt via Hebbian learning, allowing initially unselective V1 neurons to exhibit the range of response types seen experimentally, by differentially weighting each of the subcortical inputs (from the ON and OFF sheets) and inputs from neighboring V1 neurons. Reprinted from [Bednar(2013)].

Properties and mechanisms not necessary to explain the phenomena considered in this chapter, such as spiking, spike-timing dependent plasticity, detailed neuronal morphology, feedback from higher areas, neuromodulation, reinforcement learning, and supervised learning, have all been omitted, to clearly focus on the aspects of the system most relevant to the observed phenomena. The overall hypothesis is that much of the complex structure and properties observed in V1 emerges from interactions between relatively simple but highly interconnected computing elements, with connection strengths and patterns self-organizing in response to visual input and other sources of neural activity. Through visual experience, the geometry and statistical regularities of the visual world become encoded into the structure and connectivity of the visual cortex, leading to a complex functional cortical architecture that reflects the physical and statistical properties of the visual world.

At present, many of the results have been obtained independently in a wide variety of separate projects performed with different collaborators at different times. However, all of the models share the same underlying principles outlined above, and all are implemented using the same simulator and a small number of underlying components. See [Bednar(2012)] for an overview of each of the different models and how they fit together; here we present details for a simple but representative model simulating the development of orientation preferences and orientation maps for a single eye (figure 8.4), and describe a more complex but still incomplete "unified model" [Bednar(2012)] covering the other phenomena, so far published only in separate models [Bednar(2012), Antolik(2010), Ball and Bednar(2009), Ramtohul(2006), Palmer(2009), Bednar and Miikkulainen(2006)].

The goal for each of these models is the same — to explain how a cortical network can start from an initially undifferentiated state, to wire itself into a collection of neurons that behave, at a first approximation, like those in V1. Because such a model starts with no specializations (at the cortical level) specific to vision and would organize very differently when given different inputs, it also represents a general explanation for the development and function of any sensory or motor area in the cortex.

8.3 GCAL Architecture

All of the models whose results are presented here are implemented in the Topographica simulator, and are freely available along with the simulator from `www.topographica.org`. Both the basic and unified models are described using the same equations shown below, previously presented in refs. [Stevens et al.(2013)Stevens, Law, Antolik, and Bednar, Bednar(2012)]. The model is intended to represent the visual system of the macaque monkey, but relies on data from studies of cats, ferrets, tree shrews, or other mammalian species where clear results are not yet available from monkeys.

8.3.1 Sheets and Projections

Each Topographica model consists of a set of sheets of neurons and projections (sets of topographically mapped connections) between them. A model has sheets representing the visual input (as a set of activations in photoreceptor cells), sheets implementing the transformation from the photoreceptors to inputs driving V1 (expressed as a set of ON and OFF RGC/LGN cell activations), and sheets representing neurons in V1. The simple GCAL model (figure 8.4) has 4 such sheets, while the complete unified model described in [Bednar(2012)] has 29, each representing different topographically organized populations of cells in a particular region.

Each sheet is implemented as a two-dimensional array of firing-rate neurons. The Topographica simulator allows parameters for sheets and projections to be specified in measurement units that are independent of the specific grid sizes used in a particular run of the simulation. To achieve this, Topographica sheets provide multiple spatial coordinate systems, called *sheet* and *matrix* coordinates. Where possible, parameters (e.g. sheet dimensions or connection radii) are expressed in sheet coordinates, expressed as if the sheet were a continuous neural field rather than a finite grid. In practice, of course, sheets are always simulated using some finite matrix of units. Each sheet has a parameter called its density, which specifies how many units (matrix elements) in the matrix correspond to a length of 1.0 in continuous sheet coordinates, which allows conversion between sheet and matrix coordinates. When sizes are scaled appropriately [Bednar et al.(2004)Bednar, Kelkar, and Miikkulainen], results are independent of the density used, except at very low densities or for simulations with complex cells, where complexity increases with density [Antolik and Bednar(2011)]. Larger areas can be simulated easily [Bednar et al.(2004)Bednar, Kelkar, and Miikkulainen], but require more memory and simulation time.

A projection to an $m \times m$ sheet of neurons consists of m^2 separate *connection fields*, one per target neuron, each of which is a spatially localized set of connections from the neurons in one input sheet that are near the location corresponding topographically to the target neuron. Figure 8.4 shows one sample connection field (CF) for each projection, visualized as an oval of the corresponding radius on the input sheet (drawn to scale), connected by a cone to the neuron on the target sheet (if different). The connections and their weights determine the specific properties of each neuron in the network, by differentially weighting inputs from neurons of different types and/or spatial locations. Each of the specific types of sheets and projections is described in the following sections.

8.3.2 Images and Photoreceptor Sheets

The basic GCAL model (figure 8.4) has one input sheet, representing responses of photoreceptors of one cone class in one retina. The full unified GCAL model of all the input dimensions includes six input sheets (three different cone types in each eye; not shown here). For the full unified model, inputs were generated by choosing one calibrated-color image randomly from a database of single calibrated images, selecting a random patch within the image, a random direction of motion translation with a fixed speed (described in ref. [Bednar and Miikkulainen(2003)]), and a random brightness difference between the two eyes (described in ref. [Miikkulainen et al.(2005)Miikkulainen, Bednar, Choe, and Sirosh]). These modifications are intended as a simple model of motion and eye differences, to allow development of direction preference, ocular dominance, disparity, and color maps, until suitable full-motion stereo calibrated-color video datasets of natural scenes are available. Simulated retinal waves can also be used as inputs, to provide initial receptive-field and map structure before eye opening, but are not required for receptive-field or map development in the model [Bednar and Miikkulainen(2004)].

8.3.3 Subcortical Sheets

The subcortical pathway from the photoreceptors to the LGN and then to V1 is represented as a set of hardwired subcortical cells with fixed connection fields (CFs) that determine the response properties of each cell. These cells represent the complete processing pathway to V1, including circuitry in the retina (including the retinal ganglion cells), the optic nerve, the lateral geniculate nucleus, and the optic radiations to V1. Because the focus of the model is to explain cortical development given its thalamic input, the properties of these ON/OFF cells are kept fixed throughout development, for simplicity and clarity of analysis.

Each distinct ON/OFF cell type is grouped into a separate sheet, each of which contains a topographically organized set of cells with identical properties but responding to a different topographically mapped region of the retinal photoreceptor input sheet. Figure 8.4 shows the two main different spatial response types used in the GCAL models illustrated here, ON (with an excitatory center) and OFF (with an excitatory surround). All of these cells have Difference-of-Gaussian (DoG) receptive fields, and thus perform edge enhancement at a particular size scale. Additional cell classes can easily be added as needed for spatial frequency (with multiple DoG sizes) or color (with separate cone types for the center and surround Gaussians) simulations.

For the ON and OFF cells in the unified model, there are multiple copies with different delays from the retina. These delays represent the different latencies in the lagged vs. non-lagged cells found in cat LGN [Saul and Humphrey(1992), Wolfe and Palmer(1998)], and allow V1 neurons to become selective for the direction of motion. Many other sources of temporal delays would also lead to direction preferences, but have not been tested specifically.

8.3.4 Cortical Sheets

Unless otherwise stated, the simulations reported in this chapter use only a single V1 sheet for simplicity, but in the full unified model, V1 is represented by multiple cortical sheets representing different cell types and different V1 layers [Bednar(2012), Antolik(2010)]. In this simplified version, cells make both excitatory and inhibitory connections (unlike actual V1 neurons), and all cells receive direct input from LGN cells (unlike many V1 neurons). Even so, the single-sheet V1 can demonstrate most of the phenomena described above, except for complex cells (which can be obtained by adding a separate population of cells without direct thalamic input [Antolik and Bednar(2011)]) and contrast-dependent surround modulation effects (which require separate populations of inhibitory and excitatory cells [Antolik(2010), Law(2009)]).

The behavior of the cortical sheet is primarily determined by the projections to and within it. Each of these projections is initially non-specific other than the initial rough topography, and becomes selective only through the process of self-organization (described below), which increases some connection weights at the expense of others.

8.3.5 Activation

The model is simulated in a series of discrete time steps with step size $\delta t = 0.05$ (roughly corresponding to 12.5 milliseconds of real time). At time 0.0, the first image is drawn on the retina, and the activation of each unit in each sheet is updated for the remaining 19 steps before time 1.0, when a new pattern is selected and drawn on the retina (and similarly until the last input pattern is drawn at time 10,000). Each image patch on the retina represents one visual fixation (for natural images) or a snapshot of the relatively slowly changing spatial pattern of spontaneous activity (such as the well-documented retinal waves [Wong(1999)]). Thus the training process consists of a constant retinal activation, followed by recurrent processing at the LGN and cortical levels. For one input pattern, assume that the input is drawn on the photoreceptors at time t and the connection delay (constant for all projections) is defined as 0.05. Then at $t + 0.05$ the ON and OFF cells compute their responses, and at $t + 0.010$ the thalamic output is delivered to V1, where it similarly propagates recurrently through the intracortical projections to the cortical sheet(s) every 0.05 time steps. A much smaller step size of $\delta t = 0.002$ allows replication of the detailed time course of responses to individual patterns [Stevens(2011)], but this relatively coarse step size of 0.05 is more practical for simulations of long-term processes like neural development.

Images are presented to the model by activating the retinal photoreceptor units. The activation value $\Psi_{i,P}$ of unit i in photoreceptor sheet P is given by the brightness of that pixel in the training image.

For each model neuron in the other sheets, the activation value is computed based on the combined *activity contributions* to that neuron from each of the sheet's incoming projections. The activity contribution from a projection is recalculated whenever its input sheet activity changes, after the corresponding connection delay. For each unit j in a target sheet and an incoming projection p from sheet s_p, the activity contribution is computed from activations in the corresponding connection field F_{jp}. F_{jp} consists of the local neighborhood around j (for lateral connections), or the local neighborhood of the topographically mapped location of j on s_p (for a projection from another sheet); see examples in figures 8.4. The activity contribution C_{jp} to j from projection p is then a dot product of the relevant input with the weights in each connection field:

$$C_{jp}(t + \delta t) = \sum_{i \in F_{jp}} \eta_i(t)\omega_{ij,p} \tag{8.1}$$

where X_{is} is the activation of unit i on this projection's input sheet s_p, unit i is taken from the connection field F_{jp} of unit j, and $\omega_{ij,p}$ is the connection weight from i to j in that projection. Across all projections, multiple direct connections between the same pair of neurons are possible, but each projection p contains at most one connection between i and j, denoted by $\omega_{ij,p}$.

For a given cortical unit j, the activity $\eta_j(t + \delta t)$ is calculated from a rectified weighted sum of the activity contributions $C_{jp}(t + \delta t)$:

$$\eta_{jV}(t+\delta t) = f\left(\sum_{p} \gamma_p C_{jp}(t+\delta t)\right) \tag{8.2}$$

where f is a half-wave rectifying function with a variable threshold point (θ) dependent on the average activity of the unit, as described in the next subsection, and V denotes one of the cortical sheets.

Each γ_p is an arbitrary multiplier for the overall strength of connections in projection p. The γ_p values are set in the approximate range 0.5 to 3.0 for excitatory projections and -0.5 to -3.0 for inhibitory projections. For afferent connections, the γ_p value is chosen to map average V1 activation levels into the range 0 to 1.0 by convention, for ease of interconnecting and analyzing multiple sheets. For lateral and feedback connections, the γ_p values are then chosen to provide a balance between feedforward, lateral, and feedback drive, and between excitation and inhibition; these parameters are crucial for making the network operate in a useful regime.

ON/OFF neuron activity is computed similarly to equation 8.2, except to add divisive normalization and to fix the threshold θ at zero:

$$\eta_{jL}(t+\delta t) = f\left(\frac{\sum_p \gamma_p C_{jp}(t+\delta t)}{\gamma_S C_{jS}(t+\delta t)+k}\right) \tag{8.3}$$

where L stands for one of the ON/OFF sheets. Projection S here consists of a set of isotropic Gaussian-shaped lateral inhibitory connections (see equation 8.6, evaluated with $u=1$), and p ranges over all the other projections to that sheet. k is a small constant to make the output well-defined for weak inputs. The divisive inhibition implements the contrast gain control mechanisms found in RGC and LGN neurons [Antolik(2010), Felisberti and Derrington(1999), Bonin et al.(2005)Bonin, Mante, and Carandini, Alitto and Usrey(2008)].

Each time the activity is computed using equation 8.2 or 8.3, the new activity values are sent to each of the outgoing projections, where they arrive after the projection delay. The process of activity computation then begins again, with a new contribution C computed as in equation 8.1, leading to new activation values by equation 8.2 or 8.3. Activity thus spreads recurrently throughout the network, and can change, die out, or be strengthened, depending on the parameters.

With typical parameters that lead to realistic topographic map patterns, initially blurry patterns of afferent-driven cortical activity are sharpened into well-defined "activity bubbles" through locally cooperative and more distantly competitive lateral interactions [Miikkulainen et al.(2005)Miikkulainen, Bednar, Choe, and Sirosh]. Nearby neurons are thus influenced to respond more similarly, while more distant neurons receive net inhibition and thus learn to respond to different input patterns. The competitive interactions "sparsify" the cortical response into patches, in a process that can be compared to the explicit sparseness constraints in non-mechanistic models [Olshausen and Field(1996), Hyvärinen and Hoyer(2001)], while the local facilitatory interactions encourage spatial locality so that smooth topographic maps will be developed.

As described in more detail below, the initially random weights to cortical neurons are updated in response to each input pattern, via Hebbian learning. Because the settling (sparsification) process typically leaves only small patches of the cortical neurons responding strongly, those neurons will be the ones that learn the current input pattern, while other nearby neurons will learn other input patterns, eventually covering the complete range of typical input variation. Overall, through a combination of the network dynamics that achieve sparsification along with local similarity, plus homeostatic adaptation that keeps neurons operating in a useful regime, plus Hebbian learning that leads to feature preferences, the network will learn smooth, topographic maps with good coverage of the space of input patterns, thereby developing into a functioning system for processing patterns of visual input without explicit specification or top-down control of this process.

8.3.6 Homeostatic Adaptation

For this model, the threshold for activation of each cortical neuron is a very important quantity, because it directly determines how much the neuron will fire in response to a given input. Mammalian neurons appear to regulate such thresholds automatically, a process known as homeostatic plasticity or homeostatic adaptation [Turrigiano(1999)] (where homeostatic means to keep similar and stable). To set the threshold automatically, each neural unit j in V1 calculates a smoothed exponential average of its own activity ($\overline{\eta_j}$):

$$\overline{\eta_j}(t) = (1-\beta)\eta_j(t) + \beta\overline{\eta_j}(t-1) \tag{8.4}$$

The smoothing parameter ($\beta = 0.999$) determines the degree of smoothing in the calculation of the average. $\overline{\eta_j}$ is initialized to the target average V1 unit activity (μ), which for all simulations is $\overline{\eta_{jA}}(0) = \mu = 0.024$. The threshold is updated as follows:

$$\theta(t) = \theta(t-1) + \kappa(\overline{\eta_j}(t) - \mu) \tag{8.5}$$

where $\kappa = 0.0001$ is the homeostatic learning rate. The effect of this scaling mechanism is to bring the average activity of each V1 unit closer to the specified target. If the average activity of a V1 unit moves away from the target during training, the threshold for activation is thus automatically raised or lowered in order to bring it closer to the target.

8.3.7 Learning

Initial connection field weights are random within a two-dimensional Gaussian envelope. E.g., for a postsynaptic (target) neuron j located at sheet coordinate (0,0), the weight $\omega_{ij,p}$ from presynaptic unit i in projection p is:

$$\omega_{ij,p} = \frac{1}{Z_{\omega p}} u \exp\left(-\frac{x^2+y^2}{2\sigma_p^2}\right) \tag{8.6}$$

where (x,y) is the sheet-coordinate location of the presynaptic neuron i, u is a scalar value drawn from a uniform random distribution for the afferent and lateral inhibitory projections ($p = A, I$), σ_p determines the width of the Gaussian in sheet coordinates, and $Z_{\omega}p$ is a constant normalizing term that ensures that the total of all weights $\omega_{ij,p}$ to neuron j in projection p is 1.0, where all afferent projections are treated together as a single projection so that their sum total is 1.0. Weights for each projection are only defined within a specific maximum circular radius r_p; they are considered zero outside that radius.

Once per input pattern (after activity has settled), each connection weight ω_{ij} from unit i to unit j is adjusted using a simple Hebbian learning rule. (Learning could instead be performed at every simulation time step, but doing so would require significantly more computation time.) This rule results in connections that reflect correlations between the presynaptic activity and the postsynaptic response. Hebbian connection weight adjustment for unit j is dependent on the presynaptic activity η_i, the post-synaptic response η_j, and the Hebbian learning rate α:

$$\omega_{ij,p}(t) = \frac{\omega_{ij,p}(t-1) + \alpha\eta_j\eta_i}{\sum_k \left(\omega_{kj,p}(t-1) + \alpha\eta_j\eta_k\right)} \tag{8.7}$$

Unless it is constrained, Hebbian learning will lead to ever-increasing (and thus unstable) values of the weights. The weights are constrained using divisive postsynaptic weight normalization (denominator of equation 8.7), which is a simple and well understood mechanism. All afferent connection weights from ON/OFF sheets are normalized together in the model, which allows V1 neurons to become selective for any subset of the ON/OFF inputs. Weights are normalized separately for each of the other projections, to ensure that Hebbian learning does not disrupt the balance between feedforward drive, lateral and feedback excitation, and lateral and feedback inhibition. Subtractive normalization with upper and lower bounds could be used instead, but it would lead to binary weights [Miller and MacKay(1994),Miller(1994)], which is not desirable for a firing-rate model whose connections represent averages over multiple physical connections. More biologically motivated homeostatic mechanisms for normalization such as multiplicative synaptic scaling [Turrigiano(1999)] or a sliding threshold for plasticity [Bienenstock et al.(1982)Bienenstock, Cooper, and Munro] could be implemented instead, but these have not been tested so far.

Note that some of the results below use the earlier LISSOM model [Miikkulainen et al.(2005)Miikkulainen, Bednar, Choe, and Sirosh], which follows the same equations but lacks gain control and homeostatic adaptation (equivalent to setting $\gamma_S = 0$ and $k = 1$ in equation 8.3 and $\kappa = 0$ in equation 8.5). Without these automatic mechanisms, LISSOM requires the modeller to set the input strength and activation thresholds separately for each dataset and to adjust them as learning progresses. As long as these values have been set appropriately, previous LISSOM results can be treated equivalently to GCAL results, but GCAL is significantly simpler to use and describe, while being more robust to changes in the input distributions [Stevens et al.(2013)Stevens, Law, Antolik, and Bednar], so only GCAL is described here.

8.4 Results

In the following sections, we review a series of model results that account for anatomical, electrophysiological, imaging, psychophysical, and behavioral results from studies of experimental animals. Each of the results arises from the neural architecture and self-organizing mechanisms outlined in the previous section, operating on the statistical properties of the inputs, which reflect geometrical properties both of the world and of the visual system itself.

8.4.1 Maps and Connection Patterns

Figure 8.5 shows the pattern of orientation selectivity that emerges in the basic GCAL model from figure 8.4, whose subcortical pathway consists of a single set of non-lagged monochromatic ON and OFF LGN inputs for a single eye. This model robustly develops orientation maps, when given training inputs that have elongated patterns. In the model, the maps emerge as the consequence of the series of activity bubbles in response to each input pattern, which cause different regions of the cortex to learn weights corresponding to different input patterns.

Over the course of development, initially unspecific connections thus become selective for specific patterns of LGN activity, including particular orientations. Hebbian learning ensures that each afferent connection field shown represents the average pattern of LGN activity that has driven that neuron to a strong response; each neuron prefers a different pattern at a specific location on the retinal surface. Preferences from the set of all V1 neurons form a smooth topographic map covering the range of orientations present in the input patterns, yielding an orientation map similar to those from monkeys [Blasdel(1992b)]. For instance, the map shows iso-feature domains, pinwheel centers, fractures, saddle points, and linear zones, with a ring-shaped Fourier transform. As in animals [Sclar and Freeman(1982)], orientation selectivity is preserved over a very wide range of contrasts, due to the effect of lateral inhibitory connections in the LGN and in V1 that normalize responses to be relative to activation of neighboring neurons rather than absolute levels of contrast [Stevens et al.(2013)Stevens, Law, Antolik, and Bednar].

Figure 8.6 shows that the specific map pattern observed is a consequence not of the initial random weight patterns, but of the series of randomly chosen inputs over time. The overall properties of each map will be the same for any inputs drawn from the same distribution, but the specific map pattern depends crucially on the arbitrary location and order of inputs received during self-organization. The overall type of organization primarily emerges from geometric constraints on smoothly mapping the range of values for the indicated feature, within a two-dimensional retinotopic map [Miikkulainen et al.(2005)Miikkulainen, Bednar, Choe, and Sirosh, Kaschube et al.(2010)Kaschube, Schnabel, Löwel, Coppola, White, and Wolf].

The map patterns are also affected by the relative amount by which each feature varies in the input dataset, how often each feature appears, and other aspects of the input image statistics [Bednar and Miikkulainen(2004)]. For instance, orientation maps trained on natural image inputs develop a preponderance of neurons with

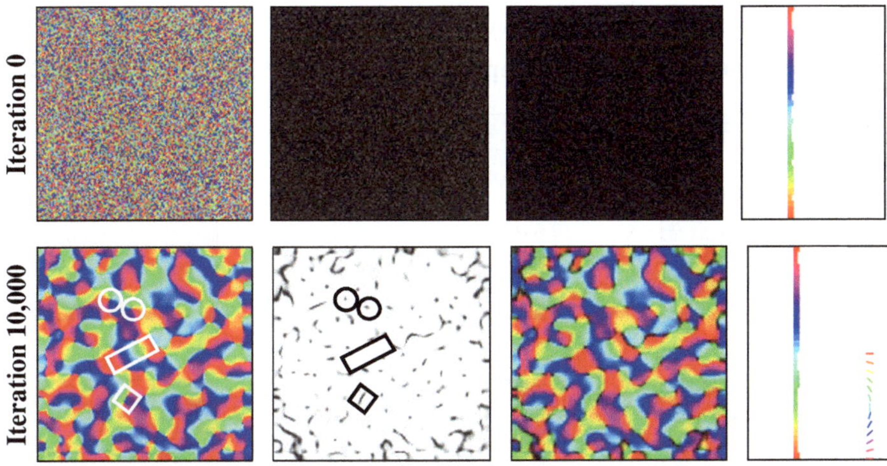

(*a*) Orientation map (*b*) Selectivity map (*c*) OR+selectivity (*d*) (*e*)

Fig. 8.5 Orientation maps trained on abstract stimuli (*color figure*). These plots show the orientation preference measured for each model neuron before (top row, iteration 0) and after self-organization (bottom row, iteration 10,000) based on artificially generated two-dimensional oriented Gaussian patterns. Each neuron in the map is colored according to the orientation it prefers, using color key (*e*). (*a*) The preferences are initially random (top). Through self-organization, the network developed a smoothly varying orientation map (bottom). Apart from the overall retinotopic mapping that was enforced at initialization, the map contains local geometric features found in maps from experimental animals, such as pinwheels (two are circled in white in *a* and black in *b*), linear zones (one is marked with a long white or black rectangle), and fractures (one between green and blue/purple is marked with a white or black square). (*b*) Before self-organization, the selectivity of each neuron for its (random) preferred orientation is very low (black in *b*, top). In contrast, nearly all of the self-organized neurons are highly selective for orientation (white in *b*, bottom). (*c*) Overlaying the orientation and selectivity plots shows that regions of lower selectivity in the self-organized map tend to occur near pinwheel centers and along fractures. Histograms of the number of neurons preferring each orientation are shown in (*d*), and are essentially flat because the initial weight patterns were unbiased and subsequent training inputs represented all orientations equally. These plots show that LISSOM (with GCAL getting essentially identical results as well; [Stevens et al.(2013)Stevens, Law, Antolik, and Bednar]) can develop realistic orientation maps through self-organization based on abstract input patterns. Reprinted from [Bednar(2002)].

horizontal and vertical orientation preferences, as seen in ferret maps and in natural images [Bednar and Miikkulainen(2004), Coppola et al.(1998)Coppola, White, Fitzpatrick, and Purves]. Figure 8.7 shows results from maps trained first on a model of spontaneous retinal activity (to account for maps present at eye opening in ferrets and cats), and then on natural images from different datasets. For natural image inputs, the map's histogram of orientation preferences will no longer be flat as it was for the artificial inputs in figure 8.5; instead it reflects the statistics of orientations present in the image dataset. Figure 8.7 shows that the model has successfully

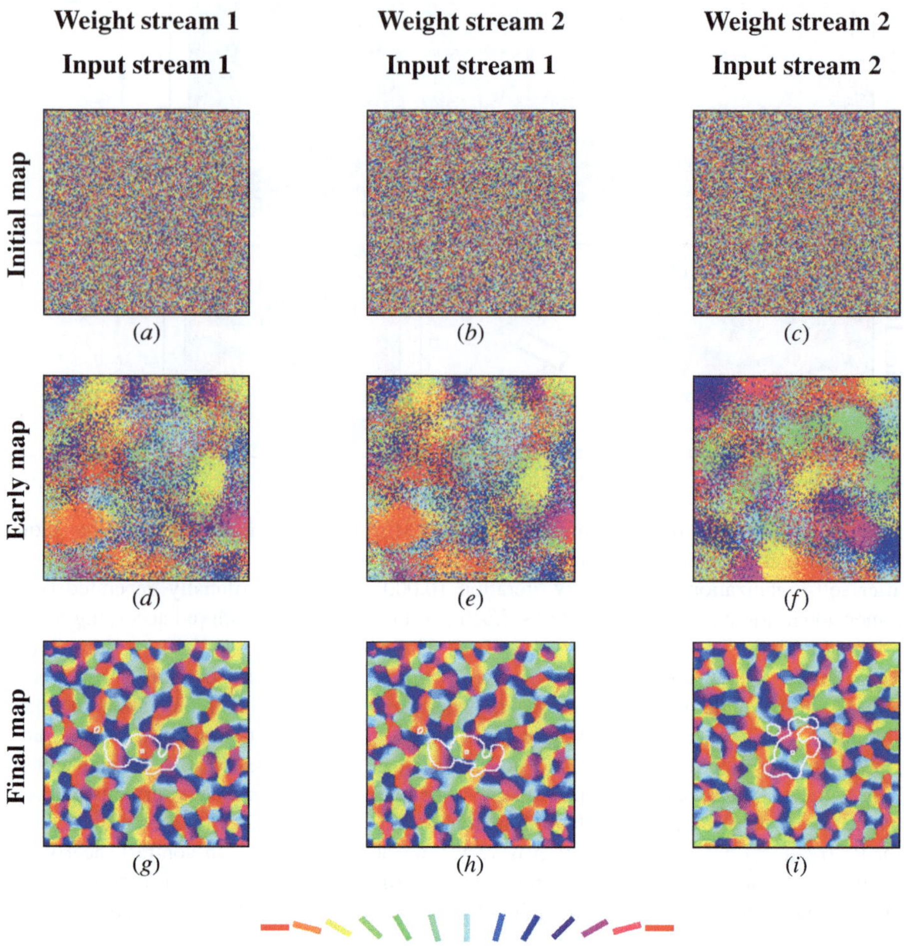

Fig. 8.6 Input stream determines map pattern (*color figure*). This figure shows that the self-organized orientation map patterns (e.g. in figure 8.5) do not depend on the random initial values of the weights. They are instead driven by the stream of input patterns presented during training. Using a different stream of random numbers for the weights results in different initial orientation maps (*a* and *b*), but has almost no effect on the final self-organized maps (compare *g* to *h*). In (*g*-*i*), the lateral inhibitory connections of one sample neuron are outlined in white, and are not affected by changing the weight stream. The final result is the same because lateral excitation smooths out differences in the initial weight values, and leads to similar large-scale patterns of activation at each iteration. (Compare maps *d* and *e* measured at iteration 100; the same large-scale features are emerging in both maps despite locally different patterns of noise caused by the different initial weights.) In contrast, changing the input stream produces very different early and final map patterns (compare *e* to *f* and *h* to *i*), even when the initial weight patterns (and therefore the initial orientation maps) are identical (*b* and *c*). Thus the input patterns are the crucial source of variation, not the initial weights. Reprinted from [Bednar(2002)].

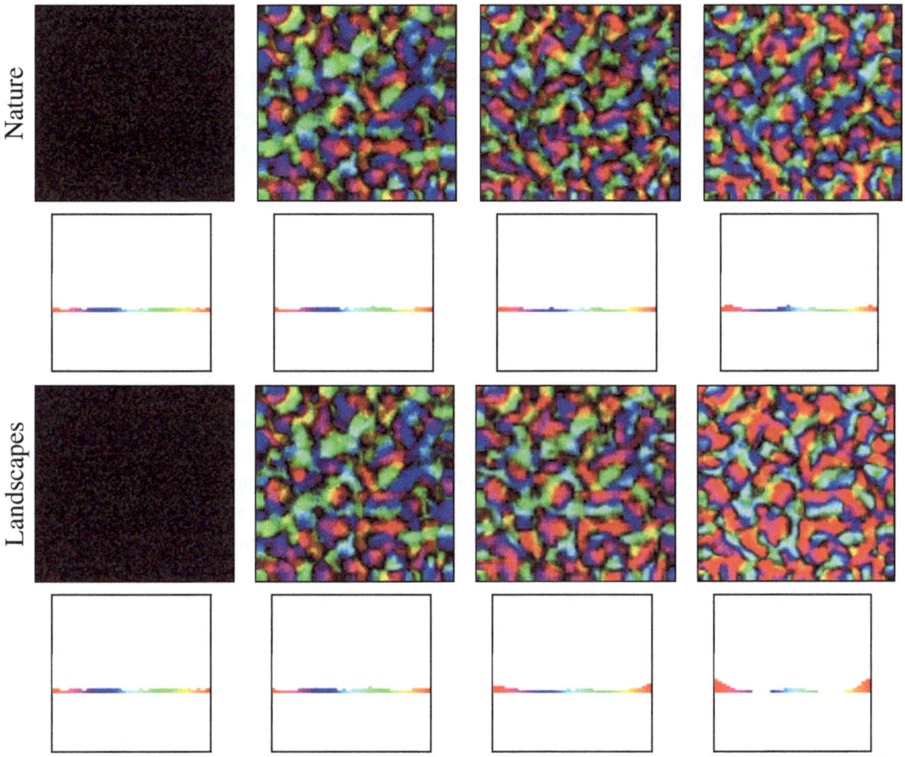

(*a*) 0: Initial map (*b*) 1000: End of prenatal training (*c*) 2500: During postnatal training (*d*) 10,000: End of postnatal training

Fig. 8.7 Postnatal training makes orientation map match statistics of the environment (*color figure*). Each row shows results from a network trained for 1000 iterations on a model of internally generated activity [Bednar and Miikkulainen(2004)], then trained for 9000 further iterations using natural images [Shouval et al.(1996)Shouval, Intrator, Law, and Cooper] to model postnatal visual experience. The orientation map plots (*b-d*) show selectivity as a brightness level, so that the postnatal improvement in selectivity will be visible. (*a*) and (*b*) are the same in each row. The top row shows the effect of postnatal training on natural images. With these images, more neurons become sensitive to horizontal and vertical contours, and the overall selectivity increases. However, the overall map shape remains similar, as found in laboratory animals ([Chapman et al.(1996)Chapman, Stryker, and Bonhoeffer]; compare individual blobs between maps right to left or left to right). The postnatal changes when trained on a different database consisting primarily of landscape images are similar but much more pronounced. With these images, the network smoothly develops strong biases for vertical and horizontal contours, within the pre-determined map shape. These results show that postnatal learning can gradually adapt the prenatally developed map to match the statistics of an animal's natural environment, as shown in figure 8.3, while explaining how an orientation map can be present already at eye opening. Reprinted from [Bednar(2002)].

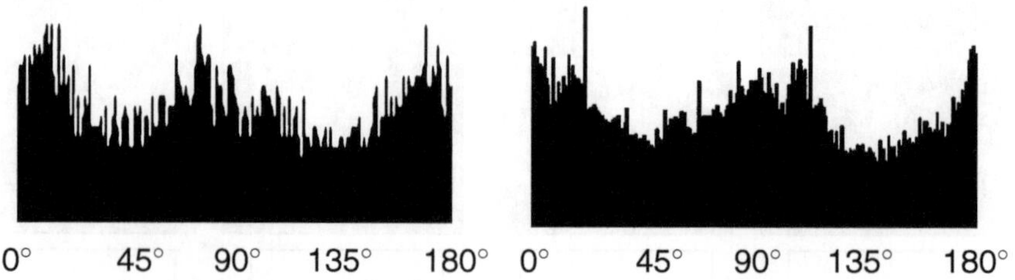

(*a*) GCAL/LISSOM model (*b*) Adult ferret

Fig. 8.8 Training on natural images gives matching orientation histograms. Looking more closely at the histogram for the network trained postnatally on images of natural scenes [Shouval et al.(1996)Shouval, Intrator, Law, and Cooper] shows that the resulting histograms are a close match to those found in adult ferret V1 (reprinted from [Coppola et al.(1998)Coppola, White, Fitzpatrick, and Purves]; copyright National Academy of Sciences, U.S.A.). The model and animals both model trained on natural images have more neurons representing horizontal or vertical than oblique contours, which reflects the statistics of the natural environment. However, the natural images were chosen specifically to exclude manmade contours, while the ferrets were raised in a laboratory environment that presumably had many long edges and sharp corners, and so it may be surprising to find such a close match for these images. Work is ongoing to identify the actual pattern of first and second order statistics in natural and laboratory environments so that these results can be interpreted clearly. Reprinted from [Bednar(2002)].

extracted the horizontal and vertical biases of natural image databases, developing many more horizontal-selective cells when trained on images with a preponderance of horizons and other horizontal patterns. This increase occurs within the context of the map already established at eye opening, with areas responding to horizontal growing larger over time, as they are activated more often than the neighboring stimuli that activate nearby regions. Figure 8.8 shows that the histogram of orientation preferences obtained in response to close-up natural images is a good match to that obtained for ferrets, which is intriguing because the ferrets have presumably been raised in a laboratory environment different from the forest and nature images used to train the model.

Figure 8.9 shows the color, motion direction, ocular dominance, spatial frequency, and disparity preferences and maps that develop when appropriate information is made available to V1 through additional ON/OFF sheets [Bednar(2012), Ball and Bednar(2009), Palmer(2009), Ramtohul(2006)]. As described in the original source for each model, the model results for each dimension have been evaluated against the available animal data, and capture the main aspects of the feature value coverage and the spatial organization of the maps [Miikkulainen et al.(2005)Miikkulainen, Bednar, Choe, and Sirosh, Palmer(2009)]. The maps simulated together (e.g. orientation and ocular dominance) also tend to intersect at right angles, such that high-gradient regions in one map avoid high-gradient regions in others [Bednar and Miikkulainen(2006)]. Each neuron becomes selective for some portion of the multidimensional feature space, and together they account for the

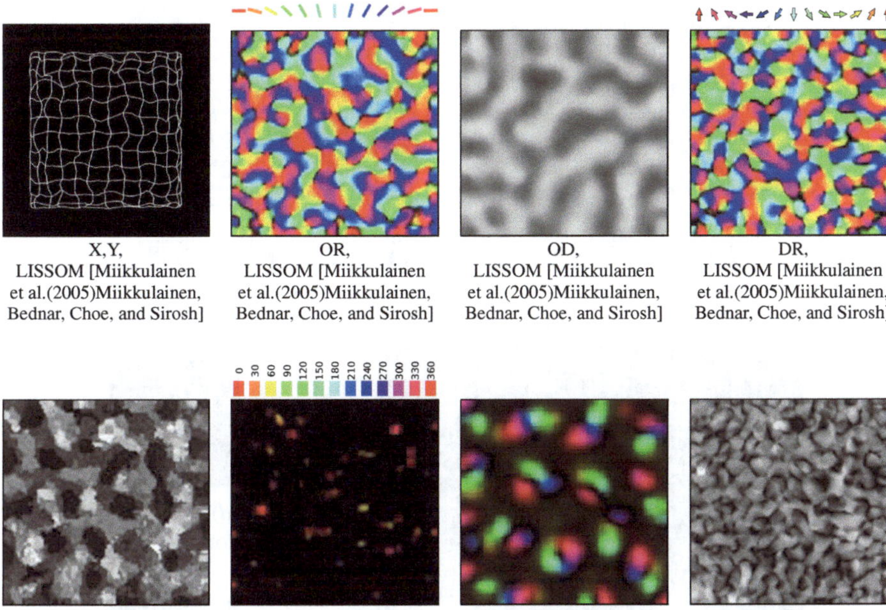

X,Y, LISSOM [Miikkulainen et al.(2005)Miikkulainen, Bednar, Choe, and Sirosh]

OR, LISSOM [Miikkulainen et al.(2005)Miikkulainen, Bednar, Choe, and Sirosh]

OD, LISSOM [Miikkulainen et al.(2005)Miikkulainen, Bednar, Choe, and Sirosh]

DR, LISSOM [Miikkulainen et al.(2005)Miikkulainen, Bednar, Choe, and Sirosh]

SF, LISSOM [Palmer(2009)]

DY, LISSOM [Ramtohul(2006)]

CR, LISSOM [Ball and Bednar(2009)]

TF, LISSOM [Miikkulainen et al.(2005)Miikkulainen, Bednar, Choe, and Sirosh]

Fig. 8.9 Model maps for other feature dimensions. Imaging results for 4mm×4mm of model V1 from the LISSOM models of retinotopy (X,Y), orientation (OR), ocular dominance (OD), motion direction (DR), spatial frequency (SF), temporal frequency (TF), disparity (DY), and color (CR). For each spatial dimension (TF has not yet been analyzed), the model develops maps that are a close match to the experimental results. All of the maps share a property of local smoothness, which results from the short-range lateral connections in the model, but the overall patterns differ with each feature depending on how those features varied during training for that simulation. Reprinted from references indicated.

variation across this space that was seen during self-organization [Bednar and Miikkulainen(2006)].

In animals, the only large-scale information available about neural properties is from imaging techniques at the map level. In the model, it is possible to see what connectivity patterns systematically lead to the observed map preferences. Figure 8.10 shows these connectivity patterns for a GCAL OR map simulation with simple and complex cells, illustrating how the neurons achieve coverage of the various possible input feature values. Lateral connections, in turn, store patterns of correlation between each neuron that represent larger-scale structure and correlations. Figure 8.11 shows the pattern of lateral connectivity for a neuron embedded in an orientation, ocular dominance, and motion direction map. Because the lateral connections are also modified by Hebbian learning, they represent correlations between neurons, and are thus strong for short-range connections (due to the

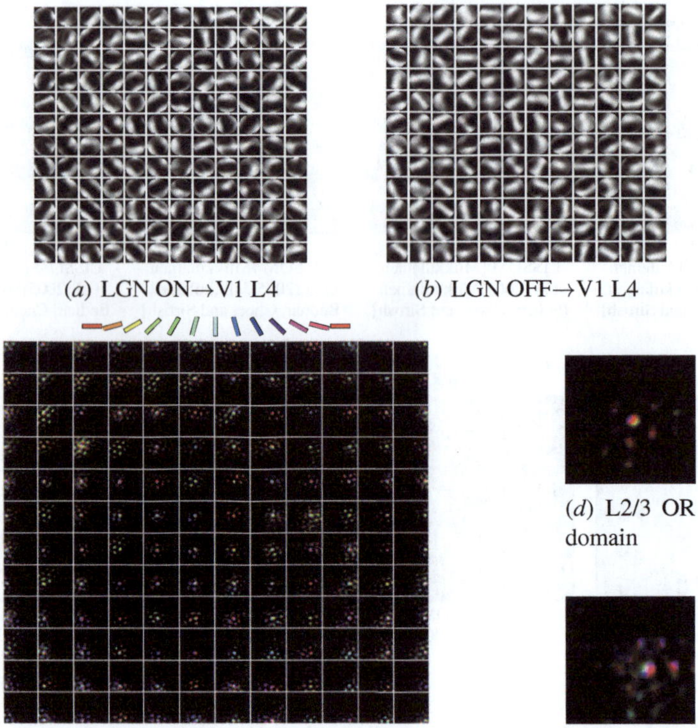

(*a*) LGN ON→V1 L4 (*b*) LGN OFF→V1 L4

(*c*) V1 L2/3→V1 L2/3

(*d*) L2/3 OR domain

(*e*) L2/3 OR pinwheel

Fig. 8.10 Self-organized projections to V1 L2/3. Unlike purely geometric models where the maps are represented directly in the model, the maps plotted in the previous figures are just summaries of the properties conferred on neurons by their connectivity patterns. These plots show the underlying connectivity patterns that lead to an orientation map, from a simulation with separate V1 L4 and L2/3 regions allowing the emergence of complex cells. (*a,b*) Connection fields from the LGN ON and OFF channels to every 20th neuron in the model L4 show that orientation preferences are reflected in the afferent connectivity to the neurons in that area. (*c*) Long-range excitatory lateral connections to those neurons preferentially come from neurons with similar OR preferences. Here strong weights are colored with the OR preference of the source neuron. Strong weights occur in clumps (appearing as small dots here) corresponding to an iso-orientation domain (each approximately 0.2–0.3mm wide); the fact that most of the dots are similar in color for any given neuron shows that the connections are orientation specific. Comparison of corresponding plots from (*c*) and (*a*) or (*b*) shows that the OR preferences of the afferent and lateral CFs are very similar. (*d*) Enlarged plot from (*c*) for a typical OR domain neuron that prefers horizontal patterns and receives connections primarily from other horizontal-preferring neurons (appearing as blobs of red or nearly red colors). (*e*) OR pinwheel neurons receive connections from neurons with many different OR preferences, because they are less selective in their responses and thus correlated with a wide range of orientation preferences. Overall, the lateral connectivity patterns reflect the patterns of co-occurrence statistics of each pair of neurons over time, due to Hebbian learning; these patterns then lead to phenomena such as orientation-specific surround modulation (figure 8.12). Reprinted from [Antolik(2010)].

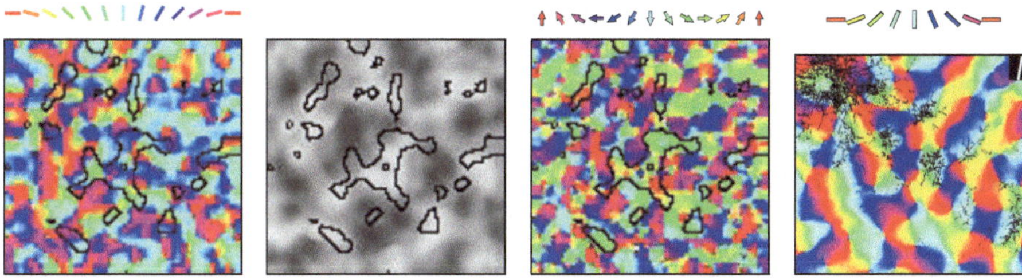

(*a*) OR+lateral [Bednar and Miikkulainen(2006)]

(*b*) OD+lateral [Bednar and Miikkulainen(2006)]

(*c*) DR+lateral [Bednar and Miikkulainen(2006)]

(*d*) Tree shrew; [Bosking et al.(1997)Bosking, Zhang, Schofield, and Fitzpatrick]

Fig. 8.11 Lateral connections across maps. LISSOM/GCAL neurons each participate in multiple functional maps, but have only a single set of lateral connections. Connections are strongest from other neurons with similar properties, respecting each of the maps to the degree to which that map affects correlation between neurons. Maps for a combined LISSOM OR/OD/DR simulation are shown above, with the black outlines indicating the connections to the central neuron (marked with a small black square outline) that remain after weak connections have been pruned. Model neurons receive connections from other model neurons with similar orientation preference (*a*) (as in tree shrew, (*d*)) but connections even more strongly respect the direction map (*c*). This highly monocular unit also connects strongly to the same eye (*b*), but the more typical binocular cells have wider connection distributions. Reprinted from refs. [Bednar and Miikkulainen(2006), Bosking et al.(1997)Bosking, Zhang, Schofield, and Fitzpatrick] as indicated.

shared retinotopic preference of those neurons) and between other neurons often coactivated during self-organization (e.g. those sharing orientation, direction, and eye preferences). The lateral connections are thus patchy and orientation and direction specific, as found in animals [Bosking et al.(1997)Bosking, Zhang, Schofield, and Fitzpatrick, Sincich and Blasdel(2001), Roerig and Kao(1999)]. Neurons with low levels of selectivity for any of those dimensions (e.g. binocular neurons) receive connections from a wide range of feature preferences, while highly selective neurons receive more specific connections, reflecting the different patterns of correlation in those cases. These connection patterns represent predictions, as only a few of these relationships have been tested so far in animals. The model strongly predicts that lateral connection patterns will respect all maps that account for a significant fraction of the response variance of the neurons, because each of those features will affect the correlation between neurons.

Overall, where it has been possible to make comparisons, these models have been shown to reproduce the main features of the experimental data, using a small set of assumptions. In each case, the model demonstrates how the experimentally measured map can emerge from Hebbian learning of corresponding patterns of subcortical and cortical activity. The models thus illustrate how the same basic, general-purpose adaptive mechanism will lead to very different organizations, depending on the geometrical and statistical properties of that feature. Future work will focus on showing how all the results so far could emerge simultaneously in a single model (as outlined in ref. [Bednar(2012)]).

8.4.2 Surround Modulation

Given a model with realistically patchy, specific lateral connectivity and realistic single-neuron properties, as described above, the patterns of interaction between neurons can be compared with neurophysiological evidence for surround modulation—influences on neural responses from distant patterns in the visual field. These studies can help validate the underlying model circuit, while helping understand how the visual cortex will respond to complicated patterns such as natural images.

For instance, figure 8.12 shows how the response to a sine grating patch can be modulated by a surrounding annulus. In animals, complicated patterns of interaction with the surround are seen depending on orientation and contrast [Sengpiel et al.(1997)Sengpiel, Sen, and Blakemore, Jones et al.(2002)Jones, Wang, and Sillito]. The model reproduces these patterns due to the orientation-specific self-organized lateral connection patterns, accounting not only for the most commonly reported and analyzed effects, but also a variety of other effects depending on the location of the neuron in the map (which affects its pattern of lateral connectivity as shown in figure 8.10). The model thus accounts both for the typical pattern of orientation contrast interactions, and explains why such a diversity of patterns is observed in animals. The results from these studies and related studies of size-dependent effects [Antolik(2010)] suggest both that lateral interactions may underlie many of the observed surround modulation effects, and also that the diversity of observed effects can at least in part be traced to the diversity of lateral connection patterns, which in turn is a result of the various sequences of activations of the neurons during development.

Although the preceding results all focused on the primary visual cortex, the mechanisms involved in these models are general purpose, based only on processing statistical regularities in input patterns to reveal the underlying geometry and properties of the external world. As a demonstration, figure 8.13 shows that the same model can be applied to a completely non-visual input modality, rodent whiskers. The same principle of activity-bubble formation due to local cooperation and more distant competition leads to very different results for this type of input, with pinwheels that develop in a strictly aligned global organization, unlike the scattered pinwheels seen in model V1 maps. But again the results are a good match to animal data, suggesting that these general principles apply across the sensory cortex, and potentially to other cortical and subcortical regions that process patterned stimuli.

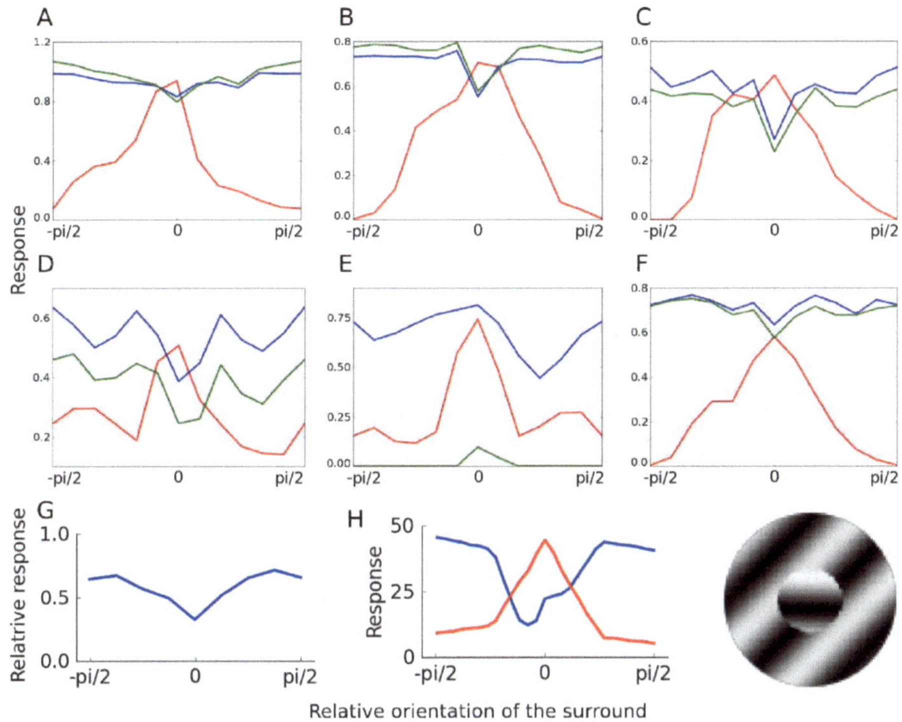

Fig. 8.12 Orientation-contrast tuning curves (OCTCs). For the OR model whose connection fields are shown in figure 8.10, the effect of the orientation-specific lateral connections can be tested using center-surround annulus stimulus like the example at the bottom right. Here the center patch is chosen to be a good match to the afferent RF of a specific V1 model neuron, and then responses are collected as the orientation of the surround is varied. In each graph A-F reprinted from ref. [Antolik(2010)], red is the orientation tuning curve for the given neuron (with just the center grating patch), blue is for surround contrast 50%, and green is for surround contrast 100%. Top row: typically (51% of model neurons tested), a collinear surround is suppressive for these contrasts, but the surround becomes less suppressive as the surround orientation is varied (as for cat [Sengpiel et al.(1997)Sengpiel, Sen, and Blakemore], G and macaque [Jones et al.(2002)Jones, Wang, and Sillito], H). Middle row: Other patterns seen in the model include high responses at diagonals (D, 20%, as seen in ref. [Sengpiel et al.(1997)Sengpiel, Sen, and Blakemore]), strongest suppression not collinear (E, as seen in ref. [Jones et al.(2002)Jones, Wang, and Sillito]), and facilitation for all orientations (F, 5%). The relatively rare pattern in F has not been reported in existing studies, and thus constitutes a prediction. In each case the observed variability is a consequence of the model's Hebbian learning that leads to a diversity of patterns of lateral connectivity, rather than noise or experimental artifacts.

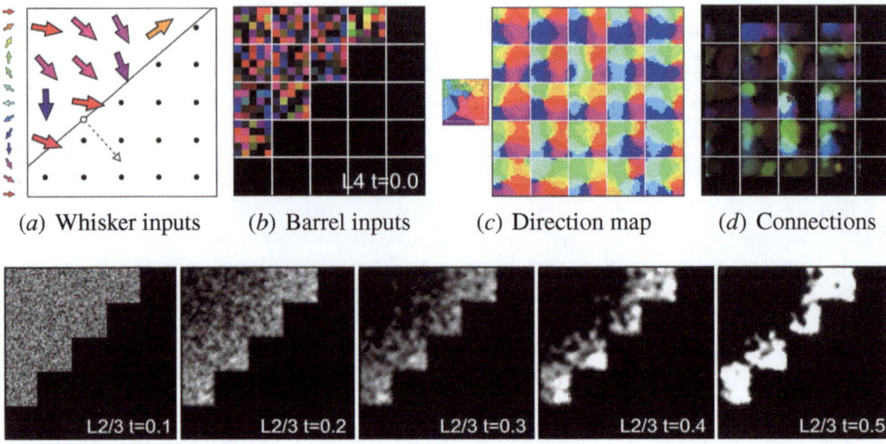

(*a*) Whisker inputs (*b*) Barrel inputs (*c*) Direction map (*d*) Connections

(*e*) Time course of activation

Fig. 8.13 Rodent barrel cortex direction map. The V1 model in GCAL and LISSOM is completely general, and contains no vision-specific components or assumptions. As a demonstration, this figure shows results from a GCAL-based model of the rat barrel cortex, which is a primary sensory area driven by the thalamus, like V1, but with inputs ultimately from rodent whiskers instead of photoreceptors (reprinted from [Wilson et al.(2010)Wilson, Law, Mitchinson, Prescott, and Bednar]). The model has a 5×5 array of whiskers that can be deflected in any direction with different strengths; (*a*) shows a typical assumed pattern of deflection, with whiskers deflected roughly in the direction perpendicular to a moving edge (e.g. an obstacle encountered by the whiskers). The corresponding thalamic input to the barrel cortex is shown in (*b*), computed using hard-wired cosine-shaped RFs analogous to the ON and OFF channels of the LGN; bright colors indicate high activation for a unit with that direction preference. The cortical response to this pattern is initially broad (*e*), as for visual stimuli to model V1, but within a few settling iterations converges into a stable pattern of activity bubbles. Due to the geometrical arrangement of the activated whiskers, the bubbles reliably form on the leading edge of the activity pattern, which causes an immediate and strong correlation between the input patterns and the neurons that respond in barrel cortex. The result is the robust emergence of globally aligned pinwheel patterns, one per whisker barrel (*c*), which is very different from the arbitrary pinwheel patterns observed for V1 development. These patterns are a close match to experimental data from rats ([Andermann and Moore(2006)]; see small map next to (*c*), showing how the map pattern emerges from the geometrical arrangement of the input stimuli and their receptors. Just as for the visual cortex models, the long-range lateral connections come from neurons with similar direction preference, due to Hebbian learning; see example for the neuron marked with a * in (*d*). For any modality, the model results reflect the geometric and statistical properties of the input, subject to constraints from the initial wiring of the cortex.

8.5 Discussion and Future Work

The results reviewed above illustrate a general approach to understanding the large-scale development, organization, and function of cortical areas, illustrating how the geometry and statistics of the external inputs interact with the geometry and architecture of the cortical architecture to determine the observed organization and operation of the visual cortex. The models show that a relatively small number of basic and largely uncontroversial assumptions and principles may be sufficient to explain a very wide range of experimental results from the visual cortex. Even very simple neural units, i.e., firing-rate point neurons, generically connected into topographic maps with initially random or isotropic weights, can form a wide range of specific feature preferences and maps via unsupervised normalized Hebbian learning of natural images and spontaneous activity patterns. The resulting maps consist of neurons with realistic spatial response properties, with variability due to visual context and recent history that explains significant aspects of surround modulation. Combining the existing models into a single, runnable visual system is very much a work in progress, but the results so far suggest that doing so will be both feasible and valuable. The simulator and example simulations are freely downloadable from `www.topographica.org`, allowing any interested researcher to build on this work.

It is important to note that many of the individual results found with GCAL can also be obtained using other modelling approaches, which can be complementary to the processes modeled by GCAL. For instance, it is possible to generate orientation maps without any activity-dependent plasticity, through the initial wiring pattern between the retina and the cortex [Ringach(2007), Paik and Ringach(2011)] or within the cortex itself [Grabska-Barwinska and von der Malsburg(2008)]. Such an approach cannot explain subsequent experience-dependent development, whereas the Hebbian approach of GCAL can explain both the initial map and later plasticity, but it is of course possible that the initial map and the subsequent plasticity occur via different mechanisms. Other models are based on abstractions of some of the mechanisms in GCAL [Yu et al.(2005)Yu, Farley, Jin, and Sur, Farley et al.(2007)Farley, Yu, Jin, and Sur, Obermayer et al.(1990)Obermayer, Ritter, and Schulten, Wolf and Geisel(2003)], operating similarly but at a higher level. GCAL is not meant as a competitor to such models, but as a concrete, physically realizable implementation of those ideas.

As discussed throughout, the main focus of this modelling work has been on replicating experimental data using a small number of computational primitives and mechanisms, with a goal of providing a concise, concrete, and relatively simple explanation for a wide and complex range of experimental findings. A complete explanation of visual cortex development and function would go even further, demonstrating more clearly *why* the cortex should be built in this way, and precisely what information-processing purpose this circuit performs. For instance, realistic receptive fields can be obtained from "normative" models embodying the idea that the cortex is developing a set of basis functions to represent input patterns faithfully, with only a few active neurons [Olshausen and

Field(1996), Bell and Sejnowski(1997), Hyvärinen and Hoyer(2001), Rehn and Sommer(2007)], maps can emerge by minimizing connection lengths in the cortex [Koulakov and Chklovskii(2001)], and lateral connections can be modelled as decorrelating the input patterns [Barlow and Földiák(1989), Dong(1995)]. The GCAL model can be seen as a concrete, mechanistic implementation of these ideas, showing how a physically realizable local circuit could develop receptive fields with good coverage of the input space, via lateral interactions that also implement sparsification via decorrelation [Miikkulainen et al.(2005)Miikkulainen, Bednar, Choe, and Sirosh]. Making more explicit links between mechanistic models like GCAL and normative theories is an important goal for future work. Meanwhile, there are many aspects of cortical function not explained by current normative models. The focus of the current line of research is on first capturing those phenomena in a general-purpose mechanistic model, so that researchers can then build deeper explanations for why these computations are useful for the organism.

8.6 Conclusions

The GCAL model results suggest that it will soon be feasible to build a single model visual system that will account for a very large fraction of the visual response properties, at the firing rate level, of V1 neurons in a particular species. Such a model will help researchers make testable predictions to drive future experiments to understand cortical processing, as well as determine which properties require more complex approaches, such as feedback, attention, and detailed neural geometry and dynamics. The model suggests that cortical neurons develop to cover the typical range of variation in their thalamic inputs, within the context of a smooth, multidimensional topographic map, and that lateral connections store pairwise correlations and use this information to modulate responses to natural scenes, dynamically adapting to both long-term and short-term visual input statistics.

Acknowledgements. Thanks to all of the collaborators whose modelling work is reviewed here, and to the members of the Institute for Adaptive and Neural Computation, and the Doctoral Training Centre in Neuroinformatics, at the University of Edinburgh, for discussions and feedback on many of the models. This work was supported in part by the UK EPSRC and BBSRC Doctoral Training Centre in Neuroinformatics, under grants EP/F500385/1 and BB/F529254/1, and by the US NIMH grant R01-MH66991. Computational resources were provided by the Edinburgh Compute and Data Facility (ECDF).

References

[Alitto and Usrey(2008)] Alitto, H.J., Usrey, W.M.: Origin and dynamics of extraclassical suppression in the lateral geniculate nucleus of the macaque monkey. Neuron 57(1), 135–146 (2008), `http://dx.doi.org/10.1016/j.neuron.2007.11.019`

[Andermann and Moore(2006)] Andermann, M.L., Moore, C.I.: A somatotopic map of vibrissa motion direction within a barrel column. Nature Neuroscience 9, 543–551 (2006), http://dx.doi.org/10.1038/nn1671

[Anderson and Rosenfeld(1988)] Anderson, J.A., Rosenfeld, E. (eds.): Neurocomputing: Foundations of Research. MIT Press, Cambridge (1988), http://mitpress.mit.edu/book-home.tcl?isbn=0262510480

[Antolik(2010)] Antolik, J.: Unified Developmental Model of Maps, Complex Cells and Surround Modulation in the Primary Visual Cortex. PhD thesis, School of Informatics, The University of Edinburgh, Edinburgh, UK (2010), http://hdl.handle.net/1842/4875

[Antolik and Bednar(2011)] Antolik, J., Bednar, J.A.: Development of maps of simple and complex cells in the primary visual cortex. Frontiers in Computational Neuroscience 5, 17 (2011), http://dx.doi.org/10.3389/fncom.2011.00017

[Ball and Bednar(2009)] Ball, C.E., Bednar, J.A.: A self-organizing model of color, ocular dominance, and orientation selectivity in the primary visual cortex. Society for Neuroscience Abstracts (2009), http://www.sfn.org, Program No. 756.9

[Barlow and Földiák(1989)] Barlow, H.B., Földiák, P.: Adaptation and decorrelation in the cortex. In: Durbin, R., Miall, C., Mitchison, G. (eds.) The Computing Neuron, pp. 54–72. Addison-Wesley, Reading (1989)

[Bednar(2002)] Bednar, J.A.: Learning to See: Genetic and Environmental Influences on Visual Development. PhD thesis, Department of Computer Sciences, The University of Texas at Austin, Austin, TX, 2002. Technical Report AI-TR-02-294 (2002), http://nn.cs.utexas.edu/keyword?bednar:phd02

[Bednar(2012)] Bednar, J.A.: Building a mechanistic model of the development and function of the primary visual cortex. Journal of Physiology (Paris) 106, 194–211 (2012), http://dx.doi.org/10.1016/j.jphysparis.2011.12.001

[Bednar(2013)] Bednar, J.A.: Constructing complex systems via activity-driven unsupervised Hebbian self-organization. In: Growing Adaptive Machines: Combining Development and Learning in Artificial Neural Networks. Studies in Computational Intelligence. Springer, Berlin (2013) (in press)

[Bednar et al.(2004)Bednar, Kelkar, and Miikkulainen] Bednar, J.A., Kelkar, A., Miikkulainen, R.: Scaling self-organizing maps to model large cortical networks. Neuroinformatics 2, 275–302 (2004), http://nn.cs.utexas.edu/keyword?bednar:neuroinformatics04

[Bednar and Miikkulainen(2003)] Bednar, J.A., Miikkulainen, R.: Self-organization of spatiotemporal receptive fields and laterally connected direction and orientation maps. Neurocomputing 52-54, 473–480 (2003), http://nn.cs.utexas.edu/keyword?bednar:neurocomputing03

[Bednar and Miikkulainen(2004)] Bednar, J.A., Miikkulainen, R.: Prenatal and postnatal development of laterally connected orientation maps. Neurocomputing 58-60, 985–992 (2004), http://nn.cs.utexas.edu/keyword?bednar:neurocomputing04-or

[Bednar and Miikkulainen(2006)] Bednar, J.A., Miikkulainen, R.: Joint maps for orientation, eye, and direction preference in a self-organizing model of V1. Neurocomputing 69(10-12), 1272–1276 (2006), http://nn.cs.utexas.edu/keyword?bednar:neurocomputing06

[Bell and Sejnowski(1997)] Bell, A.J., Sejnowski, T.J.: The "independent components" of natural scenes are edge filters. Vision Research 37, 3327 (1997), http://citeseer.nj.nec.com/bell97independent.html

[Bienenstock et al.(1982)Bienenstock, Cooper, and Munro] Bienenstock, E.L., Cooper, L.N., Munro, P.W.: Theory for the development of neuron selectivity: Orientation specificity and binocular interaction in visual cortex. The Journal of Neuroscience 2, 32–48 (1982),
http://www.ncbi.nlm.nih.gov/entrez/query.fcgi?cmd=retrieve&db=pubmed&dopt=abstract&list_uids=7054394

[Blasdel(1992a)] Blasdel, G.G.: Differential imaging of ocular dominance columns and orientation selectivity in monkey striate cortex. The Journal of Neuroscience 12, 3115–3138 (1992a),
http://www.jneurosci.org/cgi/content/abstract/12/8/3115

[Blasdel(1992b)] Blasdel, G.G.: Orientation selectivity, preference, and continuity in monkey striate cortex. The Journal of Neuroscience 12, 3139–3161 (1992b),
http://www.jneurosci.org/cgi/content/abstract/12/8/3139

[Bonin et al.(2005)Bonin, Mante, and Carandini] Bonin, V., Mante, V., Carandini, M.: The suppressive field of neurons in lateral geniculate nucleus. Journal of Neuroscience 25, 10844–10856 (2005),
http://dx.doi.org/10.1523/JNEUROSCI.3562-05.2005

[Bosking et al.(1997)Bosking, Zhang, Schofield, and Fitzpatrick] Bosking, W.H., Zhang, Y., Schofield, B.R., Fitzpatrick, D.: Orientation selectivity and the arrangement of horizontal connections in tree shrew striate cortex. The Journal of Neuroscience 17(6), 2112–2127 (1997),
http://www.jneurosci.org/cgi/content/full/17/6/2112

[Bosking et al.(2002)Bosking, Crowley, and Fitzpatrick] Bosking, W.H., Crowley, J.C., Fitzpatrick, D.: Spatial coding of position and orientation in primary visual cortex. Nature Neuroscience 5(9), 874–882 (2002),
http://dx.doi.org/10.1038/nn908

[Carreira-Perpiñán et al.(2005)Carreira-Perpiñán, Lister, and Goodhill] Carreira-Perpiñán, M.A., Lister, R.J., Goodhill, G.J.: A computational model for the development of multiple maps in primary visual cortex. Cerebral Cortex 15(8), 1222–1233 (2005),
http://dx.doi.org/10.1093/cercor/bhi004

[Chapman et al.(1996)Chapman, Stryker, and Bonhoeffer] Chapman, B., Stryker, M.P., Bonhoeffer, T.: Development of orientation preference maps in ferret primary visual cortex. The Journal of Neuroscience 16(20), 6443–6453 (1996),
http://www.jneurosci.org/cgi/content/abstract/16/20/6443

[Coppola et al.(1998)Coppola, White, Fitzpatrick, and Purves] Coppola, D.M., White, L.E., Fitzpatrick, D., Purves, D.: Unequal representation of cardinal and oblique contours in ferret visual cortex. Proceedings of the National Academy of Sciences, USA 95(5), 2621–2623 (1998),
http://www.ncbi.nlm.nih.gov/entrez/query.fcgi?cmd=retrieve&db=pubmed&dopt=abstract&list_uids=9482936

[Dong(1995)] Dong, D.W.: Associative decorrelation dynamics: A theory of self-organization and optimization in feedback networks. In: Tesauro, G., Touretzky, D.S., Leen, T.K. (eds.) Advances in Neural Information Processing Systems, vol. 7, pp. 925–932. MIT Press, Cambridge (1995),
ftp://ftp.ci.tuwien.ac.at/pub/texmf/bibtex/nips-7.bib

[Durbin and Mitchison(1990)] Durbin, R., Mitchison, G.: A dimension reduction framework for understanding cortical maps. Nature 343, 644–647 (1990),
http://www.ncbi.nlm.nih.gov/entrez/query.fcgi?cmd=retrieve&db=pubmed&dopt=abstract&list_uids=2304536

[Farley et al.(2007)Farley, Yu, Jin, and Sur] Farley, B.J., Yu, H., Jin, D.Z., Sur, M.: Alteration of visual input results in a coordinated reorganization of multiple visual cortex maps. The Journal of Neuroscience 27(38), 10299–10310 (2007), http://dx.doi.org/10.1523/JNEUROSCI.2257-07.2007

[Felisberti and Derrington(1999)] Felisberti, F., Derrington, A.M.: Long-range interactions modulate the contrast gain in the lateral geniculate nucleus of cats. Visual Neuroscience 16, 943–956 (1999)

[Grabska-Barwinska and von der Malsburg(2008)] Grabska-Barwinska, A., von der Malsburg, C.: Establishment of a scaffold for orientation maps in primary visual cortex of higher mammals. The Journal of Neuroscience 28(1), 249–257 (2008), http://dx.doi.org/10.1523/JNEUROSCI.5514-06.2008

[Hyvärinen and Hoyer(2001)] Hyvärinen, A., Hoyer, P.O.: A two-layer sparse coding model learns simple and complex cell receptive fields and topography from natural images. Vision Research 41(18), 2413–2423 (2001), http://www.sciencedirect.com/science/article/B6T0W-43GCBN2-B/1/94ef0f6b0d8

[Jones et al.(2002)Jones, Wang, and Sillito] Jones, H.E., Wang, W., Sillito, A.M.: Spatial organization and magnitude of orientation contrast interactions in primate V1. Journal of Neurophysiology 88(5), 2796–2808 (2002), http://dx.doi.org/10.1152/jn.00403.2001

[Kaschube et al.(2010)Kaschube, Schnabel, Löwel, Coppola, White, and Wolf] Kaschube, M., Schnabel, M., Löwel, S., Coppola, D.M., White, L.E., Wolf, F.: Universality in the evolution of orientation columns in the visual cortex. Science 330(6007), 1113–1116 (2010)

[Koulakov and Chklovskii(2001)] Koulakov, A.A., Chklovskii, D.B.: Orientation preference patterns in mammalian visual cortex: A wire length minimization approach. Neuron 29, 519–527 (2001), http://www.ncbi.nlm.nih.gov/entrez/query.fcgi?cmd=retrieve&db=pubmed&dopt=abstract&list_uids=11239440

[Law(2009)] Law, J.S.: Modeling the Development of Organization for Orientation Preference in Primary Visual Cortex. PhD thesis, School of Informatics, The University of Edinburgh, Edinburgh, UK (2009), http://hdl.handle.net/1842/3935

[Miikkulainen et al.(2005)Miikkulainen, Bednar, Choe, and Sirosh] Miikkulainen, R., Bednar, J.A., Choe, Y., Sirosh, J.: Computational Maps in the Visual Cortex. Springer, Berlin (2005)

[Miller(1994)] Miller, K.D.: A model for the development of simple cell receptive fields and the ordered arrangement of orientation columns through activity-dependent competition between ON- and OFF-center inputs. The Journal of Neuroscience 14, 409–441 (1994), http://www.ncbi.nlm.nih.gov/entrez/query.fcgi?cmd=retrieve&db=pubmed&dopt=abstract&list_uids=8283248

[Miller and MacKay(1994)] Miller, K.D., MacKay, D.J.C.: The role of constraints in Hebbian learning. Neural Computation 6, 100–126 (1994), http://wol.ra.phy.cam.ac.uk/mackay/abstracts/constraints.html

[Obermayer et al.(1990)Obermayer, Ritter, and Schulten] Obermayer, K., Ritter, H., Schulten, K.J.: A principle for the formation of the spatial structure of cortical feature maps. Proceedings of the National Academy of Sciences, USA 87, 8345–8349 (1990), http://www.pnas.org/cgi/content/abstract/87/21/8345

[Olshausen and Field(1996)] Olshausen, B.A., Field, D.J.: Emergence of simple-cell receptive field properties by learning a sparse code for natural images. Nature 381, 607–609 (1996), http://www.ncbi.nlm.nih.gov/entrez/query.fcgi?cmd=retrieve&db=pubmed&dopt=abstract&list_uids=8637596

[Paik and Ringach(2011)] Paik, S.-B., Ringach, D.L.: Retinal origin of orientation maps in visual cortex. Nature Neuroscience 14(7), 919–925 (2011), http://dx.doi.org/10.1038/nn.2824

[Palmer(2009)] Palmer, C.M.: Topographic and Laminar Models for the Development and Organisation of Spatial Frequency and Orientation in V1. PhD thesis, School of Informatics, The University of Edinburgh, Edinburgh, UK (2009), http://hdl.handle.net/1842/4114

[Ramtohul(2006)] Ramtohul, T.: A self-organizing model of disparity maps in the primary visual cortex. Master's thesis, The University of Edinburgh, Scotland, UK (2006), http://www.inf.ed.ac.uk/publications/thesis/online/IM060400.pdf

[Rehn and Sommer(2007)] Rehn, M., Sommer, F.T.: A network that uses few active neurones to code visual input predicts the diverse shapes of cortical receptive fields. Journal of Computational Neuroscience 22(2), 135–146 (2007), http://dx.doi.org/10.1007/s10827-006-0003-9

[Ringach(2007)] Ringach, D.L.: On the origin of the functional architecture of the cortex. PLoS One 2(2), e251 (2007), http://dx.doi.org/10.1371/journal.pone.0000251

[Ritter et al.(1992)Ritter, Martinetz, and Schulten] Ritter, H., Martinetz, T., Schulten, K.J.: Neural Computation and Self-Organizing Maps: An Introduction. Addison-Wesley, Reading (1992)

[Ritter et al.(1991)Ritter, Obermayer, Schulten, and Rubner] Ritter, H., Obermayer, K., Schulten, K.J., Rubner, J.: Self-organizing maps and adaptive filters. In: Models of Neural Networks, pp. 281–306. Springer, Berlin (1991)

[Roerig and Kao(1999)] Roerig, B., Kao, J.P.: Organization of intracortical circuits in relation to direction preference maps in ferret visual cortex. The Journal of Neuroscience 19(24), RC44 (1999), http://www.jneurosci.org/content/19/24/RC44.long

[Saul and Humphrey(1992)] Saul, A.B., Humphrey, A.L.: Evidence of input from lagged cells in the lateral geniculate nucleus to simple cells in cortical area 17 of the cat. Journal of Neurophysiology 68(4), 1190–1208 (1992), http://www.ncbi.nlm.nih.gov/entrez/query.fcgi?cmd=retrieve&db=pubmed&dopt=abstract&list_uids=1432077

[Sclar and Freeman(1982)] Sclar, G., Freeman, R.D.: Orientation selectivity in the cat's striate cortex is invariant with stimulus contrast. Experimental Brain Research 46, 457–461 (1982), http://www.ncbi.nlm.nih.gov/entrez/query.fcgi?cmd=retrieve&db=pubmed&dopt=abstract&list_uids=7095050

[Sengpiel et al.(1997)Sengpiel, Sen, and Blakemore] Sengpiel, F., Sen, A., Blakemore, C.: Characteristics of surround inhibition in cat area 17. Experimental Brain Research 116(2), 216–228 (1997), http://dx.doi.org/10.1007/PL00005751

[Shouval et al.(1996)Shouval, Intrator, Law, and Cooper] Shouval, H.Z., Intrator, N., Law, C.C., Cooper, L.N.: Effect of binocular cortical misalignment on ocular dominance and orientation selectivity. Neural Computation 8(5), 1021–1040 (1996), http://www.ncbi.nlm.nih.gov/entrez/query.fcgi?cmd=retrieve&db=pubmed&dopt=abstract&list_uids=8697227

[Sincich and Blasdel(2001)] Sincich, L.C., Blasdel, G.G.: Oriented axon projections in primary visual cortex of the monkey. The Journal of Neuroscience 21, 4416–4426 (2001), http://www.jneurosci.org/cgi/content/abstract/21/12/4416

[Stevens(2011)] Stevens, J.-L.: A temporal model of neural activity and VSD response in the primary visual cortex. Master's thesis, The University of Edinburgh, Scotland, UK (2011), http://www.inf.ed.ac.uk/publications/thesis/online/IT111096.pdf

[Stevens et al.(2013)Stevens, Law, Antolik, and Bednar] Stevens, J.-L.R., Law, J.S., Antolik, J., Bednar, J.A.: Mechanisms for stable, robust, and adaptive development of orientation maps in the primary visual cortex. Journal of Neuroscience 33, 15747–15766 (2013), http://dx.doi.org/10.1523/JNEUROSCI.1037-13.2013

[Swindale et al.(2000)Swindale, Shoham, Grinvald, Bonhoeffer, and Hubener] Swindale, N.V., Shoham, D., Grinvald, A., Bonhoeffer, T., Hubener, M.: Visual cortex maps are optimized for uniform coverage. Nature Neuroscience 3(8), 822–826 (2000), http://www.neurosci.info/courses/vision2/V1FuncOrg/swindale00.pdf

[Tanaka et al.(2006)Tanaka, Ribot, Imamura, and Tani] Tanaka, S., Ribot, J., Imamura, K., Tani, T.: Orientation-restricted continuous visual exposure induces marked reorganization of orientation maps in early life. Neuroimage 30(2), 462–477 (2006), http://dx.doi.org/10.1016/j.neuroimage.2005.09.056

[Turrigiano(1999)] Turrigiano, G.G.: Homeostatic plasticity in neuronal networks: The more things change, the more they stay the same. Trends in Neurosciences 22(5), 221–227 (1999), http://www.ncbi.nlm.nih.gov/entrez/query.fcgi?cmd=retrieve&db=pubmed&dopt=abstract&list_uids=10322495

[von der Malsburg(1973)] von der Malsburg, C.: Self-organization of orientation-sensitive cells in the striate cortex. Kybernetik 15, 85–100 (1973) http://www.ncbi.nlm.nih.gov/entrez/query.fcgi?cmd=retrieve&db=pubmed&dopt=abstract&list_uids=4786750, Reprinted in Anderson and Rosenfeld [3], 212–227

[Wiesel(1982)] Wiesel, T.N.: Postnatal development of the visual cortex and the influence of the environment. Nature 299, 583–591 (1982), http://www.ncbi.nlm.nih.gov/entrez/query.fcgi?cmd=retrieve&db=pubmed&dopt=abstract&list_uids=6811951

[Wilson et al.(2010)Wilson, Law, Mitchinson, Prescott, and Bednar] Wilson, S.P., Law, J.S., Mitchinson, B., Prescott, T.J., Bednar, J.A.: Modeling the emergence of whisker direction maps in rat barrel cortex. PLoS One 5(1), e8778 (2010), http://dx.doi.org/10.1371/journal.pone.0008778

[Wolf and Geisel(2003)] Wolf, F., Geisel, T.: Universality in visual cortical pattern formation. J. Physiol. Paris 97(2-3), 253–264 (2003), http://dx.doi.org/10.1016/j.jphysparis.2003.09.018

[Wolfe and Palmer(1998)] Wolfe, J., Palmer, L.A.: Temporal diversity in the lateral geniculate nucleus of cat. Visual Neuroscience 15(4), 653–675 (1998), http://www.journals.cambridge.org/issue_VisualNeuroscience/Vol15No04

[Wong(1999)] Wong, R.O.L.: Retinal waves and visual system development. Annual Review of Neuroscience 22, 29–47 (1999), http://www.ncbi.nlm.nih.gov/entrez/query.fcgi?cmd=retrieve&db=pubmed&dopt=abstract&list_uids=10202531

[Yu et al.(2005)Yu, Farley, Jin, and Sur] Yu, H., Farley, B.J., Jin, D.Z., Sur, M.: The coordinated mapping of visual space and response features in visual cortex. Neuron 47(2), 267–280 (2005), http://dx.doi.org/10.1016/j.neuron.2005.06.011

Index

MIX
Papier aus verantwortungsvollen Quellen
Paper from responsible sources
FSC® C105338

If you have any concerns about our products,
you can contact us on
ProductSafety@springernature.com

In case Publisher is established outside the EU,
the EU authorized representative is:
Springer Nature Customer Service Center GmbH
Europaplatz 3, 69115 Heidelberg, Germany

Printed by Libri Plureos GmbH
in Hamburg, Germany